ENGINEERING
and
OPERATIONS
in the
BELL
SYSTEM

ENGINEERING
and
OPERATIONS
in the
BELL
SYSTEM

Second Edition
Reorganized and Rewritten
Telecommunications in the
Bell System in 1982 - 1983

Prepared by Members of the Technical Staff
and the Technical Publication Department
AT&T Bell Laboratories

R.F. Rey, Technical Editor

AT&T Bell Laboratories
Murray Hill, N.J.

Foreword

The first edition of this book, published in 1977, presented a comprehensive view of the Bell System as seen from AT&T Bell Laboratories. Its primary purpose was to provide new members of technical staff of Bell Laboratories with a basic understanding of the Bell System from the standpoint of the services provided, the equipment and systems constituting the nationwide network, the planning and engineering considerations involved in the evolution of the network, and the many activities required for day-to-day operation. However, the book has been used more widely than expected as a general reference and as a primer for others unfamiliar with telecommunications.

This revision was prompted by the significant changes in technology, services, and the environment that have occurred since 1977. Consequently, material from many sections of the first edition has been updated, and the organization and presentation of material have been improved. The level of detail has been reduced in some places and more emphasis placed on explaining important concepts and defining terminology. In addition, an attempt was made to present the material in a manner suitable for a mix of academic backgrounds.

The material for this edition was almost complete when agreement was reached between AT&T and the United States Department of Justice to settle an antitrust suit by divesting the Bell operating companies from AT&T. Recognizing that it would be a long time before the massive change associated with divestiture could be recorded and that most of the information on services and systems would remain valid, revision of the first edition continued as planned. Only the first chapter was revised to indicate the major provisions related to divestiture and to provide an overview of the postdivestiture corporate units. The rest of the book portrays the Bell System as it was near the end of 1982 and early 1983 and reflects changes resulting from the Federal Communications Commission's Computer Inquiry II order that were effective January 1, 1983. Since the existence of the Bell System ends with divestiture, this second edition of *Engineering and Operations in the Bell System* is also the final edition. It is perhaps a fitting closure to an era in which the Bell System fulfilled its historic mission of providing universal high-quality, economical telephone service and also provided much of the technology for the future.

The book is divided into five parts. Part 1 presents an overview of the Bell System (and the postdivestiture configuration) in terms of the corporate units and their responsibilities, the services provided, and basic communications via the network. Part 2 deals with concepts, principles, and engineering considerations related to the components of telecommunications. In part 3, specific systems and equipment are described with emphasis on applications, distinguishing characteristics and features, and major design considerations. Part 4 describes telephone company operations on a functional level, presents a description of selected computer-based operations systems, and discusses operations planning and the evaluation of performance and services. Finally, Part 5 traces the events that have shaped the telecommunications environment and discusses factors related to the evolution of products and services.

Authors and coordinators of the material are identified at the end of each chapter. A committee consisting of R. A. Bruce, L. E. Gallaher, W. S. Hayward, Jr., J. R. Harris, M. M. Irvine, and J. A. McCarthy reviewed the plan for the book and each chapter, providing many corrections and comments. J. W. Falk and D. E. Snedeker reviewed the material from a legal and regulatory viewpoint. J. R. Harris deserves special recognition for his contributions to the planning and early development of the book and for his guidance and support throughout the project. The Technical Publication Department of Bell Laboratories provided editorial assistance throughout the project and prepared the material for publication. Many others assisted by reviewing portions of the text or serving as information sources. The principal credit, however, belongs to the editor, R. F. Rey, whose technical and managerial talents were wisely used in this undertaking.

Tom Powers

Tom L. Powers
Executive Director
Network Planning Division
Bell Laboratories

Contents

PART FOUR
OPERATIONS   569

PART ONE

INTRODUCTION TO
THE BELL SYSTEM

The first three chapters of this book provide an overview of the Bell System and prepare the reader for the more detailed treatment of topics to follow. Each chapter views the Bell System from a different perspective. Chapter 1 discusses the overall corporate structure, activities, and responsibilities of the Bell System and its constituent companies and suggests the size and complexity of the business. Material added to the chapter late in 1983 describes structural changes associated with the divestiture of the Bell System on January 1, 1984, and summarizes major provisions of the Modification of Final Judgment. Chapter 2 describes the services the Bell System makes available to customers in terms of customers' needs and uses. Although this chapter includes a discussion of terminal products, it describes the service provided by the product rather than the product itself. (Chapter 11 addresses equipment characteristics and design considerations for customer products and services.) Chapter 3 completes the overview of the Bell System: It introduces the telecommunications network and a set of terms and concepts related to network components and functions. These concepts are then applied in a discussion of the procedures involved in a typical telephone call.

1

Structure and Activities

1.1 INTRODUCTION

The Bell System led the development and use of communications equipment and techniques in the United States throughout most of this century. It became the nation's major supplier of telecommunications products and services ranging from basic residence telephones to increasingly sophisticated information services.

From its beginnings, the Bell System matched its organizational structure to the environment in which it operated. Early in the century, universal service—providing basic telephone service at an affordable price anywhere in the nation—became the Bell System goal. The Bell System approached this goal by creating a functional organization: Each of the local telephone companies and the American Telephone and Telegraph Company (AT&T) itself were organized along the lines of the job that had to be done. Tasks in each functional area were performed by specialists in that area to maximize efficiency. The local companies were responsible for responding to the particular needs of the communities they served, but they all used standard technology and operating methods. Thus, AT&T and the telephone companies achieved coordination on a national scale, while responding to local needs. As a result, the goal of universal service has been met—nearly everyone in the United States has a telephone that is connected to a single nationwide network. This **public switched telephone network**[1] is available to the general public and to other carriers and networks.

The functional organization that made providing universal service efficient and practical was possible because, for most of its history, the

---

[1] Sections 3.3.1 and 4.2.1 discuss the public switched telephone network in detail.

Bell System was almost the sole source of telecommunications service—
although under terms and conditions approved by federal and state regulators. AT&T and the telephone companies have changed their organizational structure to match environmental changes such as new and diverse
customer needs and, more recently, new markets.

Now, in the 1980s, the way in which telecommunications services are
provided is changing entirely. The 1982 Modification of Final Judgment
(MFJ), which terminated a 1974 Department of Justice antitrust suit
against AT&T, ordered the breakup (divestiture) of the integrated Bell
System. AT&T set January 1, 1984, as the target date for completion of
the massive job of restructuring a business involving about 1 million
employees and about $160 billion in assets.

Subsequent sections of Chapter 1 describe the organization of the Bell
System in 1982, the major provisions of the MFJ, and the postdivestiture
structure. The chapter was revised late in 1983 to provide an introductory account of the impact of divestiture. The rest of the book describes
the engineering and operations of the Bell System at the end of 1982 and
in early 1983 and does not reflect divestiture because much of the
material was prepared before the antitrust suit was settled in 1982.[2] However, the primary effects of divestiture are on the structure of the corporate units and the allocation of the roles in providing telecommunications services. The technology in the network, the considerations
involved in its design, and the nature of functions required to operate
the network and to provide service to customers remain essentially
unchanged. Thus, this book constitutes a valid description of telecommunications engineering and operations and meets a growing need for
an update to the previous book (Bell Laboratories 1977).

1.2 THE BELL SYSTEM IN 1982

1.2.1 AMERICAN TELEPHONE AND TELEGRAPH COMPANY

In 1982, the Bell System, serving more than 80 percent of the nation's
telephones, had long been the largest of hundreds of companies providing communications services in the United States. Before divestiture, the
Bell System operated as a partnership among AT&T; a number of Bell-
owned telephone companies (known as *operating companies*); the Western
Electric Company, Incorporated; and Bell Telephone Laboratories, Incorporated. The product of this partnership was service, provided through a
dynamic and dependable communications network designed, built, and
operated as a single system.

---

[2] Because the text describes the Bell System in 1982 and early 1983 and was written in that
time frame and earlier, the reader will find the Bell System referred to extensively in the
present tense.

Figure 1-1 shows the structure of the Bell System as it was in 1982. AT&T, the parent company, was publicly owned by 3.1 million stockholders. In turn, AT&T owned Western Electric and—totally or partially— each of the Bell operating companies.[3] AT&T and Western Electric jointly owned Bell Laboratories. Both AT&T and Western Electric also had subsidiary companies (shown on the figure); some of which supported Bell System operations, others that provided domestic and international communications services.

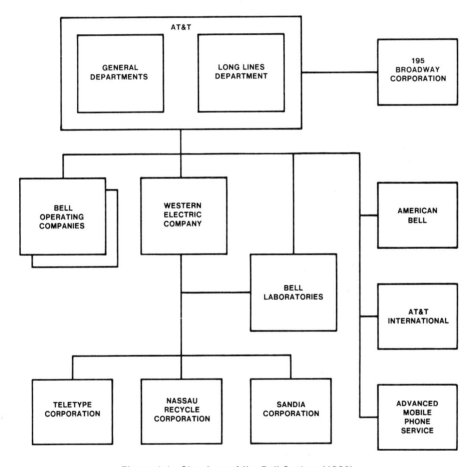

Figure 1-1. Structure of the Bell System (1982).

---

[3] AT&T was sole stockholder in twenty-one operating companies and a minority stockholder in two: the Southern New England Telephone Company and Cincinnati Bell, Inc. Bell Telephone Company of Nevada is wholly owned by the Pacific Telephone and Telegraph Company. Four Chesapeake and Potomac Telephone Companies offer service in Washington, D.C.; Maryland; Virginia; and West Virginia.

AT&T, the parent company of the Bell System, had its headquarters in New York City.[4]

Corporate Functions

General Departments. The General Departments of AT&T set the major goals and large-scale programs for the Bell System. They advised and assisted the Bell operating companies on such matters as finance, operations, personnel, legal, accounting, marketing, planning, public relations, and employee information, thereby providing continuity of direction and consolidating the particular specialities of each Bell System partner. AT&T, through its General Departments, coordinated pricing activity and represented the Bell System before federal regulatory agencies. It determined price structures and recommended costing and pricing matters through federal agencies.

The General Departments ensured that new developments, solutions to existing problems, and provisions for the future needs of customers became part of the entire Bell System. This involved directing the work of Bell Laboratories and Western Electric and coordinating the integration of new technology into the network.

The General Departments established and maintained standards and procedures for the Bell System and for the interconnection of non–Bell System equipment and facilities with the Bell System network. They served as an information clearinghouse for associations of independent telephone companies (such as the United States Independent Telephone Association) and for general-trade (that is, other than Western Electric) suppliers.

Long Lines Department. The Long Lines Department, with headquarters in Bedminster, New Jersey, operated the long-distance network. Many of its activities were similar to those of the Bell operating telephone companies. Long Lines built, operated, and maintained most of the interstate network of long-distance lines, thereby providing interstate and international communications services for people throughout the United States. It directed the overall design and management of the network and coordinated the teamwork among the various Bell and independent companies who jointly own and operate this complex, widespread system of microwave radio, coaxial cable, optical fibers, satellites, and intricate switching systems.

To handle the network efficiently, Long Lines was divided into six territorial regions (see Figure 1-2). Each region took care of engineering, sales, and network operations in its territory.

---

[4] A new headquarters building at 550 Madison Avenue has replaced the building at 195 Broadway, which was the headquarters location for many years.

Figure 1-2. Long Lines regions and regional headquarters locations.

Corporate Structure

Figure 1-3 is a block diagram of the AT&T organization as it existed in 1982. The office of the assistant to the president reviewed all aspects of the organization and ensured that each unit's plans, budgets, and operations were consistent with system requirements. It also maintained liaison with Bell Laboratories and with Western Electric. The figure shows the network function and how AT&T's customer-related operations and marketing functions were structured to reflect the Bell System's market segments: business, residence, directory, and public services.

- **Business** organizations coordinated the response of the Bell System to the needs of business customers; assisted telephone companies in the areas of marketing, pricing, costing, forecasting, training, and budget matters related to serving business customers; and supported telephone companies with installations, repairs, maintenance, customer contacts, engineering, and measurements required for customer services.

  — *Business Marketing* provided leadership and support for Bell System business marketing efforts.

  — *Business Services* combined under common management all the closely related delivery functions that flow from marketing and sales.

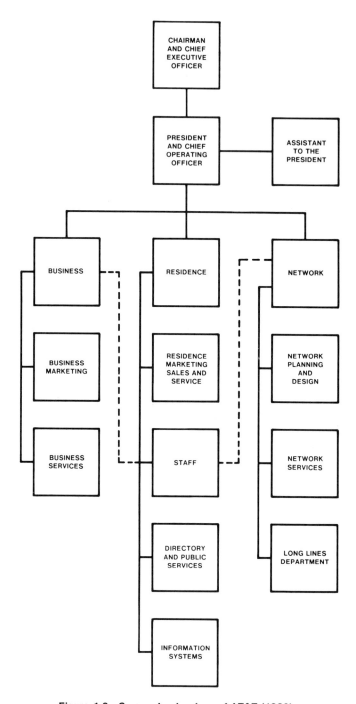

Figure 1-3. Corporate structure of AT&T (1982).

- **Residence** organizations coordinated the response of the Bell System to the needs of residential customers for telecommunications products and services.

  — *Residence Marketing Sales and Service* offered telephone companies help in marketing, pricing, costing, forecasting, training, and budget matters related to serving residence customers. It also supported telephone companies with installations, repairs, maintenance, customer contacts, engineering, and measurements required for customer service.

  In addition, for organizational purposes, several other units reported to the executive vice-president-residence.

  — *Staff* supplied support services within AT&T itself and coordinated support services such as inventory management, automotive operations, building planning, real-estate management, energy conservation, and environmental protection offered by other Bell System companies.

  — The *Directory* organization assisted telephone companies with marketing, costing, pricing, forecasting, training, and budget matters for both white pages and Yellow Pages as well as with producing and distributing directories.

  — *Public services* coordinated Bell System activities involved in providing communications services for users who are away from home or office. Public services comprised public and semipublic telephone service, including Charge-a-Call and *DIAL-IT*[5] network communications services such as Public Announcement Service and Media Stimulated Calling (see Section 2.5.1).

  — *Information Systems* provided planning, design, and development of functional accounting and information systems for use by the operating companies.

- **Network** organizations supported the business, residence, directory, and public-service markets by guiding and coordinating the operation of the network and the activities that provide telecommunications services between customers' locations.

  — *Network Planning and Design* oversaw the provision of reliable and innovative interpremises communications services, ensured that existing Bell System services were continually improved, coordinated the development of the national and international network, guided the efforts of Bell Laboratories in the area of interpremises

---

[5] Service mark of AT&T Co.

services, guided the technical planning of the operating companies, and maintained technical liaison with both independent telephone companies and international and other domestic carriers.

— *Network Services* provided methods and guidance to operating telephone company network organizations. Supported functions included the administration and maintenance of network switching systems and transmission facilities; operator services; and the engineering, construction, maintenance, and administration of distribution facilities and services.

— The *Long Lines Department* was included in the network segment. This integration made it easier to combine the planning, design, and management of the interstate network with the intrastate networks for improved overall network service.

Subsidiary Companies

AT&T owned several subsidiary companies that supported Bell System operations or provided domestic and international services. The primary subsidiaries were:

- **195 Broadway Corporation**, which provided real-estate management services for AT&T corporate locations. These services included constructing, owning, and leasing buildings; administering office space, facilities, and equipment; and providing related building and housekeeping support services such as transportation, maintenance, and cafeterias for corporate buildings.

- **AT&T International Inc.**, which was formed in August 1980 to sell Bell System products worldwide and apply Bell System technology, products, and experience to the needs of telephone administrations overseas and international business customers. It also provides technical and advisory services and directory and informations systems.

- **American Bell Inc.**, which was formed in June 1982 in response to Computer Inquiry II (see Section 17.4.3). As a "separate" subsidiary, American Bell could sell its products and services to customers without government approval and had certain limitations in the way it dealt with other Bell System companies.

- **Advanced Mobile Phone Service, Inc.**, which was responsible for planning and developing a nationwide cellular radio system to provide communications from moving customers to the land-line telecommunications system. Section 11.4.1 discusses cellular radio.

1.2.2 BELL OPERATING COMPANIES

Before divestiture, the Bell operating companies built, operated, and maintained the local and intrastate networks and provided most of the day-to-day service for customers in their individual communities. Chapter 13 discusses the many functional activities performed by the operating companies. Long-distance calls also traveled over individual company facilities, but those that went from the territory of one company to that of another were carried by Long Lines or another common carrier. (Figure 1-4 shows the Bell operating companies as they existed in 1982 and the territories they served.) The operating companies also joined with Long Lines to furnish certain interstate services such as carrying radio and television network programs to broadcasting stations throughout the country. They also cooperated with the hundreds of independent telephone companies so that the public had access to a unified national telephone network.

The operating companies differed from one another in size and organization. Geographically, the smallest was that part of the Chesapeake and Potomac Telephone Companies that offered service in the 61 square miles of the District of Columbia. The largest was the Mountain States Telephone and Telegraph Company, which operated in seven states and had a service area of more than 300,000 square miles.

The difference in size was one reason for differences in organization. For example, a function that might have been centralized in a single-state company might have had separate organizations for each state in a multi-state company. There were other reasons for differences as well. For example, the operating problems and priorities of rural areas differ from those of urban areas. Traditionally, each company worked out operational methods most suited to its own needs, within guidelines and standards provided by AT&T.

As sole or part owner of the operating companies, AT&T derived a large portion of its earnings from those companies. The relationship between AT&T and an operating company was traditionally governed by an agreement called the *license contract* (which terminated with divestiture). Each license contract described the reciprocal services, licenses, and privileges that existed between the parties. The operating company was charged a fee for the services supplied by the AT&T General Departments. The fee was based on services the company received, but it could not exceed 2.5 percent of the company's annual revenues. The licensed company agreed to certain policies and procedures defined by the parent company.

The term *license contract* goes back to the early days of the business when local companies were first licensed to use Bell telephones. For many years, the contract guaranteed that the operating companies would

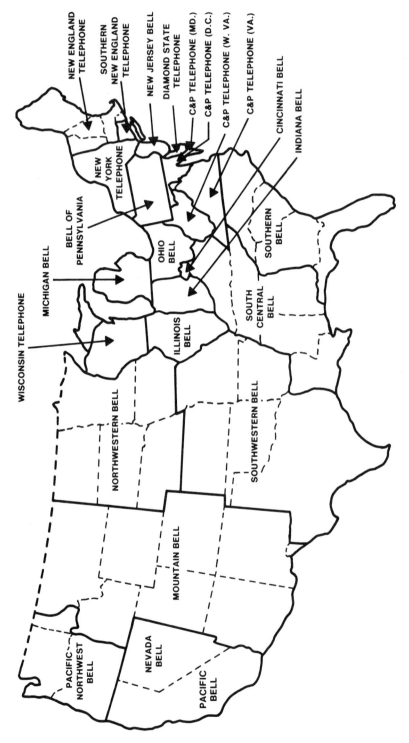

Figure 1-4. Bell operating companies and their territories (1982).

benefit from research, financing, engineering, and other important services offered by the parent company. It assured the manufacture of telephones and other devices and apparatus needed by the licensees for their business. AT&T accomplished this through Western Electric.

Corporate Structure

As with AT&T, the original organizational structure of the operating companies was defined by the jobs that needed to be done. This functional organization, shown in Figure 1-5,[6] later evolved into a network orientation. Still later, marketing became the driving force in shaping the Bell System and its operations. When AT&T reorganized around market segments, it recommended that the operating companies do the same by

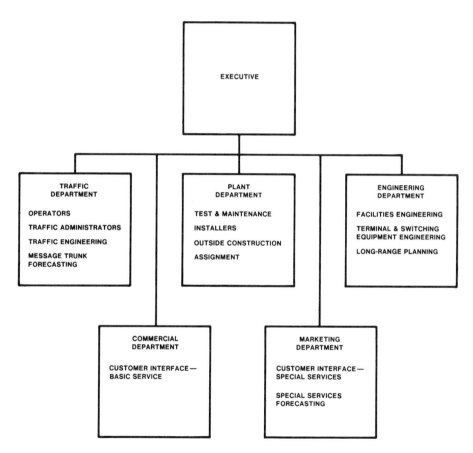

**Figure 1-5. Traditional functional organization
of the Bell operating companies.**

[6] Chapter 13 describes functional activities such as those shown in Figure 1-5.

forming business, residence, directory, public service, and network organizations.

The business, residence, directory, and public services organizations were to be responsible for marketing, sales, and delivery of products directly to customers. The network organization was to provide services between customer locations. It did the planning and engineering, provided the facilities and equipment, and operated the network. This restructuring of traditional lines of managerial authority did not reflect a difference in overall goals, however. Rather, it was done to keep pace with the emerging technological, business, and regulatory environment.

1.2.3 WESTERN ELECTRIC

Western Electric, with headquarters at 222 Broadway in New York City,[7] was the manufacturing and supply unit of the Bell System. In 1982, with about $12.6 billion in sales, Western Electric ranked 22nd on *Fortune* magazine's list of the nation's 500 largest industrial corporations. The company's almost 150,000 employees worked in nearly every state.

Before divestiture, Western Electric made a variety of customer-premises equipment, including millions of telephones each year. It also manufactured much of the other equipment that made up the telephone network. These products were designed by Bell Laboratories, manufactured by one of Western Electric's manufacturing divisions, and distributed to the telephone companies by Western Electric's Bell Sales division. Table 1-1 lists Western Electric's manufacturing divisions and their locations. Figure 1-6 shows the twenty-two manufacturing plants. The structure of Western Electric included a number of divisions responsible for major functional areas.

The **Corporate Engineering Division** coordinated the work of the manufacturing divisions to ensure that the products were compatible with the network. The division also provided research and development support for all Western Electric's engineering activities including manufacturing, equipment engineering, distribution, installation, and repair of products. In addition, it coordinated the company's quality assurance program, which required that inspectors check products to ensure that they met Bell Laboratories standards. Engineering also assisted in planning for the acquisition, leasing, and development of company facilities. It evaluated and verified company-wide cost reductions and monitored energy use at all company locations.

Western Electric's Engineering Research Center near Princeton, New Jersey, developed and improved manufacturing processes. Examples of

---

[7] In October 1983, plans were announced for the sale of 222 Broadway and the establishment of a new headquarters facility in Berkeley Heights, New Jersey.

innovations that emerged from the Research Center include industrial applications of the laser and a technique for sensing minute abnormalities in ceramics.

TABLE 1-1

WESTERN ELECTRIC MANUFACTURING (1982)

Products by Manufacturing Division	Locations
Cable and Wire Products	Atlanta Works (Norcross, GA) Baltimore Works Omaha Works Phoenix Works
Electronic Components	Allentown Works Kansas City Works (Lee's Summit, MO) Reading Works New River Valley Plant (Fairlawn, VA)
Business and Residence Products	Denver Works Indianapolis Works Kearney Works Montgomery Works (Aurora, IL) Shreveport Works
Switching Equipment	Columbus Works Dallas Works (Mesquite, TX) Hawthorne Works (Chicago) Lisle Plant (Lisle, IL) Oklahoma City Works
Transmission Equipment	Merrimack Valley Works (North Andover, MA) North Carolina Works (Winston-Salem) Burlington Works Richmond Works

The **Network Systems and Product Planning Division** ensured that products from the different manufacturing divisions were compatible with the network and met the needs of customers.

The principal points of contact between Western Electric and its Bell System customers were the Material and Account Management Division and two Bell Sales divisions. These organizations were responsible for the delivery of products and services to the customer. To facilitate this, seven Bell Sales regions were established (see Figure 1-6). Regional account management teams assisted operating companies in planning applications for Western Electric products and services and helped Western Electric identify emerging telephone company needs and develop marketing strategies to meet those needs.

- The **Material and Account Management Division** developed plans and support for the regional account management teams. It also forecast the demand for products, placed orders with the factories, and controlled stock supplies in all seven regions. The division established prices and administered the *standard supply contracts* (see below). It prepared brochures, handbooks, and customer instruction booklets on products.

- Two **Bell Sales Divisions** (East and West) provided regional account management. Their responsibilities also included systems equipment engineering, installation and repair of switching and transmission equipment, warehousing, and distribution for the Bell System. Systems equipment engineers tailored complicated Western Electric equipment to the exact needs of the customer and ensured its compatibility with existing equipment. The Bell Sales divisions operated a distribution network consisting of material management centers (huge warehouses) in each of the seven regions; thirty-one smaller service centers, which usually combined stocks of the most frequently needed items and repair facilities; and strategically located local distribution centers (see Figure 1-6).

Through its **Purchasing and Transportation Division**, Western Electric coordinated the purchase of over $4.5 billion in supplies and services from other manufacturers both for its own manufacturing needs and for resale to Bell System companies. Western Electric used more than 47,000 suppliers and transportation carriers and delivered raw materials, parts, and finished products to more than 100 company locations and to Bell customers around the country. Purchases included telecommunications equipment, computers, power equipment, telephone booths, telephone poles, office machines, maintenance items, and stationery supplies. An important part of this work, the engineering and inspection of purchased products to ensure their compatibility and quality, was the responsibility

of Purchased Products Engineering and Purchased Products Inspection, which were located in Springfield, New Jersey.

Western Electric's responsibilities in this area were defined by the *standard supply contract*, an agreement it had with the Bell operating companies. The supply contract, which terminated at divestiture, required Western Electric either to manufacture or to purchase materials that the operating companies might reasonably require, which they then might order from Western Electric. However, the supply contract did not *obligate* the operating companies to purchase these materials from Western Electric. They were free to buy from anyone.

The **Government and Commercial Sales Division** was responsible for the sale of Western Electric products and services to the United States government and other non—Bell System customers.

In addition to its role as the Bell System manufacturing and supply unit, Western Electric responded to the government's needs for both specific design projects and telecommunications systems. During World War II, Western Electric provided communications and radar equipment to the armed forces. After the war, the company did pioneering work in early-warning defense systems such as the Distant Early Warning (DEW) Line, extending from Iceland to the Aleutians. Later, Western Electric and Bell Laboratories developed the Nike-Ajax and Nike-Hercules ground-to-air missile defense systems. More recently, in the early 1970s, Western Electric was prime contractor for the Safeguard antiballistic missile system.

Western Electric has been a major technological contributor to the space program. The company provided the tracking and communications system for the United States' first manned space flight, Project Mercury, and headed the industrial team that designed and built tracking and communications systems for the Gemini and Apollo programs. Bellcomm, a subsidiary of Western Electric, was formed to carry out the systems engineering work on these programs under contract to the National Aeronautics and Space Administration. The United States Air Force and the National Aeronautics and Space Administration use a Western Electric command guidance system and missile-borne guidance equipment to support satellite launches.[8]

Western Electric has also provided complete telecommunications facilities for various government agencies both in the United States and at military bases and embassies abroad. Western Electric recently modified the Nike-Hercules air defense system for the North Atlantic Treaty Organization. The company has been engaged, with Bell Laboratories, in United States Navy submarine sonar and underwater surveillance projects

---

[8] For further details on Bell System contributions to military and space programs, see Fagen 1975/1978, vol. 2.

involving the application of acoustic technology and oceanography. These designs include underwater sensor components, cable systems, associated data-processing equipment, displays, data transmission, and communications links.

Subsidiary Companies

Western Electric owns several subsidiary companies that have supported Bell System operations or provided services. The subsidiaries include:

- **Teletype Corporation**, which develops and manufactures data terminals for the Bell operating companies, other companies, and the United States government. This equipment is used in news and wire-service operations, in data communications, and in computer systems. Teletype Corporation, with headquarters and engineering operations in Skokie, Illinois, maintains two manufacturing plants and a nation-wide network of service centers.

- **Nassau Recycle Corporation**, which reclaims and recycles nonferrous metals such as copper and zinc from scrap equipment and cable. About one-third of the copper Western Electric uses in manufacturing is provided by Nassau Recycle. The company has plants in South Carolina and New York.

- **Sandia Corporation**, which is managed by Western Electric for the United States Department of Energy under a no-profit, no-fee contract. Sandia's principal functions are research and development of nuclear ordnance, research on energy projects, and various other programs in the national interest. Sandia has laboratories in Albuquerque, New Mexico, and Livermore, California.

1.2.4 BELL LABORATORIES

Before divestiture, Bell Laboratories was the Bell System's research and development organization. Recognized worldwide as a prestigious scientific and technical institution, it was the driving influence behind the Bell System's contributions to telecommunications science and technology. The broad scope of these contributions is reflected in Table 1-2. In 1982, engineers and scientists at Bell Laboratories received 310 patents, bringing the total number of patents issued to the company since its founding in 1925 to 19,833. In 1982, they also originated 3823 technical talks to outside organizations and 2087 papers and received more than 87 scientific and engineering awards. Seven scientists from Bell Laboratories have been awarded the Nobel Prize in physics (see Table 1-3).

TABLE 1-2

A SAMPLING OF BELL LABORATORIES
CONTRIBUTIONS TO TELECOMMUNICATIONS
SCIENCE AND TECHNOLOGY*

Microelectronics

Transistor effect†
Silicon gate technology
Molecular beam epitaxy
Charge-coupled devices
Microprocessors and
  microcomputers

Software Systems

Error-correcting code
Computer languages
Computer graphics
Computer operating
  systems
Operations systems
Centralized
  maintenance
  systems
Stored-program control
  network

Digital Technology

Electrical digital
  computer
Digital switching system
Digital transmission
Packet switching
Echo canceler chip

Photonics

Lasers
Lightwave
  communications
  systems
Ultra-transparent glass
  fibers
Light-emitting diode

General Science and Engineering

Single-sideband
  transmission
Network theory
Quality control
Systems engineering
  concept
Negative feedback
Wave nature of matter†
Thermal noise
Speech synthesis
Radio astronomy
Traveling-wave tube
Microwave technology
Information theory
Solar cell
Cellular radio concept
Communications
  satellite
Supercurrent junctions†
Cosmic background
  noise†

*For a more complete list and discussion of Bell Laboratories' contributions, see Bell Laboratories 1982, Lustig 1981, and Mueser 1979.

†Nobel Prize.

TABLE 1-3

BELL LABORATORIES
NOBEL LAUREATES IN PHYSICS

1937	C. J. Davisson	Demonstration of wave nature of matter
1956	John Bardeen Walter Brattain William Shockley	Discovery of transistor effect
1977	Philip Anderson	Study of electronic structure of magnetic and disordered materials
1978	Arno Penzias Robert Wilson	Detection of cosmic microwave background radiation

NOTE: Arthur Schawlow, co-inventor (with C. H. Townes) of the laser while at Bell Laboratories from 1951 to 1962, shared the 1981 prize for work done later at Stanford on Doppler-free spectroscopy.

Purpose

Before divestiture, the purpose of Bell Laboratories was to provide the knowledge and technology essential to meeting the current and future communications needs of Bell System customers. Its activities were divided into two categories: **Research and Systems Engineering** (R&SE) and **Specific Development and Design** (SD&D).

In undertaking research, Bell Laboratories sought new knowledge relevant to communications, explored the potential usefulness of that knowledge, and looked for new modes of communication based on that knowledge. The aim of Bell Laboratories research was to improve the services provided by the Bell operating companies and to reduce the cost of providing those services. Fields of research included the physical and mathematical sciences, computer science, economics, communications principles, communications technology, engineering, and the behavioral sciences. To maintain its leadership in telecommunications, Bell Laboratories has, in recent years, devoted ever more of its efforts and resources to certain fundamental information-age technologies, especially microelectronics, software systems, digital systems, and photonics.

Systems engineers planned the nationwide telephone network and its operations. They considered the entire network rather than just one part of the *plant*[9] or one phase of operations. This included studying performance objectives, evaluating service quality, planning network configurations, generating operations plans and methods, forming requirements for equipment to provide service, and defining plans and procedures for introducing new equipment and services into the network. Systems engineering at Bell Laboratories ensured that the entire telecommunications network worked efficiently to provide continuous service while new technologies, operations, equipment, and services were introduced as they became available. Research and Systems Engineering at Bell Laboratories was funded by AT&T, primarily as a part of the service the parent company provided to the operating companies under the license contracts.

Specific Development and Design was funded by Western Electric. It was directed toward components, devices, and specific products (often involving both hardware and software) to be manufactured and furnished by Western Electric. It was concerned with designing new telecommunications applications using existing types of devices, designing completely new circuits and equipment arrangements, and preparing manufacturing information and test specifications. It involved building and testing equipment designs both in the laboratory and under field conditions, dealing with the problems of early manufacture and use of a product, and making changes as indicated by experience.

The operating companies directly funded certain other work that Bell Laboratories undertook at their specific request, for example, developing computerized operations systems (see Chapter 14) for use in telephone company business operations.

Corporate Structure

Figure 1-7 shows the corporate structure of Bell Laboratories as it existed in January 1983 following the transfer of certain customer products and services organizations to American Bell Inc.

- **Research** included divisions devoted to physics, the information sciences, the communications sciences, and materials science and engineering.

- **Legal** comprised general law, patents, regulatory matters, and corporate studies.

- **Research and Development Planning** was concerned with the organizational structure of Bell Laboratories as it was affected by the changing regulatory and business environment of the Bell System.

---

[9] All of the facilities (such as land, buildings, machinery, apparatus, instruments, and fixtures) needed to provide telecommunications services. Plant is usually divided into outside plant and inside plant.

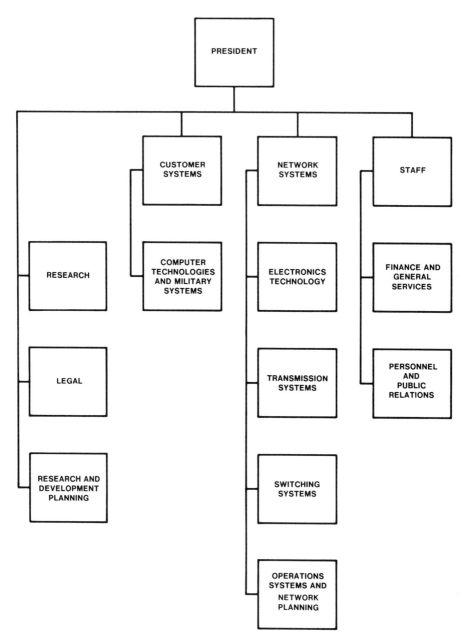

Figure 1-7. Corporate structure of Bell Laboratories (January 1983).

- **Computer Technologies and Military Systems** concentrated on research and development of computer software and hardware and on military systems work. Software development recently became an ever increasing part of the work at Bell Laboratories, and varying amounts were done in other organizational areas as well. The efforts of the Military Systems Division were devoted exclusively to Western Electric's government contracts.

- **Electronics Technology** developed electronic components for systems of all kinds. It was involved in broad areas of research and development ranging from the design and processing of integrated circuits to lightwave communications subsystems to interconnection hardware and power-conversion systems.

- **Transmission Systems** provided the systems engineering, design, and development of systems to meet needs ranging from loop transmission to long-haul land, satellite, and undersea transmission. Research and development of digital transmission was an important aspect of work in this area.

- **Switching Systems** performed research, planning, and development to maintain, improve, and offer new services on existing switching systems and for planning and developing future switching systems.

- **Operations Systems and Network Planning** developed and designed computer-based systems to support telephone company operations and formulated plans for the effective integration of these systems with each other and with the people these systems support. Network planning encompassed functions involved in the evolution and implementation of the network, determining its configuration to best meet future service needs. The Bell Laboratories' Quality Assurance Center was part of this area. Bell Laboratories set the quality standards for products, and it worked with Western Electric quality assurance organizations to monitor manufacturing operations.[10]

- **Finance and General Services** had a wide range of company responsibilities including corporate auditing and finance, internal communications services, security, and managing buildings and grounds.

- **Personnel and Public Relations** included salary administration, personnel matters, affirmative action, public and employee communications, education, and medicine and environmental health.

In 1982, Bell Laboratories' main facilities for research and development were in New Jersey and Naperville (Indian Hill), Illinois (see Table 1-4). In addition, about 15 percent of the staff of nearly 26,000 people was located at seven Western Electric manufacturing plants. These *branch laboratories* (see Table 1-5) helped coordinate and implement Bell Laboratories' specific development and design functions that resulted in drawings and specifications for the telecommunications and software products

----

[10] Section 18.7 discusses quality assurance.

TABLE 1-4

BELL LABORATORIES LOCATIONS (1982)

Location	Activities
Murray Hill, NJ	Administrative headquarters, electronics technology, basic research in various fields
Holmdel, NJ	Systems planning, network planning, operations systems planning, operations research, quality assurance, switching, transmission, customer equipment, research in communications sciences
Whippany, NJ	Loop transmission, mobile radio systems, interconnection, computing technology, engineering information, electronic power systems, military systems
Indian Hill, IL	Electronic switching, computer technology
Piscataway, NJ	Operations and network systems
Chester, NJ	Field laboratory for outside plant equipment and materials and constructed equipment
Crawford Hill, NJ	Radio and guided wave research
Freehold, NJ*	Business services operations and communications systems engineering
Neptune, NJ*	Engineering for facility networking operations and residence systems, *DATAPHONE*† digital service field support and exploratory development
Short Hills, NJ	Personnel, public relations, legal, finance and general services
South Plainfield, NJ	Computer program development, new services planning, quality assurance, network performance planning, education
Warren, NJ	Service center for New Jersey locations, stock supply center for all Bell Laboratories locations
West Long Branch, NJ	Switching and transmission engineering, network planning, operations research, quality assurance administration

* Location became part of American Bell Inc. in 1983.
† Registered service mark of AT&T Co.

TABLE 1-5

BELL LABORATORIES
BRANCH LABORATORIES (1982)

Location	Activities
Allentown, PA	Electronic devices, integrated circuits
Atlanta, GA	Wire, cable, glass lightguides, systems for joining media
Columbus, OH	Switching systems
Denver, CO*	Customer switching systems for PBX services
Indianapolis, IN	Telephones, residential terminals and home communications systems
Merrimack Valley, MA	Microwave radio, carrier transmission
Reading, PA	Electronic devices, integrated circuits

*Location became part of American Bell Inc. in 1983.

that Western Electric manufactures. Bell Laboratories also had field representatives at the headquarters locations of the Bell operating companies. They provided designers with rapid feedback on the quality and performance of new and existing telecommunications equipment.

1.2.5 RELATIONSHIPS WITH NON–BELL SYSTEM COMPANIES

Before divestiture, the Bell System served over 80 percent of the 180 million telephones in the United States, encompassing 30 percent of its geographical area. The remaining 36 million telephones were served by more than 1400 telephone companies that were not part of the Bell System. These *independents*, as they are called, worked with each other and interfaced with the Bell System through the United States Independent Telephone Association (USITA).

Through committees representing different aspects of the telecommunications business, USITA served as a focal point for agreements with the Bell System on issues such as routing the long-distance network and sharing revenues. On technical issues, the Bell System prepared network planning information and equipment compatibility specifications and released them to the independent telephone companies through USITA's Equipment Compatibility Committee and Subcommittee on Network

Planning. These releases, called *technical advisories*, ensured that the equipment and systems installed by the independents were compatible with the Bell System network.

Overseas and other international services use high-frequency radio, undersea cable, and satellite links. AT&T, other United States carriers, and foreign carriers share the ownership of some transmission facilities such as undersea cable; and they lease overseas voice circuits from satellite channels provided by the Communications Satellite Corporation (COMSAT). Service agreements with each foreign agency set up the type and extent of service and procedures for dividing revenues, and they establish the criteria for operations such as circuit engineering and quality of service.

The Bell System has participated in technical planning for international coordination through the Comité Consultatif International Télégraphique et Téléphonique (CCITT) and the Comité Consultatif International des Radio-communications (CCIR). These are the technical organs of the United Nations' specialized agency for telecommunications, the International Telecommunication Union (ITU). They function through international committees of telephone administrations and private operating agencies. Their recommendations, although not carrying the force of regulations, are generally observed, and more and more, they are becoming a consideration in system design, particularly for digital transmission and switching.

The Bell System has also worked with manufacturers other than Western Electric (general trade) who sell their products to Bell System companies. Information essential to general-trade manufacturers, specialized carriers, and other communications companies was made available in various documents. These include technical descriptions, technical manuals for Western Electric products, technical references containing interface information and technical standards for all aspects of the network and its operation, and textbooks and manuals used by Bell System designers. The public has also been able to subscribe to periodicals such as *The Bell System Technical Journal*, the *Bell Laboratories Record*, and the *Bell Journal of Economics*.

1.2.6 RESOURCES AND VOLUME OF SERVICE

After decades of growth, the goal of universal service has been achieved, and telecommunications services have become an increasingly important part of personal and business activities. Almost 1-½ billion miles of wire and radio paths interconnect almost every home and office in the United States. Over 180 million telephones have immediate, real-time access to each other and to 98 percent of another 315 million telephones in other countries.

As a result of such growth, the Bell System became a very large enterprise. Its size can be viewed from several perspectives: financial measures such as its revenues and the amount of capital investment or plant, service measures such as the number of telephones and number of calls handled, and a measure of the vast amount of effort required to build and operate the network and deliver its services—the number of employees. Table 1-6 summarizes Bell System resources and volume of service at the end of 1982. To complete the picture, the corresponding figures for the independent telephone companies are also shown.

TABLE 1-6

RESOURCES AND VOLUME OF SERVICE (1982)

	Bell System	Independents	Total
Operating companies	22*	1,432	1,454
Employees	1,009,817	192,100	1,201,917
Plant ($ billions)	158.0	41.5	199.5
Construction ($ billions)	16.8	4.7	21.5
Access lines (millions)	84.7	21.7	106.4
Central office codes	19,660	11,742	31,402
Local calls (billions)	178.9	67.1	246.0
Long-distance calls (billions)	25.9	6.6	32.5
Average calls/day (millions)	561.1	201.8	762.9
Revenues (billions)	65.1	13.9	79.0

SOURCES: AT&T 1983 and USITA 1983.

*Excluding Southern New England Telephone Company and Cincinnati Bell, Inc.

1.3 THE POSTDIVESTITURE VIEW

1.3.1 SUMMARY OF MODIFICATION OF FINAL JUDGMENT PROVISIONS

The 1982 Modification of Final Judgment (MFJ) requires AT&T to divest itself of the twenty-two Bell operating companies (BOCs). The major provisions of the MFJ are summarized in Table 1-7. Those provisions that significantly affect the conduct of engineering and operations are explained in more detail in the following paragraphs.

The nationwide Bell System network, which was designed, built, and operated as a single unit prior to divestiture, is now divided into two

TABLE 1-7

MAJOR PROVISIONS OF THE
MODIFICATION OF FINAL JUDGMENT

1. Sufficient facilities, personnel, systems, and rights to technical information must be transferred from AT&T to the BOCs, or to a new entity owned by the BOCs, to allow the BOCs to provide exchange and exchange access services independent of AT&T.

2. Facilities, personnel, and accounts used to provide interexchange services or customer-premises equipment (CPE) must be transferred from the BOCs to AT&T.

3. License contracts between AT&T and the BOCs and the standard supply contracts between the BOCs and Western Electric must be terminated.

4. BOCs may create and support a centralized organization for the provision of those services that can be most efficiently provided on a centralized basis. The BOCs shall provide, through a centralized organization, a single point of contact for coordination of the BOCs for national security and emergency preparedness.

5. BOCs must provide all interexchange carriers with exchange access services equal in type, quality, and price to those provided to AT&T. This "equal access" must be provided on a gradual basis over a 2-year period beginning September 1, 1984. By September 1, 1986, all BOC switching systems must provide equal access, although exceptions may be made for electromechanical switches or switches serving fewer than 10,000 access lines where costs of providing equal access are prohibitive. Any such exceptions shall be for the minimum divergence in access and minimum time necessary.

6. BOC procedures for procurement of products and services, dissemination of technical information and standards, interconnection and use of BOC facilities and services, and planning and implementation of new services or facilities must not discriminate between AT&T and its affiliates and their competitors.

7. BOCs may provide, but not manufacture, CPE after divestiture.

8. BOCs may produce and distribute printed directories after divestiture.

9. With the permission of the court, the BOCs may provide products or services in addition to exchange and exchange access services upon showing that there is no substantial possibility they could use their monopoly power to impede competition in the additional markets.

components: an exchange and exchange access portion provided by the divested BOCs and an interexchange portion provided by AT&T. This division does not correspond to the predivestiture distinctions between AT&T Long Lines and BOC operations, between intrastate and interstate jurisdictions, or between toll and local services. It is based instead on a definition of an exchange used in the MFJ.

Prior to divestiture, the term *exchange area* was used to describe an area in which there was a single, uniform set of charges for telephone service. Calls between points in an exchange area were *local* calls. The MFJ defines an exchange area or exchange to be generally equivalent to a Standard Metropolitan Statistical Area (SMSA), which is a geographic area defined by the United States government for statistical purposes. The MFJ concept is to group large segments of population with common social and economic interests within an exchange.

The territory served by the Bell System has been divided into approximately 160 of these exchanges, which are also referred to as *local access and transport areas* (LATAs). Depending on population densities and other factors, most LATAs serve territories ranging from major metropolitan areas to entire states. Accordingly, LATAs generally contain a number of predivestiture exchange areas.

The predivestiture BOCs performed functions that now represent both inter-LATA and intra-LATA functions. The MFJ specifies that, after divestiture, BOCs offer regulated telecommunications services within LATAs, while AT&T and other interexchange carriers (ICs) offer services between LATAs. Some examples of exchange services offered by BOCs are basic local telephone service for residence and business customers, public telephone services, and intra-LATA operator services. In addition, BOCs offer exchange access services that allow inter-LATA networks provided by ICs to access customers within a LATA and allow end-users to access inter-LATA services. Examples of IC services include inter-LATA and international telephone service and inter-LATA operator services.

In addition to reconfiguring their operations to accommodate the transfer of inter-LATA functions to AT&T, the BOCs must modify the intra-LATA networks to provide all other ICs, at their option, with exchange access equal in type, quality, and price to that provided to AT&T. The quality of exchange access is measured in terms of traffic blocking criteria (see Chapter 5) and transmission performance (see Chapter 6). In addition, BOCs must implement a new national numbering plan that provides exchange access to every IC through a uniform number of digits.

The MFJ prohibits joint ownership of switching systems, transmission facilities, and operations systems (see Chapter 14) by the BOCs and AT&T. All Bell System assets are assigned to one or the other. The MFJ does, however, allow sharing, "through leasing or otherwise," of facilities that support both BOC and AT&T services. Sharing of such multifunctional facilities for a period after divestiture is necessary because of the impracticality and enormous cost associated with the immediate reconfiguration and separation of the predivestiture Bell System network. After this transition period, BOC and AT&T facilities will be completely separated.

In its Computer Inquiry II decision (see Section 17.4.3), the Federal Communications Commission (FCC) required that all new customer-premises equipment (CPE) be provided by a separate subsidiary on a detariffed basis effective January 1, 1983. Installed CPE and remaining BOC inventories of CPE as of that date were sold or leased by the BOCs during 1983. The MFJ requires that leased CPE be transferred to AT&T at divestiture. After divestiture, BOCs are allowed to provide, but not manufacture, new CPE.

The provisions of Computer Inquiry II and the MFJ have a major impact on the BOC organizations that directly interface with customers for the provision of service and equipment, billing and collections, trouble referral, and other matters. Prior to divestiture, the BOCs provided a single point of customer contact for local and toll services as well as CPE. Under the MFJ, the BOC personnel and associated customer support responsibilities for these services are divided between the AT&T and BOC units responsible for regulated network services or CPE.

Figure 1-1, presented earlier, depicts the predivestiture relationship of the components of the Bell System. Figure 1-8 shows the new corporate structures resulting from divestiture. Sections 1.3.2 and 1.3.3 provide additional information on the organization and functions of the post-divestiture AT&T and BOCs, respectively.

1.3.2 THE NEW AT&T

The structure of the new AT&T is shown in Figure 1-8. AT&T Corporate Headquarters sets overall corporate policy and strategy for the other six entities shown. Five of these entities are divided into two sectors: **AT&T Communications** and **AT&T Technologies**, which are responsible for essentially regulated and unregulated activities, respectively. As AT&T gains experience in the new telecommunications environment, organizational structures and activities will be evolving to improve operating effectiveness. Therefore, only a brief summary of each entity is provided in this section.

AT&T Communications

The business of AT&T Communications is moving information electronically, from customer premises to customer premises, domestically and internationally. Initially, employees were drawn from BOCs, the AT&T General Departments, and Long Lines. At its inception, the company served sixty million residence customers and nearly six million

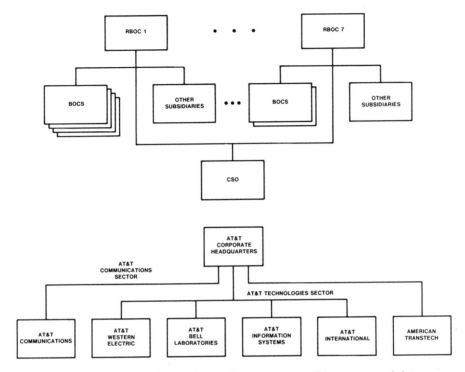

Figure 1-8. Corporate structures after divestiture. *Top,* structure of the divested operating companies (the RBOCs and their associated BOCs are identified in Figure 1-9); *bottom,* structures of AT&T.

businesses. As a result of the new regulatory environment, AT&T Communications will provide inter-LATA long-distance services, which include the interstate services previously provided by AT&T Long Lines and intrastate, inter-LATA services. Its goals, highlighted at the beginning of its mission statement, are: "To provide high-quality, innovative, widely available communications services that satisfy customers' needs to move information electronically throughout the United States and the world."

AT&T Western Electric

AT&T Western Electric continues in its leadership role as a provider of technologically advanced, high-quality products and services in the telecommunications and information systems markets. These include equipment and systems for telephone companies, consumer products, electronic components, and processors. Further, according to its mission statement, "within this role, AT&T Western Electric has a responsibility to expand its telecommunications business and address new market opportunities...."

AT&T Information Systems

AT&T Information Systems, derived initially from American Bell Inc., develops, sells, and services leading-edge communications products, information management systems, and enhanced services to business customers. It also distributes products for residential and small business customers. Its products, services, and systems reflect the rapid convergence of what were three distinct industries: telecommunications, office equipment, and data processing.

AT&T International

AT&T International, as before divestiture, markets, sells, and services current and future products and services of the AT&T Technologies sector outside the United States.

AT&T Bell Laboratories

AT&T Bell Laboratories provides the technology base for AT&T's future and designs and develops the systems and services needed by AT&T enterprises. This includes basic research and the engineering and design of components, devices, and information and operations systems and services. It also conducts systems engineering work to help identify the best solution to customers' needs. It aids the national defense by making its special capabilities and expertise available to the government.

American Transtech

American Transtech provides and/or packages quality stock transfer and related services for AT&T and regional company shareowners and customers at the lowest reasonable cost. Its mission statement further states: "American Transtech will enter new business opportunities that maximize existing functions and capacity to produce an attractive rate of return and growth."

1.3.3 THE DIVESTED OPERATING COMPANIES

The MFJ allows the BOCs considerable latitude regarding their choice of corporate structure and organization after divestiture. It explicitly states that "nothing in this Modification of Final Judgment shall require or prohibit the consolidation of ownership of the BOCs into any particular number of entities."

After study by a committee of AT&T and BOC officers, the structure adopted (see Figure 1-8) organizes the postdivestiture BOCs into seven regional BOCs (RBOCs). These seven independent corporations wholly own and are supported by a separate central services organization (CSO). These organizational units are described in the following paragraphs. They have no corporate connection with the new AT&T or its affiliates.

Regional Bell Operating Companies

RBOCs were designed to form corporate units with roughly equivalent assets and financial strength. Each RBOC contains from one to seven BOCs that serve the same general region of the country. Figure 1-9 is a map showing postdivestiture RBOC boundaries. The table below the map presents statistics that demonstrate the relative equivalence in size of the seven new corporate units.

The RBOCs operate as holding companies that hold the stock of the BOCs in their respective regions. The RBOCs are free to enter other, unregulated lines of business through the creation of separate subsidiaries. For example, each region has already formed a subsidiary to provide cellular mobile telephone services. Finally, the RBOCs jointly direct the work of the CSO.

Bell Operating Companies

The BOCs offer regulated intra-LATA telecommunications services and exchange access services within their predivestiture operating territories. While the BOCs within each region remain as separate corporations with their own boards of directors and officers, they cooperate to achieve economies of scale possible within the larger, regional framework.

The BOCs in each region have already consolidated certain functions, such as procurement and staff services. In most cases, this has been accomplished through the formation of a Regional Services Company, staffed and managed by the BOCs in the region. The BOCs within a region may also cooperate in financial, strategic, and network planning to meet their obligations to the RBOC that is their parent company.

BOC operations are modified considerably as a result of divestiture. Personnel and organizations that support the provision and maintenance of CPE, inter-LATA telecommunications services, and directory services move to AT&T entities or to other subsidiaries of the RBOCs. While most of the operations functions and organizations described in Chapters 13 through 16 are still present in the BOCs, the details of operations change considerably. The creation of LATAs defines a new set of boundaries for the operation and engineering of the network. Similarly, management and administration of the network are no longer centrally directed by AT&T. Massive transfers of assets and personnel, changes to records and support systems, and modifications to operating procedures were required to accommodate the physical rearrangement of the Bell System and the breakup of its corporate structure.

Central Services Organization

While the MFJ ends the centralized ownership and management of the Bell System and breaks up the vertically integrated structure that combined operations, research, and manufacturing, it recognizes the possibility that the BOCs might still choose to provide certain support functions

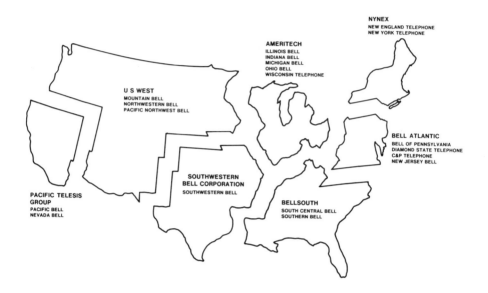

	Estimated Population Served (Millions)	Estimated Number of Employees (Thousands) 1/1/84	Total Assets ($ Billions) 6/30/83	Estimated Total Access Lines (Millions) 1/1/84	Total Number of Telephones (Millions)
NYNEX Corporation	25	98.2	17.4	12.8	17.4
Bell Atlantic Corporation	27	80.0	16.3	14.2	23.2
BellSouth Corporation	30	99.1	20.8	13.6	23.1
American Information Technologies Corporation (Ameritech)	30	79.0	16.3	14.0	23.6
Southwestern Bell Corporation	21	74.7	15.5	10.3	16.9
U S WEST, Inc.	22	75.0	15.1	10.6	16.7
Pacific Telesis Group	20	82.0	16.2	10.9	15.1

NOTE: Based on Bell System figures for December 1982 except as noted.

Figure 1-9. Regional Bell operating companies.

on a centralized basis. The CSO provides this support. The economies realized by not duplicating certain technical and support functions in each RBOC enhance the financial position of the new companies and reduce the requirements for staffing these highly technical activities. Also, the centralization of certain activities in support of the BOC networks promotes technical compatibility and supports high-quality, nationwide telecommunications service. These activities also facilitate the introduction of new technology into a national network managed by a number of totally independent corporations.

Consistent with the intent of the MFJ, the work of the CSO is directed by the RBOCs. The CSO is owned equally by each of the seven RBOCs, and its board of directors is composed of BOC and RBOC officers. CSO work plans and budgets are determined by a committee structure that includes RBOC representatives at all levels of management.

During 1982, the BOCs determined the initial set of functions to be performed by the CSO. The vast majority of these functions are technical in nature. These technical support functions include:

- *network planning*, which includes analysis and planning for new technologies and services, participation in the development of network standards, and operations systems planning

- *engineering and operations support* related to procedures and standards for network operations and quality assurance implementation

- *information systems development* for the many current operations systems whose development is assigned to the CSO and for future systems

- *technology systems*, which includes generation of generic requirements for equipment, technical analysis of products for the network, and quality assurance methods

- *applied research* in the physical sciences, mathematics, computer science, network technology, and new services.

The CSO also provides support for the BOCs in areas such as market research, regulatory matters, and other financial and administrative matters. The initial staff of the CSO is drawn primarily from Bell Laboratories, AT&T, Western Electric, and the BOCs.

AUTHORS

J. A. Civarra
B. R. Eichenbaum
J. A. McCarthy
J. A. Schelke

2

Services

2.1 INTRODUCTION

Like other aspects of the Bell System, services are dynamic: New services are introduced frequently in response to evolving customer needs and the capabilities of new technology. Consequently, at any given time, there are many new services in various stages of planning and development. This chapter covers only services currently available and some services with an announced availability date. The next few paragraphs discuss some general concepts and terminology related to services, followed by descriptions of specific services.

2.1.1 BASIC AND VERTICAL SERVICES

The concept of basic and vertical services has been central to the Bell System's role as a regulated monopoly providing telecommunications services. For several decades, **basic service** has been coupled to the Bell System's goal of universal service (see Section 17.4.1), that is, that telephone services should be available to every home in America at an affordable price. Basic service represents that universally available and affordable service.

Basic residence service, for example, generally implies that, for a fixed monthly charge, a customer receives the following:

- a standard, rotary-dial telephone set

- on-premises wiring

- a network access line—the connection to a local switching system for local calling and for access to the network

- a listing in the white pages of a directory.

All other services have been classified as **vertical services** for which customers pay a charge that is additional to the cost of basic service. Vertical services may provide greater convenience, more attractive telephone sets, or additional functions or features beyond basic service. Traditionally, revenues from vertical services have helped to maintain an affordable price for basic service. However, as competition replaces regulation in the telecommunications business, the price of a particular service will become more closely related to the cost of providing that service.

Section 2.2 describes vertical services related to station equipment. Section 2.3 describes customer switching services, vertical services used almost entirely by business customers. Sections 2.4 and 2.5 describe vertical services available through access to the network.

2.1.2 ORDINARY AND SPECIAL SERVICES

From the telephone company viewpoint, services are classified as **ordinary** or **special**. *Ordinary services* usually include residence service, public telephone service, mobile telephone service, and basic individual-line business services. All other services are considered *special services* (often called *specials*). Special services require special treatment with respect to transmission, signaling, switching, billing, or customer use and are used mostly by business customers. The overall demand for special services is growing twice as fast as the demand for ordinary telephone service.

Examples of special services described in this chapter include foreign exchange service, Wide Area Telecommunications Services, private branch exchange, and centrex services, and private-line and private network services. There are about twenty-five major categories of special services.

Facilities provided by the Bell System for other communications firms form an important and rapidly growing class of special services. These offerings include local distribution capability, interoffice facility sections, and access to the Bell System network for resale carriers and other common carriers (OCCs), including domestic satellite carriers, international record carriers,[1] and local radio common carriers. To provide communication services to their customers, these competing firms use Bell System facilities in conjunction with their own facilities or with facilities rented from independent telephone companies. Facilities offered to other carriers parallel the wide variety of general special services that the Bell System provides to its own customers: voice, data, telegraph, and television.

---

[1] *International record carriers*, such as International Telephone and Telegraph, RCA Global Communications, and Western Union, offer data and message services (like telex) internationally.

2.2 TERMINAL PRODUCTS

Terminal equipment is the principal interface between the customer and the nationwide telephone network. It ranges from the basic telephone set, which provides voice services, to the more versatile and specialized equipment that can link with computers and provide additional services such as data and graphics transmission.

This section describes terminal equipment available from the Bell System. In the past, all Bell System terminal equipment was leased to customers, but now, as the result of an order issued by the Federal Communications Commission (FCC) in 1975, customers have the option of purchasing it (see Section 2.7). Customers may purchase and install equipment from any manufacturer, provided the sets meet certain registration requirements (see Sections 8.7 and 11.1.2).

2.2.1 TELEPHONE SETS AND VERTICAL SERVICES

The telephone set is an important element of the communications network and provides access to a variety of services. The traditional rotary-dial set is available in desk and wall models and in various colors. It represented approximately 50 percent of all residential sets in service in 1980.

A number of premium telephone sets and decorator telephones are also available. (Figure 2-1 shows a group of *DESIGN LINE*[2] and other decorator telephones.) Premium sets have special features that make the use of the telephone more convenient or pleasant and provide the same access to the network as standard sets. There is an additional charge for premium sets, an example of a vertical service. The *PRINCESS*[2] telephone, a premium telephone introduced in 1959 as the "bedroom set," offers special features such as a lighted dial and a night light. The *TRIMLINE*[2] telephone features the dial in the handset and a "recall" button that allows the user to make consecutive calls without hanging up the handset.

DESIGN LINE decorator telephones use the same internal components as standard rotary or *TOUCH-TONE*[3] telephones but have distinctive housings. These sets are sold outright to the customer—there is no monthly service charge—and are covered for a limited warranty period. After this period, customers can purchase a maintenance contract for the internal mechanism.

---

[2] Registered trademark of AT&T Co.

[3] Trademark of AT&T Co.

Figure 2-1. Some *DESIGN LINE* and other decorator
telephones, including designs from non-Bell manufacturers.

Other vertical services related to telephone sets are *TOUCH-TONE*[4]
dialing and extension phones. *TOUCH-TONE* dialing is faster and more
convenient for the customer. Faster dialing also offers advantages to the
Bell System, because switching equipment is more quickly available for
other calls. *TOUCH-TONE* dialing will be required for many telephone

[4] Registered service mark of AT&T Co.

services that are expected to grow in the future, for instance, banking by telephone and other services that require access to a computer.

Extension phones make single-line telephone service in the home more convenient. In addition, the use of a second telephone line is becoming more prevalent. A series of 2-line sets that provide hold and signaling features is now available. A customer can place one call on "hold" and answer a second call, or the customer can signal another person in the house that there is an incoming call on the second line.

2.2.2 MODULAR TELEPHONES AND RETAIL SALES

The telephone sets now available from the Bell System are *modular*. A modular set has plug-ended cords that connect the telephone base to the handset and wall connector, permitting installation and removal by the customer. Modular sets provide the customer with faster and more convenient service because customers can pick up these sets at retail sales locations and do not need to schedule installation (see Section 2.7.1). This also avoids the cost of a home visit.

2.2.3 DATA PRODUCTS

As computers and other sophisticated business machines become more commonplace, data transmission is becoming an increasingly larger part of almost all business communications. For many years, the Bell System has offered a wide variety of data products to satisfy the needs of customers. This section describes two general categories of data products: **data sets** and **data terminals**.

Data Sets

Digital computers and various types of data terminal equipment produce data in digital form, that is, as a sequence of discrete electrical pulses. While digital transmission facilities, which transport data in digital form, are rapidly being deployed in the telecommunications network, analog transmission facilities, which transport data as continuous electrical waves,[5] still represent a greater share of the total network. *Data sets* (also called *modems*) provide the conversion and control functions required to transmit digital data over analog facilities. The Bell System has a number of *DATAPHONE*[6] data sets available with different capabilities and

---

[5] Chapters 6 and 9 discuss digital and analog transmission.

[6] Registered trademark of AT&T Co.

features suitable for a wide range of applications. Section 11.1.2 contains a functional description of data sets and some of the specific types and characteristics.

In general terms, the *DATAPHONE* data communications service categories associated with *DATAPHONE* data sets are based on the type of analog facilities used: narrowband, voiceband, or broadband;[7] private line or switched. Data sets are primarily categorized by the data transfer rates they provide:

- low speed — up to 300 bits per second (bps)

- medium speed — 300 bps to 9600 bps (9.6 kbps)

- high speed — over 9.6 kbps.

In the case of the public switched telephone network (PSTN), including Wide Area Telecommunications Services and foreign exchange lines (see Section 2.5.1), *DATAPHONE* data sets permit customers to send data between any two locations served by the network at speeds up to 4.8 kbps. A telephone set is used for setting up the channel and for alternate voice communications. The calling and answering can also be controlled automatically by the customer's computer or data terminal. Automatic calling and answering take place through appropriate interaction with the data sets and associated *automatic calling units*. These units dial, connect, and terminate calls.

DATAPHONE data communications service is also provided on voiceband private lines. In this service, as with *DATAPHONE* data communications service for PSTN applications, a *DATAPHONE* data set is used on the customer's premises. Arrangements can be added to permit voice communications on the private lines and to permit access to the PSTN for service backup in case of a private-line outage. Voiceband private-line configurations can be point to point or multipoint; the latter is more prevalent. Speeds up to 9.6 kbps are offered.

DATAPHONE II data communications service, the most recent offering on voiceband private lines, employs a series of advanced microprocessor-based data sets. These sets provide considerable built-in diagnostic, testing, and on-line monitoring capabilities for a data network using diagnostic control devices at a customer's central computer site. (See Figure 2-2.) This capability is particularly important to customers such as airlines and banks who have data networks with real-time applications. These customers purchase terminals, computers, etc., from many suppliers, and they must quickly identify the vendor responsible for fixing a trouble.

---

[7] See Section 6.2.1.

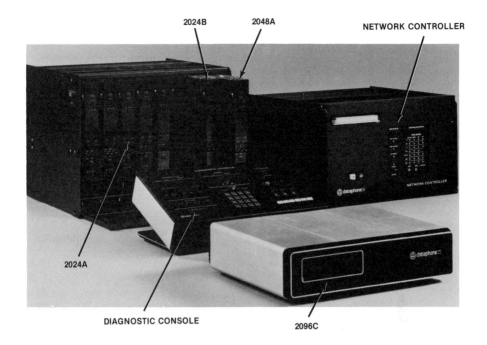

Figure 2-2. The *DATAPHONE* II data communications service new family of data products includes data sets, a diagnostic console, and a network controller.

High-speed data sets are used with analog broadband services. Both point-to-point private-line and switched offerings are provided. Speeds on broadband private-line channels range from 19.2 to 230.4 kbps. Broadband services find the greatest use in applications involving computer-to-computer transmission of large amounts of information.

The Bell System also provides several auxiliary sets for use as adjuncts to data sets. The automatic calling unit mentioned earlier is a commonly used auxiliary set. Other data auxiliary sets perform functions such as signal conversion and testing.

Data Terminals

Data terminals are the end points in data communications. They originate and/or receive data transfers. Teletypewriter, telegraph, and remote metering terminals can be used with low-speed data sets to transmit data on narrow-bandwidth analog circuits. Cathode-ray tube (CRT) terminals, certain teletypewriters, and line printers along with medium-speed data sets use voiceband private lines or the PSTN for data transfer. Some facsimile terminals, CRT terminals, and high-speed line printers are connected to high-speed data sets to transmit and/or receive data over broadband analog channels.[8]

The outputs from a number of terminals, each connected to a *data service unit*, may be multiplexed (combined) for transmission over the **Digital Data System** (DDS) network, which is described in Section 11.6.1. The data service units perform functions similar to those performed by data sets, except that conversion between analog and digital formats is not required since the DDS provides end-to-end (terminal-to-terminal) digital transmission. Computers may also appear as high-speed data terminals connected to the DDS network via an appropriate data service unit. The service provided the DDS network is called *DATAPHONE digital service* (see Section 2.5.4).

The Bell System offers many different types of terminals. Some of the more recent offerings are in the *DATASPEED*[9] terminal set line (shown in Figure 2-3), a collection of high-speed data communications terminals that includes CRT terminals, line printers, and intelligent interactive input/output terminals. (Intelligent terminals typically contain logic and memory capability.) New disk storage devices, offered with certain

Figure 2-3. *DATASPEED* **4540 line of data terminals.**
(*Courtesy of Teletype Corporation*)

[8] Table 2-1 in Section 2.5.2 lists narrowband, voiceband, and broadband offerings.

[9] Registered trademark of AT&T Co.

teleprinter and CRT terminals, permit message preparation and storage to improve interaction with a remote host computer. Section 11.1.3 provides more details on these data terminals.

2.2.4 SPECIAL-PURPOSE TERMINALS

Special telephone sets and auxiliary equipment make other services available to both residential and business customers. One such service is automatic dialing from a directory of stored numbers. Some telephones use punch-coded cards (card dialers). Others, like the *TOUCH-A-MATIC*[10] S repertory dialer, automatically dial a number at the touch of one button. The repertory of numbers is stored in electronic memory (see Figure 2-4). Many of these telephones automatically store the last

Figure 2-4. *TOUCH-A-MATIC* **S series telephone, the Bell System's first microprocessor-based telephone for the home.** Important or frequently called numbers can be dialed at the touch of a button. Red and green buttons at the top of the panel identify emergency numbers.

[10] Registered trademark of AT&T Co.

manually dialed number and redial it when the user pushes a button. Some designs have lights on their panels to help locate emergency numbers.

Auxiliary equipment, including hands-free telephone equipment together with electronic graphics, is also available for teleconferencing between two or more groups. The **4A speakerphone**, which is used by both residence and business customers, has an omnidirectional microphone and an adjustable loudspeaker to permit hands-free conversation and allow a group of people to participate in a conversation (see Figure 2-5). Portable conference telephone units are available for conferences or for communication between a student confined at home and a classroom.

Figure 2-5. 4A speakerphone. The loudspeaker is at the left. The transmitter unit has a volume control, an ON/QUIET-OFF switch, and a solid-state ON/OFF indicator light. The microphone is located under the switches.

The *GEMINI*[11] 100 electronic blackboard system uses a graphics transmission terminal that can send handwriting over conventional voice-grade telephone lines. It can be used for remote teaching and teleconferencing at a number of different locations simultaneously. (See Figure 2-6.)

[11] Registered trademark of AT&T Co.

Figure 2-6. A demonstration of the *GEMINI* 100 electronic blackboard.

Special-purpose terminals for business customers, like the **Transaction** telephone, can automatically dial a non—Bell System credit service center or data base for credit authorization or check verification. In a typical transaction, a cashier slides two cards (the merchant's identification card and the customer's credit card) through a magnetic strip reader in the telephone set (see Figure 2-7) and then keys in the sale price on a *TOUCH-TONE* telephone dial pad. The terminal automatically dials a computer in a bank or credit agency and obtains a purchase authorization as an audio response, a light, or a visual display.

2.2.5 AIDS FOR THE HANDICAPPED

Special Bell System equipment gives disabled persons access to basic telephone service. Someone with impaired mobility may find automatic dialing telephones and speakerphones easier to use than standard sets (see Section 2.2.4). For people who have lost the use of their larynx, or voice box, there is the **artificial larynx** (invented at Bell Laboratories in 1929), which replaces the vibrations of normal vocal cords with electronically controlled vibrations that can be formed into words (see Figure 2-8).

Several different aids are available to persons with impaired hearing. Among these are amplifying handsets, sets with tone ringers, and sets

Figure 2-7. The Transaction III terminal with a Transaction printer.

Figure 2-8. The artificial larynx in use.

that can be equipped with loud bells. The *CODE-COM*[12] set converts sound into either visual signals—a flashing light—or tactile signals—a vibrating disk (see Figure 2-9). To alert a deaf person to an incoming call, an ordinary household lamp may be plugged into the *SIGNALMAN*[12] relay switch, which causes the lamp to flash on and off. Alternatively, an electric fan can be plugged into the unit to signal someone who is both deaf and blind by blowing air on the person.

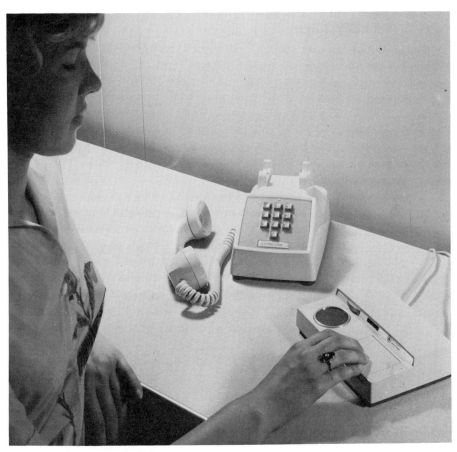

Figure 2-9. *COM-CODE* **set for the handicapped.** When connected to conventional telephones, this device allows a deaf person to receive messages via flashes of light or vibrations. The circular vibrating pad is shown on the left of the device, the sending key, which is used like a telegraph key, on the right. Light flashes come from a recess (the black rectangle) to the left of the bell symbol.

[12] Trademark of AT&T Co.

2.3 CUSTOMER SWITCHING SERVICES

The term *customer switching service* describes an arrangement that permits flexibility in the connections between one or more telephone lines and one or more station instruments.[13] Services can be tailored to the individual needs of the customer and may involve additional equipment either on the customer's premises or in a central office.

2.3.1 CUSTOMER NEEDS

Customer switching services are provided primarily for business customers, although the communications needs of some very large residences may overlap those of a small business. This section discusses only business applications.

While the specific communications needs of business customers depend on the size and nature of their businesses, for discussion purposes, these requirements may be divided into four broad categories: **intralocation calling, incoming calls, outgoing calls,** and **communications management**.

The first requirement is *intralocation calling*. Many businesses need to communicate between stations on the same premises. This is known as *intercom calling*. For voice communication, intercom calling may mean calls between people in different offices, access to an on-premises paging system, or dial access to a customer-owned recorded dictation system. For data communication, it may involve communications between a terminal and a computer or another terminal on the same premises.

The second category involves varying requirements for handling *incoming calls*. For business customers, incoming calls are important since they often represent new or additional business. Most customers, therefore, want to present a good telephone image to calling parties. This usually means having enough lines and attendants to ensure that incoming calls are promptly answered and efficiently passed to someone who can help the calling party.

The third requirements category concerns *outgoing calls*. The importance and nature of outgoing communications are, of course, a function of the customer's business. When employees spend a significant part of their workday making telephone calls, it is important to make those communications as convenient and friendly as possible. Station equipment with button-operated features and switching systems with automatic call-processing routines meet these needs. Control of outgoing call possibilities from selected lines may also be important to avoid abuses.

---

[13] Two or more telephone instruments permanently connected to a single telephone line is *not* a customer switching system.

The fourth category concerns the different needs of businesses in *managing communications*. Since communications can be a major expense, many customers have one or more managers in charge of communications facilities. To help these people do an effective job, modern equipment furnishes them with management information (data) about the use and performance of the communications service it provides.

The rest of Section 2.3 discusses the service needs of small-, medium-, and large-size business customers as well as the needs of some special customers.

2.3.2 SMALL-SIZE BUSINESS CUSTOMERS

The small business customer may be a doctor, a realtor, or a car dealer, for example. The needs of these customers are relatively simple. For incoming service, they need one or a few incoming lines. Depending on the nature of the business, these lines may be in one group accessed by a single listed directory number, or they may be individually listed so that the calling party can dial a specific number to reach a specific line. On the customer's premises, calls will be either answered at a central location and then passed to the desired person or answered at the specific dialed locations. In the latter case, there will be less need for call passing.

Call passing can be done in several ways. The desired person may be paged and requested to pick up a specific line. Another method uses switching system features that permit the transfer of the call to a telephone near the person.

In very small customer applications, where face-to-face communication is usually quite convenient, station-to-station intercom calls are often not required. These applications, however, may need paging or intercom calling to accomplish call passing. Some means may be required to hold an ongoing call temporarily so that a person may talk privately to someone else or answer another call.

The small business customer's outgoing call requirements are also simple. Like residence customers, most small business customers use their line(s) for both incoming and outgoing calls and are billed for each toll call they originate. Control of outgoing calls is also of minimal concern to the small customer because the person responsible for telephone charges can personally discuss problems with the individuals involved.

The needs of small business customers are often met by *key telephone systems*, which allow the station user to originate or answer a call on one of several lines by operating a button (*key*), and provide features such as hold, intercom calling, and message-waiting lights. Modern small communications systems (see Section 11.2.2) using integrated circuit technology, multibutton electronic telephones, and microcomputer control provide numerous additional features under software control.

2.3.3 MEDIUM-SIZE BUSINESS CUSTOMERS

The medium-size customer may be a business occupying a small office building, a manufacturer, or a municipal office such as a police station. An attendant at a central answering position most often handles incoming calls to the 20 to 200 on-premises lines representative of a medium-size customer. The physical size of this customer's organization makes convenient intralocation station-to-station calling more important, and outgoing traffic can range from a modest communications activity to a major one. Typically, management of communications services for this size customer is a part-time assignment for one person.

The customer switching service most often provided for medium-size business customers is known as a *private branch exchange* (PBX) service.[14] A PBX is a relatively small telephone switching system (*exchange*) located on the customer's premises (*private*) and connected to a central office (as a *branch*). The basic features of a PBX provide for the central answering position, convenient station-to-station intercom calling, and whatever special incoming and outgoing call features are necessary.

Customers in this size category often make and/or receive a substantial volume of calls to or from one or more distant areas of the country. To meet the needs of these customers, the Bell System offers **Wide Area Telecommunications Services** (WATS), which are described in Section 2.5.1. With WATS, calls to relatively large geographic areas are billed on a bulk basis rather than individually. The actual cost of a given call can therefore vary considerably depending on both the area called and the type of outgoing line—WATS or regular direct distance dialing (DDD)—selected.

In most cases, a medium-size customer will have a small number of different outgoing lines to access various geographic areas. Station users can then usually select the most economical line (that is, the most economical route or service) for their calls, depending primarily on the call's destination. Station users can be provided with a map or a list of area codes as the basis for selection. Other factors that may influence route selection include the time of day and whether any outgoing lines, such as WATS lines, are already busy.

In more complex installations, route selection by the individual station user is often not practical. Installations like these with calls to many geographic locations and with a greater number of different outgoing lines usually require outgoing calls to be placed through attendants. The attendants are specially trained to select the optimal route. In the most modern PBXs, this function can be performed by a computer within the PBX using software programs to provide a feature known as *automatic route selection* (see Section 2.3.4).

[14] To distinguish from early manual cord boards, current automatic systems are sometimes called *private automatic branch exchanges* (PABXs).

PBX service also meets other needs of the medium-size business customer. For example, it is usually possible to transfer a call from one station to another or to have a secretary pick up a telephone that is ringing in another office. As an alternative to PBX service, which is a customer switching service, a medium-size customer may select an exchange service (see Section 2.4), known as *centrex* service, that provides a PBX-type service.

2.3.4 LARGE-SIZE BUSINESS CUSTOMERS

Large customers have from about 200 to 10,000 or more lines. Typical locations are headquarters buildings, large banks, and combined design and manufacturing locations. At this size, centralized answering becomes less attractive, and a means of direct inward dialing[15] (DID) to station lines is often provided. Intercom calling is more important, and outgoing calls become a major concern. Most large customers also have some requirement for digital data communications on the same premises. All of these needs are satisfied by large PBXs.

The number of outgoing calls from a large system usually justifies the purchase of special long-distance facilities such as WATS lines and **foreign exchange** (FX) lines (see Section 2.5.1) that go directly to frequently called areas served by a distant central office. To use these special facilities in the most efficient way, modern PBX systems provide *automatic route selection,* a service that automatically analyzes a dialed number and selects the least-cost type of line (that is, type service) available for the call. Special arrangements are also available to distribute a relatively large number of incoming calls efficiently to a special group of lines (such as a customer service department).

In large companies, millions of dollars per year may be spent on communications. For this reason, large companies assign a small staff to manage communications services. The PBX system must provide management information for use by this staff. Modern systems not only can collect general data on system traffic (calls) but also can make a detailed record of each outgoing call, so that the costs of the communications facilities can be allocated to users.

2.3.5 SPECIAL CUSTOMERS

An airline reservation center is an example of a special customer. This type of business has a large number of incoming calls that must be evenly distributed among the reservations agents. *Automatic call distributors* (ACDs) provide this service. The operating costs and performance of an ACD greatly depend on having the correct number of agents on hand

---

[15] With *direct inward dialing,* the telephone number of the called party can be dialed directly; the call need not be passed by an attendant.

at all times. For this reason, modern ACDs also provide the customer with traffic data (in either raw or processed form) that can be used to manage the work force.

Another special customer is the telephone answering bureau, which answers other people's (clients') telephone calls. Historically, these customers have been served by providing the bureau with an extension of the client's phone. However, the service can now be implemented by means of a *Call Forwarding* feature available in modern electronic switching systems. With this feature, calls can be redirected (or "forwarded") from the original number to a special line at the answering bureau, where messages are taken for relay to the client. (See Section 11.2.7.)

2.4 EXCHANGE SERVICES

The term *exchange services* describes those services provided through the local or exchange area network. Access to the network and its services is obtained by one or more lines that connect the customer's station set(s) to the central office. Customers can choose from exchange services that range from basic local calling with standard rotary-dial telephones to PBX-like business services.

2.4.1 EXCHANGE LINES AND LOCAL CALLING

A single business or residence line connected to a rotary-dial telephone provides basic exchange services such as **local calling** and the **911 Emergency Service** described in the next section. *Local calls* are calls to any customer in the local calling area of the calling customer's central office. A *local calling area*, or *exchange area*,[16] is a geographic area within which a strong community of interest exists (that is, heavy calling volume among customers within the area). It may be served by several central offices. Basic exchange service also includes operator assistance on local calls and directory assistance[17] (see Section 2.5.1). This basic service is typically provided under a tariff[18] that allows the customer either flat-rate or measured-rate billing. With flat-rate billing, the customer can make an unlimited number of local calls for a fixed monthly charge. With measured-rate billing (also called *measured service*), the customer pays a lower fixed rate plus an additional charge for all local calls in excess of a

---

[16] This discussion is based on the traditional definition of exchange areas. The term was used with a different meaning in the 1982 Modification of Final Judgment.

[17] Some companies now charge for directory assistance calls when they exceed a fixed monthly allowance.

[18] *Tariffs* are rates and conditions for various telephone services (see Section 17.3).

specified monthly allotment. The trend in the Bell System is toward measured service.

In some locations, a customer may have the option of subscribing to single-party or multiparty service. With multiparty service, the line is shared by two or more customers; however, only one party may use the line at a time except when a call is between two customers on the line. Various ringing combinations are provided to indicate the destination of incoming calls. In 1981, about 97 percent of the Bell System was single party, with the trend toward total single-party service.

2.4.2 CUSTOM CALLING SERVICES I

Custom Calling Services I is a group of four features that take advantage of the stored-program control of electronic switching systems (see Section 10.3.1): **Call Waiting, Call Forwarding, Three-Way Calling,** and **Speed Calling**.

Call Waiting

This feature allows a customer engaged in a call to be reached by another caller. A short tone informs the customer that another call is waiting to be accepted. The tone is heard only by the Call Waiting customer; the caller hears the regular audible ringing. The customer can place the first call on "hold" and answer the second call by momentarily depressing the switchhook ("flashing"). By subsequent flashes of the switchhook, the customer can alternate between the two calls.[19]

Call Forwarding

This feature allows customers to "forward" their calls to another telephone number, which they designate by dialing a special code sequence. While Call Forwarding is activated, all incoming calls to a customer's telephone line are automatically transferred to the designated telephone.

Three-Way Calling

This feature allows a customer involved in an existing 2-way connection to place the other party on hold and dial a third party for a 3-way connection. When the third party answers, a 2-way conversation can be held before the earlier connection is re-established for the 3-way conference.

---

[19] In some business services provided by electronic switching systems, the procedures related to Call Waiting may be different.

Speed Calling

This feature allows a customer to use abbreviated codes to dial frequently called numbers. Repertories of eight (using a 1-digit abbreviated code) and thirty (using a 2-digit abbreviated code) stored numbers are available.[20] Speed Calling customers who also have the Customer Changeable Speed Calling option can assign their own Speed Calling codes to telephone numbers directly and immediately from their own telephones.

2.4.3 *TOUCH-TONE* **SERVICE**

TOUCH-TONE service replaces the customary dial pulses with tones for network signaling. As a pushbutton is depressed, two tones are generated simultaneously and the combined signal, which is clearly audible to the caller, is transmitted to the central office. Special receivers located in the central office convert the signals into a form that can be used by the switching system. *TOUCH-TONE* service provides customers with improved speed and convenience in dialing, reduces the number of digit receivers required by the central office because faster dialing uses the digit receivers for a shorter time per call, and provides the capability for end-to-end signaling once a call is established. End-to-end signaling (a capability that does not exist with rotary-dial service) allows a customer at one end of a connection to control dictation units and access data bases at the other end of the connection. The use of *TOUCH-TONE* service is increasing and will be the dominant method of customer signaling in the future.

2.4.4 **EXCHANGE BUSINESS SERVICES**

Business services offered from the exchange network satisfy many of the same business customer communication needs served by the PBX and automatic call distributor (ACD) services described in Section 2.3. With PBX and ACD services, individual customer lines connect to switching equipment on the customer's premises, and the switching equipment connects to a switching system in a local central office. With the corresponding exchange services, all of the customer's subscriber lines are directly connected to the central office, thus reducing the amount of equipment required at the customer's premises. At the central office, software or wired logic indicates which subscriber lines are part of the customer's group. By providing special features in the central office for this group of lines, both PBX-type and ACD-type services can be emulated.

---

[20] These numbers may be different in some electronic switching system business services.

The primary business services offered from the exchange network are centrex, ESS^{21}X-1, and ESS-ACD.[22] Centrex and ESSX-1 provide the same basic service elements:

- A member of a centrex or ESSX-1 group can dial another telephone number within the same group using only one to five digits.

- A member can dial calls outside the group directly, typically after dialing an access code, such as the digit 9.

- A member can receive calls that originate outside the group directly. No attendant is needed.

- Attendant positions can also be provided to allow central answering positions on the customer's premises to answer, hold, and route incoming calls to the group when the main centrex or ESSX-1 telephone number has been called.

With ESSX-1 service, the number of simultaneous incoming and outgoing calls and the number of simultaneous intragroup calls are limited by software to sizes specified by the customer. For centrex, however, the only limit is the call-handling capacity of the switching system.

The exchange service counterpart to ACD service (see Section 2.3.5) is called ESS-ACD. It is a specialized form of centrex service in which central office equipment, specifically an electronic switching system, distributes incoming calls to attendant lines. Typically, ACD attendants work full time receiving and servicing incoming calls (for example, making airline reservations). Therefore, in order to keep attendant positions efficiently loaded, there are generally fewer active attendants than the maximum number of simultaneous incoming calls.

With ESS-ACD, the central office distributes calls uniformly to the attendants, thus spreading the workload to minimize caller delay and maintain attendant efficiency. If no attendant is available, the central office will queue calls in order of arrival (see Section 5.2) and distribute them as attendants become available.

Additionally, specially designed customer-premises equipment and data links between the central office and the customer location make a large variety of management information, control, and status display features available. A customer can use statistical performance information and control capabilities to adjust the number of active positions and thus the average time a caller waits before reaching an attendant.

---

[21] Trademark of Western Electric Co.

[22] ESSX-1 and ESS-ACD are services provided by the 1ESS switching equipment.

2.4.5 911 EMERGENCY SERVICE

911 Emergency Service is designed to provide free emergency calling capability to the general public and is used in conjunction with dial-tone-first service.[23] The cost of implementing and maintaining the service is typically paid by county and state governments. With 911 Emergency Service, a single, easily remembered telephone number accesses a variety of emergency agencies. The service was established by the Bell System in 1968 in response to a recommendation by a Presidential Commission on Law Enforcement and Justice. The Commission had recommended that "wherever practical, a single (police emergency) number should be established."

Originally, 911 systems simply routed emergency calls to a centralized answering point. Later, features were added to this **Basic 911** (B911) service to provide for forced disconnect of the calling line (to prevent tying up the emergency center with nonemergency calls); holding the connection regardless of the calling party's action; emergency ringback to the calling station; and a visual and audible indication of the switchhook status of an established 911 call.

The major difficulty in implementing B911 systems is that, in many places, the boundaries of emergency agencies do not coincide with the boundaries of the local areas served by a telephone company. In some places, one local area may have twenty or more different combinations of emergency jurisdictions. When this happens, emergency calls must be selectively routed to the correct emergency agency based on the location of the calling party. The **Enhanced 911** (E911) provides the routing logic required to solve this problem (see Figure 2-10). Other features available with E911 include the ability to display the telephone number and the address of the calling party at the public safety answering point (PSAP), generally, a police station.

Approximately 800 B911 systems covering 25 percent of the population of the United States were in service by 1980, along with a total of 8 E911 systems covering a population of nine million. The potential exists for over 100 E911 systems to be in service by 1986. More efficient handling of emergency calls with these 911 systems will undoubtedly result in significant savings in life and property.

2.5 NETWORK SERVICES

In addition to exchange services, which are limited to the capabilities and resources of the exchange area network, the Bell System offers network services that make use of the broad capabilities of the **PSTN**, including

---

[23] Section 2.6 describes public telephones and dial-tone-first service.

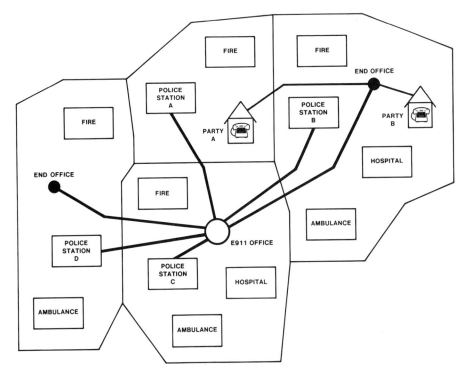

Figure 2-10. Role of an E911 office in routing calls directly to public safety answering points (PSAPs), police stations in this figure. The area shown above has four emergency agency jurisdictions. A 911 call from Party A would be routed to police station A, while a call from Party B would be routed to police station B, even though both callers are served by the same end office.

the stored-program control network (see Section 11.3.1); **private networks**; and **data networks**. The following sections describe these three types of network services.

2.5.1 PUBLIC SWITCHED TELEPHONE NETWORK SERVICES

A PSTN service, **toll service**, is used whenever calls are placed to points outside the local calling area. These calls are referred to as *toll calls*, and customers are ordinarily charged for each call. Other major PSTN services include **operator services, foreign exchange service, Wide Area Telecommunications Services, services provided by the direct services dialing capability, Automated Calling Card Service,** *DIAL-IT* **network communications service,** and **services provided by the circuit-switched digital capability.**

Operator Services

Telephone company operators provide a variety of services to PSTN customers.[24]

- **Toll-and-assistance operators** directly assist in the completion of calls.

  — **Toll-and-assistance operators** interact with customers making calling card, collect, and person-to-person calls. They may also assess charges and control the collection and return of coins for some coin calls, and place calls to points that cannot be directly dialed, such as certain mobile radios, marine stations, and certain foreign countries. These operators provide special services such as conference and call-back calls and perform manual switching, where needed. (Very few calls are switched manually because the use of direct distance dialing is widespread.) Toll-and-assistance operators assist customers who require emergency help or who are having trouble with the network. They also verify the busy-idle status of lines, accept requests for credits, provide dialing instructions, and complete calls when the customer cannot or when a customer experiences transmission problems. In today's environment, many of the operator functions on a toll call have been automated.[25]

  — **Centralized automatic message accounting – operator number identification operators** (CAMA-ONI operators) obtain the calling customer's number where the switching system does not include automatic number identification (ANI) equipment. The calling number is supplied to CAMA equipment (see Section 10.5.4) for billing purposes. The ONI function, while included for completeness, is not an actual service to the customer.

- **Number-service operators** provide information necessary for the completion of calls.

  — **Directory assistance operators** respond to customers dialing 411 for the local area code and 555-1212 for nonlocal area codes.

  — **Intercept operators** handle calls to unassigned or changed numbers.

  — **Rate-and-route operators** assist toll operators.

---

[24] Section 4.2.2 describes how operator services are integrated into the network.

[25] Section 10.4 discusses automated systems in greater detail.

Foreign Exchange Service

Foreign exchange (FX) service enables a customer to be served by a distant or "foreign" central office rather than by the nearby central office. Calls to other customers in the distant exchange area are then treated as local calls instead of toll calls. For customers who make enough calls to a particular distant exchange area, the monthly charge for FX service is less than the sum of the toll charges they would otherwise pay. Customers who find FX service economical include residence customers who often call friends or relatives in towns outside their local calling area and businesses such as firms in New Jersey who often call companies in New York City.

Wide Area Telecommunications Services

There are two types of Wide Area Telecommunications Services (WATS): **inward WATS** (INWATS), also called *800 Service*, and **outward WATS** (OUTWATS). They permit a customer, respectively, to receive from or originate to selected service areas long-distance calls that are billed to the customer on a bulk basis rather than on an individual basis. Both of these services are available on an intrastate or interstate basis. Subscribers are predominantly businesses with a substantial volume of long-distance calls to or from a wide geographical area.

For an interstate WATS customer, the United States is divided into six service areas, or bands, that extend outward from, but do not include, the customer's home state. Service Area One contains the states contiguous to the home state (but not including it) and sometimes one or two nearby states. Service Area Two includes Service Area One plus certain other states. Each successive service area includes the previous service area plus additional states. Service Area Six encompasses the entire United States (including Alaska, Hawaii, and Puerto Rico) but not the home state. Intrastate WATS is also available in most states. Under present tariff provisions, customers must purchase separate dedicated access lines to terminate interstate and intrastate WATS calls.

Expanded 800 Service, an improvement over 800 Service, uses common-channel interoffice signaling,[26] to provide three features that give customers greater flexibility in defining service areas and determining the treatment an incoming call receives. *Single-Number Service* provides subscribers with one nationwide 800 number for both interstate and intrastate calls at one or more customer locations. With *Customized Call Routing*, customers can control their call distribution based on callers' area

---

[26] Sections 8.4.2 and 8.5.5 discuss common-channel interoffice signaling.

codes. *Variable Call Routing* allows the customer to specify call distribution based on the time of day and day of week. Customers with 800 Service can add any or all of these features to meet their needs.

Services Using the Direct Services Dialing Capability

Traditionally, the Bell System has designed services to meet specific customer needs. Recently, however, there has been a trend to meet customer demand for new network services by providing a collection of service-independent network capabilities called the *direct services dialing capability*. This approach has many advantages. Customers can modify and control their services to a degree not previously possible. These capabilities are in the form of *primitives* in switching systems that can be summoned into use for various service applications. Some useful primitives might be "route the call" and "play an announcement." Services that would use the "route the call" primitive include:

- routing calls to different locations specified by the customer based on the location of the calling party, the time of day, the day of the week, the digits dialed by the caller (in response to a verbal prompt), and the busy-idle status of the customer's destination numbers. One application routes calls to the nearest retail store when there are several located in a city or town.

- routing incoming calls to different locations specified by the customer. One application has calls follow a salesperson who is moving from location to location, based on input from the salesperson.

Automated Calling Card Service

Automated Calling Card Service offers customers the ability to charge telephone calls to a number other than that of the originating station without operator assistance. This service, available to business and residence customers, automates calling card, bill-to-third-number, and collect calls. Automated Calling Card Service uses the direct services dialing capability of the stored-program control network.

A feature that accompanies Automated Calling Card Service, but which is not strictly part of it, is *Billed Number Screening*. This feature is active on any bill-to-third-number call or collect call attempt and identifies numbers that do not accept any bill-to-third-number or collect calls. This applies to business or residence customers who have requested such screening and also prevents those types of billing to public telephones.

DIAL-IT **Network Communications Service**

DIAL-IT service is a name for any of several services in which callers dial advertised telephone numbers to reach an announcement, a live answer, or both. The services fall into two categories: **Public Announcement Service** (PAS) and **Media Stimulated Calling** (MSC).

PAS plays up-to-date recorded announcements for such services as Sports-Phone, Dial-a-Joke, and Horoscope. For these services, the Bell System provides access to the announcement; the announcements themselves are provided by other companies.

Services provided by MSC include media promotion, telethons, and telephone voting service. Media promotion and telethon services give the customer the ability to connect selected callers to a live answer. Typically, callers not selected for a live answer receive a recorded announcement thanking them for their participation. The telephone voting service allows callers to respond to questions presented to them by radio or television. A caller dials one of two telephone numbers corresponding to the caller's choice or opinion and is connected to a brief acknowledgement. The calls to each telephone number are tallied, and the result is provided to the sponsor.

DIAL-IT service is available nationwide on a standard basis through the use of area code 900.

Services Provided by the Circuit-Switched Digital Capability

The circuit-switched digital capability (CSDC) will provide end-to-end digital connectivity. The CSDC, which is expected to be available as a tariffed service around the end of 1983, is an important step toward an **integrated services digital network** (ISDN), a public end-to-end digital telecommunications network capable of supporting a wide spectrum of present and emerging user needs. Like the ISDN, the CSDC will be service independent. The CSDC will provide a 56-kbps digital path over the PSTN to customers whose lines are terminated on electronic switching systems with the CSDC feature.

One of the first applications of the CSDC will be the transport of bulk data. Since the CSDC will operate at a speed of 56 kbps, it will have ten times the capacity of the 4.8-kbps data sets that are currently used on the switched network. This feature can be useful to customers like banks who must transfer large quantities (for example, tens of megabytes) of information during a limited period, perhaps overnight.

New technologies and new capabilities will help make the integrated services digital network a reality. Applications of new technology will provide terminal equipment capable of integrating voice and data into a single information flow. New service-independent capabilities, such as those to be provided by the CSDC, will route this flow over the PSTN and make many new services widely available.

2.5.2 PRIVATE-LINE SERVICES

Private-line services provide point-to-point and multipoint communication channels that are separate from channels of the PSTN. Private-line circuits are usually used for talking and signaling, but other offerings are available. These include teletypewriter services, telemetry,[27] wired music, video and television transmission, the connection of computers to other computers or input/output devices for data transmission, the extension of alarm or power control circuits from unattended to attended locations, and the connection of radio or television studios to remote transmitters.

While many private-line services can be approximated using services available on the PSTN, private lines offer the following advantages:

- Where the traffic is heavy enough and the geographic pattern lends itself to such use, private lines may be more economical.

- A private line incurs a specified charge that is independent of the amount of use.

- The time needed to establish a connection can be shorter with a private line than with the PSTN.

- Private-line services are dedicated to the customer and not shared, as in the PSTN, thereby ensuring a through (nonblocking) connection at all times (see Section 5.2).

Private lines are offered in several designated series, which serve as a basis for service negotiations between marketing representatives and customers. Different series lines have different uses and electrical characteristics. Table 2-1 lists the series numbers and types of service.

2.5.3 PRIVATE NETWORK SERVICES

Large business customers with geographically dispersed locations subscribe to private network services. Each of the customer locations is usually served by a PBX, centrex, or ESSX-1. As long as the calling volume over the private facility is such that the toll charges for equivalent PSTN calls are higher than the monthly charge for the dedicated facility, a private network is cost effective.

---

[27] Low-speed transmission of measured quantities. Generally, *telemetry* (or *telemetering*) refers to an arrangement in which measurements taken in one place are recorded in another place.

TABLE 2-1

PRIVATE-LINE OFFERINGS

Series	Examples of Service
1000	Low-speed (narrowband*) data, for example, private-line telegraph, teletypewriter, teletypesetter, and remote metering (telemetering)
2000	Voice
3000	Medium-speed (voiceband*) data
4000	Telephoto/facsimile
6000	Audio (music transmission)
7000	Television
8000	High-speed (broadband*) data

*Section 6.2.1 discusses voiceband, narrowband, and broadband channels.

The simplest kind of private network would be a transmission facility dedicated to a customer and interconnecting two geographically separated customer PBXs or centrex/ESSX-1 locations (in Los Angeles and New York, for example). One PBX/centrex location calls the other by dialing a code to access the other location and then dialing the extension number of the station at the distant location. This example is often not considered to be a true network and is usually referred to as *tie-line service*—the transmission facility ties together the customer locations.

If there are three customer PBXs located in Los Angeles, Chicago, and New York, the customer might acquire tie lines between Los Angeles and Chicago and Chicago and New York. A caller at the Los Angeles PBX who wishes to call the New York PBX first dials an access code to reach the Chicago PBX and then dials another access code to instruct the Chicago PBX to connect the Los Angeles tie line to the New York tie line. After the Chicago PBX makes the connection, the Los Angeles customer dials the called person's extension number to complete the call through the New York PBX. This type of service is known as *Tandem Tie-Line Service* because the Chicago PBX must be able to connect or "tandem" the call between the two tie lines. With this type of service arrangement, a different access code is required from each originating location to reach a particular location. In addition, the customer must not only pay for the dedicated transmission facilities but must also pay for PBX or centrex switching capabilities to make direct or tandem connection to tie lines (see Section 3.3.2).

To establish a network of tie lines with a uniform numbering plan similar to that which exists in the PSTN, the customer must subscribe to private network services like **common-control switching arrangement** (CCSA), **Enhanced Private Switched Communication Service** (EPSCS), or **electronic tandem switching** (ETS). These services are described in the next few paragraphs. Each of them allows interlocation dialing on a 7-digit basis, where the first three digits uniquely identify each location and the last four digits identify that location's PBX or centrex stations. The first three digits do not correspond to the station's normal telephone number and are only used for private network calls. The result is that the private network customer has a unique 7-digit dialing plan that is uniform for all locations on the network.

All of these services—tie lines, tandem tie lines, CCSA, EPSCS, and ETS—allow the private network the option of carrying calls that go off the network, that is, calls that do not terminate at one of the customer's PBX or centrex systems. To enter the public network to complete a call, tie-line customers dial "9" plus a PSTN number. The CCSA, EPSCS, and ETS networks recognize 10-digit calls as off-network calls (where the ten digits are PSTN numbers). The CCSA service, EPSCS, and ETS carry the call over the dedicated facilities of the private network to a point close to the desired location where the call enters the PSTN. (Chapter 4 contains more information about private network configurations and various call-routing arrangements.)

Common-Control Switching Arrangement

CCSA service was the first private network service to offer a customer with geographically dispersed locations uniform dialing over dedicated private facilities. The CCSA is primarily an interstate service regulated by the FCC. Any station within a CCSA network may directly dial any other station by using a uniform 7-digit dialing plan. The first three digits identify the location, and the last four digits identify a location's PBX stations or centrex stations. The network switching systems that perform the routing function are selected by the Bell System or an independent telephone company, depending on the location, and are never on the customer's premises; that is, CCSA routing switches cannot be PBXs. Dedicated access lines from the PBX or centrex provide access to the private network and to the selected network switches.

To use a CCSA private network, the customer dials an access digit at the PBX or centrex and, after being connected to the Bell System CCSA switch, dials the 7-digit on-network number or a 10-digit off-network number.

Customers with extensive tie-line networks find that the costs of adding CCSA switches are justified by the convenience of the uniform dialing plan. Costs are still lower than they would be using the PSTN.

Enhanced Private Switched Communication Service

EPSCS is an improved CCSA-like service introduced in 1978. Like CCSA, EPSCS is an interstate service regulated by the FCC and uses uniform 7-digit dialing. It, too, utilizes switching systems selected by the Bell System with dedicated access lines to customer PBXs and centrex switching systems to accomplish network routing functions. However, EPSCS offers features in addition to those available in CCSA both as part of the standard EPSCS offering and as options at extra cost.

Two unique standard features of EPSCS are 4-wire transmission (to improve transmission quality)[28] within the private network and a Customer Network Control Center, which customers can use to control some network operations and to obtain private network usage and status information automatically and on demand. Other features include:

- automatic route selection of FX and WATS facilities for off-network calling

- automatic alternate routing

- time-varying routing to accommodate expected changes in traffic loads

- call queuing when a network, FX, or WATS facility is busy

- authorization code entry when placing a call to provide controlled use of expensive facilities

- special recorded announcements

- "meet me" conferencing with 6-station capability[29]

- automatic dialing.

Electronic Tandem Switching

ETS is another recently introduced (in 1979) private network service. ETS is regulated by state commissions and is not in itself an interstate service. It is a collection of features offered by the same switching equipment that provides PBX and centrex service. There are no special Bell System

---

[28] See Section 6.2.2.

[29] "Meet-me" conferencing allows a maximum of six people to participate in a conference call. The participants dial a special network number (called a *conference dial code*) at a prearranged time. Only those people who dial the assigned code have access to the conference.

switches. To obtain ETS, the customer must be served by PBXs and centrex switches that are capable of being equipped with ETS features. Once equipped, these PBX and centrex switches offer the basic uniform dialing plan for dedicated private facilities characteristic of private networks.

Many of the features available to EPSCS customers are also available to ETS customers, including automatic route selection, call queuing, and authorization codes. ETS also offers a Customer Administration and Control Center, which is similar to the Customer Network Control Center used with EPSCS but less sophisticated. ETS offers the customer less sophistication than EPSCS, but generally costs less.

2.5.4 DATA SERVICES

Section 2.2.3 discussed data products for use on customer premises and described some data services derived from the use of *DATAPHONE* data sets and their inherent capabilities. In *DATAPHONE* II data communications service, for example, monitoring and control of a data network is accomplished through the capabilities of *DATAPHONE* II data communications service equipment; the telecommunications network provides only transport functions; that is, it serves only as a communications path or channel.

This section describes data services that require specific network implementations or functions, such as synchronization, multiplexing, and switching. These network services have evolved from the differing needs of users. *DATAPHONE* digital service, for example, responds to the need for very high-quality data transmission. *DATAPHONE* Select-a-station satisfies a special class of applications involving the interconnection of a large number of low-speed data stations on analog private-line facilities. The other services described in this section are emerging to meet rapidly growing, diverse needs and are expected to be available in 1983 or 1984.

DATAPHONE **Digital Service**

DATAPHONE digital service offers data communications over point-to-point or multipoint private lines at data rates of 2.4, 4.8, 9.6, and 56 kbps. The objectives of this service are high performance and excellent availability. Availability is provided by a network with high reliability and rapid restoration of service when failures do occur. It is attractive to customers such as on-line reservation services for airlines, who require low error rates and high network availability. In general, error rates and network availability are substantially better with *DATAPHONE* digital service than with other private-line services or on the PSTN.

The service is provided by the **Digital Data System** (DDS), a synchronized data network (see Section 11.6.1). Data transmission is full duplex (that is, 2-way simultaneous transmission) and remains digital end to end (terminal to terminal). There are no restrictions on the data format; that is, the service is transparent to any data sequence. The DDS is designed for data only—there is no provision for voice transmission.

In 1982, *DATAPHONE* digital service was available in about 100 metropolitan areas; it is expected to be available in over 125 areas in the next several years.

DATAPHONE Select-A-Station

DATAPHONE Select-a-station service allows customers to establish a series of point-to-point connections rapidly between a master location and a large number of remote locations. It is suitable for users who need remote telemetry from a large number of remote locations. A typical customer might be a central station alarm company that would use the service to provide security and fire protection by monitoring business and residential premises. The service is provided over voiceband private lines and is supported by high-speed switching equipment designed especially for this application. The equipment is located in central offices but is controlled by the customer's master station equipment. Section 11.6.3 provides more information on the operation of the system.

Basic Packet-Switching Service[30]

The Basic Packet-Switching Service (BPSS) is a private-line switching arrangement for switching data packets among a customer's various locations.[31] BPSS is particularly suited for customers with the following requirements:

- large numbers of data calls

- large quantities of data that must be transmitted between various locations

- data flow occurring in bursts, with a peak data rate that is high compared to the average data rate. (The higher the peak-to-average ratio, the more efficient the transport of data over BPSS.)

---

[30] Basic Packet-Switching Service has been renamed *ACCUNET* Packet Service. (*ACCUNET* is a service mark of AT&T Co.) Shortly before publication of this book, the names of several services were changed. Time did not permit changing the names throughout the book; however, the new names are indicated in footnotes the first time an applicable service is mentioned.

[31] With packet switching, the data are divided into *packets*, each of which includes destination and control information (see Section 5.8.1) in addition to data. Section 11.6.2 discusses packet-switching systems.

An airline company, which might use BPSS for reservations and other operations, is an example of this type of customer.

BPSS is furnished at a telephone company central office. It is accessed through ports that operate at transmission speeds of 9.6 and 56 kbps.[32] Each port is dedicated to a single customer, and a customer may combine ports from more than one BPSS packet switch with access lines and trunks to form private packet-switching networks. The design of BPSS permits different customers to obtain ports on the same switching arrangement and therefore share its common switching capability. Although a switching arrangement may be shared by many customers, its inherent design ensures privacy of communication between different customer networks. When the access lines to BPSS ports and the trunks between ports on different switching arrangements are provided by the Digital Data System, BPSS offers high availability and reliability.

BPSS provides two types of fundamental capabilities: **virtual call** and **permanent virtual circuit** capabilities. *Virtual call capability* allows setup and clearing on a per-call basis. Once a call is set up, it appears to have a dedicated connection for its duration, that is, until cleared. *Permanent virtual circuit capability* provides the same functions as virtual call capability, except that call-related procedures (setup and clearing) are eliminated. Permanent virtual circuits are permanently defined in tables in the switching arrangement(s) when service is established. Data terminals connected by a permanent virtual circuit appear to have a full-time, dedicated connection. A customer must place an order with the telephone company to establish, change, or discontinue a permanent virtual circuit.

Customers for the virtual call capability of BPSS may include retailers who need access to several different data bases periodically during business hours. An automotive parts or plumbing supply outlet may have to place orders with several different distributors. Customers for the permanent virtual circuit capability of BPSS may include retail stores whose checkout clerks routinely obtain clearances for credit-card purchases by customers.

The switching arrangement available with BPSS has a nominal switching capacity[33] of up to 500 data packets per second. It supports maximum packet sizes of either 128 or 256 data octets (8-bit characters) as specified by the customer. Various software functions are available, with which, for example, customers may create their own logical subnetworks.[34]

---

[32] BPSS supports the 1980 access protocol Comité Consultatif International Télégraphique et Téléphonique (CCITT) Recommendation X.25 (see Section 8.8).

[33] Actual switching capacity depends on traffic characteristics (see Section 5.8).

[34] The capabilities available in BPSS are described in detail in AT&T Long Lines 1982.

High-Speed Switched Digital Service[35]

High-Speed Switched Digital Service (HSSDS) provides a 1.544-megabits-per-second (Mbps) network for transmission of voice, data, or video within or between HSSDS switching nodes. Customers call an 800 number to reserve HSSDS facilities between nodes. Access between customer premises and an HSSDS switching node is provided through **High-Capacity Terrestrial Digital Service.**[36] The network, which is composed of terrestrial and satellite digital facilities, is planned to have nodes in forty-two cities by the end of 1983. It will support 2-point, multipoint, and broadcast connections.

High-Capacity Digital Transport Services

These services, first available in 1983, provide customers with 1.544-Mbps digital circuits on a full-time (24 hours a day) basis. Two services are included, one using terrestrial facilities, the other satellite facilities.

High-Capacity Terrestrial Digital Service (HCTDS) can be used to connect two customer locations or to connect a customer location to a telephone company central office. Equipment may be provided at the central office that enables the digital circuit to carry twenty-four voiceband channels, each of which can be terminated on the switching system.

High-Capacity Satellite Digital Service (HCSDS)[37] use dedicated earth stations at the customer's location or shared earth stations at four locations in the United States. Customers using shared earth stations obtain dedicated terrestrial links via HCTDS to one of the four shared stations. Point-to-point and multipoint communications among shared and dedicated earth stations are permitted. Features such as echo cancellation, elastic stores,[38] and earth station control are also available.

2.5.5 MOBILE TELEPHONE SERVICES

Mobile telephone services utilize radio transmission to provide telephone service to customers on the move. Until recently, development of these services has been constrained by limited radio-frequency assignments and

---

[35] High-Speed Switched Digital Service has been renamed *ACCUNET* Reserved 1.5 Service.

[36] High-Capacity Terrestrial Digital Service has been renamed *ACCUNET* T1.5 Service.

[37] High-Capacity Satellite Digital Service has been renamed *SKYNET* 1.5 Service. (*SKYNET* is a service mark of AT&T Co.)

[38] A buffer memory that can hold a variable amount of data. The length of time that specific data items remain in the store depends on the amount of data it contains.

technological complications. Mobile telephone services include: **land mobile telephone service,** BELLBOY[39] **personal signaling set paging service, air/ground service, marine radiotelephone services,** and **high-speed train telephone service.**

Land Mobile Telephone Service

Land mobile telephone service provides 2-way voice communications, through designated central offices, between mobile units and land telephones or between two mobile units. Users in the mobile serving area have full access to the PSTN either on a manual (operator-handled) or a direct-dial basis. In 1981, the FCC approved a new type of system, called *cellular*. Beginning in 1983, the Bell companies will serve a much larger number of customers through this new design, which makes more efficient use of frequency assignments. Section 11.4.1 discusses this system and other land mobile telephone systems, present and future.

BELLBOY **Personal Signaling Set Paging Service**

The *BELLBOY* personal paging service notifies customers when someone wants to talk with them. Customers carry a cigarette-pack-sized radio receiver that emits an audible tone when the number assigned to that unit is called. (A new receiver being offered by some telephone companies also incorporates a visual display of the calling number.) The receiver is activated by an array of radio transmitters that provide coverage for an urban area. In 1980, *BELLBOY* personal signaling set paging service had about one hundred thousand customers nationwide (see Section 11.4.2).

Air/Ground Service

Air/ground service provides 2-way telephone service between customers flying in private aircraft and customers on the PSTN. The service uses radio base stations connected to control terminals and mobile service switchboards that interconnect with the PSTN. All radio equipment mounted on aircraft is customer owned and customer maintained. In 1980, approximately two thousand aircraft were equipped with this service, and sixty thousand calls were placed during the year.

Marine Radiotelephone Services

Marine radio telephone services include **very high frequency (VHF) maritime service, coastal-harbor service,** and **high-seas maritime radiotelephone service**. These services provide 2-way telephone service to water

---

[39] Registered service mark of AT&T Co.

craft. The three services differ in the range of distances over which they operate. VHF maritime service offers reliable communications up to 50 miles offshore and on inland waterways. Coastal-harbor service communications can range up to 200 miles offshore, and high-seas service is intended for ships engaged in oceanic operations and transoceanic passages. The radio equipment for all three services mounted on board ships is customer owned and maintained.

High-Speed Train Telephone Service

High-speed train telephone service provides telephone service between a passenger train and the PSTN. Operator-handled train telephone service was inaugurated in 1947, and by 1952, service was provided to nineteen trains on five railroads. These installations are now out of service, in most cases because of the demise of the equipped trains. More recently, in 1968, train telephone service was installed aboard the Metroliner trains operating between New York and Washington, D.C. The service provides the public with coinless *TOUCH-TONE* telephones. Approximately forty-five thousand calls were handled yearly during the 1970s. This service is being phased out in anticipation of the new cellular service.

2.5.6 VIDEO TELECONFERENCING SERVICE

In 1964, AT&T offered *PICTUREPHONE*[40] visual telephone service using small desktop units that contained both a special camera tube and a small black and white receiver. Public booths were also established, and service was offered between locations in Chicago, New York, and Washington, D.C. The offering was continued until 1975 as a market trial.

The experience gained in the trial led to a reorientation of the service. In 1975, *PICTUREPHONE meeting service* was introduced (also on a trial basis), using rooms equipped for conferences of various sizes. The trial of the reoriented service continued until 1981, when AT&T announced its plans to offer video teleconferencing as a standard service in forty-two cities. Long-distance transmission for the standard service is provided by the High-Speed Switched Digital Service (see Section 2.5.4). Access from conference rooms to the nodes is offered under full-period High-Capacity Terrestrial Digital Service (see Section 2.5.4) tariffs. Customers may have private conference rooms on their own premises or may use public rooms offered by AT&T.

The objective of the video teleconferencing service is to make conferences both effective and pleasant. As shown in Figure 2-11, six people

---

[40] Registered service mark of AT&T Co.

Figure 2-11. Conference room equipped for
PICTUREPHONE **meeting service.**

are accommodated at a conference table. Each person is within range of one of three cameras. The picture selected for transmission depends on which person is speaking; thus the speaker is on camera. Other cameras are provided for overviews, various graphic displays, and for a speaker at an easel. Monitors show both the incoming and outgoing video.

Video teleconferencing service will prove to be economical when measured against the rising cost of travel and the time lost in travel that could be applied to other business or personal responsibilities.

2.6 PUBLIC COMMUNICATIONS SERVICES

By the end of 1981, there were approximately 1.6 million telephones providing Public Communications Services and generating about $3 billion in annual revenues. Most public telephones are installed at locations where a public need exists such as airports, bus depots, train stations, hotel lobbies, large office buildings, and on public streets and highways. They provide the general public with access to all United States and international telephones, generally on a customer-dialed basis.

About 30 percent of the telephones that provide Public Communications Services are called *semipublic* because they are not always available to the public. Semipublic telephones are most often found in service stations, delicatessens, self-service laundries, and similar businesses. The proprietor can use the line for business purposes and make a coin telephone available to the public during business hours. Many businesses also install semipublic telephones to control outgoing calls by employees. The semipublic station can also be equipped with a noncoin, answer-only extension to allow the proprietor to answer incoming calls without having to pick up the coin telephone. While most of these have a nondial telephone bridged across the line without any privacy protection for the user of either station set, arrangements can be made to provide privacy to users of the coin telephone. A business with a semipublic telephone can be listed in both the white pages and the Yellow Pages of telephone directories.

Revenues from public telephones are generally shared with the owners of the premises where the sets are located. For telephones located on public property, most state regulatory agencies have set up some formula to provide a commission to the political unit that grants the franchise. For example, urban sidewalk installations usually yield a commission to the city. Similarly, commissions on airport locations are paid to the government agency operating the airport.

Charging arrangements for semipublic telephones vary considerably from state to state. Some states require a minimum amount of revenue to be generated. If that amount is not achieved, the proprietor is billed for the amount needed to reach the minimum. Other states require a flat monthly fee with no guarantee. Rarely are any commissions paid on semipublic telephone service.

There are two types of public telephones—coin and Charge-a-Call, or coinless (see Figure 2-12). With a coin telephone, a customer may place either sent-paid calls (paid for at the time the call is made by depositing one or more coins) or non-sent-paid calls (where payment is not made at the time of the call). The latter include collect calls, calls charged to a third number, and calling card calls (that is, calls billed to another telephone). Charge-a-Call stations can complete only non-sent-paid calls.

Coin telephones can provide either coin-first or dial-tone-first service. Coin-first telephones require a specific deposit (ten, fifteen, twenty, or twenty-five cents) before the receipt of dial tone. Dial-tone-first telephones allow the completion of service calls (for example, directory assistance, 911, and repair service) and access to Traffic Service Position System (TSPS) operators or TSPS/Automated Coin Toll Service (ACTS) equipment without coin deposits. In addition to providing coin-free emergency access, the dial-tone-first station gives the customer some assurance that it is working before any coins are deposited.

For toll calls, a toll-and-assistance operator informs the caller of the charges and confirms the correct deposit of coins. Most toll service operators are supported by TSPS (see Section 10.4.1). Where TSPS has been enhanced by ACTS equipment, the operator function is automated, and synthesized voice messages are used for interaction with the caller.

Figure 2-12. Public telephones. *Left*, a coin set; *right*, a Charge-a-Call set.

About two-thirds of the public telephone revenues are from non-sent-paid calls. At locations such as airports and railroad terminals, a substantial percentage of the calls are non-sent-paid. These locations are attractive candidates for the installation of Charge-a-Call telephones, which can handle such calls, and for the provision of Automated Calling Card Service (see Section 2.5.1), which automatically bills them.

2.7 CUSTOMER SUPPORT SERVICES

The preceding sections describe many of the telecommunications services available to Bell System customers. These services are provided through various types of terminal equipment and the capabilities of the network. This section discusses a different class of services—those provided by an operating company to support customer needs in the areas of acquisition, use, and maintenance of telecommunications services.

2.7.1 RETAIL SALES AND SERVICE

In 1980, over 20 million requests for telephone service (for example, to initiate, terminate, or change service), representing more than half of the Bell System activity, were handled through the 1700 Bell PhoneCenters[41] then maintained by Bell operating companies. Before the introduction of Bell PhoneCenters, residential customers had to contact a telephone company business office to order service and then wait for an installer to bring the telephone set(s) to the home.

Until recently, telephones and other terminal equipment were leased from the local operating company. On January 1, 1983, under provisions of the FCC's Computer Inquiry II decision, sale of new terminal equipment at Bell PhoneCenters was transferred to AT&T.[42] Operating companies may sell terminal equipment that was in their inventory on that date, and they continue to operate service centers where customers may conveniently order service and replace faulty equipment bought or leased from the operating company.

Customers can select from a variety of basic and *DESIGN LINE* decorator telephones. They can also obtain the lengths of handset and mounting cords they desire and the particular adapters they need to make their inside wiring compatible with modular telephone technology (see Section 2.2.2). Once the wiring inside a residence has been adapted, station sets may be plugged in or removed quite conveniently.

2.7.2 BUSINESS OFFICE SERVICES

Where telephone company customer service centers exist, service representatives in business offices refer requests for new residential service to them. Business offices continue to initiate service orders for new phone installations in some cases, for example, when inside wiring records indicate that customers will be unable to install the phones they desire. In addition, they answer customers' questions about billing for residential services.

2.7.3 INSTALLATION AND MAINTENANCE SERVICES

Installation and maintenance services include the inside wiring and connection of station equipment. This may be required by business customers and by those residence customers who either cannot or do not wish to

---

[41] Also called PhoneCenter Stores.

[42] Terminal equipment from several manufacturers may be purchased at many retail stores.

install their own equipment. The installer may perform the inside wiring job for new installations and make changes to existing wiring, as required.

Service problems are handled by a call to Repair Service. The nature of the customer's complaint is recorded, and after the probable cause is determined by testing, repair action is initiated. Section 13.2.3 further discusses maintenance operations.

2.7.4 DIRECTORY SERVICES

Customers are entitled to have their names, addresses, and telephone numbers listed in white pages directories, but listings will be withheld if a subscriber so desires. Subscribers not listed in the white pages may also specify that their addresses and telephone numbers not be published in Information Services directories for public disclosure. Arrangements may be made to print listings in bold-face type or in association with special instructions, such as, "After 5 o'clock, call 555-5555."

An additional fee is required for some white pages services. All Yellow Pages listings require additional fees. Costs depend on the details of the special listing, in particular, the size of the listing and whether or not it is accompanied by an advertisement.

To make Directory Services more helpful to users, innovations are being introduced. For example, in some white pages directories, federal, state, and county governmental listings are printed on blue paper in a special section. Upon request, partial addresses or no addresses may be listed with names and telephone numbers. Four-color advertisements are being introduced in Yellow Pages directories. New Yellow Pages books specialize in user-market needs such as Medical Directories, Tourist Directories, etc. White and Yellow Pages data bases have been developed for user access via the emerging electronic information technology.

AUTHORS

C. E. Betta
H. J. Bouma
P. S. Browning
W. R. Byrne
R. Carlsen

D. C. Franke
W. G. Heffron
T. B. Morawski
K. J. Pfeffer
P. T. Porter

S. H. Richman
J. F. Ritchey
A. T. Vitenas
C. H. Zima

3

Introduction to the Network

3.1 WHAT IS A TELECOMMUNICATIONS NETWORK?

As a starting point in defining a telecommunications network, a general definition of a network may be helpful. In a broad sense, a *network* is a system of interconnected elements. Topologically, it can be represented by a set of nodes and a set of links that interconnect pairs of nodes. A network is needed when certain types of services must be provided to many, widely dispersed customers. Depending on the types of services, the characteristics of the network elements may differ greatly.

Another concept necessary for the definition of a telecommunications network is the notion of telecommunications traffic or, simply, traffic. *Traffic* is the flow of information or messages through the network. Traffic may be generated by simple telephone conversations, or it may be the result of complex data, video, and audio services. (Chapter 5 deals with traffic theory and its application to engineering the network.)

A *telecommunications network*, then, is a system of interconnected facilities designed to carry the traffic that results from a variety of telecommunications services. (Chapter 2 discusses the various telecommunications services available.) The telecommunications network as a whole has two different but interrelated aspects. In terms of its physical components, it is a **facilities network**. In terms of the variety of telecommunications services that it provides, it is a set of many **traffic networks**, each representing a particular interconnection of facilities. The distinction between traffic networks and the facilities network is discussed in more detail later in this chapter.

As stated earlier, a network can be represented by nodes and links. In the telecommunications network, the nodes represent switching offices and facility junction points, and the links represent transmission facilities. Traffic is the flow of information in the network.

Telecommunications has three characteristics that dictate the basic nature of the network. First, traffic must be carried among customers dispersed over large geographic areas. Second, traffic may be generated between any pair of customers at virtually any time, although the duration of each call may be fairly short. Third, the ability to exchange information between any pair of customers is expected to be available with relatively short delay.

Figure 3-1 illustrates some key concepts in the design of a telecommunications network.[1] Figure 3-1A shows a highly oversimplified situation in which no switching is used and telephones at all four end points (customer locations) are directly interconnected in pairs by transmission paths. A telecommunications network designed in this way would be inefficient and prohibitively expensive because it would require many telephones at each end point and many transmission paths as well.

The design depicted in Figure 3-1A can be improved considerably by the introduction of switching. For example, the use of a switch at each location would eliminate the need for all but one telephone at each location. This situation is shown in Figure 3-1B. In this case, although the total number of telephones has decreased, the number of required transmission paths remains the same, and the implementation of switching at each end point would be expensive.

Figure 3-1C shows that a much more efficient use of network elements results from the introduction of switching at a central location to interconnect transmission paths emanating from the end points. Both the number of switches and the number of transmission paths are substantially reduced. In a network with no switching (as in Figure 3-1A), if there are n end points ($n=4$ in Figure 3-1A), $n(n-1)$ telephones and $n(n-1)/2$ transmission paths are needed. However, the network design shown in Figure 3-1C requires only n telephones, n transmission paths, and one central switch.

More efficient network design and lower cost to the customer are achieved at a certain price: The number of simultaneous connections through the network is limited. Thus, while the network depicted in Figure 3-1A would allow the simultaneous connection of all pairs of end points, thereby supporting six calls at the same time, the centrally switched network configuration of Figure 3-1C can support only two simultaneous connections. This limited potential for simultaneous calls is not a serious drawback, however, because the concurrent use of the network by all or even most users is unlikely. Chapter 5 includes a further discussion of topics related to network usage.

---

[1] For an in-depth analysis of the design and cost characteristics of telecommunications networks, see Skoog 1980.

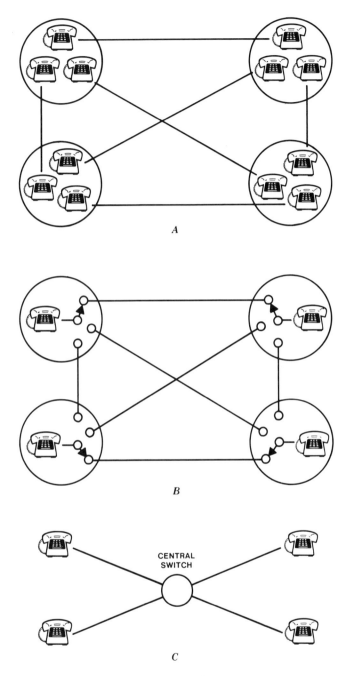

Figure 3-1. Networking and the tradeoff between transmission and switching. *A*, direct interconnection of telephones; *B*, interconnection through switches at end points; *C*, interconnection through a centralized switch.

3.2 THE FACILITIES NETWORK

The telecommunications network was defined in Section 3.1 as a system of interconnected facilities designed to carry traffic that results from a variety of telecommunications services. When viewed from the perspective of its physical components, or facilities, the network may be referred to as the *facilities network*.

The components of the facilities network may be divided into three broad categories.

- **Station equipment** is generally located on the customer's premises.[2] Its primary functions are to transmit and receive the information flow and required control signals between customers and the network.

- **Transmission facilities** provide the communications paths that "carry" the information between customers. In general, transmission facilities consist of some sort of transmission medium (for example, the atmosphere, paired cable,[3] coaxial cable, lightguide cable)[4] and various types of electronic equipment located at different points along the transmission medium. This equipment amplifies and, sometimes, regenerates the transmitted signals. In addition, various types of facility terminal equipment provide functions needed where transmission facilities connect to switching systems and at facility junction points.[5]

- **Switching systems** interconnect the transmission facilities at various key locations and route traffic through the network. As mentioned in Section 3.1, the introduction of central switching into the network yields cost savings in station equipment and transmission facilities.

In addition to the functions just described, transmission facilities and switching systems provide for **signaling** in the network. (Chapter 8 describes signaling, a major network function.)

The Bell System provides a large percentage of the telecommunications facilities in the United States nationwide network. However, numerous independent telephone companies and other common carriers also own both transmission facilities and switching systems.

The following sections give an overview of the three basic categories of facilities, which are discussed in detail in subsequent chapters.

---

[2] *Customer-premises equipment* (CPE), a broader term, includes more than station or terminal equipment. For example, private branch exchange equipment located on a customer's premises performs customer switching functions.

[3] Paired or multipair cable contains a number of twisted pairs of wires.

[4] Section 6.3 describes these transmission media.

[5] Chapters 6 and 9, respectively, describe some of the functions and equipment.

3.2.1 STATION EQUIPMENT

Station equipment is the user's interface with the rest of the network and the available services. The most common station equipment is the ordinary single-line telephone set. The functional components of a telephone set include a transmitter and a receiver (most often combined in a handset), a rotary or pushbutton dial, a switchhook mechanism, and a bell or other alerting device. The telephone converts an acoustic signal (which may be generated by a human voice or by a device that transforms data into a series of tones) into an electrical signal, which it sends over a transmission facility. For reception, the telephone converts an incoming electrical signal back into an acoustic signal.

In addition to transmitting and receiving information in the form of electrical signals, the telephone also provides for two kinds of signaling functions: **supervision** and **addressing**. *Supervision* includes the constant monitoring by the local switching system or by a private branch exchange (PBX) of the status (idle or busy) of the telephone and alerting the user that a call is being made by providing an audio (or visual) signal. *Addressing* refers to the task of specifying to the network the destination of a call.

The telephone switchhook is used to signal idle or busy status. When the telephone is idle, or "on-hook," the switchhook contacts are open. When the telephone is busy, or "off-hook," the switchhook contacts are closed. These supervisory signals allow certain equipment at the central office (or PBX) to recognize origination, answer, and termination of a call. A customer is alerted to an incoming call by a bell or tone ringer (or, in some cases, a light).

Addressing is done by either a rotary dial or a set of pushbuttons, producing a signal that corresponds to the called number. With a rotary dial, a series of pulses, equivalent to alternate on-hook, off-hook conditions, represents each digit dialed. With a pushbutton set, a different pair of tones represents each digit.[6]

Many other kinds of station equipment convey various types of information. Some enable computers to communicate directly with one another over the telecommunications network. Others are used for visual information such as video, facsimile, and graphics. (Section 11.1 discusses station equipment in detail.)

3.2.2 TRANSMISSION FACILITIES

Transmission facilities provide the communications paths that carry the traffic between any two points in the network. These communications

---

[6] There are also telephone sets with pushbuttons that produce dial pulses.

paths are referred to as *channels* or *circuits*[7] and may be classified as follows into three broad categories.

- Ordinary channels (circuits) that connect customers' station equipment to a switching system are called *lines* or *loops*.[7] *Loop* is derived historically from the pair of wires that form a loop between the customer's location and the switching system. Originally, all loops were wire pairs. Today, however, many loops do not have an associated physical pair of wires for the entire path. Instead, loop carrier systems[8] are used.

- Ordinary channels (circuits) that connect two switching systems are called *trunks*. There may be several switching nodes with trunks connecting them between the calling customer's telephone and the called customer's telephone. The trunks and switching systems carry traffic generated by many customers, while loops are dedicated to individual customers.[9]

- Finally, channels (circuits) dedicated to a specific customer to provide special services are called *special-services circuits*. They encompass both circuits to a customer's equipment and circuits between network switching systems.

Based on the above classification of channels, it is possible to divide the transmission facilities into two general categories.

- **Loop transmission systems, subscriber loop systems,** or, simply, **loop systems** carry loops and special-services circuits to a customer's premises. Typically, loop transmission systems are paired cable (also called *multipair cable*) that is suspended from telephone poles, buried directly in the ground, or placed in underground cable ducts (often called *conduit*). The average length of customer loops is about two miles.

- **Interoffice transmission facilities** carry trunks and special-services circuits between network switching systems. These facilities vary greatly, but the majority are carrier systems. Interoffice transmission facilities range in length from less than a mile to several thousand miles.

---

[7] For the most part, the terms *channel* and *circuit* are used interchangeably, as are the terms *loop* and *line*. Where differences are perceived, it is likely to be the result of traditional usage in a particular technology area.

[8] Carrier systems combine a number of circuits on two physical pairs of wires using a technique called *multiplexing* (discussed in Section 6.5).

[9] Except in the case of party-line service, where loops are shared by two or more customers.

3.2.3 SWITCHING SYSTEMS

This section provides a broad characterization of switching systems as elements of the facilities network. (Chapter 7 discusses switching functions and concepts, and Chapter 10 describes a number of systems.)

loops,
trunks,
special
service

The primary function of switching systems is to interconnect circuits. Depending on the types of circuits involved, switching systems fall into two functional categories: **local** and **tandem**. *Local* switching systems connect customer loops directly to other customer loops or customer loops to trunks. Local switching systems, which may serve many thousands of customer loops, are also referred to as *central offices*. The central office building contains one or more switching systems (central offices),[10] certain transmission and signaling equipment, and other equipment necessary to provide telephone service to customers in the nearby geographic area. A central office may be divided into two or more 3-digit central office codes.[11] The last four digits of a telephone number provide up to 10,000 line numbers within each central office code.

The term *tandem* is used generically for any switching system that connects trunks to trunks. In a more limited sense, it is often used to denote systems typically found in metropolitan networks within the public switched telephone network (PSTN). These *local tandem* switching systems connect local switching systems to each other or to other systems in the PSTN or interconnect other metropolitan tandem systems. Generic tandem (that is, trunk-to-trunk) systems that perform class 1 through class 4 functions in the toll switching hierarchy (described in Section 4.2) are called *toll* systems. The terms *local* and *toll* reflect the tariff distinction between local and toll traffic.

In addition, some switching systems perform both local and tandem switching functions. (Chapter 4 discusses the implications of combined local/tandem switching systems.)

A recent development in local switching is the use of **remote switching systems** that serve small population centers typical of rural areas. Customer loops are connected to these systems, which, in turn, are connected to central offices in larger population centers. The remote systems are essentially extensions of the host central office. (Section 10.3 discusses these systems.)

In addition to network switching systems located in telephone company buildings, another type of switching system, the **PBX**, is typically located on a customer's premises. As described in Section 2.3, a PBX connects a localized community of users to each other and, through special-

---

[10] The term *central office* is sometimes applied loosely to a central office building and its equipment.

[11] Also called *exchanges*.

Private Branch Exchange

services circuits, to a network switching system. An attendant may handle calls involving an off-premises party. The circuits connecting a PBX to a central office are called *PBX trunks* since they interconnect two switching systems. In most cases, the circuits appear as lines to the central office. PBXs and associated PBX trunks are part of the total facilities network.

Finally, operator services (such as directory assistance, special charging on toll calls, and intercept of calls to nonworking numbers) are provided by arrangements of equipment that are considered to be switching systems. (Section 4.2.2 describes how operator services are provided within the network, and Section 10.4 describes the related equipment.)

3.3 TRAFFIC NETWORKS

The description of the telecommunications network in the previous section emphasized the physical components of the network, namely, station equipment, transmission facilities, and switching systems. In that context, the network was referred to as the facilities network. It is also important to consider the manner in which this network provides the various telecommunications services. From this perspective, the network may be thought of as a set of traffic networks sharing common facilities. For example, the PSTN, which provides public switched telephone network services, is the largest and best-known traffic network. Many other traffic networks provide a variety of special services such as private-line voice and data and audio and video program services. Each traffic network is designed to meet a particular set of requirements related to transmission performance, reliability, maintenance, and the ability to handle the expected traffic volume. The following sections describe some of these traffic networks and show how they use common elements and, in some cases, share them.

3.3.1 PUBLIC SWITCHED TELEPHONE NETWORK

Because of the large volume of business and residential telephone traffic that it carries, the PSTN is probably the most familiar of the traffic networks. This network provides the public switched telephone network services described previously in Section 2.5.1.

The various types of traffic in the PSTN represent communications between any two end points in the network. Traffic is switched through each switching office, or node, it encounters and travels between nodes on trunk groups. The offices and trunk groups are arranged in a hierarchical routing structure, as described in Section 4.2. Circuits 1 and 2 in Figure 3-2 carry this type of traffic. The foreign exchange (FX) line shown as circuit 5 is noteworthy because it appears as a line at distant

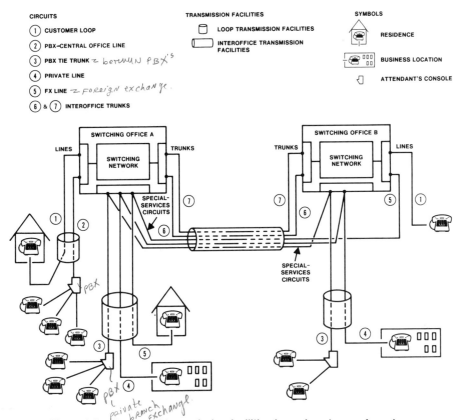

CIRCUITS

(1) CUSTOMER LOOP

(2) PBX-CENTRAL OFFICE LINE

(3) PBX TIE TRUNK ← between PBX's

(4) PRIVATE LINE

(5) FX LINE ← foreign exchange.

(6) & (7) INTEROFFICE TRUNKS

TRANSMISSION FACILITIES

LOOP TRANSMISSION FACILITIES

INTEROFFICE TRANSMISSION FACILITIES

SYMBOLS

RESIDENCE

BUSINESS LOCATION

ATTENDANT'S CONSOLE

Figure 3-2. The use of transmission facilities by various types of services.

switching office B. It is generally used primarily for traffic to and from other lines in the local calling area of office B. It can also use office B as the access point for the entire PSTN.

The PSTN is, by far, the largest traffic network in terms of both equipment utilization and traffic volume. In 1981, it handled about 270 billion calls, and at the end of that year, it served about 180 million telephones in the United States. (These figures include both Bell System and non—Bell System calls and telephones.)

3.3.2 PRIVATE-LINE VOICE NETWORKS

As described in Section 2.5.3, large businesses with many dispersed locations often use private-line voice networks—dedicated facilities that connect a company's various locations. The circuits used for private networks are special-services circuits.

Some of these circuits are switched; others are nonswitched. Among nonswitched private networks, many serve only two stations and are called *point-to-point networks*. Others, called *multipoint networks*, interconnect a number of stations at dispersed locations and may have signaling arrangements to alert appropriate stations when communication is desired. Private-line nonswitched networks may be interconnected at telephone company switching offices, although they are not switched through the switching systems. The PBX tie trunk, circuit 3, and the private line, circuit 4, in Figure 3-2 are examples.

In addition to the nonswitched private networks, several thousand private switched voice networks serve government organizations and large business customers. The trunks and access lines in these networks are private-line circuits that interconnect switching systems either at customer locations (for example, PBXs) or in telephone company switching offices. A typical private switched network sharing switching systems with the PSTN is shown in Figure 3-3. The switching systems are partitioned so that only private network access lines may access the private network trunks. Private switched networks range in size from large nationwide networks that serve hundreds of locations to small networks in metropolitan areas that serve fewer than ten locations.

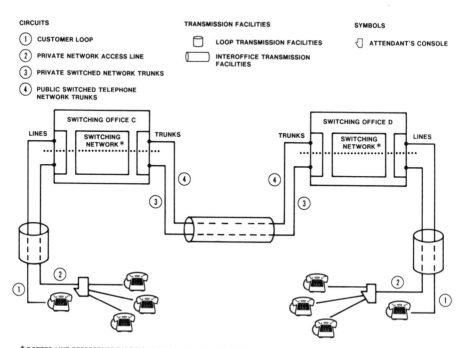

Figure 3-3. Example of a private switched network.

Examples of private switched networks are CORNET (the corporate network used in the Bell System), FTS (the Federal Telecommunications Service network serving the civil organization of the Federal Government), and AUTOVON (the automatic voice network of the United States military).

A key point illustrated by Figures 3-2 and 3-3 is that private networks, both switched and nonswitched, may share many elements of the facilities network with the PSTN. Specifically, the various circuits for the PSTN and private networks are provided by common transmission facilities. Furthermore, in switched private networks, even the switching systems can be shared, as shown in Figure 3-3; although certain private switched networks (for example, AUTOVON) have dedicated switching systems.

3.3.3 PRIVATE-LINE DATA NETWORKS

Section 2.2 describes private-line data services over analog facilities. Section 2.5.4 describes *DATAPHONE* digital service, and Section 11.6 describes the system aspects of the Digital Data System (DDS) that support the service.

Another example of private-line data networks involves telegraph channels. These channels employ a fraction of the voice bandwidth[12] and are used for services such as teletypewriter, remote metering, and burglar alarms. There are about 8000 interstate private teletypewriter networks. About half are simple 2-point networks; the rest are multipoint networks, most of them nonswitched.

In addition to private data networks provided by the telephone companies, customers may form their own switched data networks by leasing private-line circuits and interconnecting them with computers operating as data switches. Both the National Aeronautics and Space Administration and the Advanced Research Projects Agency have such private data networks. In a certain sense, the structure of private data networks is generically similar to that of private voice networks. The channels involved are dedicated to the customer, are usually not switched by network switching systems, and typically share transmission facilities with the PSTN.

Private voice and data channels differ primarily in their transmission requirements, particularly where high-speed data is involved. Furthermore, because of the current advances in both switching and transmission technologies, the distinction between voice and data networks is becoming less and less significant.

---

[12] The *voice bandwidth* is the range of frequencies necessary to transmit and receive acceptable quality speech signals. See Section 6.1.

3.3.4 PROGRAM NETWORKS

Radio and television broadcasters use program networks extensively to distribute program material simultaneously to a number of their affiliated stations. Because of different transmission requirements (see Chapter 6), both audio and video program networks exist. A key difference between program networks and private switched voice networks is the directional nature of the communications. While voice networks support simultaneous 2-way communications, the program networks provide 1-way transmission of audio and video programs from one source to many destinations. Another difference concerns the typical transmission capacities required. In a private voice network, two customer locations may be connected by a single voiceband channel; the transmission of audio and video programs, on the other hand, requires much larger bandwidths for the video information, typically, equivalent to hundreds of voice circuits. Except for these two distinguishing features, the structure of program networks is similar to that of private voice networks; hence, program networks share the transmission facilities with other private networks and the PSTN.

3.4 A TYPICAL TELEPHONE CALL

To introduce the rudimentary operation of the network, this section presents a functional description of a typical telephone call, the most familiar service provided by the PSTN. The description illustrates some of the terms defined in previous sections and introduces some new terms and concepts.

public switched telephone network

3.4.1 SETTING THE STAGE (FIGURE 3-4)

Mrs. Cooper, a local realtor, is calling Mrs. Mahon, a prospective buyer, at her home in a neighboring town. Mrs. Cooper's telephone is served by central office A, and her central office code is 747. Mrs. Mahon's telephone is served by central office code 951 in central office B. Since many calls are placed between central offices A and B, a number of trunks provide a direct route between the two offices. An alternate route through tandem office C is also available. (Chapter 5 describes traffic engineering aspects, including alternate routing.)

3.4.2 INITIATING THE CALL (FIGURE 3-5)

When Mrs. Cooper picks up her handset, the switchhook contacts of the telephone set close, signaling its off-hook status. Control equipment in the switching system at office A detects a change from on-hook to off-hook status and interprets the change as a request for service. At this time, dial tone is connected to Mrs. Cooper's telephone, assuming that a

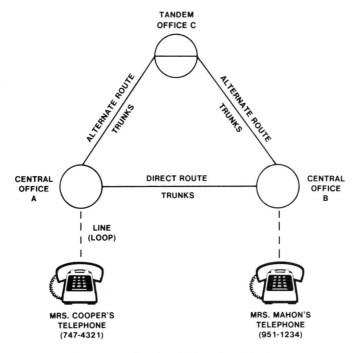

Figure 3-4. Direct and alternate routes for a
call from Mrs. Cooper to Mrs. Mahon.

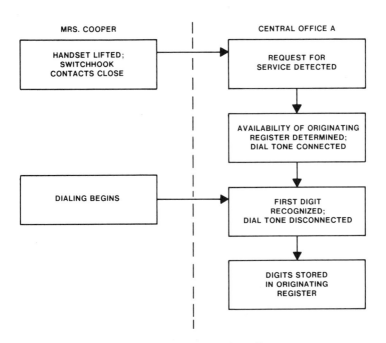

Figure 3-5. Initiating the call.

register, usually called an *originating register*,[13] is available to accept and store the digits she will dial.

After Mrs. Cooper dials the first digit, the dial tone is disconnected. The digits dialed by Mrs. Cooper (951-1234) are received and stored in the originating register.

3.4.3 CALL PROCESSING AT THE ORIGINATING CENTRAL OFFICE (FIGURE 3-6)

Next, the control equipment in central office A translates the dialed number. By examining the leading digits, usually the first three,[14] it determines that Mrs. Cooper's call is to another central office code; that is, it is not an *intraoffice* call. Her call is an *interoffice* call and must be connected to a trunk going to another office. Routing information stored in the system indicates which paths (trunk groups) are appropriate and translates the desired paths to representations of physical locations or terminations of trunks. If the call is billable, an automatic message accounting (AMA) register is requested (see Section 3.4.5). At this time, control equipment transfers the call information to a register in another storage area (the outpulsing register shown in Figure 3-7), releasing the originating register from the call. The control equipment begins scanning the outgoing trunks to find an idle trunk to office B. An idle trunk is found directly between offices A and B.

The control equipment could have found that all trunks in the trunk group(s) to office B were busy. In this case, it would have begun to scan the outgoing trunks to tandem switching office C, since the call could be routed on a trunk from office A to office C and from there to office B (as shown in Figure 3-4). If all trunks to tandem office C had also been busy, it would have been impossible to complete the call. In that case, Mrs. Cooper would have heard a reorder tone, often called a *fast busy tone* since it has 120 interruptions per minute (ipm), compared to the 60 ipm of the busy tone.

3.4.4 CALL ADVANCEMENT TO THE TERMINATING CENTRAL OFFICE (FIGURE 3-7)

The first event shown in Figure 3-7 is the seizing of an idle trunk to office B. When a trunk is seized for a particular call, it appears busy to the switching system and becomes unavailable for other calls. A 2-way

---

[13] To illustrate the functional operations involved in the call, this discussion uses generic terms for equipment. Because of the variety of switching systems in the network, the generic term may not fit all cases. For example, step-by-step switching systems, the oldest systems used (see Section 10.2.2), may not have originating registers and may complete the switching functions differently. Likewise, stored-program control systems (see Sections 10.3.1 and 11.3.1) are computer-like in their operation.

[14] In some cases, an access code, such as the digit 1, is used as a prefix to the address digits on calls outside the local area. (See Section 4.3.)

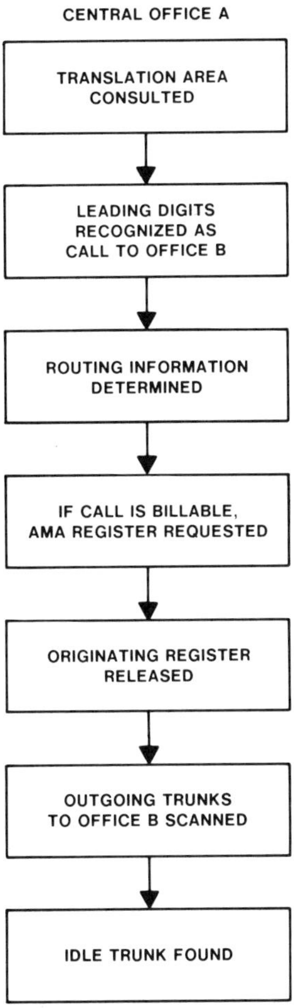

CENTRAL OFFICE A

```
┌─────────────────────────┐
│    TRANSLATION AREA      │
│       CONSULTED          │
└─────────────────────────┘
            │
            ▼
┌─────────────────────────┐
│    LEADING DIGITS        │
│    RECOGNIZED AS         │
│   CALL TO OFFICE B       │
└─────────────────────────┘
            │
            ▼
┌─────────────────────────┐
│   ROUTING INFORMATION    │
│       DETERMINED         │
└─────────────────────────┘
            │
            ▼
┌─────────────────────────┐
│   IF CALL IS BILLABLE,   │
│  AMA REGISTER REQUESTED  │
└─────────────────────────┘
            │
            ▼
┌─────────────────────────┐
│   ORIGINATING REGISTER   │
│       RELEASED           │
└─────────────────────────┘
            │
            ▼
┌─────────────────────────┐
│    OUTGOING TRUNKS       │
│   TO OFFICE B SCANNED    │
└─────────────────────────┘
            │
            ▼
┌─────────────────────────┐
│     IDLE TRUNK FOUND     │
└─────────────────────────┘
```

Figure 3-6. Processing the call at the
originating central office.

trunk may be seized by the switching system at either end to originate a
call, while a 1-way trunk may only be seized from one end. Transmission occurs in both directions on either type of trunk.

Mrs. Cooper's line is connected to the outgoing trunk through a path
in the switching network within the switching system. The identity of
this trunk, the number of digits to be transmitted, and additional information that may be necessary for call setup are recorded in an outpulsing
register.

In central office B, an incoming register of the switch will be seized[15]
and will signal readiness to receive address information. The control

---

[15] Seizing the register makes it unavailable for other incoming calls until it is released.

95

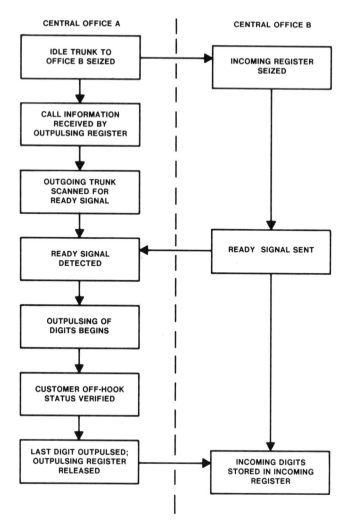

CENTRAL OFFICE A

CENTRAL OFFICE B

IDLE TRUNK TO
OFFICE B SEIZED

INCOMING REGISTER
SEIZED

CALL INFORMATION
RECEIVED BY
OUTPULSING REGISTER

OUTGOING TRUNK
SCANNED FOR
READY SIGNAL

READY SIGNAL
DETECTED

READY SIGNAL SENT

OUTPULSING OF
DIGITS BEGINS

CUSTOMER OFF-HOOK
STATUS VERIFIED

LAST DIGIT OUTPULSED;
OUTPULSING REGISTER
RELEASED

INCOMING DIGITS
STORED IN INCOMING
REGISTER

Figure 3-7. Call advancement to the terminating
central office.

equipment in Mrs. Cooper's central office will periodically scan for this "ready" signal. When this "ready" signal is detected, outpulsing of digits begins. If central office B contains a single central office code, only the last four digits of Mrs. Mahon's number will be transferred. This is because all calls on the direct trunk group will terminate at central office B. However, if office B contains more than one central office code, additional digits must be transmitted to identify the particular central office code serving Mrs. Mahon.

Before the last digit is sent, the control equipment checks to see that the calling customer's line is still off-hook. If the calling customer has hung up (abandoned the call), the control equipment will terminate the call-processing sequence and release associated equipment and circuits.

When the last digit is outpulsed, the outpulsing register is released. The digits are now stored in the incoming register at central office B. (Sections 8.4 and 8.5 discuss several techniques for sending the digits between central offices.)

3.4.5 CALL COMPLETION (FIGURE 3-8)

Once the digits are stored in an incoming register at the terminating office, many functions are initiated and supervised by the control equipment. The 4-digit line number is translated to Mrs. Mahon's physical line termination. The status of Mrs. Mahon's line is interpreted and signifies that the line is idle. (If Mrs. Mahon's line were busy, a busy signal would be returned to Mrs. Cooper.)

The incoming trunk is connected through the switching network to Mrs. Mahon's line. A ringing register is seized, the incoming register is released from this call, and Mrs. Mahon's telephone rings. An *audible ring*, a tone that has the timing of a ringing signal and that indicates that a ringing signal is being applied to Mrs. Mahon's telephone, is sent back to Mrs. Cooper at this time.[16] The control equipment at the terminating office will scan Mrs. Mahon's line status for an answer (off-hook) indication and, when it is detected, will terminate the ringing signal and return answer supervision to office A. This will be used to record answer or connect time for billable calls.

Mrs. Mahon answers the phone, and the conversation begins. As Mrs. Cooper talks into her handset, the acoustic speech signal is converted into an electrical signal by the transmitter in the handset. The signal generated by conventional transmitters is an electrical analog of the acoustic signal. This electrical analog of the speech may proceed through the switching systems and transmission facilities to Mrs. Mahon's telephone in that form, or it may proceed through part of its path in digital form. The latter would then require analog-to-digital and digital-to-analog conversions.

With conventional technology, the signal reaching Mrs. Mahon's telephone will be analog, and the receiver will convert the analog signal back to an acoustic signal. The acoustic signal from the receiver is not an exact reproduction of that at the transmitter. One reason for this is that the frequency content is limited by the transmission path (see Section 6.2). Also, impairments such as noise and loss occur, and if the call

---

[16] Although initiated at the same time, the audible ring is separate from the ringing signal that activates the ringer in the called party's telephone.

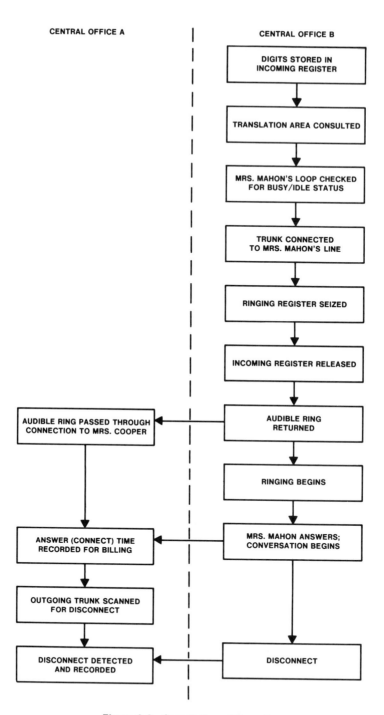

Figure 3-8. Completion of the call.

travels a long distance, an echo effect could occur. (These impairments and ways of controlling them are discussed in Section 6.6.)

During the conversation, the originating office, office A, monitors the outgoing trunk to office B for disconnect. If the calling party hangs up first, the connection is released, and disconnect supervision is sent to the terminating office. The trunk is idled when the terminating office returns on-hook supervision.

If the called party (Mrs. Mahon, in this example) hangs up first, a timed-release period of 10 to 11 seconds is initiated. The connection is released after this time—or earlier if the calling party hangs up.

Completion of the call is detected and recorded at central office A for accounting purposes if there is a charge for the call; that is, if it is not covered by a fixed monthly charge or a flat rate. When the call is first dialed, the control equipment in central office A determines whether the call is billable by the routing information associated with the first three digits (see Figure 3-6). If the call is billable, a register is requested from an automatic message accounting system to receive information that is to be recorded about the call. For Mrs. Cooper's call, the information recorded includes the number of Mrs. Cooper's telephone, the number dialed, the time Mrs. Mahon answered, and the time the connection was released. Data on this call and other billed calls from central office A are forwarded to a data-processing accounting center where they are periodically processed to compute customer charges. If the call is billable, Mrs. Cooper's next monthly telephone bill will include a charge for the call. (Section 10.5 describes how the data are processed to determine the charge, and Section 13.2.2 discusses operations at the accounting center.)

Thus, a basic telecommunications service—the simple telephone call—requires a relatively complex sequence of events.

AUTHORS

J. L. Bazley
T. Pecsvaradi

PART TWO

NETWORK AND SYSTEMS CONSIDERATIONS

The five chapters in this second part present the basics of network structure and planning and explain engineering considerations applicable to network and customer-services systems. Network functions such as transmission, switching, and signaling are discussed, and a background of telecommunications concepts, principles, and technology is presented with emphasis on relevance to the related network and system topics. This should provide a general foundation for the description of specific network and customer-services systems in Part Three.

Chapter 4 expands on the introductory material in Chapter 3 and describes the structure of traffic and facilities networks. It explains the difference between local and toll networks, discusses the role of network planning, and describes the unified numbering plan. Chapter 5 presents basic traffic concepts through a discussion of topics such as traffic theory and its application to the engineering of switching systems and trunk groups. Interrelated traffic considerations that affect network engineering, the management of traffic networks, and considerations in the design of data networks are also discussed.

Chapter 6 presents fundamental transmission principles and technology. Concepts such as signals, channels, media, modulation, and multiplexing are examined, and transmission impairments and objectives are discussed. Beginning with a discussion of the basic role of switching, Chapter 7 explains switching functions, networks, and control. In addition to a description of the basic switching functions involved in establishing a connection, the various auxiliary functions are also discussed. Part Two concludes with Chapter 8, which explains the fundamentals of signaling and interfaces in terms of the basic functions they provide in the network and some of the methods and concepts involved in fulfilling their roles.

4

Network Structures and Planning

4.1 INTRODUCTION

The telecommunications network, as defined in the last chapter, has two interrelated aspects. In terms of its physical components—station equipment, transmission facilities, and switching systems—the telecommunications network is a facilities network. To provide various services, those facilities are interconnected in many ways; in this sense, the telecommunications network is a set of traffic networks that share the facilities.

This chapter continues the discussion of the telecommunications network with descriptions of the structures of the facilities and traffic networks. For traffic networks, the emphasis is on the largest and most complex—the public switched telephone network (PSTN)—although private networks are covered briefly.[1] The discussion also includes a description of the PSTN worldwide numbering plan. The last section of the chapter discusses considerations and approaches used in planning the configuration of the telecommunications network so that it continuously meets constantly growing and changing demands.

4.2 STRUCTURE OF TRAFFIC NETWORKS

4.2.1 THE PUBLIC SWITCHED TELEPHONE NETWORK

The PSTN actually consists of two strongly interdependent networks: the **local network** (sometimes called the *exchange area network*) and the **toll network**. The interdependence results from extensive integration and

---

[1] As noted in Section 2.1.2, special services (which include private networks) have been growing rapidly. One result of this growth is that about 5.2 million (43 percent) of the approximately 12 million interbuilding circuits in service in the Bell System in 1982 were special-services circuits.

sharing of functions to reduce overall network costs. The following discussion, which is designed to convey basic concepts, presents a simplified view of the local and toll network structures. The actual structure is far more complex due to the variety of ways in which network functions are integrated to meet diverse needs in particular segments of the network.

Local Network Structure

The structure of the local network begins with customer station equipment connected by loops to local switching systems. All customers connected to a local switching system (central office) in a particular central office building[2] are said to be located in a *wire center area*, and the location of the building is called the *wire center*. These concepts are illustrated in Figure 4-1. Customers located within a wire center area communicate with each other through the local switching system, or systems, at the wire center. As indicated in Section 3.1, this arrangement reduces network costs by adding some switching costs in return for a large reduction in transmission costs.

Figure 4-2 illustrates the concept of judiciously combining switching and transmission to minimize overall costs in the local network. In the 2-level switching hierarchy shown, which is typical of most metropolitan areas, the switching systems at adjacent or nearby wire centers are connected by trunks, either directly or through one or two tandem switching systems. Thus, customers in adjacent or nearby wire center areas communicate with each other using their dedicated loops and the trunks interconnecting their local and tandem switching systems.

Whether it is more economical to provide direct trunks between two adjacent wire centers, to interconnect them indirectly using tandem trunks and tandem switching systems, or to use a combination of both depends on the traffic volumes, the distances involved, and the opportunities for sharing the facilities among many customers.

In Figure 4-2, there is a strong community of interest (high traffic volume) between offices at wire centers A and B, justifying a direct trunk group (represented by the dashed line). Traffic between wire center C and the other two wire centers does not warrant direct trunk groups and is carried by tandem groups (represented by solid lines) through tandem office T. Using tandem trunk groups and switching systems to provide service in a local area usually involves longer transmission paths and more switching but proves to be more economical when the traffic volumes between pairs of local switching systems are very low.

For intermediate traffic volumes, the most economical solution may be a combination of direct and tandem trunks. The routing technique that

---

[2] As noted in Chapter 3, a central office building may contain one or more switching systems.

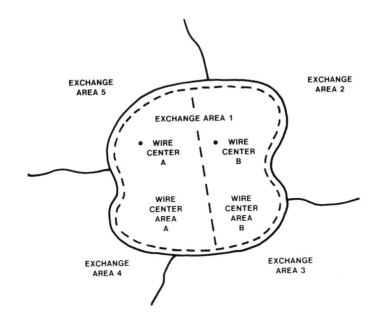

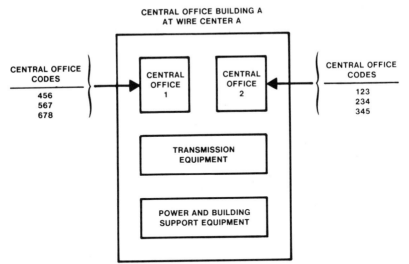

Figure 4-1. Local network topology. *Top*, wire center areas and exchange areas. Traditionally, a single, uniform set of charges exists for telephone service in an exchange area, and a call between any two points in the area is a local call. (Different meanings will apply after divestiture.) *Bottom*, central office building terminology.

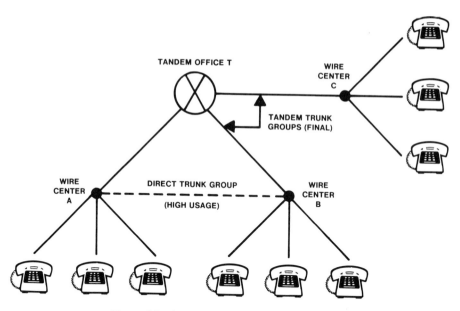

Figure 4-2. A typical local network configuration.

takes advantage of this network structure is called *automatic alternate routing*. With automatic alternate routing, a switching office first offers a call to a *high-usage (HU) trunk group* (a primary direct route between two switching systems). If all trunks in the HU group are busy, the call is routed via a tandem office using a *final trunk group*. Final trunk groups are the final routes traffic can take. When all transmission paths on a final trunk group are busy, the calling customer is sent a reorder tone and must try again later.

It is important to note that the terms *direct trunk group* and *tandem trunk group* describe the topological structure of the network, while *high-usage trunk group* and *final trunk group* refer to the manner in which the trunk groups are used when routing traffic on them. Consequently, a direct trunk group is not necessarily an HU trunk group. If, for example, the traffic between offices A and B in Figure 4-2 were not allowed to overflow to the tandem trunk groups, then the direct trunk group would not be an HU group. HU and final groups are traffic-engineered differently, as described in Section 5.4.

Section 5.5 describes the method of determining the most economical apportionment of HU and final trunks. As discussed there, an important factor in the determination is trunk group *efficiency*—the concept that average traffic carried per trunk increases with trunk group size. This concept is the basis for efforts to concentrate traffic into larger parcels, a major consideration in the design of traffic networks.

When the number of central offices in a metropolitan area is large, it may be advantageous, in terms of trunking arrangements, to group central offices in sectors that reflect communities of interest. Each sector is served by a tandem office located near the traffic-weighted center of the sector. This reduces the total length of tandem trunking. The attendant penalty is that some calls may require three consecutive tandem trunks[3] and, consequently, may have a slightly longer set-up time. A more complicated structure with more routing choices may also result if certain office pairs, such as A and D in Figure 4-3, reside in different sectors but have enough traffic between them to justify a direct trunk group.

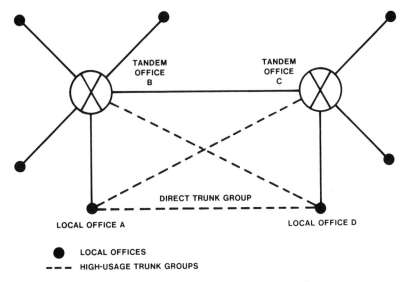

Figure 4-3. A 2-tandem-office local network.

Toll Network Structure

As mentioned earlier, the switching configuration shown in Figure 4-2 is a 2-level hierarchy. While it provides an economical tradeoff between switching and transmission costs within most local metropolitan areas, it is not the most economical structure for interconnecting all switching offices in the nationwide network. The volume of traffic between widely spaced local offices is usually small, and direct trunks to serve this traffic would be too expensive. Therefore, the switching hierarchy now used in the United States has five levels (illustrated in Figure 4-4), in which successively higher level offices (also called *classes*) concentrate traffic from increasingly larger geographical areas.

---

[3] Trunks interconnecting tandem offices are sometimes called *intertandem trunks*.

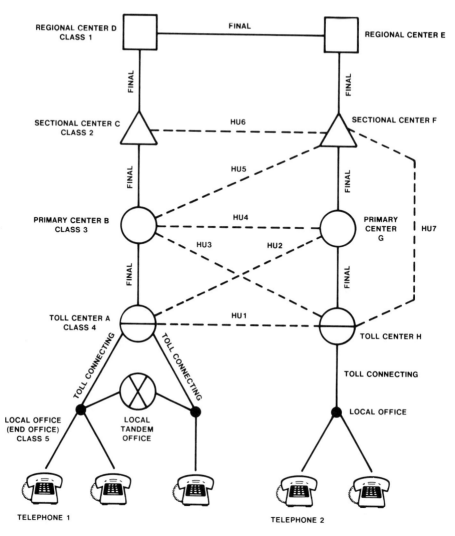

Figure 4-4. The switching hierarchy.

Class 5 offices (also called *local switching offices* and *end offices*) are part of the local network. The toll network consists of the class 4 and higher switching offices (toll centers, primary centers, sectional centers, and regional centers as shown in Figure 4-4) and the trunks interconnecting them. Table 4-1 gives the number of offices at each level of the PSTN hierarchy. All trunks within the toll network are called *intertoll trunks*. Class 5 end offices are connected to the class 4 toll centers by *toll connecting trunks*.

TABLE 4-1

DISTRIBUTION OF OFFICES IN
THE PSTN HIERARCHY (1982)

	Class				
	1	2	3	4	5
Bell operating companies	10	52	148	508	9803
Independent telephone companies			20	425	9000

An office connected by a final trunk group to a higher class office is said to "home" on that office, although it should be noted that offices do not always home on the next higher class office. For example, while most class 5 offices (end offices) home on class 4 offices (toll centers), some class 5 offices home on class 3, 2, or 1 offices.

Typically, offices in the same homing chain are located relatively near each other. Consequently, many final trunk groups are only a few miles long. The final and HU trunk groups that interconnect two different regions (that is, two different homing chains) are the *long-haul trunks* of the toll network. If the volume of traffic between offices of the same class or differing in class by one is high enough, it is more economical to connect them directly by HU trunks (HU1 through HU6 on Figure 4-4) than to send traffic by some indirect path. Sometimes traffic between offices in the same homing chain justifies interconnection by an HU group, as illustrated by HU7 in Figure 4-4.

The basic rule for routing a toll call is to complete the connection at the lowest possible level of the hierarchy, thus using the fewest trunks in tandem. In Figure 4-4, a call from telephone 1 to telephone 2 first goes to the local office (a class 5, or end office). That office recognizes the call as a toll call and sends it over toll connecting trunks to toll center A. Toll center A searches for an idle HU trunk, first in the HU1 group, then in the HU2 group. If all trunks in those groups are busy, the call overflows to the final trunk group connecting toll center A and primary center B. If all trunks in this final group are busy, the call is blocked and the customer receives a reorder tone. Otherwise, the call reaches primary center B, and the sequential search procedure continues up through the hierarchy, searching HU3 through HU6 and related final groups until either an idle path (sequence of trunks in tandem) between the two end offices is found or all possible routing alternatives are exhausted. The average toll call uses slightly over three trunks, including toll connecting trunks. The maximum number of trunks that may be used in a connection is nine.

The actual structure of the network is far more integrated and complex than the simplified local and toll network structures just described. For example, advances in switching technology allow the introduction of local switching systems that can record billing information and perform alternate routing, thus combining local, local tandem, and toll functions. Figure 4-5 shows the distribution of the different switching functions over the Bell System switching systems. The net effect of sharing among switching systems is a reduction in switching and transmission costs.

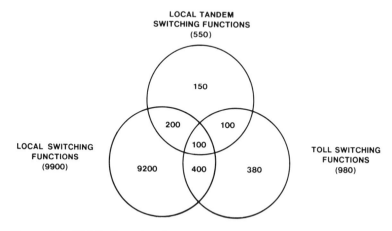

Figure 4-5. Distribution of switching functions over Bell System switching systems. The numbers represent the number of switching systems performing the indicated functions at the end of 1978. The union of any two (or all three) circles represents systems providing a combination of functions. For example, 400 systems provide a combination of local and toll functions.

One manifestation of the ability to combine or integrate local and toll functions in one switching system is that toll calls between certain pairs of end offices are not carried on toll connecting trunks up to the class 4 toll center on which the end office homes. Rather, they are carried on *end office toll trunks.* Many of these are direct trunk groups that carry toll traffic only between two end offices. Other end office toll trunks may carry toll traffic from one end office to the class 4 toll office on which a distant end office homes. These trunk groups are HU groups, so that any overflow from them would be routed through the toll hierarchy in the usual manner. Because they save on toll switching costs, end office toll trunks are economical when there is a sufficiently high community of interest between two end offices that are a considerable distance apart. In this arrangement, the switching systems involved must be able to record billing information and perform alternate routing. This type of trunking is prevalent in certain areas of the country. In New Jersey, for example, nearly 36,000 of the approximately 168,000 trunks carrying toll traffic at the end of 1982 were end office toll trunks.

4.2.2 OPERATOR FUNCTIONS

Section 2.5.1 describes the operator services provided for PSTN customers. The related operator functions are performed primarily in conjunction with computer-based operator systems and automatic call distributors
(ACDs).[4] Toll switchboards account for a small and declining percentage
of operator positions. Two systems that automate certain operator functions appear in Figure 4-6. The Traffic Service Position System (TSPS)
automates many functions of the toll-and-assistance operator related to
billing calling card, collect, coin, and person-to-person calls. Most intercept functions are now provided by the Automatic Intercept System (AIS).
With AIS, most intercept calls are handled automatically but operators are
available to handle unusual problems and provide additional clarification
to customers, if necessary.[5]

Connections to the PSTN switching hierarchy from the switchboards
and automated systems are shown in Figure 4-6 and are described below.

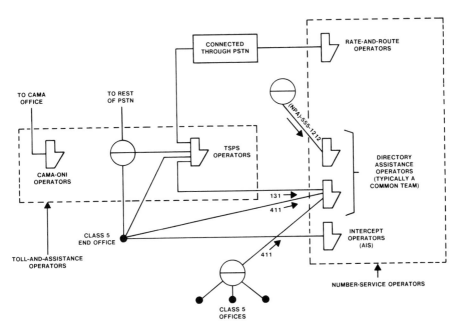

CLASS 4 OFFICE (OR HIGHER LEVEL OFFICE IN TOLL HIERARCHY)
NPA   NUMBERING PLAN AREA

Figure 4-6. Typical topology of operator functions. *Note*: Connections to
directory assistance operators show numbers dialed by originator of request.

[4] Section 11.2.6 describes ACDs.

[5] Section 10.4 describes TSPS and AIS.

- Operators at TSPS consoles (or one of the remaining toll switchboards) are interconnected with the PSTN in several ways. The most common function, providing assistance on outward calls, requires connection between class 5 offices and a toll office. Other functions may require connection to dedicated facilities—such as specialized trunks to class 5 offices for busy-idle testing. These facilities are usually accessed via toll offices. To perform some of their functions, operators at TSPS consoles require connections to directory assistance and rate-and-route operators.

- Centralized automatic message accounting — operator number identification (CAMA-ONI) operators are temporarily connected to customer-dialed, station-to-station calls to identify the caller's telephone number to CAMA equipment when the local office does not have AMA and automatic number identification (ANI) capability. These operators can be connected to any toll office (class 4 or higher) that provides a CAMA function.[6]

- Directory assistance operators respond to customers' requests for unknown telephone numbers. In most cases, the customer dials different numbers for directory assistance depending on the location of the requested number relative to the calling customer as described below.

  — For numbers in the local calling area, 411 is dialed. Local offices close to the ACD handling directory assistance are connected by direct trunk groups; more distant offices typically connect through a toll office where directory assistance traffic is concentrated on a trunk group to the ACD.

  — For numbers in the same or *home numbering plan area* (HNPA)[7] but not in the same local calling area, 555-1212 is dialed. These calls are typically routed to a toll office where they are concentrated on a trunk group to the ACD.

  — For numbers in another or *foreign NPA* (FNPA), NPA-555-1212 is dialed. These calls are routed through the toll network to a terminating toll office near the ACD and concentrated on a trunk group to the ACD.

- Intercept operators (or AISs) respond to calls made to nonworking numbers (changed or unassigned numbers). These operator positions are connected on direct trunks from terminating local offices.

- Rate-and-route operators provide toll-and-assistance operators with routing codes, rate information, and lists of numbers that may be coin

---

[6] Section 10.5.4 describes CAMA operation.

[7] Section 4.3 discusses the worldwide numbering plan.

telephone lines. The rate-and-route operators are connected through the toll network to toll-and-assistance operators.

4.2.3 STRUCTURE OF PRIVATE SWITCHED NETWORKS

The communications needs of large corporate customers can often be satisfied at lower cost by private networks. Private networks may also provide features or capabilities that are not available with the PSTN. (See Section 2.5.)

In many respects, the structure of private networks is similar to that of the PSTN. For instance, a typical Enhanced Private Switched Communications Service (EPSCS) network provides large corporate customers with an advanced, electronically switched voice and data communications service. It links thousands of telephones, private branch exchanges (PBXs), and data sets in hundreds of geographically dispersed customer locations throughout the country.[8] Customer-premises equipment is connected by access lines to EPSCS switching systems, which are 1ESS switching systems (see Section 10.3.3) especially modified to provide customers with high-quality transmission of voice and data over long distances. The EPSCS switching systems, in turn, are connected by trunks.

The EPSCS network has a 2-level hierarchy, which means that most EPSCS calls are completed through only one or two switching systems and usually involve no more than two or three trunks in tandem. For example, in Figure 4-7, a call made from the San Diego PBX to the Miami PBX of a large corporation is routed most directly and economically through the EPSCS switching systems at Los Angeles and Atlanta over the high-usage trunks between them. If all trunks on this route are busy, the call overflows to a less direct route. The trunking arrangement shown in the figure suggests several alternative routes, with the final route through all four EPSCS switching systems.

It is important to note that parts of the EPSCS networks are shared among different EPSCS customers and parts are dedicated to individual customers. For example, the 1ESS systems are usually shared because most customers do not have enough traffic to justify a dedicated switching system. But the access lines from the customer's PBX to an EPSCS switching system, the trunks between EPSCS switching systems, and the switch terminations are dedicated to a single EPSCS customer.

Private and public networks also share facilities. Thus, most EPSCS switching systems also provide ordinary PSTN services. Furthermore, even the dedicated trunks between EPSCS switching systems are provided by transmission facilities that may contain many other private and public circuits. It is through such pervasive sharing of facilities that economies of scale can be realized in the telecommunications network.

---

[8] Sections 11.1.2 and 11.2 describe the operation of data sets and PBXs, respectively.

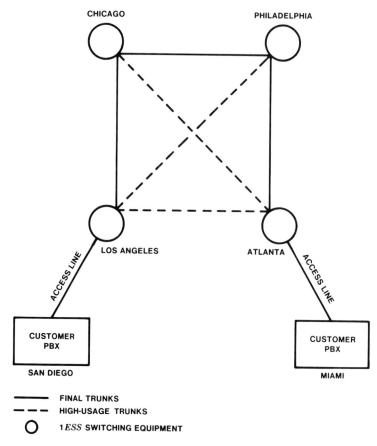

CHICAGO   PHILADELPHIA

LOS ANGELES   ATLANTA

ACCESS LINE   ACCESS LINE

CUSTOMER PBX   CUSTOMER PBX

SAN DIEGO   MIAMI

—————— FINAL TRUNKS
— — — HIGH-USAGE TRUNKS
O 1*ESS* SWITCHING EQUIPMENT

Figure 4-7. A typical EPSCS network for a large corporation with offices in San Diego and Miami. The network has a 2-level hierarchy with the switching systems at Los Angeles and Atlanta homing on systems in Chicago and Philadelphia, respectively. This structure creates an efficient alternate-routing private network.

4.3 NUMBERING PLAN

A fundamental requirement for a switched telephone network is a numbering system that identifies each main station[9] by a unique address that is convenient, readily understandable, and similar in format to those

_____

[9] A telephone that is connected directly to the central office either by an individual or shared line. Main stations include the principal telephone of each party on a party line. They do *not* include telephones that are manually or automatically connected to the central office through a PBX or extension telephones (telephones that have been added to an individual or shared line to extend the telephone service to other parts of the subscriber's home or business premises).

of other main stations connected to the network. Since 1947, this need has been met for the PSTN in most of North America by an integrated, unified numbering plan. Within the plan, each main station is assigned a unique 10-digit address consisting of a 3-digit area code (representing a numbering plan area), a 3-digit central office code, and a 4-digit station number. In some instances, a prefix, a suffix, or both may be added. In addition to providing telephone numbers for use in all customer-to-customer traffic (including direct distance dialing [DDD] calls), the numbering plan provides for:

- calls from a customer to a telephone company service (for example, calls to directory assistance)

- calls from an operator to a customer (for example, calls established by an operator at a prearranged [deferred] time)

- calls from one operator to another (for example, calls that cannot be dialed directly by a customer).

When international DDD (IDDD) became possible, the numbering plan was expanded to provide access to countries outside North America.

The following paragraphs describe the numbering plan for the national and international PSTN. Numbering plans used in other traffic networks (for example, private networks) are outside the scope of this discussion.

4.3.1 NOMENCLATURE

The following set of symbols is commonly used in discussing the numbering plan and dialing procedures:

$$N \ = \text{Any digit 2 through 9.}$$
$$X \ = \text{Any digit 0 through 9.}$$
$$0/1 = \text{Either 0 or 1.}$$

Some related terms are:

- *Address* — A unique 10-digit number assigned to a main station.

- *Central Office Code* — A 3-digit identification number preceding the station or line number. There may be up to 10,000 station numbers per central office code. Several central office codes may be served by a central office, as noted in Section 3.2.3.

- *Dual-Tone Multifrequency (DTMF) Symbol* * — The star (or asterisk) button, located at the lower left in the standard 4-row, 3-column pushbutton array. (Section 11.1.1 discusses DTMF signaling.)

- *DTMF Symbol* # — The number sign (or pound sign) button, located at the lower right in the standard pushbutton array.

- *Listed Directory Number* — The 7-digit number composed of the 3-digit central office code and the 4-digit station or line number.

- *Numbering Plan Area (NPA)* — A geographical division defined by the familiar "area code," within which telephone directory numbers are subgroups. In North America, a 3-digit $N0/1X$ code is assigned to denote each NPA or area code.

  > *Home NPA (HNPA)* — The NPA within which the calling line appears at a local (class 5) switching office.

  > *Foreign NPA (FNPA)* — Any NPA other than the home NPA.

- *Prefix* — Any signal dialed prior to the address. Prefixes are used to place an address in proper context, to indicate service options, or both. With international calls, for example, the prefix 011 marks a simple IDDD call. The prefix 01 calls for special handling such as that required for a person-to-person call.

- *Service Code* — A code, typically of the form $N11$, that defines a connection for a service. Examples are 411 for directory assistance and 611 for repair.

- *Station (or Line) Number* — The final four digits of a standard 7- or 10-digit address. These digits define a connection to a specific customer's line within a central office code.

- *Suffix* — Any signal dialed after the address. Operators use suffixes, for example, to indicate the end of dialing.

- *System Code* — A 3-digit code, usually of the form $1XX$ but including $0XX$ assignments, available only to operators or to switching equipment for use as part of a special or modified address to influence route selection.

- *Toll Center Code* — A 3-digit code of the form $0XX$ that identifies a specific toll center and is available only for operating company use. It is always preceded explicitly or implicitly by an area code; therefore, toll center codes need not be used the same way in all NPAs.

4.3.2 INPUT DEVICES AND DIALING PROCEDURES

The numbering plan must accommodate all authorized address input devices, for example, customers typically use rotary-dial or pushbutton telephone sets; operators rely largely on multifrequency keysets. The plan must also accommodate the switching equipment that interacts with the address input device and all authorized users, including system test and maintenance personnel.

To reach a particular destination, dialing must follow a prescribed sequence that may depend on the originating point. First, one or more prefix digits may be needed; next, the address must be transmitted; and finally, a suffix may apply. A customer served by a PBX, for example, may dial the prefix 9 to access the PSTN. A DDD call may require the additional prefix 1 and a 10-digit address. To reach the same number, an operator would begin with the prefix KP (key pulse), a keyset signal that unlocks the device provided to register the address digits to follow. (The risk of dialing before dial tone and thus reaching a wrong number under certain circumstances is thereby avoided in operator dialing since nothing is accepted in the absence of a KP.) The same 10-digit address is used by both customers and operators, but the operator indicates end of dialing by the suffix ST (start).

In some cases, a dialing procedure is subject to *critical timing*. Critical timing applies when a switching system cannot tell from the digits dialed whether dialing is complete, as in 0+ (0 followed by other digits) dialing sequences, since 0 alone may be the complete dialed input. Long-range plans seek to minimize such timing. For cases where format alone is not definitive, as in the "dial 0" case cited, the *TOUCH-TONE* telephone input # will be increasingly available as an end-of-dialing (cancel timing) indication. Applications in international dialing and custom calling services already exist.

4.3.3 HISTORY AND EVOLUTION

As noted in Section 4.3, the basic address format used in most of North America for *customer identification* consists of ten digits, divided into a 3-digit area code, a 3-digit central office code, and a 4-digit station number.[10] To avoid a difficult and traumatic transition for both telephone companies and customers, growth in telephone service must be accommodated within the limits of this 10-digit format. The evolution of the address format to accommodate growth is described in the following paragraphs and the changes are summarized in Table 4-2.

---

[10] In the toll network, many calls are carried in formats that differ from the standard 10-digit DDD address. An operator call to another operator may require only the digits 121. Conversely, an operator-dialed call to Mexico may be served most conveniently with an 11-digit address. Most toll switching systems can process calls with addresses ranging from three to eleven digits.

TABLE 4-2

EVOLUTION OF 10-DIGIT CODE

Period of Application	Area Code	Listed Directory Number	
		Office Code	Line Number
Original	N0/1X	NNX	XXXX
Present	N0/1X	NXX	XXXX
Future	NXX	NXX	XXXX

N	=	2 through 9
X	=	0 through 9
0/1	=	0 or 1

Originally, 10-digit addresses were of the form N0/1X-NNX-XXXX.[11] (Table 4-3 shows the allocations of the one thousand 3-digit codes that characterized the original plan.) In most NPAs, the NNX-XXXX format is still used for the 7-digit directory number, but the NXX-XXXX format has been introduced on a limited basis and will find increasing application where additional central office codes are needed. This less restrictive NXX format embodies the concept of interchangeable office and area codes (that is, the formats of area codes and central office codes will no longer be mutually exclusive).

The concept of interchangeable codes was incorporated into the plan in 1962 when it appeared that the basic set of 152 area codes possible using the N0/1X format (shown in Table 4-2) would be exhausted by the mid-1970s.[12] (Careful code management has postponed this date to about the turn of the century.) A more immediate concern in many geographic areas was the limit of 640 central office codes (NNX) per NPA. Accordingly, network preparation for the additional 152 office codes provided by the NXX format (less N11) was completed in 1971, with Los Angeles and

[11] Although all-number calling is now the system standard, telephone numbers have an alphanumeric tradition. Despite the personal appeal of names (which often had local geographical significance, for example, MUrray Hill 7-1234) rather than all-number codes, letters were a basic barrier to the use of the full range of dial-code sequences and to the uniformity of international addresses. Prior to all-number calling, directory numbers were commonly referred to as "2L+5N" to call attention to the alphanumeric usage. It should be noted, though, that the alphanumeric format also used the "3-4" character subgrouping.

[12] When there is no longer an economic alternative (for example, boundary realignment), the normal procedure to accommodate growth is to split an NPA when it requires relief. A recent split occurred in 1973 in Virginia, when NPA 703 was divided to make room for NPA 804. Since the process of splitting involves substantial expense, requires changes likely to produce customer annoyance, and imposes a need for networkwide coordination, decisions to split are subject to close scrutiny. Transition plans are a principal consideration.

TABLE 4-3

CODE PARTITIONING

200 Toll Center and System Codes			

000. 099

100. 199

152 Area Codes

8 Service Codes   640 Central Office Codes

200 . . . 210	211	212 . . . 219	220 299
300 . . . 310	311	312 . . . 319	320 399
400 . . . 410	411	412 . . . 419	420 499
500 . . . 510	511	512 . . . 519	520 599
600 . . . 610	611	612 . . . 619	620 699
700 . . . 710	711	712 . . . 719	720 799
800 . . . 810	811	812 . . . 819	820 899
900 . . . 910	911	912 . . . 919	920 999

New York City as early application sites commencing in 1974. With the $N0/1X\text{-}NXX\text{-}XXXX$ format, mandatory use of the prefix 1 indicates that a 10-digit number will follow, while an initial digit N indicates that a 7-digit number will follow. $N11$ calls are recognized individually.

When DDD was introduced, areas with step-by-step switching equipment (see Section 10.2.2) required a DDD prefix. Since step-by-step equipment acts on each digit as it is received, a unique steering prefix was required to identify toll calls. Calls with the local $NNX\text{-}XXXX$ formats were subject to immediate local switching, so prefix choices were limited to digits or digit combinations starting with 0 or 1. The digit 0 was already being used for operator access, leaving 1 as an apparent choice. But service codes of the form $11X$ for repair service and directory assistance were in common use and had to be accommodated. While prefixes such as 112 and 115 were among early choices, the DDD prefix now in use is predominantly 1. Conflicts with $11X$ service codes were avoided by changing to codes chosen from the $N11$ series or by using standard 7-digit numbers as an alternative.

Eventually, a need for a second prefix was established. The purpose was to permit not only ordinary DDD but a combination of customer dialing and options for operator assistance. The prefix 0, the only remaining single-digit choice, was pressed into service with the understanding that 0+ calls would typically be associated with the "dial 0" assistance calls that operators were handling in any event. (These calls are listed in Table 10-1 in Section 10.4.1.)

4.3.4 THE PRINCIPAL CITY CONCEPT

Both the toll hierarchy and the numbering plan are clearly linked to geography. But how these plans are linked to one another is not obvious. The answer lies in the principal city concept.

In seeking suitable routes to a destination switching node, a switching system examines part of the address or destination code. If every 3-digit area code defines such a destination without violating hierarchical access rules, administrative functions are substantially simplified. The principal city concept provides the needed association of a hierarchical node with an area code by providing a common point, a principal city, to which all calls to a given area code may be routed. The principal city, which is usually located within the numbering plan area served, must be prepared to route all valid calls delivered to it. Thus, a new central office destination code may be activated anywhere without coordinated preparation at all possible call sources. It follows that any call for area code $N0/1X$ may be sent to its one principal city. On the other hand, selected calls for area code $N0/1X$ may bypass the principal city if analysis of the 6-digit code (the area code and central office code) confirms the suitability of such routing.

4.3.5 INTERNATIONAL NUMBERING

IDDD depends as directly on numbering as does national DDD, but standards are now a matter of global concern. The Comité Consultatif International Télégraphique et Téléphonique (CCITT) foresaw this need and organized a study to determine how to satisfy it. The standard, approved in 1964, establishes eleven digits as a preferred maximum length for international numbers but allows twelve.

The international number is flagged by a dialed prefix, not internationally standardized, that alerts the switching equipment. The international number itself consists of a country code and a national number. Country codes are standardized and vary in length from one to three digits, the first digit of which constitutes the world zone number. National numbers are the familiar telephone numbers used for domestic long-distance service. (Table 4-4 gives world zone number assignments.)

The countries or zones anticipating the greatest telephone population by the year 2000 were assigned the shortest country codes to allow for longer national numbers. Specifically, the unified North American world zone is 1 and the Soviet Union zone is 7. In these cases, world zone and country code are the same. Other zones contain a mix of 2-digit and 3-digit country codes (for example, 52 for Mexico and 502 for Guatemala). Countries where the number of telephones to be served can be handled by nine or fewer digits are assigned 3-digit codes. Certain combinations of the initial two digits (22, 23, 24, 25, and 26 for world zone 2; 35 for

TABLE 4-4

WORLD ZONE ASSIGNMENTS

World Zone	Principal Areas Covered
1	Canada, United States
2	Africa
3, 4	Europe
5	South and Central America, Mexico
6	South Pacific
7	U.S.S.R.
8	North Pacific
9	Far and Middle East
0	Spare

NOTE: Specific country code assignments (see AT&T Long Lines 1982, p. 122) tend to be stable but may be changed by mutual agreement.

world zone 3; 50 and 59 for world zone 5) are selected in forming 3-digit codes. The other pairs are assigned as 2-digit country codes. Thus, from the initial two digits, switching systems can determine whether the country code is two or three digits long.

The dialing sequence for an IDDD call is illustrated by the following call from England to the United States. The customer in England dials 010-1-*NXX-NXX-XXXX*, where 010 is the international subscriber dialing prefix used in England; 1 identifies North America as the world zone (and, in this case, is the country code); and the remaining digits are the familiar 10-digit address or national number used in North America.

The Bell System has authorized two prefixes for outwardbound IDDD. The prefix 011 indicates simple coin or noncoin automatic calls. The prefix 01 indicates a desire for operator assistance.

4.3.6 OTHER SPECIAL NUMBERING

International dialing, of course, is only one of a number of services dependent on numbering. A selection of representative formats for other services indicates that numbering is not static and that new services invariably require some adaptation of numbering.

- *Custom Calling Services* — for Call Forwarding: *72 with *TOUCH-TONE* dialing or 1172 with rotary dialing (followed after dial tone by the 7-digit number of the telephone to which calls will be sent)

- *Toll-free service* (area code 800) — for directory assistance: 1+800-555-1212

- **DIAL-IT** *network communications service* (area code 900) — for Sports-Phone: 1+900-976-1313

- *Calling card service* — 0 followed by a 10-digit destination address, followed after signal by a 14-digit calling card number.

4.4 STRUCTURE OF THE FACILITIES NETWORK

Section 3.2 introduced the three major elements of the facilities network—station equipment, transmission facilities, and switching systems. From that discussion, it is evident that station equipment is dedicated to particular customers; and therefore, the quantity, type, and configuration are dictated by the services a customer chooses. Transmission and switching facilities, on the other hand, are shared. For this reason, station equipment is omitted from the following discussion, and the term *facilities network* refers only to transmission and switching facilities.

Many factors determine the structure of the facilities network. Some of the more important ones are customer location, telecommunications services desired, performance objectives, available communications technologies and their costs and performance characteristics, and the need for redundancy in the network to protect against major service interruptions.

For discussion purposes, it is convenient to divide the facilities network into the **local facilities network** and the **interoffice facilities network,** according to the two major classes of transmission facilities: loop and interoffice. No physical dividing lines separate these facilities; the distinction is merely an aid in describing the characteristics of the overall facilities network.

4.4.1 LOCAL FACILITIES NETWORK

The local facilities network consists of the local switching systems located at wire centers and the loop transmission facilities through which customers are connected to local switching systems. The local switching systems are located at what can be considered the conceptual boundary between the local facilities network and the interoffice facilities network. In this sense, they belong to both networks.

A wire center area is divided into several (usually four) distinct geographical areas called *feeder route areas.* Feeder route areas are further divided into *allocation areas,* and each allocation area is divided into one to five *distribution areas.* Figure 4-8 shows a portion of the structure of the local facilities network in a typical suburban area, including a wire center building and one feeder route area. The inset shows the structure of a distribution area in detail.

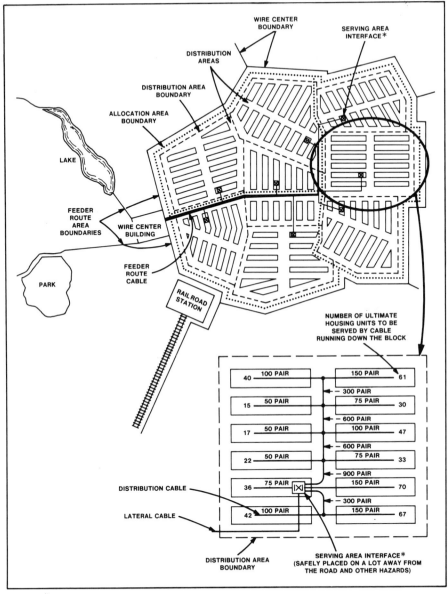

WIRE CENTER
BOUNDARY

SERVING AREA
INTERFACE*

DISTRIBUTION
AREAS

DISTRIBUTION AREA
BOUNDARY

ALLOCATION AREA
BOUNDARY

LAKE

FEEDER
ROUTE
AREA
BOUNDARIES

WIRE CENTER
BUILDING

FEEDER
ROUTE
CABLE

PARK

RAILROAD
STATION

NUMBER OF ULTIMATE
HOUSING UNITS TO BE
SERVED BY CABLE
RUNNING DOWN THE BLOCK

40	100 PAIR		150 PAIR	61
		— 300 PAIR		
15	50 PAIR		75 PAIR	30
		— 600 PAIR		
17	50 PAIR		100 PAIR	47
		— 600 PAIR		
22	50 PAIR		75 PAIR	33
		— 900 PAIR		
36	75 PAIR		150 PAIR	70
		— 300 PAIR		
42	100 PAIR		150 PAIR	67

DISTRIBUTION CABLE

LATERAL CABLE

DISTRIBUTION AREA
BOUNDARY

SERVING AREA INTERFACE*
(SAFELY PLACED ON A LOT AWAY FROM
THE ROAD AND OTHER HAZARDS)

* A rearrangeable cross-connect point between feeder and distribution cables
in the loop plant.

Figure 4-8. A suburban feeder route area.

Loop transmission facilities have a tree-like structure: A feeder route (typically, a large paired cable[13]) extends out from the wire center into each feeder route area; lateral cables are spliced to the feeder cables at various points along the route; distribution cables are joined to the lateral cables; and, finally, customers' station equipment is connected to the distribution cable. The feeder cables near the wire center are generally quite large (containing 1200 to 3600 wire pairs). At each lateral cable splice, feeder cables become progressively smaller as the distance along the route from the wire center increases (indicated by the progressively narrow feeder route line in Figure 4-8).

The primary transmission facility in the loop network is paired cable suspended on telephone poles, placed in underground conduit, or buried directly in the ground. In recent years, the application of pair-gain systems has become an important consideration in the design of the loop network. To reduce the number of cables required, subscriber pair-gain systems use carrier and concentrator techniques (see Chapter 9), thereby enabling a number of customer loops to share the same wire pair and electronics.

Local switching systems may be either electromechanical or electronic.[14] Switching system size depends on the number of customers in the area served and the area's growth rate. System type depends on area characteristics and the switching technology available at the time of installation. Table 4-5 lists the average values of some parameters of the facilities network in urban, suburban, and rural environments. Trunks are provided on interoffice transmission facilities, and thus, they are a part of the interoffice facilities network discussed in the next section. However, trunks are relevant to wire center parameters because, as stated earlier, they terminate at local offices.

A number of observations may be made about Table 4-5. The average area served by a wire center in suburban and rural areas is about an order of magnitude larger than the average urban area served. Although greater switching economy might be achieved by serving even larger rural areas, the savings would be more than offset by the additional cost of the loop electronics necessary to maintain service quality. While about 85 percent of the customers served by small, unattended switching offices in rural areas are located within 2 miles of the wire center, the average length of those rural loops that extend beyond 2 miles greatly exceeds the global average. Thus, a relatively small percentage of the customers

---

[13] Paired (or multipair) cable consists of pairs of wires twisted together to form the core of a communications cable. Section 6.3 discusses transmission media, including paired cable.

[14] Chapter 7 describes the concept of switching, and Chapter 10 describes switching systems.

accounts for a disproportionately large percentage of the total loop mileage in rural areas.

In rural areas, the volume of traffic per main station is lower, and the percentage of intraoffice calls (calls between customers served by the same switching system) is higher. This aspect of traffic has a noticeable impact on the average number of lines per trunk, which increases from eight in urban wire centers to sixteen in suburban wire centers and twenty in rural wire centers.

The statistics in Table 4-5 illustrate how factors such as geography, population density, and calling patterns affect the structure of the local facilities network. The next section discusses the significance of these factors for the interoffice facilities network.

TABLE 4-5

AVERAGE WIRE CENTER PARAMETERS FOR THE PSTN

Parameter	Urban	Suburban	Rural
Number of switching systems	2.3	1.3	1.0
Area served (square miles)	12	110	130
CCS/main station*	3.1	2.7	2.1
Intraoffice calling (percent)	31	54	66
Working lines	41,000	11,000	700
Trunks	5,000	700	35
Trunk groups	600	100	5.5

*CCS/main station is a measure of traffic in which the unit of measurement is hundred call seconds (CCS) per hour. For example, a typical urban telephone line is used for 310 seconds (approximately 5 minutes) during the central office busy hour (see Section 5.2.3).

4.4.2 INTEROFFICE FACILITIES NETWORK

The interoffice facilities network consists of interoffice transmission facilities, tandem switching systems,[15] and local switching systems. (As mentioned earlier, local switching systems constitute the conceptual boundary

[15] Tandem is used here in the generic sense, that is, any system that connects trunks to trunks. Tandem switching systems may serve as tandem offices, toll offices, or combinations.

between the local and interoffice facilities networks.) The interoffice facilities network can be divided into metropolitan, rural, and intercity facilities, each of which is described below.

Metropolitan Networks

Table 4-6 shows the distribution of trunks from all class 5 end offices over various trunk group sizes and distance bands. The information in Tables 4-5 and 4-6 helps characterize the metropolitan network. According to Table 4-5, the average urban (metropolitan) wire center serves an area of 12 square miles, reflecting the fact that a metropolitan area has a large number of wire centers relatively close together. Clearly, then, the shorter trunks in Table 4-6 are mostly in metropolitan areas, and it is reasonable that distances between wire centers of up to 15 miles be used to represent the metropolitan areas. Table 4-5 shows that an average of 600 trunk groups terminate in a metropolitan (urban) wire center. Applying the 15-mile distance assumption to Table 4-6, it can be determined that most of these trunk groups are of modest size, with over half containing fewer than forty trunks.

These characteristics strongly influence the structure of the interoffice facilities network in metropolitan areas. Although the structure of the interoffice facilities network is more complex than that of the local facilities network, the same general principles of sharing facilities are applied.

TABLE 4-6

DISTRIBUTION OF TRUNKS FROM CLASS 5 OFFICES
(EXCLUDING INTERTOLL TRUNKS)

Trunk Group Size (Number of Trunks)	Distance Between Wire Centers (Mileage Bands)						Total Percentage of Trunks in Each Trunk Group Size Band
	0 to 5 (Miles)	5 to 10 (Miles)	10 to 15 (Miles)	15 to 20 (Miles)	20 to 25 (Miles)	Over 25 (Miles)	
Over 195	5.0	3.4	0.8	0.2	0.1	0	9.5
180 to 195	0.4	0.4	0	0	0	0	0.8
160 to 180	0.4	0.7	0.2	0.1	0	0	1.4
140 to 160	0.7	0.5	0.2	0	0	0	1.4
120 to 140	0.9	1.1	0.2	0.1	0	0	2.3
100 to 120	1.0	1.3	0.6	0.1	0	0	3.0
80 to 100	1.1	2.0	0.4	0.1	0	0	3.6
60 to 80	1.3	2.9	1.3	0.2	0.1	0	5.8
40 to 60	1.6	2.7	2.7	0.6	0.1	0.1	7.8
20 to 40	2.3	6.3	6.3	2.2	0.3	0.3	17.7
Under 20	3.6	10.2	14.6	10.5	5.3	2.5	46.7
Total Percentage of Trunks in Each Mileage Band	18.3	31.5	27.3	14.1	5.9	2.9	100.0

Figure 4-9 shows the interoffice facilities network in downtown Chicago. Many wire centers are located close together and are interconnected by a transmission facilities network that has a grid structure following the pattern of the streets. From this structure it is clear that a relatively small number of transmission facilities emanating from each wire center are shared by the many trunk groups terminating at each wire center. The

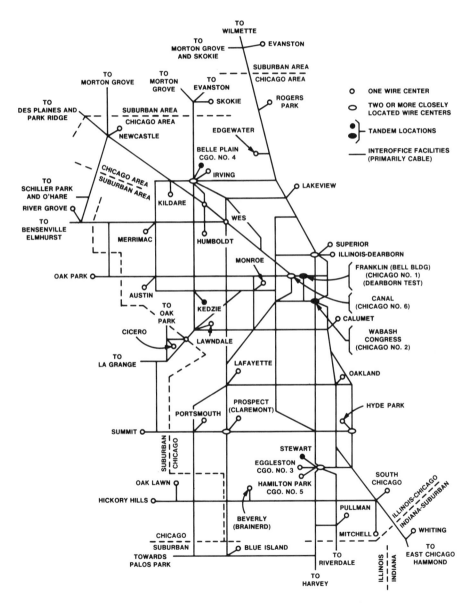

**Figure 4-9. A metropolitan interoffice facilities network—
the Chicago area.**

predominant transmission facilities are voice-frequency facilities (see Section 6.2) and digital carrier systems (see Section 9.4) on paired cable. Most of the cables are placed in conduit, which explains why the grid-like structure of the network follows the street pattern. A route section may contain both voice-frequency circuits and digital carrier channels on cable (sometimes, in the same cable) and may have a total cross section of over 20,000 equivalent voice-frequency circuits.

It should also be noted that metropolitan networks contain both local and local tandem switching offices. Several tandem offices are located in the downtown Chicago area.

Rural Networks

As shown in Table 4-5, the wire centers in rural environments serve large, sparsely populated areas and are located far apart. Consequently, the trunks are usually longer than the 15-mile limit associated with metropolitan trunks, and the trunk groups are small. (Table 4-6 shows that 80 percent of the longer trunk groups [15 miles or longer] contain fewer than twenty trunks.) Table 4-5 shows that rural wire centers are quite small in terms of both the number of lines and the number of trunks. There is also more intraoffice calling in rural areas than in metropolitan areas. This, in turn, means that fewer trunks are required to interconnect rural wire centers.

These characteristics determine the structure of the interoffice facilities network in rural areas. Figure 4-10 shows the interoffice facilities network in the Greenwood District of Mississippi. The area shown covers approximately twenty counties and contains forty-six Bell System and independent wire centers serving 120,000 telephones. The distance between wire centers, the volume of traffic, and the low percentage of interoffice calls in a rural area justify few direct trunks between wire centers. Most trunks are in toll connecting trunk groups connecting class 5 offices to class 4 and class 3 offices. This results in a tree-like facilities network in which many individual chains of transmission facilities branch out from the toll offices in routes typically following major highways. Each transmission facility along a chain serves offices that are located farther away from the toll center. Thus, in general, the closer the transmission facilities are to the toll office, the larger their channel capacities. In this sense, the interoffice transmission facilities network in a rural area is similar to the feeder network in the local facilities network.

Although open-wire[16] facilities still exist in certain areas, the primary interoffice transmission facilities in rural areas are carrier systems (see Chapter 9) on paired cable either buried directly in the ground or supported on telephone poles.

---

[16] Open-wire lines consist of uninsulated pairs of wires supported on poles. Section 6.3 discusses transmission media, including open-wire lines.

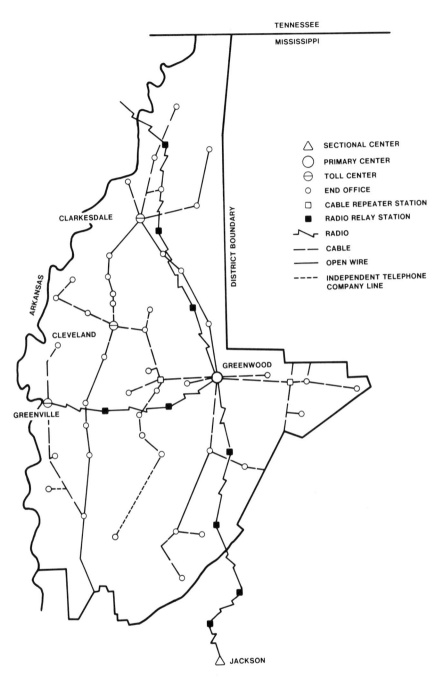

Figure 4-10. A rural interoffice facilities network—Greenwood District, Mississippi. (End offices co-located with toll and primary centers are not shown.)

Rural areas are most often served by small, unattended, electromechanical switching systems called *community dial offices* (CDOs). Recently, the Bell System has made a concerted effort to introduce electronic switching—a move made possible by the development of remote switching equipment such as the No. 10A Remote Switching System and the No. 5A Remote Switching Module.[17] Remote switching equipment operates as an extension of a host electronic switching system located at a population center as far as 100 miles away.

The introduction of a remote office in a rural area, especially when it replaces a small, slowly growing CDO, can result in a significant reconfiguration of the trunking arrangements. For example, since the remote office is connected to the rest of the facilities network through the host, trunks previously connecting the CDO to other offices (toll center, other class 5 offices, operator positions, etc.) are disconnected. On the other hand, additional trunks from the host to these other offices may be required to carry the traffic generated by the remote office. As a result of trunk group efficiencies gained at the host, however, the number of additional trunks required at the host is generally less than the total number of trunks eliminated at the remote office.

Intercity Networks

To enable customers located in different, widely dispersed geographical areas to communicate, it is necessary to interconnect the metropolitan and rural areas. This interconnection is accomplished by a network of intercity facilities.[18] Generally, intercity facilities consist of toll switching systems (class 4 through class 1) and long-haul transmission facilities (facilities that provide long-haul intertoll trunks and special-services circuits).

Intercity trunk groups are characterized by large variations in length, size (number of trunks or cross section), and number of groups originating and terminating at the various toll switching systems. For example, analysis of the Long Lines facilities provided in 1982 for the PSTN reveals that the size of intercity trunk groups ranged from fewer than 12 to over 1000 trunks, with the average group containing about 30 trunks. The length of these trunk groups ranged from less than 100 miles to over 4000 miles, with an average group length of approximately 1100 miles.

These requirements for intercity facilities result in a complex network structure. This is indicated by Figure 4-11, which shows existing high-capacity intercity facility routes.

---

[17] Section 10.3 describes remote switching equipment in more detail.

[18] The distinction between the intercity and the rural networks is less well defined than that between intercity and metropolitan networks. But, since the current trend is to use the term *intercity*, this chapter adheres to this convention.

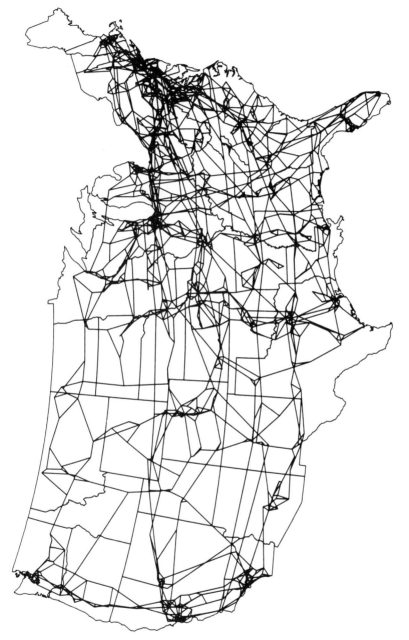

Figure 4-11. Major Bell System transmission routes.

Virtually all continental long-haul transmission uses microwave radio and coaxial cable systems.[19] One notable exception is the optical fiber transmission system that will interconnect the Boston, New York, and Washington metropolitan areas.[20]

The existing structure of intercity facilities has resulted from two fundamentally conflicting goals: to minimize the total cost of the network and to provide redundancy as a protection against major service interruptions. A well-designed network represents a judicious balance between these two objectives. Skoog (1980) shows that the total cost of the network is minimized by funneling many point-to-point circuit requirements onto a relatively small number of "backbone" routes with large cross sections. This is a direct consequence of the economy of scale exhibited by the long-haul transmission facilities (see Section 4.5.1). The major transmission routes resulting from this funneling scheme interconnect the nation's population centers (see Figure 4-11). Redundancy is provided by multiple, physically separate long-haul routes, although at an increased cost. An additional feature of intercity facilities is the existence of metropolitan junction offices surrounding some of the larger population centers. By means of these offices, circuits (and, therefore, traffic) can be routed around rather than through the more congested city centers.

4.4.3 EVOLUTION OF THE DIGITAL FACILITIES NETWORK

Until the early 1960s, the facilities network evolved as an analog network. Both transmission facilities and switching systems were designed to transmit and switch analog signals. This has changed, particularly over the last decade, as a result of the rapid development of digital facilities that can transmit and switch signals in digital form. A major advantage of digital transmission is signal durability (see Section 6.4.3). Other advantages of digital transmission are discussed later in this section.

The T1 digital carrier, a short-haul (metropolitan area) system using paired cable, was introduced into the network in 1962 and has grown rapidly because it provides low-cost and reliable transmission. The T1 system carries twenty-four voiceband channels on two pairs of wires and has a transmission range of approximately 50 miles.[21] Digital-to-analog converters are required at its end points to accept and transmit signals in

---

[19] Coaxial cable is a transmission medium that consists of a tube of electrically conducting material surrounding a core conductor held in place by insulators. Section 6.3 discusses transmission media, including coaxial cable, and Chapter 9 discusses transmission systems and their applications.

[20] Section 6.3.5 discusses lightguide cable, which consists of multiple optical fibers.

[21] Section 9.4.2 describes the T1 system in detail.

analog form for switching[22] and to allow interconnection with analog transmission facilities.

Since the introduction of the T1 system, many new digital transmission facilities have been developed and deployed in the network. These new facilities include other T-carrier systems using paired and coaxial cable, digital microwave radio systems, and, most recently, lightwave systems.[23] Lightwave systems generally provide high-capacity, high-quality transmission, which makes them attractive for both short-haul and long-haul transmission.

In addition to these terrestrial facilities, digital satellite systems are being developed for long-haul transmission. A satellite system consists of the satellite itself and the ground stations. Additional facilities are required to connect the ground stations with the terrestrial network.

Increasingly sophisticated digital technology has also affected switching systems. The first major application of time-division switching[24] and digital switching technology was the Bell System's 4ESS switching equipment (introduced in 1976) whose primary application is in the toll network. At the end of 1982, there were 91 4ESS systems in service; this number is expected to increase to 123 by the end of 1985 and to 135 by 1990.[25]

Digital switching technology is being applied to local switching systems as well. The 5ESS switching system, a local digital system introduced in 1981, is designed in a modular form that will eventually enable it to terminate up to 100,000 lines. Initially, local digital switching systems will be deployed in rural and suburban areas.

Remote digital switching systems are also being developed and deployed in the facilities network. They are generally much smaller than local switching systems and are used primarily to replace small, slowly growing CDOs and to establish new, small switching offices. An example is the No. 5A Remote Switching Module currently under development in conjunction with the 5ESS system.

The deployment of digital technology is accelerating because of the synergies resulting from combined use of digital transmission and digital switching in the network. Until digital switching systems were introduced, digital transmission facilities required analog-to-digital and digital-to-analog conversion at the ends of the digital facility. As the number of digital switching systems increases, the need for conversion decreases and, therefore, more opportunities arise where the introduction

---

[22] As will be seen later in the chapter, converters are not required where digital switching systems are used.

[23] Chapter 9 describes these systems.

[24] Section 10.3.4 describes time-division switching.

[25] Numbers include non-Bell System applications.

of digital transmission facilities becomes cost effective. In addition to the lower cost and simpler interfaces that result when digital transmission and digital switching facilities are used together, the reduced number of analog-to-digital conversions required on an end-to-end digital connection results in improved transmission (see Section 6.4.3).

Because the first digital switching system to be deployed in the facilities network was the 4ESS system, a toll switching system, the initial cost savings due to the integration of digital transmission and switching occurred in the toll portion of the network. Now, with the introduction of digital switching at the local office level, savings can also be realized in the local portion of the network. Nor has the evolution of digital facilities stopped at the local office. Digital pair-gain systems extend the kind of savings resulting from integrated transmission and switching to loop equipment. Current indications are that this rapid evolution of digital technology will continue until a highly connected digital facilities network is achieved.

4.5 NETWORK CONFIGURATION PLANNING

In a broad sense, network planning is a multifaceted discipline that encompasses the functions involved in planning the evolution and implementation of the network, designing and engineering the configuration of the network, and managing the total network investment. Its objectives are to provide network capacity and to guide the operations functions in the use of the capacity to meet customer demands in a cost-effective manner.

One result of network planning is the establishment of plans for various aspects of the organization and operation of the network. Although these plans will not be discussed here, it is important to be aware of their existence and the essential roles they play in achieving effective, high-quality network performance. Some of the organizational and operational areas affected are: the PSTN switching hierarchy, the worldwide numbering plan, analog transmission loss (see Section 6.6), and the analog and digital multiplexing hierarchies (see Sections 9.3 and 9.4, respectively).

In keeping with the general character of this chapter, this section does not address all aspects of network planning;[26] it concentrates, rather, on what may be called *network configuration planning*.

---

[26] Other chapters cover specific aspects of network planning as their main subjects or as side issues related to their main subjects. Chapter 15, for instance, deals with operations planning (ensuring that operations are assigned to people and systems in ways that realize potentials for greater efficiency and better customer service), and Chapter 16 discusses the considerations involved in setting service and performance objectives.

Network configuration planning begins with this information:

- a forecast of point-to-point (for example, central office—to—central office or wire center—to—wire center) demand for various services

- currently available and projected technology

- the existing configuration of the network

- regulatory, legal, structural, financial, and other constraints.

Its specific goal is to determine the future configuration of the facilities network as a function of time so as to meet the increased demand for a variety of services in an orderly fashion while minimizing total network costs.

Because of the special characteristics of the network, the long lead times required to expand capacity in the network, and the long-term economic impact of major facility installations, network configuration planning has a relatively long horizon, typically on the order of 20 years.

Just as the structure of the facilities network may be divided into two parts, network configuration planning may be divided into two disciplines that parallel those parts: **local network planning** and **interoffice network planning** (shown in Figure 4-12). Here and in the rest of this chapter, the term *network planning* will be used in the narrower sense of network configuration planning.

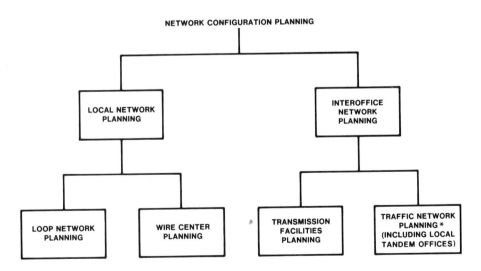

* OFTEN REFERRED TO AS TOLL OFFICE PLANNING

Figure 4-12. Major components of configuration planning.

Network planning is not a process of performing various tasks in separate, isolated planning disciplines and then combining the results into a single plan. Rather, it is a continuous process within each of the disciplines with significant interactions among them so that changes in one may affect plans in others. This is particularly true today with the introduction of new technologies that tend to blur the distinctions between planning disciplines.

In network planning, while there are issues and methodologies specific to each discipline, there are also basic approaches common to all the disciplines. Two of these basic approaches are particularly important because, together, they illustrate the essential nature of network planning. First, network planning identifies and generates options. Some areas where numerous options exist are:

- the basic *network configuration* (as a result of the tradeoff between transmission and switching)

- *traffic engineering* (determining the number and type of channels or communications paths required between switching points and the call-handling capacity of the switching points)

- *circuit routing* (determining the most efficient configuration of transmission facilities to provide the required circuits)

- *capacity expansion* in both transmission facilities and switching systems

- *multiplexing* (combining a number of transmission paths).

Second, network planning virtually always makes the selection based on economic analysis, that is, which option is the most economical over the specific interval for which the planning is being done. The following sections describe these two approaches and other generic network planning issues in more detail.

4.5.1 ALTERNATIVE PLANS[27]

Network planning is a complex problem because the possible plans have many interdependencies that must be considered. For example,

- Decisions made in one part of the network often affect adjacent and even distant parts of the network.

- Decisions made at certain times may have significant consequences at other times.

---

[27] See Skoog 1980 for a detailed discussion of most of the considerations covered in this section.

• As stated earlier, different network planning disciplines are highly interdependent. For instance, switching office planning affects transmission facility planning and vice versa.

The following sections discuss considerations in generating options.

Switching and Transmission Tradeoffs

One of the most important tradeoffs in network planning is between switching and transmission. In general, as more switching is introduced into the network, transmission costs decrease and switching costs increase. The proper balance between switching and transmission is not easy to achieve because of the large number of possible configurations. The solution depends on decisions such as where and when to introduce switching systems, what type of systems to use, the classes of offices and homing arrangements, and the desired connectivity among the systems.

Network planning must identify the reasonable choices in switching system and transmission facility configurations and select the one that minimizes the total network cost over the planning horizon.

Traffic Engineering

In broad terms, traffic engineering determines how much switching and transmission capacity is required to meet the expected demand. The many choices encountered in traffic engineering stem from the combined effects of a multilevel switching hierarchy, high-usage and final trunk groups, and the notion of trunk group *efficiency* (see Section 5.5). Clearly, traffic engineering is a complex discipline and an essential part of network planning.

Two switching systems may be connected in many ways, for example, with only direct trunks, with only tandem trunks, or with a combination. Generally, if two switching offices are connected by both direct and tandem trunks, then the direct trunks and some of the tandem trunks are contained in high-usage trunk groups, and some of the tandem trunks are in final trunk groups. In this case, the number of possibilities is increased even further because of the options in high-usage and final trunk group sizing.

In addition to strictly economic considerations, other factors often influence the manner in which switching offices are interconnected. One such factor is the requirement to provide service protection through diversity in the trunking arrangement (that is, physically separate facility paths between end points so that if transmission cannot be accomplished over one path, another path can be used).

Circuit Routing

Another major task in network planning is the routing of point-to-point circuit demands between all pairs of nodes in the network. Specifically, circuit routing assigns each point-to-point circuit demand to a set of links[28] that forms a continuous path between the two end points.[29] The point-to-point demands are converted to demands over time for transmission capacity on each of the network links. The proper choice of transmission facilities on each link can be determined only after all point-to-point demands have been routed.

The overall routing scheme that helps minimize total network cost is called *minimum-cost routing*. Minimum-cost routing involves determining a path through the network for each point-to-point demand for each year such that when the point-to-point demands are provided for on these paths and the resulting optimal capacity-expansion problem (the next topic in this section) is solved, the total cost of transmission facilities is minimized. Thus, circuit routing and capacity expansion must be performed jointly and optimally in the sense of the total network cost.

The fundamental tradeoff in minimum-cost routing is between length- and size-dependent transmission facility costs. Typical transmission facility line-haul costs are linear functions of the length and nonlinear functions of the size of the facilities. Specifically, the size-dependent part of the cost functions has the property that the higher the facility capacity, the smaller the average cost per circuit. Therefore, if the routing of a point-to-point demand is changed from one path to a longer path where it combines with other point-to-point requirements to form a larger demand, the net increase in length-dependent cost may be more than offset by the net decrease in size-dependent cost. Hence, in a more general network context, concentrating point-to-point circuit demands along nonminimum distance paths can yield cost advantages related to the size of transmission facilities.

In a large network, such as the PSTN, there are many possible ways to combine circuits. Minimum-cost routing evaluates these possibilities and establishes the proper balance between length-dependent and size-dependent transmission facility costs.

Capacity Expansion

Capacity expansion deals with determining the types, sizes, locations, and timing of switching system and transmission facility installations. If

---

[28] A link is a topological connection between two nodes; it represents one or more transmission facilities directly interconnecting the two nodes.

[29] Circuit routing is not to be confused with traffic routing, which is the process of setting up calls (routing calls) in the PSTN.

there were only a single type and size of facility available, capacity expansion would be much simpler; questions of type and size would not arise, and the timing of installations would be a direct consequence of the facility size and the growth of demand. In reality, many types and sizes of transmission facilities and switching systems exist, and as a result, capacity-expansion problems are complex.

A typical capacity-expansion problem has several characteristics. First, it is defined over a certain planning horizon during which there is a growing requirement for some type of capacity, and the equipment that provides this capacity has a relatively long service life. Second, the cost of providing the capacity can be modeled by an initial cost that occurs at the time the equipment is installed and incremental costs that occur periodically as the equipment is more fully utilized. Moreover, the initial cost of the equipment exhibits economy of scale; that is, the average initial cost per unit of capacity decreases as the capacity of the equipment increases. Finally, as mentioned earlier, the choice between two different sequences of capacity installations is based on economic analysis.

In a capacity-expansion problem, a tradeoff is made between short- and long-term economics. Because of the particular cost characteristics of the facilities involved, these are often in conflict. The installation of low-capacity facilities is generally a good capacity-expansion strategy from the point of view of short-term economics because it normally involves small initial costs. Often, though, the long-term economics of such a strategy may be highly undesirable because of the cost of successive expansion jobs at relatively short intervals. Alternatively, the installation of large facilities to satisfy demand over a longer interval (often characteristic of a long-term economic solution) may result in poor short-term economics by requiring too large an initial capital outlay. The principal goal of a capacity-expansion problem, then, is to select a sequence of facility installations that yields the proper balance between short- and long-term economics. (Sections 4.5.3 and 18.3 discuss the considerations involved in economic analysis in more detail.)

Multiplexing

Multiplexing is the process of combining many individual signals for transmission, thereby using the transmission media more efficiently.[30] Efficient use of these media permits deferral of additional facility constructions, yielding overall savings in transmission facility costs. The basic tradeoff, then, is between the cost of multiplexing equipment and the line-haul cost of transmission facilities.

---

[30] Section 6.5 discusses multiplexing in more detail.

Multiplex planning addresses the following questions:

- Which point-to-point circuit requirements in the network should be combined and met by multiplexing?

- Under what circumstances or where should the multiplexing be done?

- What sizes should the bundles of circuits be between any two points (nodes) in the network?

The number of possible options for concentrating the various point-to-point circuit requirements make these multiplexing decisions extremely complex.

In developing alternative plans, circuit-routing and capacity-expansion tasks are usually done first. Then, appropriate multiplexing alternatives are developed.

4.5.2 ECONOMIC ANALYSIS

The alternatives generated in the process of developing a long-term network plan are evaluated and compared using economic analysis. Each alternative plan is modeled as a time sequence of events. Typical events are installing a new piece of equipment, removing or retiring an existing facility, converting from one type of facility to another, and rearranging parts of the network. Associated with each event is a time sequence of capital expenditures and either 1-time or periodically recurring expenses (maintenance costs, taxes, etc.). In addition, some events generate a time sequence of revenues (payments received for providing products and services). Capital expenditures, expenses, and revenues are commonly referred to as *cash flows*. Conventionally, revenues are considered positive cash flows and capital expenditures and expenses negative cash flows.

A key notion in economic analysis is the preferential ordering of the cash flows associated with options. Skoog (1980) shows that a reasonable approach to preferential ordering of cash flows leads to the *present-worth* criterion. According to this criterion, cash flows occurring at different times are *discounted* by an appropriate rate to reflect their present worth and then added algebraically. Through this process, a cash flow stream over a planning horizon can be reduced to a single number for each option that then permits comparison on an equal basis.

Two alternative plans may reach the end of a planning horizon with different capacities for handling future growth. This is referred to as the *end-of-study effect* and indicates the need for further analysis. One way to make this problem more manageable is to append an appropriate extension period to the planning period.

Economic analysis must also consider the finite useful service lives of facilities installed to provide increased capacity. As a result of wear and tear, obsolescence, etc., each piece of equipment must be retired and replaced at some point to maintain service. Retired systems are usually replaced by larger, more modern systems that provide capacity for future growth while meeting current demands.

These are only a few of the factors that play an important role in the economic evaluation of possible plans. Section 18.3 is a more complete treatment of the entire subject.

4.5.3 OTHER NETWORK PLANNING CONSIDERATIONS

Level of Detail in Modeling

Virtually every component of a network planning study has to be a model, or approximation, of the real situation. Obvious examples of this are the network topology, the physical and cost characteristics of the available technologies, the various types of demands, different engineering constraints, and the cost and revenue streams associated with a network plan. There is an inherent conflict, however, between the desire to represent a network planning problem realistically and the limitations of the available modeling techniques. Generally, if the various components of the problem are described in very detailed, realistic terms, the sheer size of the problem may render it intractable. On the other hand, if the problem is represented by highly aggregated models (where, for example, many network nodes are replaced by one aggregate node[31]), it becomes difficult to translate the solution for the modeled problem into a solution for the real problem accurately.

One approach to resolving the conflict is to introduce a variable level of detail into the overall network planning process. Thus, the long-term evolution of the network would be accomplished by using appropriately aggregated models; for the early years of a plan, the essential realism and implementability would be achieved by using more detailed models. This approach is particularly appealing because it ensures that the long-term evolution of the network is unencumbered by unnecessary details, while the early years of the plan are sufficiently realistic.

Implementing this approach is not simple. The planner must determine which components of the problem should be modeled at which levels of detail. Perhaps even more difficult is the problem of translating the solution at an aggregated level into a solution at a finer level of detail. Often, the only guidelines available for these aggregation and disaggregation problems are the planner's experience and extensive knowledge of the network.

---

[31] Figure 4-9, the interoffice facilities network in the Chicago area, shows examples of aggregate nodes.

Subnetwork Planning

A related approach to solving a large network problem is: (1) to partition the network into subnetworks, (2) to develop a plan for each of the subnetworks, and (3) to integrate them into a total, unified network plan. There are two difficulties with this approach. First, optimum subnetwork plans do not guarantee an optimum overall plan. Second, fundamental difficulties exist in modeling the interactions between adjacent subnetworks.

An alternative approach is: (1) to create a sufficiently aggregated model for the network and develop an overall network plan at an aggregated level; (2) to translate the aggregated plan into various constraints, initial conditions, and conditions at subnetwork boundaries and to develop the plans for each one of the subnetworks; and (3) to integrate the subnetwork plans into a single overall plan. The advantage of this approach is that the subnetwork plans start with certain initial conditions, boundary conditions, and constraints that are, at least on an aggregated level, consistent with the total network plan. It is likely that even with this approach, there may be local conflicts between adjacent subnetworks. These must be resolved when the subnetwork plans are integrated into a single network plan.

There may be other feasible alternatives to solving a large network planning problem. For example, it may be possible to iterate between overall network and subnetwork planning. Another approach might be: (1) to start with a single highly aggregated network and develop an aggregated plan for it; (2) to partition the network into a small number of disjointed subnetworks and use the previously obtained aggregated solution to define initial conditions, boundary conditions, and constraints for the subnetworks to develop the plan for each but at a higher level of detail than before; and (3) to continue planning on smaller and smaller networks with more and more detail, always using the previous solutions as initial conditions.

All of these approaches are used at one time or another depending on existing conditions.

Planning Interval and Plan Monitoring

Once a network plan is being implemented, it must be monitored to ensure that it remains valid and cost effective. Assumptions, forecasts, and parameter values may change with time. These changes necessitate deciding when the plan is no longer valid and should be updated.

This is a difficult decision because of the complex mechanism through which a particular assumption or parameter affects the overall network plan. This difficulty can be alleviated to some extent by performing the appropriate sensitivity analyses (projecting results if initial assumptions and parameters were different) as part of the original plan development. Sometimes, however, changes may be large enough to cast serious doubt on the validity of the plan. In these cases, the plan should be updated.

A more structured approach to plan monitoring and updating uses regular planning intervals. Since the demand forecast is usually generated at regular intervals, it seems reasonable to tie the planning intervals to the forecasting intervals, although not necessarily on a one-to-one basis. Such regular planning cycles are particularly attractive in a highly mechanized planning environment. With mechanized data bases and planning tools, the planner may generate many options quickly. Using network economic analysis to evaluate the options, the validity of the original network plan can be checked. If, at this point, the original plan is found invalid, a new, cost-effective network plan can be developed.

Network configuration planning takes advantage of both sensitivity analyses and plan monitoring and updating.

Mechanized Planning Environment

As the complexity of the network grew, as new and advanced technologies evolved, and as the demand for many complex telecommunications services increased, traditional manual planning methods became inadequate. Consequently, the network planning environment today is becoming increasingly mechanized.

Mechanization in the planning environment affects two distinct areas: data-base development and design of planning tools. Data-base development ensures that the great variety and large volume of data needed for effective network planning are readily available and can be easily accessed and updated. Part of a mechanized data base should be a data base of record, which contains all relevant data. This data base should include information about the network topology; network initial conditions such as existing facilities by location, their spare capacities, the demands on these facilities, and their potential for conversion to other types of facilities; technology characteristics, such as modularities, capacities, and costs; and point-to-point demand forecasts identified by type, source, and other attributes.

To make such a data base useful for planning, it must be possible to extract subsets of the data, to strip away unnecessary details, and to aggregate the data to the appropriate level of detail. Moreover, the data base must be capable of being updated so that it always contains current information. These requirements clearly point to a need for an efficient, well-designed data-base management system.

Mechanized planning tools enable network planners to consider many more options than would be possible in a strictly manual planning environment. Interrelationships among the various assumptions, parameter values, and planning disciplines in these network problems are so complicated that without the aid of mechanized planning tools, it would be difficult to generate even a feasible solution; the optimum solution would be virtually impossible to obtain. Since these problems are particularly well suited for computer implementation, the optimal (or at least near optimal) solutions can be found using mechanized tools.

TABLE 4-7

SUMMARY OF MECHANIZED PLANNING TOOLS IN USE IN THE BELL SYSTEM

Acronym	Name	Application
CUCRIT	Capital Utilization Criteria	Economic analysis
EFRAP	Exchange Feeder Route Analysis Program	Loop network planning
FCAP	Facility Capacity	Intercity transmission facility network planning
LRAP	Long Route Analysis Program	Loop network planning
LRSS	Long Range Switching Studies	Traffic network planning
LSRP	Local Switching Replacement Planning System	Wire center planning
MATFAP*	Metropolitan Area Transmission Facility Analysis Program	Metropolitan transmission facility network planning
OFNPS	Outstate Facility Network Planning System	Rural transmission facility network planning
TCSP	Tandem Cross Section Program	Traffic network planning
TSORT	Transmission System Optimum Relief Tool	Intercity transmission facility network planning

*See Section 14.4.2.

Table 4-7 is a list of the more common mechanized planning tools used by Bell System network planners. These systems are constantly updated, modified, and enhanced, and new systems are being developed as the need for them arises.

AUTHORS

R. W. Hemmeter
R. J. Keevers
T. Pecsvaradi

5

Traffic

5.1 INTRODUCTION

Traffic, as defined in Chapter 3, is the flow of information through the telecommunications network. It may be generated by telephone conversations or by data, video, and audio services. The potential number of users generating traffic on the public switched telephone network (PSTN) in the continental United States is well over 100 million. However, only a small percentage of that total number actually uses the network or some portion of it at any given time. Obviously, it is more economical to design the network so that equipment can be shared—different people use the same equipment at different times—rather than dedicated to each of the millions of users. When equipment is shared, though, a particular person may not be able to complete a call because the necessary equipment is busy. Thus, a tradeoff between cost and service exists: The more equipment available for use, the better service the user will receive, but the more costly that service will be.

The same kind of traffic considerations and tradeoffs occur in many situations unrelated to the telephone environment, for example, at a checkout counter in a supermarket. The more attended counters there are, the better (that is, the faster) the service, but more checkout counters mean higher costs. Each counter requires equipment (a cash register, for example) that must be purchased and maintained. Each counter in use requires an attendant (checker), who has to be paid. A store manager has two cost−versus−quality-of-service tradeoffs to consider: How many counters are needed to meet peak business periods? and How many checkers are needed to meet daily and hourly variations in business?

The quantitative tradeoff relationship between cost and service in any system depends on the statistical parameters that describe the way users place demands on the system. Methods for determining how much ca-pacity (for a telephone network, this means switching and transmission capacity) is required as a function of the expected demand are called *traffic engineering methods.*[1]

The techniques used to develop traffic engineering methods are rooted in a branch of applied probability theory called *traffic theory*. This chapter describes some of the basic concepts of traffic theory and the mathemati-cal models of telephone network usage generated from them. Specific examples show how the models are applied to some of the many traffic engineering problems that arise in the effort to design a low-cost telecom-munications network that provides high-quality service.

The concepts and techniques presented in Sections 5.2 through 5.7 are described in terms of the PSTN, but they also apply to traffic networks generally. A mixture of traffic types (voice, data, etc.) on a traffic network does not necessarily require adjustments to the engineering process; the PSTN, for example, carries such a traffic mixture. Some special-purpose networks, however, may involve additional engineering considerations. An important example is data networks, discussed in Section 5.8.

5.2 TRAFFIC THEORY BACKGROUND

5.2.1 MEASURE OF TRAFFIC DEMAND: OFFERED LOAD

The common ingredient in engineering checkout counters and engineer-ing telephone equipment is the nature of the demand placed on both sys-tems. This demand is a function of customer decisions and is, for the most part, beyond the control of the engineer. It is also, in some sense at least, unpredictable. But, if quantitative engineering techniques are to be developed, some means of quantifying the demand must be found.

The demand on a traffic system is called the *offered load* and is defined by two parameters: the average rate at which customers request service (called the *average arrival rate*), λ, and the average length of time they require service (called the *average holding time*), τ. Cooper (1981, Sec-tion 3.1) shows that *in any system with enough servers such that all arrivals are served immediately, the average number of busy servers is always the product of λ and τ*, independent of patterns of arrival and variations in holding times.[2] Thus, offered load, a, is defined by the equation

$$a = \lambda\tau. \tag{1}$$

---

[1] To ensure consistent service throughout the telecommunications network and to avoid the duplicated effort that would result if each Bell operating company devised its own methods, Bell Laboratories, in concert with AT&T, traditionally has designed the required methods and procedures for all the companies.

[2] If, on the average, one customer per minute (λ=1) arrives at a checkout counter in a supermarket and the average service time is 3 minutes (τ=3), three checkers would be busy on the average (1×3). To ensure immediate service when bursts of customers arrive, more than three checkers would be needed.

When λ and τ are expressed in the same unit of time, load a is the mean number of arrivals per holding time and is a dimensionless quantity whose numerical values are expressed in *erlangs* in honor of the Danish mathematician A. K. Erlang, the founder of traffic theory.[3]

5.2.2 GRADE OF SERVICE

Offered load, the average number of busy servers if there are *enough* servers to handle all customers immediately, is an important quantity. However, by itself, this average does not indicate how many servers are enough. Furthermore, immediate service is often not a desirable—and may not be a possible—system objective. The analysis of traffic systems therefore requires the specification of the quality of service, or grade of service (GOS), that customers should receive. In turn, specification of the GOS depends partially on how the system is designed to handle situations in which an arriving customer finds all servers busy.

When customers find all open checkout counters in a supermarket busy, they normally form waiting lines (queues) at each counter, usually at their own discretion. Some customers will leave rather than wait, particularly if all the lines are too long. In each line, the customers are (usually) served in the order they entered the line (in queuing terminology, this particular queue discipline is called *first in, first out* [FIFO]). In the supermarket, then, it is reasonable to express the GOS in terms of the time customers spend waiting on line.

Telephone systems are more complex. When all servers are busy, the customer usually does not have the option of waiting for a server to become idle. If, for example, all the circuits between a New York City office and a Chicago office are busy, a customer calling from the New York office is not permitted to wait for an idle circuit. Instead, the New York office returns a reorder tone (a fast busy signal) to the customer, indicating that the customer should hang up and try again later. Systems like this one, in which waiting positions are not provided, are called *blocked-calls-cleared* (BCC) systems. A reasonable service criterion for BCC systems pertains to the percentage of call attempts that encounter all servers busy, that is, the *blocking probability*.

In other telephone situations, for example, where average holding times are short enough that waiting times are not likely to be excessive, waiting positions are provided. By lifting the telephone handset, for example, a customer initiates a request for a digit receiver to be attached to the line so that the customer may dial. The system sends a dial tone

---

[3] The operating telephone companies generally measure load in units of hundred call seconds (CCS) per hour. (The first C is the Roman numeral for 100.) Since there are 3600 seconds in an hour, a continuous traffic load in a 1-hour period would represent 36 CCS—one erlang. The CCS unit is convenient for measurements, particularly with devices that scan a piece of equipment every 100 seconds and record busy-idle status.

when the digit receiver connection has been made. Usually this happens so quickly that the dial tone is present even before the customer gets the handset to the customer's ear. But sometimes all the digit receivers of the type needed (dial-pulse receivers for rotary-dial telephones, dual-tone multifrequency[4] receivers for TOUCH-TONE telephones) are in use. In these cases, the customer should wait for a dial tone, that is, for a digit receiver to become available and be connected to the line. Some customers do not wait. Instead, some hang up and try again later; some depress the switchhook repeatedly to express annoyance; some dial anyway. Neither of the last two strategies works, and dialing anyway only results in an ineffective attempt or a wrong number.

Systems where waiting lines are allowed to form are called *blocked-calls-delayed* (BCD) systems. For such systems, the percentage of customer requests that encounter all servers busy (that is, the blocking probability) is as important as it is in BCC systems. But it may also be important to guard against excessive delays even if the probability of being delayed is small. It may even be economically attractive to permit many small delays. Thus, the GOS criterion for such systems usually pertains to the *delay distribution*, $[P(D>t)]$, the probability that the delay exceeds some number of time units, t. Figure 5-1 shows a possible delay distribution for a fixed offered load and a varied number of servers.

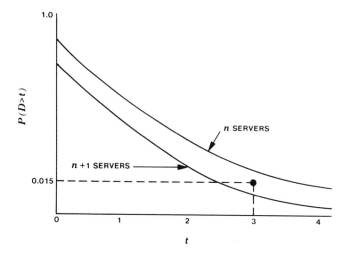

Figure 5-1. Example of delay distribution. $P(D>t)$ depends on how many servers are available to handle a given offered load. The number is chosen to meet a GOS objective. For example, the GOS objective for average dial-tone delay is typically $P(D>3 \text{ seconds}) \leqslant 0.015$. In the case shown, $n+1$ servers would be required. If n servers were used, $P(D>3 \text{ seconds})$ would be greater than 0.015.

[4] Section 11.1.1 discusses dual-tone multifrequency (DTMF) signaling.

It is also possible to define mixed BCC and BCD systems. If there are, for example, a fixed number of waiting positions, requests that find all servers busy and an idle waiting position are presumed to wait for service, while requests that find all servers busy and all waiting positions filled are presumed to be cleared from the system. A mixed system is most frequently encountered in PSTN electronic switching systems (see Section 10.3) where only a given amount of storage area is available to "remember" calls waiting for a particular equipment item.

In addition, it is not always true that waiting calls should be served in order of arrival (FIFO). For example, when an electronic switching system becomes congested (there are more calls in progress than the switching system can efficiently process), waiting lines can become long, so long that the customer may hang up (that is, defect from the queue). The customer most likely to defect is the customer who has been there the longest. During periods of extreme congestion, a last-in−first-out (LIFO) queue discipline prevents the congested system from wasting time trying to process a call that has probably defected. This makes the best use of the system, since it increases the probability that the next call processed from the queue has not defected. In any event, the shape of the delay distribution depends strongly on whether the queue discipline is FIFO, LIFO, or some other scheme.

5.2.3 ENGINEERING PERIODS

The decision to specify a GOS for a load on a given system is a good start, but to which load should the GOS be applied? The formulas developed in traffic theory typically assume that the load is stationary[5] during the period of time for which a traffic system is being analyzed. For the traffic theory formulas to be useful, then, those time periods for which the load is stationary must be determined and the GOS criteria applied to some or all of them. (This does not imply that the load must be exactly the same during each engineering period, only that the load variation must be confined to a small range within a given period.)

In the PSTN, loads do not remain constant during every period of every day. They may peak during certain periods of the day, particular days of the week, or particular seasons of the year. However, in the Bell System, the stationary period is generally taken to be an hour, since data support the assumption that offered load may be considered stationary for intervals of an hour or so. The load does vary from hour to hour; however, it usually varies in recognizable patterns. Data describing calling

---

[5] Roughly speaking, a stationary process is one in which the parameters describing the process (λ, the average arrival rate, and τ, the average holding time, in traffic theory) remain constant.

patterns between two local offices show that there are two or three distinct load peaks (*busy hours*) during the day—one in the morning, one in the afternoon, and perhaps, one in the evening. Moreover, there will usually be two or three periods during the year when the daily peaks are higher than normal (*busy seasons*). For this reason, the Bell System defines the *engineering period* (the period in which GOS criteria are to be applied) as the *busy season busy hour* (BSBH), that is, the busiest clock hour of the busiest weeks of the year. The decision to concentrate on BSBH periods is an outgrowth of the Bell System service philosophy—to provide high-quality, economical service during normal, daily use of the network.

The identity of the busy season busy hour varies by geographical area. For example, in metropolitan areas where business traffic tends to dominate, the BSBH typically occurs between nine and eleven o'clock on weekday mornings. (The particular hour may occur from nine to ten, ten to eleven, or from half past ten to half past eleven, etc.) Another peak in traffic often occurs in the afternoon, and in some local offices, this may be the BSBH. In local offices with a significant amount of residential traffic, an evening busy hour may occur. Figure 5-2 shows the traffic variation typical of a local office with business traffic predominating. Since the busy season (as mentioned before, there might be two or three in a year) normally lasts for a month or longer, loads will be measured in upwards of twenty busy hours (one particular hour in five weekdays over four weeks) two to three times a year.

Although traffic theory is useful in predicting the performance of a given system for a given load submitted to a given number of servers, considerable engineering judgment is required to select the particular

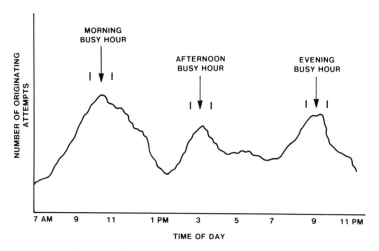

Figure 5-2. Traffic variation with time of day.

load levels (engineering periods) about which to be concerned. One consideration is the type of equipment being engineered. For trunk groups (circuits between switching systems) the average of the twenty BSBH measurements is used, giving rise to the concept of the *average busy season busy hour* (ABSBH).[6] A GOS criterion for such a trunk group might read: For the ABSBH load, a call requiring a circuit in the trunk group should encounter all trunks busy (be blocked) no more than 1 percent of the time. Or the GOS might state: The average blocking over the twenty BSBHs should not exceed 1 percent.[7] In any event, the GOS for trunk groups is always stated in terms of blocking because trunk groups are engineered as blocked-calls-cleared systems[8] (that is, there are no waiting positions for trunk circuits).

For equipment within a switching system (for example, digit receivers), the average busy season busy hour load is not always a good measure of system demand. When switching systems become congested, they tend to spread the congestion to other switching systems in the network (for example, systems awaiting signaling responses from the congested system). So, peak loads are of more concern than average loads when engineering switching equipment, and engineering periods other than the ABSBH are defined. Examples of these periods are: the highest BSBH and the average of the ten highest BSBHs. Sometimes, the engineering period is the weekly peak hour (which may not even be a BSBH).

In addition to selecting suitable engineering periods, it is necessary to express the GOS criteria in meaningful terms. Since most switching equipment items provide waiting positions, the GOS criteria are usually expressed in terms of *delay probabilities*. For example, the digit receivers in switching systems are commonly engineered in terms of dial-tone delay. The number of digit receivers to be provided is chosen such that the probability of dial-tone delay exceeding 3 seconds is less than a specified value. To reflect the concern with performance during peak periods, GOS criteria may be specified for multiple engineering periods. (For dial-tone delay, criteria for three engineering periods are often specified.) Figure 5-3 presents a simplified view of the effect of multiple criteria. It shows two of the frequently used engineering periods: the highest BSBH and the average of the ten highest BSBHs. One criterion may dominate (determine the number of digit receivers required), depending on the ratio of load values for the two engineering periods (busy hours).

[6] The ABSBH concept is limited to trunks.

[7] These two criteria are usually not equivalent (see Section 5.4.2). The blocking for the average of twenty loads (ABSBH) tends to be less than the average blocking across the twenty loads.

[8] At present, this is true for the PSTN, but there may be exceptions in private networks.

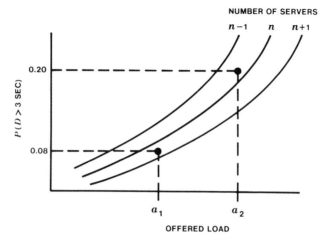

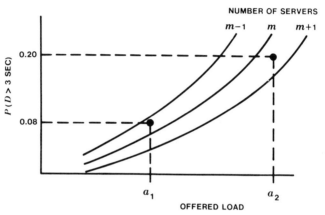

Figure 5-3. Effect of multiple criteria on engineering digit receivers. Two frequently used criteria for engineering digit receivers are that the probability of dial-tone delay greater than 3 seconds, $P(D>3$ seconds), be as follows:

$P(D>3$ seconds) $=0.08$ for the average of the ten highest BSBHs (a_1)
$P(D>3$ seconds) $=0.20$ for the highest BSBH (a_2) .

Relative values for a_1 and a_2 vary for different parts of the network. In some cases, *top graph*, a_1 dominates—requires a greater number of servers ($n+1$). In other cases, *bottom graph*, a_2 dominates—requires ($m+1$) servers.

Thus, the selection of engineering periods and statements of GOS criteria are critical components in determining the quality of service customers will receive. (Section 5.7.2 discusses Bell System service objectives.)

As mentioned previously, the Bell System service philosophy concentrates on engineering the network for the BSBH. However, the Bell System is also concerned with network performance when load levels are substantially higher than, or calling patterns are significantly different from, any BSBH. These include Mother's Day, Christmas Day, heavy snow days, and periods of natural disaster such as floods and earthquakes. Providing enough equipment to handle these infrequent dramatic peaks would be far too costly. Network management maintains network integrity at those times (see Section 5.6).

The distinction between BSBH and dramatic peak loads is important because most of the useful traffic models developed describe systems that are, in some sense, stable. The formulas resulting from these models do not predict system performance under heavy overloads well. The following sections then, assume that the traffic demands are those encountered during the busy season.

5.2.4 TRAFFIC THEORY TECHNIQUES

Telephone traffic cannot be predicted exactly, but it is reasonable and useful to view customer requests and service times as statistical processes that can be described in terms of probabilities. Thus, the number of calls that will arrive into the system during the next t seconds may not be known, but the probability that there will be some number k of them can be estimated. Similarly, the probability that a given call now being served will leave the system during the next t seconds can be estimated.

These and other probabilities relevant to traffic analysis can often be calculated using methods developed in a branch of applied probability theory known as *traffic theory*.[9] Many of the engineering methods used in the telephone system are direct applications of traffic theory models. Even for complex network operations that do not readily lend themselves to traffic theory analysis, traffic theory concepts and results often form the basis for engineering procedures.

Traffic theory models begin by ascribing some underlying probabilistic structure to the processes of call arrivals and call holding times. One common assumption about call arrivals, for example, is that the probability of an arrival during some small interval $(T, T+t)$ is proportional only to the length of the interval, namely, t. The constant of proportionality is some number λ (the average arrival rate), so that

$$Pr[call\ arrival\ in\ (T, T+t)] = \lambda t. \qquad (2)$$

Cooper (1981, pp. 50-51) shows that the simple assumption in equation (2) leads to a formula that calculates $p(k; t)$, the probability that k calls arrive in any interval of length t:

$$p(k; t) = \frac{(\lambda t)^k}{k!} e^{-\lambda t}. \qquad (3)$$

---

[9] This branch of applied probability theory is known by several other names, such as *queuing theory* and *congestion theory*.

Equation (3) is the well-studied *Poisson distribution* (Feller 1968, eq. [6.1]), and any process described by equation (3) is, therefore, called a *Poisson process*. Equation (3) is applicable to many physical situations with a large (essentially infinite) number of potential users, acting independently, such that the percentage of users actually requesting service at any time is small relative to the maximum possible. The population of telephone users in a given central office displays this property, and equation (3) *describes well the way prospective calls originate at a central office during busy periods.*

Of course, in a telephone network, the process describing calls departing from the network after they obtain a server must be considered as well as the process describing call arrivals. As with equation (2), the simple assumption can be made that, during an interval of length t, each call in progress will terminate with probability μt, where μ, the *departure rate*, is merely the reciprocal of the average holding time, that is, $\mu = 1/\tau$. If $H(t)$ denotes the probability that a given arrival requires service for t seconds or less, Cooper (1981, pp. 42ff) shows that

$$H(t) = 1 - e^{-\mu t} = 1 - e^{-t/\tau}. \tag{4}$$

Equation (4) is the well-studied *negative exponential distribution* (Feller 1966). One of its more startling properties is its lack of memory: Equation (4) describes the distribution of the call's remaining time in the system *no matter how long the call has already been in progress*. Of course, that was indirectly assumed by stating that μt completely defined the departure probability. Though this assumption may seem unrealistic, it is, nevertheless, true that the *negative exponential distribution describes the distribution of telephone conversation times reasonably well.* Numerous studies through the years have verified this result.

When arrivals follow a Poisson process and holding times are negative exponential, traffic is said to be random. Traffic theory models have been developed for other cases, but the two most useful formulas in the Bell System are based on the assumption of random traffic. These formulas, developed by A. K. Erlang, are applicable to many traffic problems. The next sections describe these models and show the resultant formulas without proof.[10] Instead, the discussion emphasizes the underlying assumptions and their implications for traffic engineering.

5.2.5 ERLANG'S BLOCKED-CALLS-DELAYED MODEL

If a random load, $a = \lambda \tau$, is submitted to a system of c servers such that blocked calls wait until served on a (possibly infinite) waiting line, waiting calls are served in order of arrival, k is an index of the number of

---

[10] Cooper 1981 gives the derivations of these and other traffic formulas.

arriving calls, and P_j denotes the probability that an arriving call finds j customers in the system (either being served or waiting); then

$$
P_j = \begin{cases} \dfrac{a^j}{j!} \, P_0 & j = 0, 1, ..., c \\[2ex] \dfrac{a^j}{c! \, c^{j-c}} \, P_0 & j = c, c+1, c+2, ..., \end{cases}
$$

(5)

where

$$
P_0 = \left[\frac{a^c}{(c-1)! \, (c-a)} + \sum_{k=0}^{c-1} \frac{a^k}{k!} \right]^{-1} .
$$

(6)

Equations (5) and (6) are valid only when the load is less than the number of servers $(0 \leqslant a < c)$. If the load is greater $(a > c)$, the system is unstable and the queue length grows without bound.

The quantity $(P_c + P_{c+1} + P_{c+2} + ...)$ represents the probability that an arriving call will be delayed. This probability is denoted by $C(c, a)$ and, from equation (5),

$$
C(c, a) = \sum_{j=c}^{\infty} P_j = P_0 \sum_{j=c}^{\infty} \frac{a^j}{c! \, c^{j-c}} = \frac{a^c}{c!} \, \frac{c}{c-a} \, P_0 .
$$

(7)

Then, from equation (6),

$$
C(c, a) = \frac{a^c}{(c-1)! \, (c-a)} \left[\frac{a^c}{(c-1)! \, (c-a)} + \sum_{k=0}^{c-1} \frac{a^k}{k!} \right]^{-1} .
$$

(8)

Equation (8) is *Erlang's Delay Formula*, also referred to as the *Erlang C* formula. This formula has been extensively tabulated since many traffic systems are modeled using the Erlang C assumptions.

Equation (8), however, is not sufficient for engineering purposes. Because the Erlang C model describes a blocked-calls-delayed system, the delay distribution is needed as well. Assuming Poisson arrivals, negative exponential holding times, and first-in—first-out queue service, it can be shown that the *conditional delay distribution* (that is, the probability that delay exceeds t, given delay occurs at all), denoted by $P(D > t | D > 0)$, is given by the formula

$$
P(D > t | D > 0) = e^{-(c-a)\mu t} .
$$

(9)

That is, the conditional delay distribution is negative exponential with mean

$$\overline{D}|D>0 = \frac{\tau}{c-a},$$ (10)

where $\tau = 1/\mu$.

From equations (8) and (9), then, the *unconditional delay distribution* $P(D>t)$ is such that

$$P(D>t) = C(c,a)e^{-(c-a)\mu t},$$ (11)

whose mean is given by

$$\overline{D} = C(c,a)\frac{\tau}{c-a}.$$ (12)

Equations (9) and (11) depend on the first-in—first-out assumption. The average delays, however, given by equations (10) and (12), are valid for *any* order of service. The negative exponential holding-time assumption in the analysis is important. Exhibiting delay formulas such as equation (11) for other holding-time distributions is complex and invariably produces formulas that are numerically difficult. This is one reason the Erlang C model is sometimes used as an approximation even in cases where the model assumptions are not valid.

5.2.6 ERLANG'S BLOCKED-CALLS-CLEARED MODEL

Assuming the same system as in the previous section (c servers, random load a), except that there are no waiting positions and blocked calls are immediately cleared from the system, then P_j (the probability that arriving calls find j customers in the system) is equivalent to the probability that an arriving call finds j servers busy. Obviously, $j \leqslant c$ for this system, and it can be shown that

$$P_j = \frac{\dfrac{a^j}{j!}}{\displaystyle\sum_{k=0}^{c} \frac{a^k}{k!}} \qquad j = 0, 1, ..., c.$$ (13)

The probability that an arriving call is blocked, then, is given by P_c but is commonly denoted by $B(c, a)$, so that

$$B(c, a) = \frac{\dfrac{a^c}{c!}}{\displaystyle\sum_{k=0}^{c} \frac{a^k}{k!}}. \tag{14}$$

Equation (14) is *Erlang's Loss Formula*,[11] referred to as the *Erlang B* formula. It holds even when the load is greater than the number of servers ($a > c$) because, unlike the blocked-calls-delayed model in which all calls are eventually served, the blocked-calls-cleared (BCC) model allows calls to be lost when all servers are busy. Therefore, the BCC system never becomes unstable.

For a BCC system, then, the offered load can be divided into the *load lost* and the *carried load*. The load lost, α, is easily calculated:

$$\alpha = aB(c, a). \tag{15}$$

The carried load, L, is given by

$$L = a - aB(c, a) = a[1 - B(c, a)]. \tag{16}$$

(In a blocked-calls-delayed system, no calls are lost [$\alpha = 0$]; the carried load equals the offered load [$l = \alpha$].)

The efficiency, ρ, of a BCC system can then be defined as the load carried per server:

$$\rho = \frac{L}{c}. \tag{17}$$

The efficiency, ρ, is commonly referred to as the *occupancy* of a group of c servers. Since a given server can never carry more than one erlang, L is always less than c, implying that $\rho \leqslant 1$.

Perhaps the most important property of an Erlang B system is that it produces an economy-of-scale effect. That is, for a fixed blocking, ρ increases with increasing load. Figure 5-4, which plots the number of servers required to produce 0.01 blocking as a function of offered load, illustrates this fact. As offered load increases, the number of *additional* servers needed to maintain blocking at 0.01 decreases. This property is extremely important in designing telephone networks (see Section 5.5).

---

[11] This formula is valid for any holding-time distribution with a finite mean.

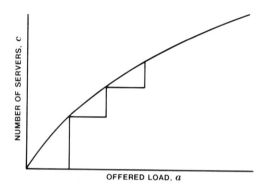

Figure 5-4. Servers versus load for $B(c, a) = 0.01$.

5.3 ENGINEERING SWITCHING SYSTEMS

A switching system includes many different elements needed during the processing of a telephone call, but not all elements of a switching system are traffic sensitive. Trunk scanners, for example, periodically check each trunk connected to a switching system to determine the trunk's busy-idle status. The workload of the trunk scanners thus depends on the number of trunks terminated and the necessary scan rate, but the workload is independent of the number of calls in progress. For most switching system elements, though, the workload *is* sensitive to traffic, and these elements must be engineered accordingly. The next few sections provide some examples of the problems involved in engineering switching system components and describe some engineering approaches.

5.3.1 DIGIT RECEIVERS

Calls encountering all digit receivers busy are, from the customer's point of view, waiting for dial tone. As long as digit receivers are engineered so that long delays are unlikely, it is reasonable to assume that the customer will wait for dial tone without defecting. Thus, a blocked-calls-delayed model is appropriate for estimating blocking probabilities. The grade-of-service (GOS) criteria for digit receivers are related to delay probabilities, however, not to blocking. Whether the Erlang C delay equation is appropriate depends on the distribution of digit-receiver holding times. (As noted previously, the Erlang C delay equations assume negative exponential holding times.)

Since the negative exponential assumption is reasonable for dial-pulse receivers, the Erlang C model performs well. Rotary dialing generates different pulsing times for different digits, ranging from short (1) to long (0); and the time between dialing successive digits is a random variable with a wide range, at times, several seconds. This behavior, combined with enough partial-dial abandons (where the customer hangs up before dialing is complete) that generate small holding times, leads to an overall dial-pulse-receiver holding-time distribution for which the negative exponential is a good fit. Strictly speaking, the lack of memory cannot, obviously, apply to dialing times. Once four or five digits have been dialed, for example, the remaining holding-time distribution cannot be the same as the initial distribution. But the discrepancy between the lack-of-memory assumption and actual dialing behavior is not great enough to render the approximation ineffective.

On the other hand, the Erlang C model cannot be expected to describe dialing behavior for dual-tone multifrequency receivers accurately. *TOUCH-TONE* dialing times are much shorter and less variable than rotary dialing times. Each digit has almost the same pulse time, and there is a tendency to dial rapidly. A constant holding-time assumption might seem more appropriate and would be, except that the number of digits per call is so variable (seven, ten, possibly three or four with centrex/private branch exchange systems or from partial-dial abandons) that the constant holding-time assumption is too optimistic. Despite the inaccuracy of its assumptions, therefore, the Erlang C model is preferable because it errs on the side of good service but not to the extent of adding significant cost.

Similarly, the Erlang C model is used to engineer the digit equipment needed for communication between switching systems (digit transmitters and incoming digit receivers, for example). Delays for these items have nothing to do with dial tone, since the customer has already received dial tone and has entered the called-party digits; rather, delays in transmitting and receiving digits between switching systems affect the time needed to establish a talking path. The considerations, however, are similar to those in the dial-tone-delay problem for originating users. For example, the sending system must wait for a signal from the receiving system before it can send the digits. The waiting time depends on the number of digit receivers in the receiving system.

5.3.2 OPERATOR FORCE

In the mathematical sense, the number of operators required (the *operator force*) can be "engineered" using the same modeling techniques as for equipment items. Since calls directed to an operator that find all operator positions busy have to wait, determining how many operators are needed for a given GOS is a traffic problem. Using assumptions of arrival rates and operator service times, traffic models can be developed to express delay probabilities, and in fact, the Erlang C model (with adjustments to account for known deviations from the model's assumptions) is used to do that. Given the offered load and average operator service times, the number of operators required is determined by the objective speed of answer (the delay in obtaining connection to an operator). Depending on the type of service and the equipment, the objective speed of answer ranges from 2 to 6 seconds.

Traffic theory considerations aside, however, operators are not machines; and this fact plays an important part in engineering procedures. In particular, it is not enough to determine the number of operators required for peak load periods, put them in service, and leave them there permanently. The problem becomes one of *forcing*, that is, scheduling the operator work force so that the desired grade of service is met while minimizing the wasted expense of providing too many operators. The forcing problem thus involves careful forecasting of load variations from hour to hour, day to day, and week to week, not just the forecasting of some peak load period.[12]

Computer-based systems developed jointly by Bell Laboratories and AT&T provide operating company personnel with data on traffic volume, measurements of time to answer, and service time. A modified Erlang C model is used to forecast operator requirements by 15-minute intervals during the week, 2 weeks in advance.

5.3.3 STORED-PROGRAM CONTROL SYSTEMS

Stored-program control (SPC) switching systems (see Section 10.3) process thousands of simultaneous calls on a time-shared basis. The performance of the stored program is traffic sensitive; that is, the more calls being processed, the longer it takes for the SPC to return to the next stage in processing a given call. It is possible, for example, for an SPC to generate excessive dial-tone delay even though digit receivers are available, because the controller may spend several seconds processing calls that are already in progress before processing the digit-receiver request for new call originations.

The SPC, then, has a definite capacity, that is, a maximum volume of calls it can process during a given interval and still satisfy all GOS requirements. This capacity depends on many factors including the stored-program organization, the speed of the logic, the amount of memory and hardware available, and the kinds of calls being processed. (Different types of calls—rotary or *TOUCH-TONE* telephone, coin or non-coin telephone, intraoffice or interoffice, etc.—require different call-processing functions or different processor work times for a given function.) Traffic theory models are not adequate for determining SPC capacity because of the complex interactions between traffic, software, and hardware.

---

[12] Since forcing is an operating company activity, Section 13.3.1 discusses it further.

In the absence of suitable traffic theory models, simulation techniques have been developed for the various SPC systems to measure system performance as a function of traffic input. Simulation results have been useful in constructing formulas that engineers can use to predict a given system's capacity before it is placed in service. Once the switching system is in service, the formulas are used with measurement data to adjust capacity estimates. This allows the engineer to predict when a given system will *exhaust* (when the forecasted traffic load will reach the estimated call-handling capacity of the system) and to re-evaluate the prediction regularly.

One important aspect of SPC traffic is the way the processors react to heavy load conditions. With many such systems, once the number of calls in progress becomes too great and the processor begins to give poor service, deterioration accelerates rapidly. This acceleration occurs because the deterioration of service in itself generates additional processor work items for each call (moving calls in and out of waiting lines, for example). In addition, the number of calls to be processed is likely to increase as customers reattempt calls that have failed to complete. To compensate for this effect, logic is included that allows the processor to monitor its performance and to take appropriate action when there are indications that the load has exceeded capacity levels. These automatic overload control actions can take many forms, such as deferring nonessential maintenance tests, reducing the rate of scanning for new call originations, or changing the order of processing various queues (see Section 5.2.2). Determining the proper overload detection logic and setting the parameters of the control strategy are traffic engineering problems for which simulation techniques have also been particularly important.

5.3.4 CAPACITY CONSIDERATIONS – LOAD BALANCING

The preceding traffic engineering examples assume that all customers can reach all idle servers at all times. In fact, depending on the traffic, an efficiently designed switching network that gives access to the servers may not provide an idle path at all times. Customer terminations are grouped as shown in Figure 5-5, and an above-average load in one group will result in more blocking than for a group with below-average load. It is a property of the blocking curves that such an imbalance in loads results in poorer average service than would result if the loads were evenly distributed. Furthermore, one of the groups may be given very poor service even when the average appears to be satisfactory. This can be illustrated in the following example for dial-tone delay where the servers are dial-tone registers. As shown in Figure 5-6, load balancing equalizes service to customers, while at the same time, giving the best

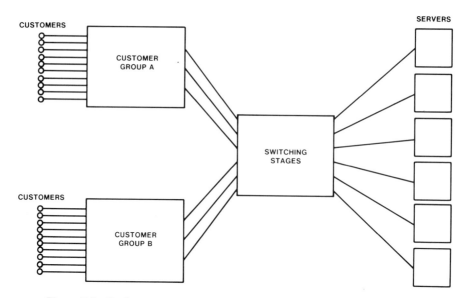

Figure 5-5. Customer access to servers. In this simplified representation, an above-average load from customers in group A may cause blocking on the links from group A that exceeds the objective value, even though there may be idle servers.

overall average service. Line finders in step-by-step systems, horizontal groups in crossbar systems, and concentrators in electronic switching systems are the most critical components for load balancing.[13]

The simplest and most economical way to achieve and maintain load balance is through control of routine line assignments: New customers are assigned to lightly loaded components, and assignments to heavily loaded components are avoided.[14] As customer disconnects occur, the load on the heavily loaded components is reduced. However, when imbalance is severe (that is, when some components are badly overloaded), correction by routine line assignment may take too long. It may be necessary to disconnect active customers from overloaded components and transfer them to underloaded components. Such *corrective action* is clearly more costly.

The existence of an official Load Balance Index Plan reflects the importance of load balancing. Without load balancing, switching systems

---

[13] Chapter 10 describes switching systems and their components.

[14] Assignment of customers to specific line equipment is a network administration function. Section 13.3.2 discusses network administration functions.

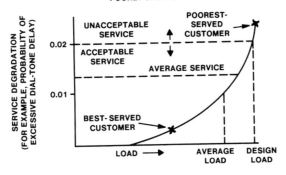

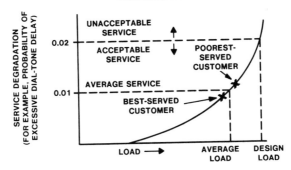

Figure 5-6. Load-service curves. Because the rate of service degradation increases with increasing load, for identical switching systems operating at the same average load, average service is better in a balanced system than in an imbalanced one. In addition, some customers in a poorly balanced system, *top*, may receive an unacceptable level of service, while service to all customers is equalized in a well-balanced system, *bottom*.

would have to operate at a load level estimated to be about 5 to 10 percent below their design capacity to maintain the same quality of service.

5.4 ENGINEERING TRUNK GROUPS

The previous section explored some of the traffic engineering problems and techniques associated with switching systems. This section focuses on some of the problems and techniques associated with the trunk groups connecting the switching systems.

Several models exist to help determine trunk requirements. Typically, an application model is used to generate trunking tables that are, in turn, used by operating companies to determine the number of trunks in a group for a given demand.

5.4.1 APPLICABILITY OF THE ERLANG B MODEL

In the PSTN, there are two kinds of trunk groups: **high-usage** (HU) groups, from which blocked calls are offered to other (alternate-route) groups for completion, and last-choice or **final** groups, from which blocked calls are given a reorder tone indicating that the caller should try again later.[15] Whether an HU group should be placed between two offices and, if so, how many trunks it should have are discussed in Section 5.5, which treats the larger question of traffic network design. Final groups, however, are identified a priori by the switching hierarchy described in Section 4.2.1 and are engineered to provide an average blocking of 0.01 during the busy season. Because calls blocked on a final group are not permitted to wait,[16] it would appear reasonable to assume that the Erlang B model is applicable to these groups. In actuality, however, there are several effects not accounted for in the Erlang B model that affect its applicability, so that modifications are needed to engineer final groups properly. The next few sections discuss these modifications.

5.4.2 DAY-TO-DAY VARIATION

The grade-of-service (GOS) criterion for final groups pertains to average blocking during the busy season, which typically comprises 20 hours of data. Generally, the offered load varies between different busy hours in the busy season. As Figure 5-7 shows, the Erlang B formula is such that the blocking for the average of loads a_1 and a_2 is less than the average blocking for the two loads. Similarly, blocking for the average busy season busy hour (ABSBH) load is less than the average blocking during the busy hours of the busy season. This day-to-day variation effect was both predicted and verified many years ago. Because of this effect, the Erlang B formula, which underestimates the group size needed to average 0.01 blocking when day-to-day variation is anything but zero, was not used to build trunk group tables. Instead, the tables were based on the *Poisson Blocking Formula*, $P(c, \bar{a})$, where

$$P(c, \bar{a}) = e^{-\bar{a}} \sum_{k=c}^{\infty} \frac{\bar{a}^k}{k!}. \tag{18}$$

---

[15] Historically, large HU groups with the potential for significant overload were engineered for 0.01 blocking, and for service protection, calls blocked on these groups were not offered to other routes. These groups were called *full* groups or *special final* groups. But engineering and overload control techniques have reduced the need for them.

[16] Trunk holding times tend to be long because, unlike digit receivers, trunks have to be held for the duration of the call. Allowing blocked calls to wait for an idle trunk would involve long delays and inefficient use of system capacity. This is not the case for traffic with short holding times, typically, certain data applications such as packet-switching networks. (Section 5.8 discusses data traffic.)

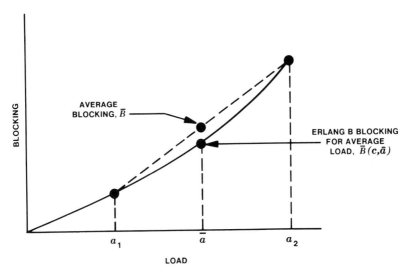

Figure 5-7. Day-to-day variation effect.

In this formula, c is the required number of servers, $\bar{a}$ is the ABSBH load, and k is an index of the number of arriving calls. $P(c, \bar{a})$ is the tail of the Poisson distribution, hence the formula's name.

Equation (18) assumes that delayed calls have a duration, including the delay, that is the same as it would be if there were no delay. This assumption is mathematically convenient, but it has no physical justification in actual customer behavior. The practical value of the Poisson Blocking Formula in equation (18) is that, by chance, it produces approximately the correct answer. It predicts a number of servers significantly smaller than the Erlang C formula (calls wait forever) but greater than the Erlang B formula (calls do not wait at all). Consequently, it serves as a suitable compromise between the two formulas.

Engineering tables that considered day-to-day variation were first introduced in the mid-1960s, based on work by R. I. Wilkinson (1956). More refined techniques were applied to provide later tables (see Hill and Neal 1976). These later tables, known as the *Neal-Wilkinson* $\bar{B}$ tables, are replacing Poisson tables in the operating companies.

5.4.3 REATTEMPTS

Customers whose attempts are blocked frequently try again. If the number of blocked attempts is small, the *reattempt effect* will be insignificant. The effect is potentially significant on trunk groups that operate at very high occupancies (see Section 5.2.6). Because these groups have very little built-in slack, blocking on them increases dramatically when an overload is present. Reattempts cause a departure from the random arrival assumption used in the traditional models and, if they are not accounted for, cause an overestimate of offered load based on measurements. In practice, the accounting has been done by assuming that only 35 percent of blocked calls are lost (do not return to the system).

The potential problem caused by reattempts is minimized because the trunking tables use the models (originally, the Poisson, now, the Neal-Wilkinson $\bar{B}$) only for groups of fewer than 250 trunks. When the $\bar{B}$ model calls for more than 250 trunks (that is, for the higher occupancy groups), the group is engineered to the same (lower) occupancy that a 250-trunk group provides.

5.4.4 NONRANDOM LOAD

Both the Erlang B and the Poisson models assume that calls arrive in accordance with the Poisson distribution. For final groups serving overflow from HU groups, however, the Poisson arrival assumption is demonstrably false. Poisson load has a variance-to-mean ratio equal to 1; overflow load has a variance-to-mean ratio significantly greater than 1. The variance-to-mean ratio of the offered load is called the *peakedness* and is denoted by z. More trunks are required to provide an equivalent GOS for peaked ($z > 1$) load than for Poisson ($z = 1$) load.

The effect of peaked (non-Poisson offered) traffic is modeled using a technique known as the *equivalent random method*. It assumes that all peaked traffic will behave the same way as traffic overflowing a single trunk group that meets the same assumptions as the Erlang B Loss Formula. The model is constructed by calculating a load, A, and a number of trunks, C, that will have an overflow traffic with mean, a, and variance, $v, (v > a)$, equal to the traffic to be modeled. The desired overflow from a group of trunks to serve the peaked traffic is expressed as *overflow load, α*. For example, the objective may be 1-percent blocking, or $\alpha/A = 0.01$. If A erlangs offered to $C + c$ trunks produce α erlangs of overflow, then the a erlangs of peaked traffic offered to c trunks will also produce α erlangs of overflow.

Calculated values of C and A for a wide range of overflow loads and peakedness are tabulated. Calculation of c, then, involves iteration of the equation:

$$(\frac{\alpha}{A}) = B(C + c, A). \tag{19}$$

The disadvantage of the equivalent random method is that it presumes that the group of c trunks is engineered using the Erlang B model. But, as mentioned previously, engineering final groups using the Erlang B model fails to account for day-to-day variations. Nor can inserting the Poisson blocking probability in equation (19) compensate for day-to-day variations, since A and C are estimated by assuming blocked calls are cleared on the C trunks.

The development of alternate-route networks, in which non-Poisson offered traffic reaches the final groups, spurred the effort to produce models that accurately account for peakedness. This effort, in turn, generated a need for models (for example, Neal-Wilkinson) that properly account for both day-to-day variation effects and non-Poisson offered traffic. The next section discusses the advantages of alternate-route networks and some factors complicating their design.

5.5 TRAFFIC NETWORK DESIGN

The previous two sections described the problems of engineering switching systems and engineering trunk groups as if they were independent activities. However, when the nodes (switching systems) and the links (trunks) are combined to form a network, optimizing overall network design becomes the key problem. This optimization involves providing satisfactory (from the customer's viewpoint) end-to-end service at the lowest possible cost. Because the telephone network is so large and complex and must be flexible enough to respond to increasing demands, new services, and changing technology, a precise formulation of the design problem is impossible. However, alternate routing has proven to be a key factor in optimizing the network. The next few sections describe one of the simple models developed to produce cost-optimal (in the static sense) alternate-route networks and how this model is used as part of a process to provide the required number of trunks in the network—a process that captures dynamic (year-to-year) effects too complex to include directly in a single optimization model.[17]

5.5.1 THE ECONOMICS OF ALTERNATE ROUTING

Every central office (CO) in the PSTN is connected by a final trunk group to some tandem[18] switching office, so that routing between COs is always possible over these "backbone" final groups (see Section 4.2). Figure 5-8, in which COs 1 and 2 have final groups to the same tandem office A, illustrates this construction. Traffic between COs 1 and 2 can always be

---

[17] Section 13.3.1 describes the process for providing trunks and special-services circuits in the network (*provisioning*).

[18] *Tandem* is used in this discussion in the generic or functional sense, that is, an office that switches trunks to trunks. In the switching hierarchy, this office may be a toll office or a local tandem office.

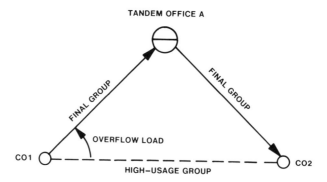

TANDEM OFFICE A

FINAL GROUP

FINAL GROUP

OVERFLOW LOAD

CO1

CO2

HIGH—USAGE GROUP

Figure 5-8. Alternate-route triangle.

routed on groups 1-A and A-2 and would encounter (roughly) 0.02 blocking on the average during the busy season, assuming 0.01 blocking on
each final group.

Where enough traffic exists, there are significant advantages in also
building a direct group between COs 1 and 2 and allowing calls blocked
on the 1-2 group to overflow to the alternate route 1-A-2. This routing
strategy effectively balances the conflicting forces of cost and service.
Trunks in the direct group, being shorter, tend to be cheaper than trunks
in the alternate route, which seems to suggest that all traffic between the
offices should be carried on a direct group. If this were done, however,
the group would have to be engineered for objective blocking, requiring
a larger number of trunks. In addition, there would be no service protection if a serious failure in the trunk group occurred. The alternate-route
network in Figure 5-8 provides better service at lower cost than either the
alternate-route-only or the direct-route-only strategies, provided the
correct number of trunks is determined for the direct group.[19]

The model for determining the correct number of HU trunks is based
on the simple triangle network of Figure 5-8. The following discussion
assumes that there is a cost per alternate-route trunk, C_A (which includes
the cost of switching through tandem office A plus the transmission cost
of trunks 1-A and A-2), and a cost per direct trunk, C_D. The *marginal
capacity*, γ, of a trunk group engineered to 0.01 blocking can be defined as
the additional load that can be offered to the group, without changing
the blocking, when one trunk is added. (Mathematically,

$$\gamma = \left. \frac{\partial a}{\partial c} \right|_{B=0.01}$$

[19] An exception occurs when the traffic between COs 1 and 2 is small. Because small trunk
groups are inefficient compared to larger trunk groups (that is, small groups operate at
low average occupancies), there are cases when no direct group can be justified
economically.

where a is the offered load and c is the number of trunks in the group.) Since the 1-A and A-2 groups are carrying other traffic (group 1-A carries traffic from CO1 to other COs; group A-2 carries traffic from other COs to CO2), they tend to be reasonably large and efficient; that is, they have a high marginal capacity. The following discussion assumes both groups have the same marginal capacity, denoted by γ_A. (In practice, γ_A is usually between 26 and 28 CCS.)[20]

If n trunks are put into the CO1-CO2 group, the cost of these trunks will be nC_D. Moreover, the overflow from this n-trunk HU group, $\alpha(n)$, will be offered to the alternate route, which will require the addition of $\alpha(n)/\gamma_A$ trunks to group 1-A and to group A-2 to maintain 0.01 blocking. The incremental[21] cost, then, of putting n trunks in the CO1-CO2 HU group, denoted by $C(n)$, is given by

$$C(n) = nC_D + \frac{\alpha(n)}{\gamma_A} C_A. \tag{20}$$

Figure 5-9 shows $C(n)$ and its two components in graph form, treating n as a continuous variable. As shown, $C(n)$ has a unique minimum, which occurs at the value of n such that

$$-\frac{\partial \alpha(n)}{\partial n} = \gamma_A \frac{C_D}{C_A} = \frac{\gamma_A}{C_R}, \tag{21}$$

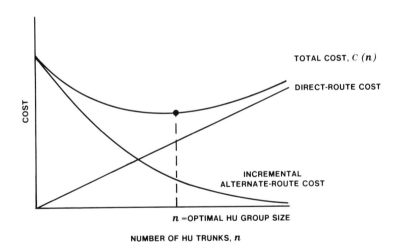

Figure 5-9. Cost function for alternate routing.

<hr/>

[20] Hundred call seconds per hour (see footnote 3).

[21] The background traffic on trunk groups 1-A and A-2 requires trunks too, of course, but the cost of those trunks is assumed to be independent of n. Thus, only the *additional* cost of routing the CO1-CO2 overflow via 1-A-2 must be considered.

where $C_R = C_A/C_D$. C_R is referred to as the *cost ratio* of the HU group. (If there is no n such that equation (21) holds, then the minimum occurs at $n=0$; that is, tandem route only is the optimal strategy.)

The quantity $-\partial\alpha(n)/\partial n$ has a practical interpretation: It is the load on the last trunk of an n-trunk group, assuming the Erlang B model applies to the HU group. Thus, the quantity γ_A/C_R is the most economical load on the last trunk, and the optimal group size is such that the last trunk carries γ_A/C_R load. Since the telephone companies measure load in CCS, the quantity γ_A/C_R is referred to as the *economic CCS* (ECCS). The technique of sizing HU groups such that the last-trunk load is γ_A/C_R is commonly called *ECCS engineering*.

5.5.2 MODIFICATIONS TO THE ECCS ENGINEERING TECHNIQUE

The cost function in equation (20) represents an approximation to the actual incremental cost of providing n trunks in the HU group. The marginal capacity, for example, is not truly a constant but depends on both the peakedness (variance-to-mean ratio) of the background traffic on the alternate route and the peakedness of the overflow from the HU group. Nevertheless, the ECCS technique yields a good approximation to the optimum HU group size. This approximation can be improved by recognizing certain cost realities that the ECCS model ignores. These realities are discussed in the following paragraphs.

Minimum Group Sizes

There is an administrative cost associated with an HU group that is not reflected in equation (20). This cost applies when $n>0$ but not when $n=0$. When this administrative cost is added to the total cost for the optimal group size as determined by the ECCS method, it may well be that $n=0$ is a cheaper solution. Empirical studies have determined that this happens when the ECCS method suggests small HU groups and can be corrected effectively by using the ECCS method in conjunction with a minimum HU group size. The recommended minimum, determined empirically, is typically three trunks per group for local networks and six trunks per group for toll networks; smaller HU groups are not built. The difference between the two minimum values is explained by the fact that cost ratios (C_A/C_D) in local networks tend to be higher than in toll networks, so that smaller groups are economical.

Modular Engineering

The cost function of equation (20) assumes that there is a unit cost, C_D, for each trunk added to the direct group. However, many of the transmission components that implement trunk groups have the capacity to handle multiple trunk circuits. For example, trunks can be added to a digital interface frame (see Section 9.4.3) in increments of twenty-four. To compensate for these effects, a technique known as *modular engineering* is used.

With modular engineering, the optimal HU group size as determined by ECCS engineering is rounded to the nearest relevant module size. For the Long Lines network, the relevant module size is twelve trunks because the basic multiplex unit (the channel group) implements twelve circuits at a time.[22] For local T-carrier (see Section 9.4.2) networks using dedicated digital terminations, the relevant module size is twenty-four trunks for 2-way groups and twelve trunks for 1-way groups. (Using twelve trunks for 1-way groups maximizes the chances that the trunk requirements of two opposing 1-way groups will add up to a multiple of twenty-four circuits.)

Noncoincidence of Busy Hours

The triangular cost model illustrated in Figure 5-9 assumes that the busy hour of the direct group coincides with the busy hour of the two alternate-route groups. This need not be true. For instance, the busy hours of the two alternate-route groups need not be the same. In that case, overflow from the direct HU group might force additional trunks to be added to group 1-A but not to group A-2 in the group 1-A busy hour. Similarly, trunks might have to be added to group A-2 but not to group 1-A for overflow during the group A-2 busy hour.

One method of compensating for possible noncoincidence effects is the *cluster busy hour* technique, shown in Figure 5-10, in which a cluster of potential HU groups overflows to a common final group. The offered loads to each group are summed for each busy hour, and the maximum hourly sum identifies the cluster busy hour. Each HU group is then sized, using the offered load for that group that corresponds to the cluster busy hour. This technique works well in many applications, but it ignores the possibility that the alternate-route busy hour may, in fact, depend on the HU group sizes selected. When this happens, no single hour can be identified a priori as the "correct" hour for all HU groups. A technique known as *multihour engineering* (see Elsner 1977) has been developed to solve the problem generally, but the algorithm is complex and must be implemented by means of a highly sophisticated computer program.

5.5.3 DYNAMICS OF NETWORK DESIGN

The ECCS cost model used to size HU groups determines an optimum group size only in the static sense, that is, for a given load or collection of hourly loads and a given alternate-route cost ratio. As trunk group

---

[22] Sections 6.5, 9.3.5, and 9.4.3 discuss multiplexing and channel groups.

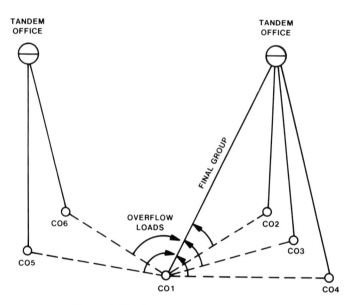

TANDEM OFFICE

TANDEM OFFICE

FINAL GROUP

OVERFLOW LOADS

CO6

CO5

CO2

CO3

CO1

CO4

Figure 5-10. Originating cluster concept.

requirements change from year to year, other costs are incurred that are not included in the ECCS model but that can strongly influence economic decisions.

The impact of trunk requirements on the facilities network layout (see Section 4.4) is particularly important. For the facility-provisioning process to position correctly the facility routes needed to implement trunk groups, estimates of trunk requirements are needed years in advance. The ECCS model is used to forecast trunk requirements using projections of offered loads and routing configurations.

Forecasting errors are to be expected and must be corrected to minimize total network cost while providing the grade of service required. The network design process therefore involves three distinct but interrelated phases: **forecasting, engineering,** and **servicing**. The following paragraphs discuss some of the traffic considerations involved in these operations. Activities involved in each operation are described in Section 13.3.

Forecasting

The key to accurate trunk forecasting is accurate projection of loads offered to the trunk groups. Load projection, in turn, requires a measurement system to obtain basic load data and an extrapolation process that translates basic data into expected loads.

There are two sources of basic load data: automatic message accounting (AMA) tapes and trunk group measurements. AMA tapes record information related to telephone calls for billing purposes (see Section 10.5). The Centralized Message Data System (CMDS) analyzes the tapes to determine traffic patterns, but the process is expensive. So, CMDS samples only 5 percent of all completed, or billable, calls (messages) to generate a point-to-point load base. Although the resultant estimate of total load is based on a small sample of calls that excludes incomplete calls, it is, nevertheless, useful because it is relatively independent of the network switching configuration. This is particularly important if it is known that the switching configuration will be changing during the period of the trunk forecast.

Trunk group measurements are more reliable than point-to-point estimates and are usually less expensive to obtain. But such measurements are only possible, obviously, on groups that exist. They become less useful where network configuration changes occur and new trunk groups come into existence.

Engineering

The network trunk requirements produced by the forecasting process are sensitive to errors in load projection. Such errors can generate significant and frequently unnecessary year-to-year "churning" (fluctuation) of trunk group requirements, the cost of which can be high. Furthermore, the decision regarding the number of trunks to place in a given group should depend on both the number of trunks already in place and the expected future of the group. For example, disconnecting trunks is expensive, but so is maintaining more working circuits than are required to meet service objectives. An optimal disconnect policy, therefore, depends on whether the reduced requirements of a given group are expected to return to their original size in the near future. If so, it is usually more economical to leave the unneeded circuit(s) in place rather than pay a disconnect cost now and a reconnect cost later.

Another important consideration in engineering is proper reserve capacity. If final trunk groups, for example, were always engineered to provide exactly 0.01 blocking, then, on the average, half of the final groups would actually generate greater than 0.01 blocking due to forecasting errors. Trunk servicers (discussed in **Servicing** below) would have to augment (add trunks to) these underprovided groups to correct for the errors, and this augmentation is expensive.

An optimal reserve capacity balances the cost of overproviding against the cost of servicing. Too much reserve is wasteful; too little can lead to severe and expensive servicing problems. It thus becomes economical to overprovide final groups, since the cost of some additional (reserve) trunks is offset by the savings in labor cost during servicing.

Using the trunk forecast as input, the engineering process attempts to smooth the forecast to eliminate churning due to statistical error and to reflect disconnect and reserve economics. Historically a manual process, engineering is now being automated to provide more accurate, stable, and economic (in the total cost sense) trunk requirements to the facility planners and engineers.

Servicing

In trunk servicing, performance measurements of final trunk groups are used to determine where circuits have to be added or deleted to meet service objectives. This process is supported by the computer-based Trunk Servicing System (TSS). TSS makes the servicing of HU groups as well as finals possible since it analyzes overall network performance and can determine when, for example, the overflow from a poorly engineered HU group is responsible for the poor performance of a given final group.

5.6 NETWORK MANAGEMENT

Telephone networks employing alternate routing and common-control switching systems (see Chapter 7) provide efficient use of facilities. This efficiency, however, can lead to operating penalties when heavy traffic overloads or major equipment failures occur. The solid curved line in Figure 5-11 illustrates this phenomenon. The figure shows simulated results for a network engineered to carry 1600 erlangs but subjected to a substantial overload. Without network management controls, as the

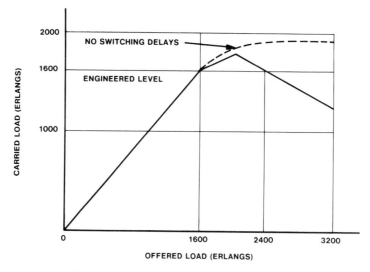

Figure 5-11. Carried load versus offered load.

offered load increases beyond engineered levels, the carried load begins to decrease and eventually drops below engineered capacity. The reduction in carried load is the result of trunk and switching congestion. It is the responsibility of network management to prevent such deterioration of service and to keep the network operating near maximum efficiency when unusual traffic patterns or equipment failures would otherwise force inefficiencies because of congestion.

5.6.1 TRUNK CONGESTION

As direct routes in a network become occupied, more and more calls are forced to follow alternate routes, which involve more trunks and switching systems per call. The efficiency of the network decreases as this occurs. If traffic levels are so high that the alternate routes also become occupied, more and more calls receive reorder tone; and the switching equipment and circuits used to handle these calls up to the point where blocking occurs will have been wasted. At very high load levels, then, otherwise efficient alternate-route networks become inefficient. Moreover, since high blocking generates more reattempts per call and still more load for the network, such congestion tends to sustain itself.

5.6.2 SWITCHING CONGESTION

A call arriving at a switching system from a preceding system will be delayed if all incoming digit receivers are in use. While waiting for an idle receiver, the digit transmitter in the preceding system is tied up. After 20 to 30 seconds, the transmitter will time out and route the call to a reorder tone. (Under normal conditions, this rarely happens.) The result is an aborted call that has used several times the normal transmitter holding time and has accessed routing equipment in the first system twice—once to set up the call and once to route it to a reorder tone. Incoming receiver congestion in the second switching system, therefore, has the potential to generate congestion in other switching systems. Switching overloads can spread throughout the network if appropriate countermeasures are not taken.

5.6.3 ANALYSIS OF CONGESTION

Standard traffic models, such as the Erlang B or Erlang C models, are not useful for analyzing network behavior under heavy overloads. They assume equilibrium conditions and, therefore, do not apply when overloads occur and congestion begins to spread. Furthermore, network measurements describing overload phenomena and the effectiveness of management techniques are difficult to obtain: Most congestion patterns do not repeat themselves with enough fidelity to permit accurate analysis, and experimentation with poor service conditions violates the operating companies' dedication to providing the best possible service.

Computer simulation techniques have proved to be the best tools for studying network congestion and evaluating network management techniques. Simulation programs can take many forms. The program used to construct Figure 5-11 is a call-by-call simulator; that is, each call entering the simulated network is followed through the network. This simulation shows that, without controls, when the offered load exceeds the engineered load (1600 erlangs) by 50 percent or more (2400 erlangs), the carried load falls below the engineered level. Moreover, as the dashed line indicates, increasing switching capacity and, thereby, eliminating queue delays would significantly improve the overload characteristics.

The results of another run on this simulator are shown in Figure 5-12 and illustrate several principles. This figure shows the time response of the Figure 5-11 network when a 100-percent overload (3200 erlangs) is offered and no controls are used. The vertical scales show both carried load in erlangs and ineffective (blocked) attempts. The carried load is initially about 1800 erlangs, this being the level at which the network was operating when the overload was introduced. As the congestion builds, the number of attempts blocked by trunk congestion increases rapidly. Ineffective attempts due to timeouts (switching congestion), on the other

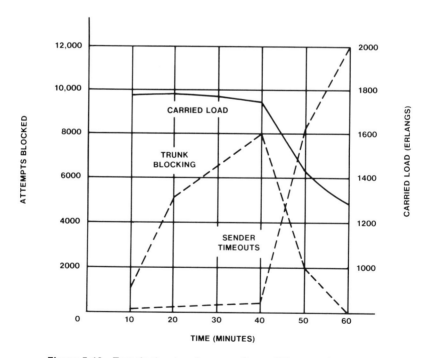

Figure 5-12. Transient network congestion—100-percent overload.

hand, are initially small. As reattempts due to trunk blocking combine with the heavy overload of initial attempts, the switching system becomes congested. Timeouts increase rapidly after 40 minutes and carried load falls off accordingly. As a result of switching congestion, the trunk congestion disappears, since so many short-holding-time calls (namely, digit transmitter timeouts) are using the trunks.

If special network management measures were not taken, the poor performance illustrated in Figure 5-12 might occur several times a year. Christmas Day, Mother's Day, and, occasionally, other holidays, for example, produce intertoll trunk congestion. While the total message volume on these days is well below that of the average business day, the calling pattern is different. Local and intrastate toll traffic is light, but long-haul interstate traffic far exceeds engineered capacity levels. Such skewed patterns of overload also occur in metropolitan areas during snowstorms, for example, when calls from the suburbs to the center city exceed normal levels.

Another type of overload that can cause service deterioration is the *focused overload*—heavy calling from many points to one point. Telephone calling after the Mount Saint Helens eruption in May 1980 is a good example. For a few days following the eruption, call attempts from the rest of the country to the Pacific Northwest Bell area were three to five times higher than they had been on Mother's Day the week before, typically the company's peak day. Similar focused overloads may occur in metropolitan areas when, for example, a radio or television station encourages mass calling, or when, during a power failure, a power company becomes the target for large call volumes.

5.6.4 DETECTION OF CONGESTION

Network congestion is detected and often predicted by interpretation of appropriate traffic data. As shown in Figure 5-12, overloads often begin with trunk congestion. Signals indicating all-busy status for trunk groups are provided in Network Management Centers. In more sophisticated systems, data are used to calculate *attempts per circuit per hour* (ACH) and *connections*[23] *per circuit per hour* (CCH) for trunk groups as frequently as every 5 minutes. ACH and CCH patterns can be revealing. For example, a higher than normal ACH coupled with a normal CCH implies that demand is heavy, but calls are being completed. If CCH becomes higher than normal, however, trunk holding times are short. This implies that ineffective attempts are being switched, and some control may be required.

---

[23] *Connection* implies only that a call attempt has been switched through a switching system to an outgoing circuit. It does not imply that the call has reached the called party.

The most effective indicators of switching congestion are dial-tone delay (for central offices) and incoming digit-receiver delay (for toll and local tandem offices). If excessive dial-tone delay is the result of heavy originating traffic, the network manager can do little. If, however, dial-tone delay at an end office is the result of heavy traffic terminating at that office, as is possible with electronic switching systems (see Section 5.3.1), controls can relieve terminating congestion. When digit-receiver delay becomes excessive, timeouts become a problem (see Figure 5-12). Controls can maintain delays at reasonable levels.

5.6.5 CONTROL OF CONGESTION

Network management controls may be *restrictive* or *expansive*. Restrictive controls eliminate or reduce routing alternatives to prevent traffic from reaching congested areas. Expansive controls permit calls to be completed using additional nonstandard routing to bypass congested trunk groups and switching systems.

Table 5-1 gives some examples of the various controls and a brief description of the function of each. As the table indicates, controls such as directionalization and code blocking are particularly effective during focused overloads. Without such controls during the Mount Saint Helens eruption, for example, incoming traffic would have made it impossible to complete emergency calls from the affected area.

For more general overloads, such as those that occur on Mother's Day and Christmas Day, changing normal routing patterns can be an effective technique. If, for example, the eastern seaboard is experiencing congestion at ten o'clock on Christmas morning, traffic between Eastern time-zone offices can be routed via Central, Mountain, or even Pacific time-zone offices. Because of time-zone differences, circuits to those areas and the offices themselves probably have unused capacity at that time. Under normal circumstances, such circuitous routing would be inefficient, but network managers can temporarily redefine alternate-route sequences when the need arises.

5.7 TRAFFIC MEASUREMENTS AND SERVICE OBJECTIVES

The preceding discussion frequently referred to the various service objectives that govern the engineering of equipment. The existence of such service objectives presumes that measurements will be taken to determine whether the network is meeting them. The interplay between traffic measurements and service criteria is important: There is no point in establishing service objectives that cannot be adequately verified by the measurement process. In fact, the service objectives must be such that

TABLE 5-1

NETWORK MANAGEMENT CONTROLS

Control	Type	Function
Directionalization of Trunk Groups by Making Trunks Appear Busy*	Restrictive	When operated at one switching system, reduces call pressure to a distant switching system. For a 2-way trunk group, additional equivalent 1-way outward trunks for the distant switching system are provided. Commonly used in focused overloads to hold traffic away from the focal point.
Cancellation of Alternate Routing*	Restrictive	Removes alternate-routed traffic from a group and, hence, reduces the load on the distant switching systems. Since alternate routing is reduced, the average number of links per call is also reduced. There are two forms: CANCEL-FROM (selectively cancels traffic overflowing from a high-usage group to another high-usage group or final group); CANCEL-TO (cancels all traffic overflowing to the controlled group). Usually used in peak-day overloads (such as Christmas Day) to increase network efficiency by reducing links per call and to control the amount of alternate-routed traffic that reaches the upper levels of the hierarchy.
Rerouting*	Expansive	Routes overflow traffic to a trunk group not in the normal route advance sequence. Usually used when all normal routes are busy.
Code Blocking*	Restrictive	Blocks calls (routes to reorder tone or to recorded message) according to destination code. Useful during focused overloads, especially if calls can be blocked at or near origination. Blocking need not be total unless the destination office is completely disabled through natural disaster or equipment failure. Switching systems equipped with code-blocking features typically can control 50 percent, 75 percent, or 100 percent of calls to a particular code. The controlled code may be as broad as the destination numbering plan area or as restricted as the office line.
Dynamic Overload Control (DOC)	Restrictive	An automatic control system that senses congestion (as measured by the number of calls waiting for a sender) at a toll or local tandem office and that sends an electrical signal to subtending offices (those that home on it). Responses of the subtending switching systems depend on how they have been programmed or wired. Commonly, subtending offices cancel alternate-routed traffic or make trunks busy. When DOC is properly deployed, about 40 percent of the subtending traffic will be controlled, and a control action may last only about 10 seconds. DOC controls are preferred over manual controls because they are activated as soon as congestion appears and removed as soon as the need disappears.
Selective Dynamic Overload Controls (SDOC)	Restrictive	DOC does not discriminate between traffic with a high completion probability and that with a low completion probability. SDOC combines DOC and code blocking to form a system in which the subtending offices either calculate or receive completion statistics by code, and, if the control office encounters congestion, the subtending offices respond to a DOC signal by canceling alternate routing for hard-to-reach codes or by making trunks appear busy for these codes.

* These controls require manual activity at a key or switch and sometimes involve wiring changes. Hence, they may be applied too slowly and are often left in effect too long.

measuréments can be developed to pinpoint the corrective action that must be taken when objectives are not met.

Intuitively, it is expected that service objectives are set first, then the network is engineered and equipped, and then measurements are taken to evaluate performance. Frequently, with switching equipment in particular, this natural progression is followed. For the traffic network as a whole, however, this progression is not feasible. Limitations inherent in the measurement process can influence both the setting of service criteria and the design of the traffic network itself. The next section discusses some traffic measurement procedures, after which it is possible to explain the motivation behind the apparently arbitrary decision to engineer every final group to 0.01 blocking rather than to design the network to some end-to-end service objective.

5.7.1 TRAFFIC MEASUREMENT PROCEDURES

A key measurement that must be taken for any group of servers is offered load. Not only are load estimates used as the basis for projecting future demands, but when a given service objective has not been met, it is important to know what the load was so that the proper number of additional servers needed can be calculated.

One method of measuring load on a trunk group requires a traffic usage recorder, which scans the trunks every 100 seconds and counts the number of busy trunks. The sum of the counts during the measurement interval, usually an hour, gives a direct estimate in hundred call seconds (CCS) per hour of the usage of the trunk group.[24] This usage measurement, however, estimates carried load, L. To convert the estimate, $\hat{L}$, to one of offered load, it is necessary to use equation (22),

$$\hat{L} = \hat{a}\left[1 - B(c,\hat{a})\right], \tag{22}$$

iteratively to solve for a, the estimate of offered load. This procedure, though, requires an assumption about the peakedness of the offered load. Further, the measurement includes usage on trunks made busy (taken out of service) because of maintenance activity.

Another method of estimating trunk group offered load uses peg count and overflow (PCO) data. *Peg count* refers to the number of attempts offered to the group during the measurement interval; *overflow* refers to the number of attempts encountering all trunks busy. The ratio of overflow to peg count directly estimates blocking,[25] from which offered

---

[24] There is, of course, statistical error because of the discrete scanning.

[25] PCO measurements do not distinguish initial attempts from reattempts; hence, the ratio is only an estimate of first-attempt blocking.

load can be deduced, assuming Erlang B behavior and an estimate of offered load peakedness. The combination of both usage and PCO data gives the best estimates of offered load and grade of service, since peakedness assumptions are not required when both data sources are available.

Similar measurements, usage and peg count, are taken for switching equipment such as incoming digit receivers. Usage data for switching items, however, are typically based on 10-second scan rates, since the holding times of these items are shorter than trunk holding times. For those items that must meet delay criteria, it is also necessary to record delay data.

Older common-control equipment, such as crossbar systems (see Section 10.2), require expensive external measuring devices to record data. For electronic switching systems (see Section 10.3), traffic measurements can be taken directly by a stored program. Such data are more accurate and complete (different types of telephone calls are easily distinguishable, for example) but require processor real time. Therefore, the amount of data taken must be limited so that the system's capacity for call processing is not restricted.

However measurements are taken, there are errors attributable to discrete sampling, maintenance activities, hardware failures, and incomplete knowledge of the underlying traffic parameters (peakedness and reattempts, for example). Also, the more data taken to minimize errors, the more expensive the process of analyzing the data and interpreting the results. The current trend is toward mechanizing data collection and using computer programs for analysis and most of the interpretation.

5.7.2 SERVICE OBJECTIVES

Some service criteria used in the Bell System, such as dial-tone delay and incoming digit-receiver delay, are self-explanatory, deriving directly from the server's function, customer expectation, and the potential effect of poor service on the network. However, setting proper service criteria for the traffic network as a whole is not so simple. Ultimately, the customer is most concerned with the probability that a telephone call will reach its destination, independent of where along the way it might be delayed or blocked. If service objectives were set to govern that probability, a reliable measurement process to estimate the total number of initial attempts between every pair of end offices and the number of those attempts that failed in the network would be needed. Given that data and some end-to-end service objective, the question would be: What to do if the objective were not met? Failed attempts could result from a problem in one or more of many trunk groups from which an end-to-end path might be formed and/or from a problem in one of several switching systems. The specific location(s) and cause(s) of the problem would have to be determined. The amount of data that would have to be collected, coordinated, and analyzed is staggering, as is the problem of deciding upon corrective actions.

From this perspective, the effectiveness of a hierarchical network with its backbone finals engineered to 0.01 blocking becomes clear. The worst service that a call can receive during the average busy hour is roughly 0.09 blocking (there are, at most, nine finals in tandem in any route). Because of the noncoincident busy hours of the network groups and the economical placement of high-usage groups to provide alternate routing, the blocking seen by the customers is substantially less. Empirical studies show that end-to-end blocking averages about 0.02, and almost all the traffic experiences blocking of 0.05 or less under normal busy season conditions.

Equally crucial, the problems of measuring performance and then taking corrective actions are simplified as much as possible. For example, attention is focused on individual network components, rather than on overall network activity with its many possible interactions. If, for example, a given final group is not giving 0.01 blocking, that fact can be determined relatively easily; and solutions are immediately known. The hierarchical design with probability-engineered final groups thus greatly simplifies the problems of measuring and servicing the network and, in so doing, significantly reduces the cost of administering and maintaining that network.

5.8 TRAFFIC CONSIDERATIONS IN DATA NETWORKS

Data communications are growing rapidly both in terms of traffic volume and the diversity of data applications. The Bell System initially met the growing need by developing equipment that allows data to be transmitted over the PSTN, private switched networks, and private-line multipoint networks. Later developments led to increased PSTN capabilities and other networks specifically designed to carry data traffic and support data applications.[26] The following sections describe characteristics of data traffic, performance parameters, and some concepts peculiar to engineering data networks.

5.8.1 THE NATURE OF DATA TRAFFIC

Section 5.2.1 defined the traffic on a telephone network in terms of average call arrival rate and average call holding time. These parameters are also applicable to data networks, although the models that describe them

---

[26] Section 11.6 describes some of the network capabilities the Bell System provides to accommodate the many different types of data traffic.

are different. A simple example of an established data call is a user logged on to a time-sharing computer. The *call rate*, as in telephone networks, is the number of call ("logon") attempts per unit time, and the *call holding time* is the duration from call establishment (logon) to call disconnect ("logoff"). Call holding time, of course, includes times when no data are being transmitted, just as a telephone call holding time includes pauses in conversation.

Call rate and call holding time can vary significantly from one data application to another. An airline reservation desk application, for instance, may require only one call per day—a connection is established at the beginning of the day and is maintained until the end of the day. A reservation clerk may send or receive data many times during that one call. Alternatively, in a time-sharing application, a data call might have to be made each time the application is run.

When data calls are transmitted over networks sensitive to the quantity of data transmitted per unit time, two additional parameters are useful—transaction rate and transaction length. Each time data is transferred to or from a customer, a transaction occurs. There may be many data transactions during a data call (as in the airline reservation desk application) or only one to and from a computer (as in the time-sharing application). Each transaction may itself contain variable quantities of data. The quantity of data is called the *transaction length* and may be described in terms of bits, bytes or characters, or packets.

Packet-switching networks,[27] in particular, are sensitive to the quantity of data in a call and take advantage of the "burstiness" (that is, the data is sent and received in bursts of high intensity interspersed among periods of little or no activity) of that data traffic. In packet switching, data are divided into packets, each of specific format (including destination and control information) and of maximum length. The transmission path is occupied by a particular packet for a time that is dependent on the packet length, and the path is then available for use by other packets that may be transferring data between different data terminals. This contrasts with circuit-switching data networks where a physical end-to-end transmission path is established at the start of the call and is dedicated to the call for its duration. The end-to-end transmission path in packet-switching networks is called a *virtual circuit*. As they traverse this virtual circuit, packets are *switched* on a per-packet basis from one network node (packet switch) to the next, using the specific control information they carry.

Each type of data network is sensitive to call rate, call holding time, transaction rate, and transaction length in a different way. Figure 5-13 depicts the relationship between the four data traffic parameters.

---

[27] Sections 2.5.4 and 11.6.2 discuss packet-switching network services and systems.

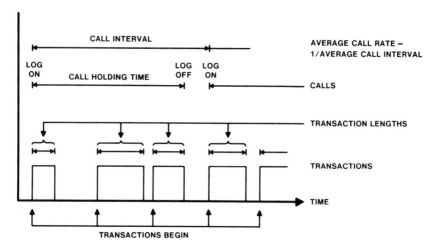

Figure 5-13. Data traffic parameters.

5.8.2 DATA TRAFFIC MODELS

Telephone networks are dominated by voice traffic, facilitating the modeling of call rate and holding time. Modeling traffic on data networks is more difficult. No general models for data traffic exist because of the diversity of data applications and the short experience in measuring data traffic compared to many years of measuring voice traffic. The data source for a data network may be another network or a customer-provided node that concentrates its own end-users. In these cases, the individual data calls would not necessarily be explicitly observable. Call rate and holding time are, thus, modeled only after the particular data application is understood as well as possible.

Data transaction rate and transaction length, however, lend themselves to modeling by the telephone traffic call rate and holding time models, respectively. Transactions are generally expected to arrive according to a Poisson process. Thus, the probability that k transactions arrive in any time interval, t, is given by equation (3), where λ is now the *average transaction arrival rate*:

$$p(k;t) = \frac{(\lambda t)^k}{k!}\, e^{-\lambda t}. \qquad (3)$$

Typical transaction arrival rates are shown in Table 5-2. The rate is assumed to apply to both directions in 2-way data applications (for example, inquiry-response applications) and for one direction in 1-way data applications (for example, batch applications).

TABLE 5-2

SAMPLE DATA TRAFFIC CHARACTERISTICS

	Application			
	Time Sharing	Inquiry-Response	Message	Batch
Call rate (average number of calls per busy hour)	0.50	0.25	3.33	0.15
Call holding time (average number of minutes)	12	16	3	11
Average number of transactions per call	10	10	1	1
Transaction rate (average number per busy hour)	5.00	2.50	3.33	0.15
Transaction length (average number of characters per transaction)				
In	35	120	600	40,000
Out	200	200	–	–

Assuming the transaction has arrived, data transaction length in characters is treated as a geometric probability function (see Fuchs and Jackson 1970). As in the assumptions for equation (4), it is assumed that the probability of a call ending in a small interval, t, is a constant, μt, and that the probability that the character last received is the last one in the transaction is a constant, $1/c$. Under these assumptions, the probability of a transaction having m characters is:

$$P(m) = \frac{1}{c}(1-\frac{1}{c})^{m-1} \qquad m = 1, 2, 3, ..., \qquad (23)$$

where c is the average number of characters in a transaction.

For packet-switching networks, the data transaction length is stated in terms of the number of packets, since packet-switching systems are more sensitive to the number of packets than to the number of characters. Assuming a geometric distribution of transaction lengths, if the characters of a transaction are loaded into packets D at a time, except for the last packet, which may have any number from 1 to D, the distribution of packets per transaction will also be geometric. The probability, given a packet contains at least one character, that this packet will be the last one in the transaction is one minus the probability that D or more characters remain in the transaction. From equation (23), this probability is $(1-1/c)^D$. Therefore, the probability that any packet, once stated, will be the last one is constant—the condition for the geometric probability function. The number of packets per transaction is, therefore, a geometric distribution with mean N:

$$N = \frac{1}{1-(1-\frac{1}{c})^D}. \tag{24}$$

The data traffic models are useful as inputs to other models of the data network and network components. They can be used directly for engineering access lines and switch access-line ports and as direct inputs, for example, to load-service models of access lines and switch access-line ports. Traffic loads at other data network components (for example, at interswitch trunks or switch trunk ports) can be derived from the data traffic models and models of the network components through which the data traffic flows.

There is no universal flow model applicable to all networks, since the model depends on the network and packet-switching system architecture, method of operation, and traffic-handling capabilities. The models of data traffic and related models for data network components can be used to design, for example, new data network control strategies and to specify data network engineering rules. The engineering process includes adjustments to network and component sizing that are based on data network measurements.

5.8.3 DATA PERFORMANCE CONCERNS

Telephone traffic performance concerns, such as call blocking, are also relevant to data traffic. In addition, data accuracy, throughput, and network delay are important in data network performance. Data accuracy and throughput are concerns in both circuit switching and packet switching, whereas network delay is usually a concern in packet switching.[28]

[28] The recommendation of categories and levels of performance for packet-switching networks is a current activity of the Comité Consultatif International Télégraphique et Téléphonique (CCITT) 1980-1984 plenary period. Performance categories and levels are reasonably well understood at this time, although the 1980 recommendations treat only a few performance categories.

Accuracy is only indirectly related to traffic engineering through objectives such as lost packet rate. Throughput and network delay, allocated on a network component basis, are closely coupled to traffic engineering.

Data Accuracy

Data accuracy objectives are typically bit- or character-related or are reflected in a specified number of error-free seconds as in the Digital Data System (see Section 11.6.1). On packet-switching networks, accuracy performance objectives also include errored, misdelivered, lost, duplicated, and out-of-sequence packet rates. Typically, rates for each of these are on the order of 1 in 10^9 to 1 in 10^7 under normal service conditions.

Throughput

Throughput is the quantity of data a user can transfer during a time interval. On packet-switching networks, the end-to-end transmission path (hence, channel capacity) is not dedicated. Contention for the channel capacity occurs when multiple users attempt data transmission. Thus, throughput and the related performance objectives are functions of the network parameters that control the sharing of the channel capacity. Once a call is established on circuit-switching data networks, the end-to-end transmission path, or channel capacity,[29] is dedicated to the data call. In both types of switching networks, the achievement of throughput objectives also depends on the round-trip network delay.

Network Delay

Network delay is usually measured from the last character input to the network to the first character received from the network. In both circuit-switching and packet-switching networks, network delay is a function of data transmission speed, propagation time, and distance through the network. In packet-switching networks, network delay is also a function of packet-switch processing time and queuing delay, both of which can be influenced by traffic engineering. Other factors related to network delay either are under customer control (for example, transmitting at different data rates) or are determined by network design (for example, limiting end-to-end transmission paths to a small number of packet switches).

---

[29] Bandwidth for analog channels, bit rate for digital channels. Section 6.2 describes analog and digital channels.

Network delays are usually divided into call-related times (for example, call establishment) and data transfer times.[30] Typical packet call-related times are approximately 500 milliseconds. Packet data transfer times typically range from 150 to 500 milliseconds.

5.8.4 TRAFFIC-ENGINEERING CONCEPTS

The quantity of data traffic carried by the PSTN is small compared to the voice telephone traffic it carries. Therefore, little or no explicit planning is done for data traffic. If it is known that a significant quantity of data traffic will be present (for example, when a large data user moves into a new area), planning may include accommodating the data traffic on the PSTN, engineering special data networks, or both. This section addresses traffic-engineering concepts for data networks. Moreover, since engineering circuit-switching data networks is similar to engineering telephone networks, only concepts related to packet-switching data networks are described.

Typically, data networks are engineered to support some *peak* busy hour condition. *Extreme value engineering* (see Barnes 1976) is sometimes used. Another method models the highest traffic hour of the busiest day considered by excluding some number of busiest days of the year and then multiplying average busy hour traffic by a factor from 1.2 to 1.5.

Estimates of peak busy hour traffic and performance are used to engineer data network transmission facilities and packet switches. The data traffic models in Section 5.8.2 are a basis for the engineering rules and models. For example, peak busy hour traffic and performance estimates are applied to the rules and/or models to predict time of exhaust for packet switches and other network components and to determine appropriately sequenced network growth steps.

Network access lines are engineered using the total traffic entering the network at the access points. Network trunks and switches are engineered using the total traffic entering the network at the access points and the traffic flow patterns. Network access lines and trunks are engineered to a utilization level, typically from 50 to 70 percent in packet-switching networks, to achieve allocated network performance objectives.

Packet-switch engineering involves determining the appropriate switch configuration so that the allocated network performance can be achieved. Typical resource limits on state-of-the-art packet switches include maximum number of ports, throughput of each port, switch real-time processing, and port-and-switch memory. Ideally, all of these

---

[30] *Data transfer time* is the network delay for switching data and does not include, for example, routing and address translation time as does call establishment.

resources should exhaust simultaneously when switching system traffic reaches a certain level. This rarely happens, though, and some limits are never reached. Typically, packet-switch real-time processing is most critical and is used as the engineering load-versus-service (for example, cross-switch delay) relationship.

AUTHORS

N. Farber
R. A. Farel
W. S. Hayward
V. S. Mummert
S. H. Richman

6

Transmission

Chapters 3 and 4 introduced the network elements (for example, switching systems and transmission facilities) involved in transmitting information over the telecommunications network. This chapter introduces some of the fundamental transmission concepts and principles underlying the way those physical facilities carry information. Sections 6.1 through 6.3 describe types of **signals,** the concept of a **channel**, and important **transmission media**, respectively. Sections 6.4 and 6.5 discuss two processes used to prepare signals for transmission (**modulation** and **multiplexing**) and some of the techniques used in the Bell System. The last section describes some **impairments** that degrade the performance of transmission facilities, how the impairments are controlled to meet performance **objectives**, and how transmission performance objectives are allocated to the network components.

6.1 SIGNAL TYPES AND CHARACTERISTICS

The telecommunications network can transmit a variety of information, which originates in either of two basic forms: **analog signals** and **digital signals.** *Analog signals* are in the form of a continuous, varying physical quantity (for example, voltage) that reflects the variations of the signal source in time (for example, changes in the amplitude and pitch of the human voice during speech). Speech and video signals are examples of analog signals. *Digital signals*, on the other hand, are discrete and have well-defined states at any point in time. Telegraph signals and pulses transmitted from a dial-pulse telephone are examples of digital signals. An increasingly common digital signal is that originated by computers and data terminals. It is important to note that signals may originate in one form, but may be converted to the other form for transmission over the network (see Sections 6.4 and 6.5). The following sections describe three types of analog signals (**speech, program,** and **video**) and one type of digital signal (**data**).

6.1.1 SPEECH SIGNALS

The speech signal is the most common signal transmitted over Bell System facilities. The transmitter in a telephone set transforms sound waves (*acoustic signals*) generated in the speaker's larynx into electrical analog signals. The analog waveform[1] of a typical speech signal can be defined in terms of frequencies over a band—from 30 hertz (Hz) to 10,000 Hz, or 10 kilohertz (kHz)—with most of the energy between 200 Hz and 3.5 kHz. It is not necessary, however, to reproduce speech waveforms precisely to achieve acceptable transmission quality because the human ear is not sensitive to fine distinctions in frequency, and the human brain can correct for inaccuracies such as missing syllables. Since transmission costs are directly related to the range of frequencies transmitted (*bandwidth*), the electrical signal used to provide commercially acceptable quality for telephone communications is, thus, *bandlimited* to the range of frequencies in which most of the energy occurs—between 200 Hz and 3.5 kHz (see Figure 6-1).

The characteristics of the speech signal that vary with time are not easily described. For example, the dynamic range is large; that is, the amplitude varies rapidly and widely from moment to moment and from speaker to speaker. There are talk spurts followed by silent intervals during which the speaker pauses for breath or stops to listen to the other person.[2] Because of these variations, it is difficult to evaluate the speech signal precisely, and the design of transmission systems capable of achieving acceptable voiceband channel performance relies on measured survey data. (Section 6.2 discusses channels.)

6.1.2 PROGRAM SIGNALS

Program signals include high-fidelity radio broadcasts, the audio portion of television programs transmitted between a broadcaster's studio and the station transmitters, and "wired music system" material that is distributed to subscribing customers. The average volume, dynamic range, duration, and bandwidth of program signals are usually greater than for ordinary telephone speech signals. The bandwidth of a program signal, for instance, may be as narrow as that required when speech alone is transmitted, such as during a newscast; or it may be as wide as 35 Hz to 15 kHz when high-quality music for frequency modulation (FM) stereo

---

[1] Green 1962 provides a good introduction to the concept of waves and their characteristics.

[2] The Time Assignment Speech Interpolation (TASI) System exploits this characteristic of human conversation. TASI switches one conversation onto the transmission facility during the idle time that occurs in another conversation.

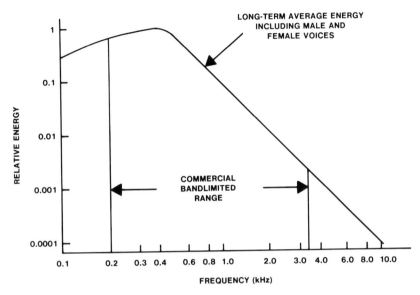

Figure 6-1. Average speech signal energy. Experiments have shown that frequencies below 1 kHz contain the intelligibility information and the higher frequencies convey the articulation and improve the naturalness of the received signal. There is a significant amount of energy below 200 Hz, but these frequencies are excluded because they are subject to noise and inductive interference (see Section 6.6.1).

broadcasts is distributed. The energy distribution is also difficult to specify because of the variety of program material transmitted (for example, speech, music, sound effects).

6.1.3 VIDEO SIGNALS

Most video signals carried over Bell System facilities today are color television signals. The color information (hue and saturation) is combined with the black and white (luminance) information in the television picture and encoded into a highly complex electrical signal for transmission. The picture image is scanned rapidly at the transmitter in a systematic manner and then reproduced at the television receiver.

The synchronous scanning process used in the United States is illustrated functionally in Figure 6-2. The basic process consists of a sequence of nearly horizontal scanning lines that run from left to right, beginning at the top of the picture image. When one scan, or field, reaches the bottom of the image, the process is repeated, the next set of scanning lines falling between the lines from the previous field. Two successive fields total 525 scan lines and make a complete picture frame. At a rate of 30 frames (15,750 scan lines) per second, the received picture appears not to flicker because of the persistence of the human eye.

To permit decoding of the video signal at the receiver, it is necessary to indicate the beginning of horizontal scanning lines and fields. This is done by transmitting synchronizing pulses interleaved with the picture information. To prevent the synchronizing pulses and traces of returns to the next horizontal scan line from being visible on the screen, the picture signal is driven to a very low brightness level during the intervals when pulses are transmitted. Since synchronizing pulses are essentially digital signals and picture information is an analog waveform, video signals may be thought of as *hybrid signals.*

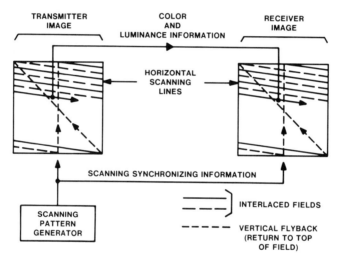

Figure 6-2. Broadcast television scanning process. Horizontal picture resolution is determined by the finest detail that can be resolved along a line and vertical resolution by the number of horizontal lines per frame. Subjective viewing tests have determined that a signal bandwidth of about 4.3 megahertz (MHz) is required to achieve horizontal resolution about equal to the vertical resolution.

6.1.4 DATA SIGNALS

The signal produced by a computer or data terminal usually consists of a stream of pulses that represents information coded into binary digits, or *bits* (for example, 0 and 1). Unlike the speech signal, which contains much inherently redundant information, the basic data signal has little or no redundancy. For protection, bits are often added to the basic signal to permit error detection (error-detecting codes). By adding more bits (error-correcting codes), it is possible to correct many errors (that is, to reconstruct the original signal without retransmitting the data).

Since data signals are often transmitted over systems that were originally designed for speech signals alone, the data signals must be processed at the source in a manner that ensures compatibility with the system. A train of "rectangular" digital pulses (see Figure 6-3, *top*) is not directly suitable for transmission over analog facilities because it has frequency components extending from 0 frequency (dc) to a high frequency (determined by the sharpness of the pulses and the rate at which they are transmitted). To fit the bandwidths available in the analog transmission channels, high-frequency components are restricted by modifying the shape of the pulses (see Figure 6-3, *bottom*) and limiting the pulse rate. The pulse rates most commonly used in the Bell System are those compatible with transmission over voiceband channels and range up to several thousand pulses per second.

The characteristics of actual analog channels are accommodated by changing the form of the signal through a process called *modulation*. An example of modulation is *frequency-shift keying* (FSK), a technique in which the transmitted frequency is shifted back and forth between 1.2 kHz (corresponding to the presence of a pulse or a binary 1) and 2.2 kHz (corresponding to the absence of a pulse or a binary 0). (Section 6.4.4 contains more information on FSK.)

Digital facilities, coming into increasing use, can be a more effective means for transmitting data signals. Even so, the digital pulse stream requires processing at the source so that it fits into the format of the Digital Data System (see Sections 6.6.2 and 11.6.1) or the time-division multiplex hierarchy (see Sections 6.5.2 and 9.4.3).

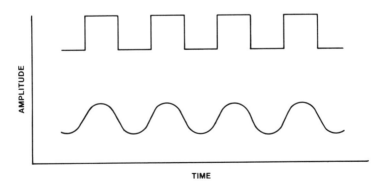

Figure 6-3. Digital pulse streams. *Top,* rectangular pulses; *bottom,* modified pulses.

6.1.5 OVERVIEW

In summary, signals may be characteristic of the producer or the receiver of the signal (for example, a person or a teletypewriter). They may be analog, digital, or, to some degree, hybrid (for example, video signals). And signals may be converted from one form to another to make them compatible with the transmission path over which they will travel with no significant loss of information.

6.2 CHANNELS

6.2.1 BASIC CONCEPT

A *channel* is a transmission path for providing communications between two or more points within the network or between the end points of a connection (for example, from customer to customer or from one customer to several customers).[3] A channel may be dedicated to a particular customer full time, may be switchable, or may even be changed in assignment during a call. A *channel* is the *means* by which information is transmitted over the network; the *signal* transmitted is that *information*. This distinction is important to remember throughout the rest of this discussion.

Channels may be classified by their intended use and the corresponding kind of signal transmitted, for example, voiceband channels, program channels, telegraph channels, and video channels. Signal requirements may dictate the channel characteristics required to meet transmission objectives; for example, 1-way broadcast television requires wideband channels to carry the signal bandwidth described earlier. In other cases, the signal format that is actually transmitted over the channel may have to be tailored to meet the characteristics of the channel. For example, a data signal must be processed (coded, pulse-shaped, modulated) to be suitable for transmission over a voiceband channel in the public switched telephone network (PSTN).

Channels may also be classified by the broad nature of the transmission facility that provides the channel. If the transmission facility accepts a band of frequencies and is compatible with the transmission of analog signals, it is said to provide analog channels, and the facility is called an *analog facility*. Analog channels can be further characterized by *bandwidth*: narrowband channels (for example, 100 Hz, 200 Hz); voiceband channels (4 kHz);[4] broadband channels (for example, 48 kHz, 240 kHz).

---

[3] The term *multipoint* is used for connections involving more than two end points.

[4] The nominal voiceband channel is defined as 4 kHz although the speech signal is essentially bandlimited to between 200 Hz and 3.5 kHz. The additional bandwidth allows for a guard band on either side of the speech signal to lessen interference between channels.

If the transmission facility accepts a train of pulses and carries signals in digital form, it is said to provide digital channels and is called a *digital facility*. Digital channels can be further characterized by *pulse rate* or *bit rate*. For example, channels with a bit rate of 64,000 bits per second, or 64 kilobits per second (kbps), are often used in the Bell System. Digital signals at lesser bit rates (2.4, 4.8, 9.6, and 56.0 kbps) may be carried on these channels. For the lower bit rates, several signals may be combined (multiplexed, as described in Section 6.5) onto the 64-kbps channel.

6.2.2 VOICE-FREQUENCY AND CARRIER-DERIVED CHANNELS

Channels in the Bell System network are provided by either **voice-frequency** (VF) or **carrier** transmission facilities. A VF facility is an analog facility that provides just one voiceband channel. In this case, the voiceband channel is termed a *baseband channel* because the information signal is carried at its original frequencies. VF facilities are operated in either a 2-wire or a 4-wire mode. In the 2-wire mode, both directions of transmission are carried on the same pair of wires. Local loops and short trunks are operated in this way to save copper wire and to be compatible with local 2-wire switching systems, as shown at the top of Figure 6-4. Because long 2-wire facilities requiring excessive gain would suffer from instability[5] and singing,[6] longer VF trunks and certain VF special-services

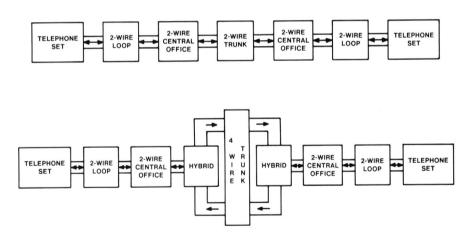

Figure 6-4. Voice-frequency 2-wire and 4-wire channels. *Top*, a 2-wire channel; *bottom*, a combination 2-wire/4-wire channel.

---

[5] The reflection of some portion of the transmitted signal back toward its source in sufficient amplitude causes gain instability (see Section 6.6.1).

[6] An undesirable self-sustained oscillation in a channel resulting from excessive positive feedback. In a telephone connection, singing manifests itself as a continuous whistle or howl.

circuits (see Section 3.2.2) are often operated on a 4-wire basis (that is, with a separate transmission path for each direction of transmission). As shown at the bottom of Figure 6-4, a transmission circuit element called a *hybrid* is necessary as an interface between short-distance 2-wire trunks and long-distance 4-wire trunks.

In carrier transmission, signals are processed and converted to a form suitable for a particular broadband medium. The facility and, correspondingly, the channels may be analog or digital. A number of channels are combined (multiplexed) in one carrier system, and they are called *carrier-derived channels* (see Figure 6-5). Providing a number of channels on one facility results in savings (for example, in copper wire) for trunk lengths greater than some prove-in length. Therefore, although the electronics required to combine channels may be expensive, carrier systems become increasingly economical as their length and channel cross section increase.

Carrier systems consist of three major functional building blocks, as shown in Figure 6-5:

- **Multiplex/demultiplex terminals** combine voiceband channels at the carrier system input and separate them at the output.

- **Carrier facility terminals** process signals into a form suitable for transmission over the high-frequency line and process signals that have been transmitted over the high-frequency line into a form suitable for separation.

- The **high-frequency line** provides a broadband facility for the simultaneous transmission of a large number of voiceband channels. In this

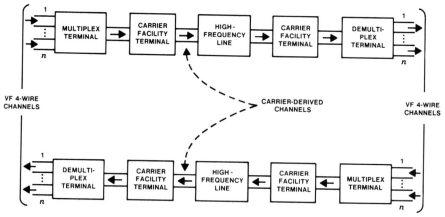

VF VOICE FREQUENCY
n ANY NUMBER

Figure 6-5. Example of carrier-derived 4-wire channels.

context, microwave radio qualifies as a high-frequency line, as does a pair of wires in the T1 carrier system (see Section 9.4.2).

Carrier-derived channels operate in the 4-wire mode with a separate path (for the combined channels) for each direction of transmission. Most multichannel carrier systems on paired cable (see Section 6.3.2) use a separate pair of wires for each direction of transmission. Some short-haul carrier systems and submarine (undersea) carrier systems, on the other hand, use an "equivalent 4-wire" mode of operation. Here, as in certain radio systems, the paths for the two directions of transmission are separated by frequency rather than physically.

6.3 TRANSMISSION MEDIA

Transmission media may constrain and guide communications signals or permit signals to be transmitted but not guide them. Five types of guided media used in the Bell System are **open-wire lines, paired cable** (also called *multipair cable*), **coaxial cable, waveguides,** and **lightguide cable**. The atmosphere and outer space are unguided media for transmitting **terrestrial microwave radio** signals and **satellite** communications, respectively. This section describes fundamental applications and basic advantages and disadvantages of each medium.

6.3.1 OPEN-WIRE LINES

Open-wire lines consist of uninsulated pairs of wires supported on poles spaced about 125 feet apart. Five open-wire pairs may be mounted on a single crossarm on a pole with a foot of space between wires to prevent momentary short circuits in high winds. Most poles are limited to about ten crossarms. The wires range in diameter from 0.08 inch (12 gauge) to 0.165 inch (6 gauge) and may be copper, copper-clad steel, or galvanized steel.

Open-wire lines have been largely supplanted by paired cable (see Section 6.3.2), except in some rural areas, because of the limited number of wire pairs that can be accommodated on a pole; the susceptibility of the wires to storm damage, atmospheric corrosion, and interference from power lines that share the poles; and the high cost of maintenance. A singular advantage of open-wire is its low attenuation[7] at voice frequencies compared with paired cable.

6.3.2 PAIRED CABLE

Paired cable is composed of wood-pulp or plastic-insulated wires twisted together into pairs. In some cable, many twisted pairs are stranded into a rope-like form called a *binder group*; several binder groups are, in turn,

---

[7] A decrease in the signal amplitude during transmission.

twisted together around a common axis to form the cable core; and a protective sheath is wrapped around the core. The amount of twist applied is varied among pairs within each binder group to reduce crosstalk.[8] Paired cable is manufactured in a number of standard sizes and may contain from 6 to 3600 wire pairs. Typically, wires range in diameter from 0.016 inch (26 gauge) to 0.036 inch (19 gauge). Figure 6-6 shows typical paired cable.

The main applications for paired cable are in the loop and exchange areas. The cable may be strung on poles, installed in underground conduit, or buried directly in the ground. To protect the cable in these

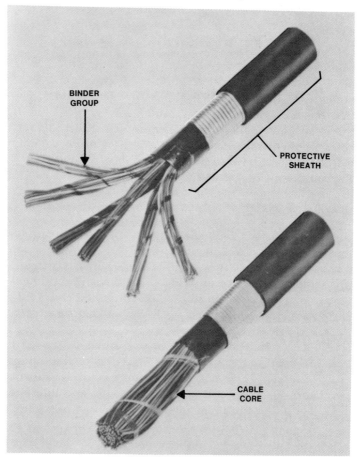

BINDER
GROUP

PROTECTIVE
SHEATH

CABLE
CORE

Figure 6-6. Typical paired cable.

[8] The interference in one communication channel caused by a signal traveling in an adjacent channel (see Section 6.6.1).

environments, sheaths of different materials or combinations of materials are used. For example, cables that may be exposed to water are sheathed in aluminum, steel, and polyethylene coated with a waterproofing compound. In addition, they are filled with a petroleum compound to keep out water if the sheath is damaged. Most older paired cable is made waterproof with lead sheathing and pressurized air within the cable core. The lead sheath also provides protection from noise, but it is subject to corrosion and ultimately may admit moisture.

On a given facility route, more circuits can be accommodated by paired cable than by open-wire lines. Increased circuit demands have also led to the extensive use of paired cable in carrier systems where, however, the higher frequencies result in greater crosstalk between pairs within a cable. To combat crosstalk, it is sometimes necessary to use different cables for the two directions of transmission, to select pairs within the same cable for opposite directions of transmission according to special engineering rules, or to use a cable with an electrical screen separating binder groups.

6.3.3 COAXIAL CABLE

Coaxial cable contains from four to twenty-two coaxial units called *tubes*. Each coaxial tube consists of a 0.100-inch copper inner conductor kept centered within a 0.375-inch cylindrical copper outer conductor by polyethylene insulating disks spaced about 1 inch apart. The outer conductor is formed into a cylinder around the disks and is held closed by interlocking serrated edges along its longitudinal seam. Two steel tapes are wound around the outer conductor for added strength. In addition to coaxial tubes, coaxial cable contains a small number of twisted wire pairs and single wires that are used for maintenance and alarm functions. Figure 6-7 shows typical coaxial cable.

An important advantage coaxial cable has over paired cable is its capability to operate at very high frequencies, which permits it to carry a relatively large number of carrier-derived channels. In addition, since the copper outer conductor is grounded and provides a shielding that improves with increasing frequency, crosstalk decreases rather than increases with frequency as is the case with paired cable.

Coaxial cable is used primarily on intercity routes in the long-haul network (see Section 4.4.2) where heavy cross sections of traffic exist. In these applications, the installed first cost (materials, manufacturing, labor, etc.) *per circuit* is substantially lower than for paired cable, although the *overall* installed first cost is much higher than for paired cable.

Coaxial cable is also used in undersea cable systems (see Section 9.3.4). However, the unique operating environment dictates design, operational, and reliability requirements different from those for cable used on land.

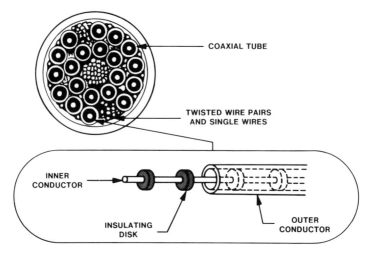

Figure 6-7. Typical coaxial cable. (*Redrawn and adapted with permission from Martin 1976, p. 162.*)

6.3.4 WAVEGUIDES

A waveguide is a rectangular or circular copper pipe that confines and guides radio waves between two locations. Its main advantage is very low attenuation at microwave frequencies compared, for instance, with coaxial cable. Very wide bandwidths can be achieved as well. The application of waveguide is limited, though, because (1) it must be manufactured to extreme uniformity and (2) extreme care is required during installation to minimize sharp bends for hilly terrain or right-of-way curvature, which result in transmission mode[9] conversion and increased attenuation.

Circular waveguides have been used to transmit 4-, 6-, and 11-gigahertz (GHz)[10] microwave signals simultaneously from the bases of microwave radio towers to the antennas on top. (The position of a circular waveguide is shown in Figure 6-10.) The waveguide diameter is chosen to achieve low attenuation across this band of frequencies, considerably lower attenuation than for the same length of rectangular pipe. In addition, the circular cross section of the waveguide permits the transmission of two signal polarizations so that two signals at the same frequency

---

[9] A transmission mode is a particular form of signal propagation within a waveguide characterized by the configuration of the electromagnetic field in the plane transverse, or perpendicular, to the direction of propagation. Major modes are the transverse electric (TE) and the transverse magnetic (TM). In the TE mode, the electric field is perpendicular to the axis of the waveguide and the magnetic field is parallel to it.

[10] One gigahertz (GHz) equals one billion hertz.

can be transmitted simultaneously, effectively doubling the channel capacity.[11]

6.3.5 LIGHTGUIDE CABLE

In lightguide transmission systems, an electrical signal is converted to a light signal by a light source (for example, a light-emitting diode or a laser) and then coupled into a glass fiber. Two types of lightguide cable are currently in use: ribbon-fiber cable and stranded cable. One form of ribbon-fiber cable consists of from one to twelve flat ribbons, each containing twelve glass fibers, thereby providing up to 144 one-way paths for optical signals. As shown in Figure 6-8, the ribbons are stacked into a rectangular array and twisted to improve the flexibility of the cable. Two sheaths, both consisting of polyethylene with steel reinforcing wires, protect the hair-thin fibers (0.005 inch) from excessive pulling tensions. The cable diameter is about 0.5 inch, regardless of the number of optical fibers it contains. Single-unit stranded cable has up to 16 fibers woven around a central strength member. A 72-fiber multiunit cable consists of six single units, each having 12 fibers.

Optical fibers (see Figure 6-9) are composed of three concentric cylinders made of dielectric materials:[12] the core, the cladding, and the jacket. The jacket is a light-absorbing plastic that prevents crosstalk and protects the surface of the cladding. The core and the cladding are transparent glass designed to guide the light within the core and to reflect it

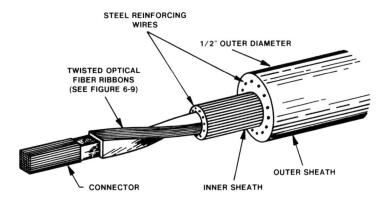

Figure 6-8. Typical ribbon-fiber lightguide cable.

[11] Signal polarization refers to the direction of a signal's electric field. The number of signals carried by a channel may be doubled if the polarizations of two signals are perpendicular to one another.

[12] Materials that do not conduct electrical current.

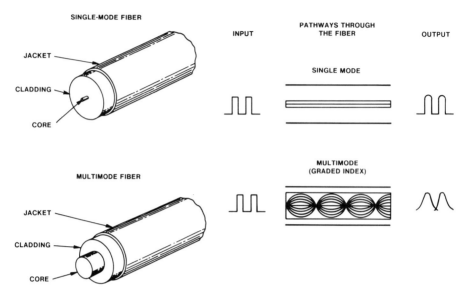

SINGLE-MODE FIBER

JACKET

CLADDING

CORE

MULTIMODE FIBER

JACKET

CLADDING

CORE

INPUT

PATHWAYS THROUGH
THE FIBER

OUTPUT

SINGLE MODE

MULTIMODE
(GRADED INDEX)

Figure 6-9. Optical fibers.

as it grazes the interface between them. The refractive index[13] of the core is greater than that of the cladding, so light in the cladding moves faster than light in the core. As a light signal in the core moves toward the cladding (from a region of lower velocity to one of higher velocity), it is bent back toward the core and is, thus, guided along the fiber.

The two basic types of optical fibers are **single mode** and **multimode**. The core diameter of a *single-mode* fiber is sufficiently narrow to restrict light to one mode of propagation (that is, light travels only one path through the fiber). Because of this restriction, transmission over single-mode fibers is not subject to modal dispersion,[14] but it is more difficult to couple light from one single-mode fiber to another. *Multimode* fibers have much wider cores that allow light to enter at various angles and travel through the fiber along different zigzag paths of varying lengths. Since the path lengths are different, so are the times required to travel them at the same speed (that is, with a uniform index of refraction). This results in modal dispersion. To reduce dispersion, the index of refraction is graded across the core (that is, the index of refraction in the core

---

[13] The ratio of the velocity of a lightwave in free space to its velocity in a given medium.

[14] Signal distortion caused by multiple modes combining at the output of a fiber and spreading out the signal in time. For example, if the signal were in the form of pulses, the pulses would spread out and, thus, overlap causing them to be less distinguishable at the receiving end of transmission.

decreases as distance from the center of the core increases). Thus, light traveling a shorter path near the center moves more slowly, while light traveling the longer path moves more quickly. The result is that the travel times tend to be equal.

The main advantages of lightguide cable are:

- immunity to electromagnetic interference and crosstalk.

- small size (0.5-inch outer diameter) and light weight (approximately 80 pounds per 1000 feet of cable). Paired cable of equivalent capacity is approximately 3 inches in diameter and weighs 100 times as much.

- low attenuation allowing repeaters[15] to be positioned far apart. Maximum spacings with current technology are 25 miles for single-mode fibers and 15 miles for multimode fibers.

- high communication capacity. With currently available systems, 144-fiber cable can accommodate approximately 250,000 voice-frequency circuits. This capacity will increase rapidly in the future as a result of new technology.

- wide bandwidth: 100 GHz for a single-mode fiber and 1 to 2 GHz for a multimode fiber at the longer wavelengths.

In many potential applications, the advantages outweigh the disadvantages. These applications include intraoffice data busing between computer peripherals and equipment frames in digital switching systems, loop carrier systems, medium- and large-capacity interoffice trunks, large-capacity long-haul intercity routes, special video hookups, and transoceanic cable systems.

6.3.6 TERRESTRIAL MICROWAVE RADIO

The terrestrial microwave radio medium consists of a line-of-sight propagation path[16] through the earth's atmosphere and associated tower-mounted transmitting and receiving antennas shown in Figure 6-10. Microwave energy can be focused into a narrow, strongly-directional beam, similar to a beam of light. Antenna towers must, therefore, be rigid enough to withstand high winds without excessive antenna deflection, which results in increased attenuation. The Federal Communications Commission (FCC) makes a portion of the radio-frequency bands of the electromagnetic spectrum available for common-carrier service.

---

[15] Devices that perform amplification or regeneration and associated functions. In transmitting digital signals (for example, lightguide signals), repeaters are regenerative; that is, they reconstruct each digital pulse to its original form.

[16] The straight path between the radio transmitting antenna and the receiving antenna.

ANTENNAS

CIRCULAR
WAVEGUIDE

Figure 6-10. A typical microwave radio station.

The Bell System applications have been primarily in the 4-, 6-, and 11-GHz bands. Under normal atmospheric conditions, the received microwave energy on a line-of-sight path decreases in proportion to the square of the distance between two radio towers. The decrease in energy is relatively constant over a wide band of frequencies.

The two main characteristics of radio transmission are that no physical facility is required to guide the microwave energy between repeater stations and that these stations may be positioned many miles apart. The spacing between repeater stations is determined by the geography of a given route, the technology used in the terminal equipment, and the transmitter power permitted by the FCC. Typical repeater spacings in the Bell System are 20 to 30 miles, but much longer spacings are possible where there is little *fading* activity. (Fading is discussed in the following paragraphs.) Microwave radio, thus, has the advantage of being able to span natural barriers such as rugged or heavily wooded terrain and large bodies of water.

However, using the earth's atmosphere as the transmission medium results in problems not experienced with other media. For example, heavy ground fog or very cold air over warm terrain can cause enough atmospheric refraction to reduce the power of the signal noticeably. This fading affects a wide band of frequencies and may last several hours. The remedy is to use higher antennas or position them closer together.

A second type of fading, called *multipath fading*, occurs mostly at night and at dawn during the summer, when there is no wind to break up atmospheric layers. Normally, microwave energy radiated outside the line-of-sight path (called *off-axis energy*) is bent by atmospheric refraction into the receiving antenna. This energy travels a longer path than the line-of-sight signal and, therefore, has a longer travel time. Depending on the amount of off-axis energy and its phase (the amount of delay relative to the primary signal), instantaneous reduction and even cancellation of the primary signal may occur. Multipath fading may recur frequently but lasts only a few seconds and, generally, affects only certain frequencies. It can be minimized by using *frequency diversity*.[17] Since the use of frequency diversity requires more of the limited radio-frequency spectrum available, it is generally not applied to low-density routes. An alternative approach, *space diversity*,[18] is used in this case and for paths that experience severe fading.

Another problem is reduction of received signal power from rain attenuation, which is particularly severe at 11 GHz and higher frequencies. Greater absorption of microwave energy occurs with increasing frequencies as the wavelength of the signal approaches the size of the rain drops. Neither frequency diversity nor space diversity is effective in countering rain attenuation. The only remedy for maintaining acceptable transmission quality is the use of shorter repeater spacings.

6.3.7 SATELLITE

The transmission medium of a satellite system consists of a line-of-sight propagation path from a ground station to a communications satellite and back to earth. The ground station includes the antennas, buildings, and electronics needed to transmit, receive, and multiplex signals. The satellite is usually placed in a geosynchronous orbit (about 22,300 miles above the earth) so that it appears stationary from any point on the earth. It

---

[17] Transmission of the same radio signal over different microwave frequencies at the same time. For a given set of atmospheric conditions, radio signals at different frequencies will experience different degrees of multipath fading.

[18] Space diversity uses two receiving antennas usually mounted on the same tower and separated vertically by several wavelengths. By switching from the regular antenna to the diversity antenna whenever the signal level drops, the received signal is maintained nearly constant.

acts like a radio relay station in the sky and uses the same frequency spectrum as terrestrial microwave radio.

The satellite's microwave antennas are directional and may cover the whole earth or just a part. A beam of energy from the satellite about 17 degrees wide will just cover the earth's surface. To cover a portion of the earth's surface only 500 miles across, a beamwidth of 1 degree would be required. While narrow beamwidths result in greater received power, they require sophisticated antenna control systems to keep the antennas oriented properly. One method is to spin the satellite in one direction and its antennas in the opposite direction at an equal rate.

One advantage of satellite transmission is the capability to send large amounts of information to almost anywhere on the earth, regardless of how remote or distant. Another is that transmission cost is independent of the distance between sending and receiving locations. Further, each satellite system requires only one satellite repeater. A comparable terrestrial microwave system between distant locations requires many repeaters placed in tandem because the curvature of the earth interferes with the line-of-sight transmission path. Amplification of the signal by each repeater increases the effects of distortion and noise. In some cases, this characteristic of satellite systems may make transmission via satellite superior to transmission via terrestrial microwave radio.

However, because of the great distance between the earth and the satellite repeater, satellite transmission produces greater attenuation and longer delays (approximately 0.25 second one way) than transmission using terrestrial systems. The effects of attenuation can be overcome by using high-gain repeaters and high path-elevation angles (above 20 degrees) for the ground antennas (to reduce the path length through the atmosphere) together with narrow-beam transmission. In the 4- and 6-GHz frequency bands, attenuation is not a significant problem. At frequencies above 10 GHz, however, rain attenuation is a problem on satellite paths. Since heavy rain is not likely to occur simultaneously at two widely separated ground stations, rain attenuation can be minimized by a form of space diversity.

The long delay is another matter. Besides impairing the quality of voice communications, long delay can seriously affect data transmission unless protocols[19] and control of user terminals are designed to match this media.

6.4 MODULATION

Modulation converts a communication signal from one form to another more appropriate form for transmission over a particular medium between two locations. *Demodulation* restores the signal to its original

---

[19] Strict procedures for initiating and maintaining communications. (See Section 8.8.)

form at the receiving end of the transmission medium. Modulation is often necessary because a signal, even when bandlimited, is rarely in the best form for direct transmission over the frequency spectrum available on a given medium.

A simple example of modulation is the transformation of an acoustic signal into an electrical analog signal by the telephone transmitter (discussed in Section 6.1.1). When the telephone is off-hook, a direct current flows over the customer's loop from the central office to the station set. This direct current can be conceived of as a 0-Hz carrier signal whose amplitude is varied by changes that occur in the sound-wave pressure when a customer speaks. The modulating signal in this case is the acoustic signal; the modulated signal is the varying amplitude direct current, and the ac component of this modulated current is usually referred to as the *speech signal*.

Some reasons for modulating a communication signal prior to transmission are:

- to enable a number of signals from voiceband channels to be combined (multiplexed, as described in Section 6.5) and transmitted simultaneously over a common broadband medium, thereby reducing the per-channel transmission cost.

- to shift the signal frequencies upward to simplify the design of repeater components and, in the case of radio communication, to translate the signal spectrum into one of the frequency bands allocated to common carriers by the FCC. Demodulation shifts the signals down.

- to convert voice or other analog signals to digital form to achieve lower noise or less distortion as well as economies in system implementation.

- to convert digital signals for transmission over analog channels that would destroy the digital waveform.

The following sections describe some commonly used modulation methods applied in the Bell System network. These include **double-sideband amplitude modulation, single-sideband amplitude modulation, pulse-code modulation,** and methods of **modulating digital data signals for analog facilities**.

6.4.1 DOUBLE-SIDEBAND AMPLITUDE MODULATION

One of the more familiar forms of amplitude modulation (used in AM broadcasting) is double-sideband amplitude modulation (DSBAM). In amplitude modulation, the amplitude of a sinusoidal carrier signal is varied continuously by the information signal, and the modulated signal is translated in frequency to occupy a distinct frequency band for transmission over a particular medium. DSBAM generally uses product modulation; that is, the product of an analog baseband information signal, $a(t)$, and a sinusoidal carrier signal, $\cos \omega_c t$, is formed. To obtain a modulated signal, $M(t)$, from which the baseband signal is easily recovered by demodulation, the carrier signal is added to the product:

$$M(t) \cdot \cos \omega_c t + a(t) \cdot \cos \omega_c t$$

$$= [1 + a(t)]\cos \omega_c t.$$

Here, $\omega_c = 2\pi f_c$ and f_c is the carrier frequency. Figure 6-11 illustrates this process.

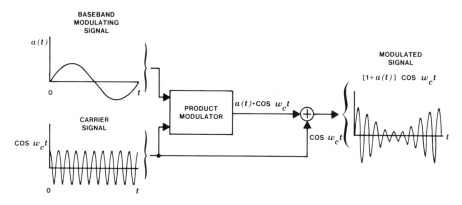

Figure 6-11. Double-sideband amplitude modulation process.

Figure 6-12 is a representation of the modulated signal in the frequency domain. DSBAM translates the frequency spectrum of the baseband signal symmetrically about the carrier frequency. The upper sideband constitutes a pure frequency translation above the carrier frequency, while the lower sideband inverts the baseband frequencies below

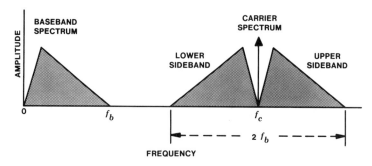

Figure 6-12. Frequency representation of DSBAM.

the carrier frequency. Both sidebands contain the same information. The carrier frequency is chosen so that the lower sideband does not overlap the baseband signal spectrum.

This form of modulation is attractive because filters are not required to eliminate an unwanted sideband as in single-sideband amplitude modulation (see Section 6.4.2), and therefore, terminals are less expensive. In addition, since the carrier signal is transmitted in these systems, no separate carrier signal needs to be supplied by the receiving terminal for demodulation.

However, since the carrier signal and the redundant sideband are both transmitted, the power-handling requirements of the transmission equipment are much greater than if only one sideband were transmitted. Also, DSBAM transmission requires twice as much bandwidth as the baseband signal. Both of these factors increase the line-haul cost (cost per mile) per voiceband channel of these systems and limit their economic application to shorter distances than those served by single-sideband systems.[20]

6.4.2 SINGLE-SIDEBAND AMPLITUDE MODULATION

To obtain a single-sideband amplitude modulated (SSBAM) signal, either sideband of the product-modulated spectrum described in Section 6.4.1 may be selected for transmission by a *bandpass filter*. In most Bell System applications, the lower sideband is selected.

Single-sideband transmission requires only one-half the DSBAM bandwidth per voiceband channel, allowing twice as many voiceband channels to be carried in a given frequency band (by multiplexing). (Figure 6-13 illustrates this.) For this reason, SSBAM is the technique usually found in the multiplex terminals used with long-haul broadband facilities incorporating frequency-division multiplexing of voiceband channels (see Sections 6.5.1 and 9.3.5). Also, since the individual carrier frequencies are not transmitted over these facilities, the ratio of information power to total transmitted power is greatly improved. Thus, the cost of line repeaters is significantly less than for DSBAM.

Suppression of the carrier frequency at the transmitting terminal improves the power efficiency of SSBAM but requires the insertion of a precise and stable carrier frequency at the receiving terminal for demodulation. This must be done with extreme phase accuracy to avoid an impairment called *quadrature distortion.*[21] Data signals are particularly sensitive to distortion. Improved control of the phase as well as the frequency of the carrier supply signals in SSBAM systems is provided by the

---

[20] A short-haul system (N2) is described in Section 9.3.3.

[21] Waveform distortion caused by a phase error in the demodulation process (see Bell Laboratories 1982, pp. 100-102, 300).

nationwide Bell System Carrier Synchronization Network, which distributes synchronization signals throughout the United States based upon two standard reference signals (see Section 6.5.3).

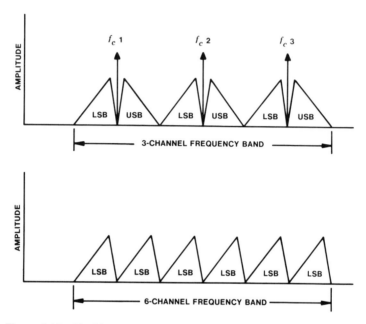

Figure 6-13. Stacking voiceband channels in DSBAM and SSBAM. *Top,* using DSBAM with transmitted carrier; *bottom,* using SSBAM.

6.4.3 PULSE-CODE MODULATION

In both *pulse-amplitude modulation* (PAM) and its extension, *pulse-code modulation* (PCM), the unmodulated carrier signal is a series of precisely timed pulses of equal amplitude. The carrier signal is modulated by samples of the amplitude of a baseband analog signal. The baseband signal may be reconstructed from the amplitude samples, providing the samples are taken often enough and the baseband signal is bandlimited.

The sampling theorem (see AT&T 1977, vol. 1, pp. 211–212) expresses the necessary conditions:

> If a baseband analog signal is sampled (that is, if instantaneous measures of signal amplitude are taken) at a rate rapid enough so that at least two samples are taken during the course of the period corresponding to the highest significant baseband

signal frequency, then the samples uniquely determine the signal, and the signal may be reconstructed (demodulated) from the samples without
distortion.

A baseband signal bandlimited to f_m Hz is completely specified by the
amplitude of its samples spaced in time,

$$T = \left[\frac{1}{2f_m} \right]$$

seconds apart. For example, analog signals bandlimited to 4 kHz require
a sampling interval (called the *Nyquist interval*) of 125 microseconds,
which corresponds to 8000 amplitude samples per second.

As shown in Figure 6-14, the sampling process can be thought of as a
form of product modulation. Since the amplitude of the sampling carrier
pulses, $s(t)$, is varied according to sample values of the baseband signal
amplitude, $m(t)$, this form of modulation is called *pulse-amplitude modulation*. A representation of the modulated signal in the frequency domain
is shown in Figure 6-15.

To demodulate a PAM signal and reconstruct the analog baseband signal, the modulated signal is passed through a *low-pass filter* with cutoff
frequency f_m. For speech signals, the low-pass filter need not have an
extremely sharp cutoff. Rather, since the speech signal is effectively limited at the transmitting end to 3.5 kHz and the sampling rate selected is
8000 samples per second, there is a guard band of $8 - 2(3.5) = 1$ kHz, over
which the filter attenuation may attain a high value.

PCM is distinguished from PAM by two additional signal-processing
steps, *quantizing* and *encoding*, that take place before the signal is transmitted. Quantizing replaces the exact amplitude of the samples with the
nearest value from a limited set of specific amplitudes. The sample

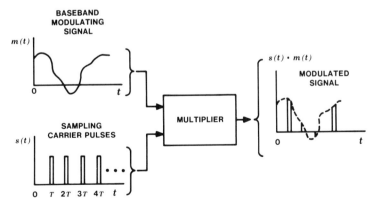

Figure 6-14. Pulse-amplitude modulation process.

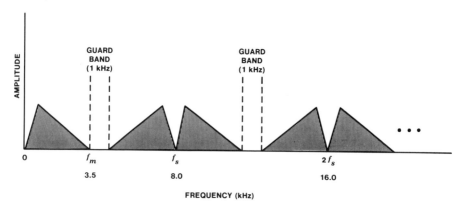

Figure 6-15. Amplitude spectrum of PAM.

amplitudes are then encoded, and the codes are transmitted. Quantized sample amplitudes are typically transmitted as binary codes. In general, a code consisting of n binary (for example, on-off) pulses is sufficient to represent 2^n quantized sample amplitudes (for example, a group of three binary digits represents $2^3 = 8$ quantized sample amplitudes).

Noise and other interference in the transmission channel alter the coded pulse stream. However, since only the presence or absence of a particular pulse need be established for PCM, detection circuitry is relatively simple, and most important, a pulse stream identical to the original coded pulse stream can be regenerated. This can be done as often as required to ensure that noise in any one segment of the transmission path is not likely to cause a pulse to appear as no pulse or vice versa. Thus, noise is not additive in a properly designed PCM system. This is an advantage over analog transmission where the noise present in the channel is amplified along with the signal because the signal cannot be separated from the noise.

Representing an exact sample amplitude by one of the 2^n predetermined, discrete amplitudes introduces an inherent error called *quantization noise*. This noise is controllable and may be made negligible by using a sufficiently large number of quantizing levels or steps. Studies have shown that eight bits (256 quantizing levels) per sample provide a satisfactory signal-to-noise ratio for speech signals. Therefore, to transmit an analog speech signal bandlimited to 4 kHz, 64 kbps are required (eight pulses per sample and 8000 samples per second). Quantization noise is further controlled by minimizing repeated encoding and decoding of the signal (analog-to-digital conversion of the signal and vice versa). Transmission planning for the network controls the number of such encodings and decodings. Typically, only two or three are permitted.

The transmission of speech signals with PCM techniques, while more effective in many cases, results in less efficient use of the bandwidth of the facility than with analog transmission. A speech signal requires a nominal 4 kHz of analog bandwidth in the transmission channel if the signal is sent in analog form.[22] An indication of the bandwidth required to send signals in digital form on an analog channel is given by a corollary to the sampling theorem stated earlier: A baseband analog channel can be used to transmit a train of independent pulses at a maximum rate that is twice the channel bandwidth. An analog voiceband channel with a nominal 4-kHz bandwidth could, therefore, carry a maximum of 8000 pulses per second.[23] Since current PCM practice converts the analog speech signal into a 64-kbps digital pulse stream, it appears that a bandwidth of 32 kHz is required to transmit speech using PCM. The apparent 8-to-1 ratio in bandwidth between transmitting speech using PCM versus analog form may be useful as an indicator, but a more meaningful comparison would require consideration of other factors. For instance, while a channel used for analog transmission of speech must be of good quality in terms of certain impairments, the robust nature of PCM transmission permits the use of a channel of significantly lower quality.

Thus, although a PCM signal requires a wider bandwidth than a baseband analog signal, this greater-than-minimal bandwidth is traded for significant economies in repeater design and a robust signal format having a high immunity to noise and other interference. The PCM signal can be completely regenerated at each repeater location, provided that noise between repeaters is not great enough to cause errors in pulse recognition. Transmission impairments, therefore, are largely controlled by the design of the PCM terminals. With proper design, a PCM signal can be transmitted almost any distance without loss in quality.

Among the applications that are attractive for PCM are loop carrier systems, metropolitan short-haul wire-pair carrier facilities, short-haul microwave radio systems, and digital lightwave transmission systems. All but the last of these applications take advantage of the rugged signal in the presence of noise. In lightwave transmission, the lightguide cable has an extremely wide bandwidth and is naturally suited for digital signals (on-off pulses of light intensity).

6.4.4 MODULATING DIGITAL DATA SIGNALS FOR ANALOG FACILITIES

Many modulation methods have evolved to enable digital data signals to be transmitted with minimal distortion over voiceband analog channels. Generally, selection of a method involves a tradeoff between simplicity of

---

[22] Of course, the quality of the analog channel in terms of impairments to speech signals must be adequate to ensure that the received speech signal is of acceptable quality.

[23] While PCM uses binary pulses, the corollary does not limit the pulses to binary form; that is, more than two values can be represented at a given pulse position.

instrumentation and efficient use of the available channel bandwidth. Efficiency is often expressed in terms of the bit rate per unit of bandwidth (that is, bits per second per Hz of bandwidth).

Frequency-shift keying (FSK) is applied where economical hardware design is more important than efficient bandwidth use. In FSK, the carrier signal is shifted back and forth between two distinct, predetermined frequencies. The channel bandwidth required is equal to nearly twice the bit rate plus the difference between the two carrier frequencies. In data applications on voiceband channels, FSK has been used at speeds up to 1.8 kbps.

Differential phase-shift keying (DPSK) uses bandwidth more efficiently than FSK by putting more information in each transmitted sample pulse, but it requires more elaborate designs for signal generation and detection. DPSK has become the worldwide standard for 2.4-kbps data transmission over voiceband channels.

In DPSK, information is transmitted in the *phase* of the received signal. Four possible signal states are represented by each of four phases: +45 degrees, −45 degrees, +135 degrees, and −135 degrees. The phases correspond to the symbols 00, 01, 10, and 11, respectively. Since each symbol carries two bits of information, the channel bandwidth required is somewhat less than the bit rate.

In principle, a transmitted symbol may carry many bits of information. For example, a symbol could carry three bits if eight distinct phases were transmitted using DPSK. Similarly, four bits could be carried with sixteen phases. Using more levels, therefore, allows for transmission of higher and higher data rates over a voiceband channel. However, as the number of phases to be transmitted increases, there is a greater likelihood that noise and other impairments will cause incorrect symbols to be received. To achieve higher bit rates (for example, 4.8 and 9.6 kbps), an adaptive equalizer that self-adjusts (based on the actual received data signal) to compensate for distortion encountered on a wide variety of voiceband channels is needed. In addition, to achieve a transmission rate of 9.6 kbps on private-line facilities, a technique called *quadrature-amplitude modulation* (QAM) is used (see Lucky, Salz, and Weldon 1968, pp. 174-178). With QAM, two independent baseband data signals can be transmitted in the same channel bandwidth through the use of two DSBAM carrier signals in quadrature phase.

6.5 MULTIPLEXING

Multiplexing is a technique that enables a number of communications channels to be combined and transmitted over a common broadband channel. At the receiving terminal, *demultiplexing* of the broadband channel separates and recovers the original channels. In general, demultiplexing involves processes that are the reverse of those used in multiplexing.

The primary purpose of multiplexing is to make efficient use of a transmission facility's bandwidth capability to achieve a low per-channel transmission cost. This involves an economic tradeoff similar to that illustrated in Figure 6-16. As more channels are bundled together using additional multiplex equipment, the per-channel cost of the line-haul portion of the transmission system is lowered (indicated in Figure 6-16 by a reduced slope), but the cost of the terminals is increased (indicated by a higher intercept). The tradeoff between a higher terminal cost and a lower mileage cost per channel results in a net savings, provided the application involves system lengths greater than some distance, L.

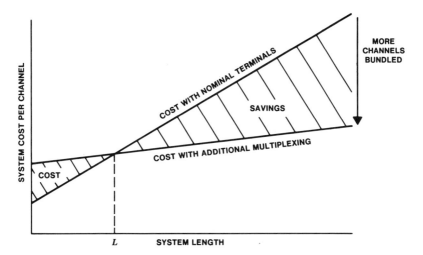

Figure 6-16. Multiplex cost tradeoff.

6.5.1 FREQUENCY-DIVISION AND TIME-DIVISION MULTIPLEXING

The two basic multiplexing methods used in the Bell System are **frequency-division multiplex** (FDM) and **time-division multiplex** (TDM). FDM divides the frequency spectrum of a broadband transmission system into many full-time communications channels. Signals from these channels are transmitted at the same time but at different carrier frequencies. In the Bell System, for example, the standard analog communications channel is the nominal 0-kHz–to–4-kHz voiceband channel. At the transmitting (multiplex) terminal (shown in Figure 6-5), voiceband channels are shifted up in frequency using amplitude modulation and occupy specific frequency slots associated with broadband transmission systems. At the receiving (demultiplex) terminal, all the voiceband channels are shifted back in frequency using amplitude demodulation, and then separated for distribution.

In TDM, a transmission facility is shared in time rather than frequency. This is accomplished as indicated in Figure 6-17, by interleaving
PCM samples from a number of voiceband channels on a common bus.
At the transmitting terminal, several slow-speed pulse streams (digital
channels) are combined into a composite high-speed pulse stream. At the
receiving terminal, the slower speed streams are separated. It is important to note that transmitting and receiving terminals must be synchronized so that the pulses can be correctly identified and kept in the proper
relation to one another. Network signaling bits and framing bits are
added to the coded information bits as noted in the figure.

6.5.2 MULTIPLEX HIERARCHIES

The time-division multiplexing of twenty-four voiceband channels illustrated in Figure 6-17 is typically accomplished in the local network. In
the intercity network, the concentration of traffic provides economic

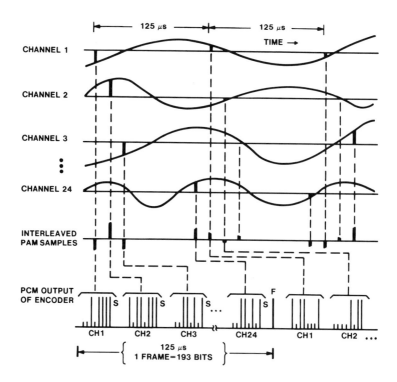

NOTES:

1. FRAMING BIT IS 193RD BIT (F).

2. SIGNALING BIT IS 8TH BIT IN EACH CHANNEL IN ONE FRAME OUT OF SIX (S).

3. THE OUTPUT PULSE RATE SHOWN IS 1.544 Mbps, THE DS1 LEVEL IN THE
   TDM HIERARCHY. (SECTIONS 6.5.2 AND 9.4.3 DISCUSS THE TDM HIERARCHY.)

Figure 6-17. Time-division multiplexing.

opportunities to combine channels into larger and larger bundles for long-haul transmission. A number of different size bundles are needed to make the most efficient use of the available transmission facilities to satisfy the range of route capacities and distances found in the network. A hierarchy of multiplex levels or steps has been developed to meet this need and to provide for orderly and efficient growth of the network. In fact, there are two hierarchies: one for analog transmission systems and one for digital transmission systems.

In the FDM hierarchy, multiplex levels correspond to increasingly wider frequency bands. The initial level is a grouping of twelve voiceband channels by an A-type channel bank. Additional levels eventually lead to the multiplexing of a large number of channels compatible with broadband transmission systems. (Section 9.3.5 includes information on A-type channel banks and the FDM hierarchy, and Figure 9-19 is a graphic representation of it.)

The TDM hierarchy consists of multiplex levels corresponding to increasingly higher pulse rates. The initial level is shown in Figure 6-17. It is called the *DS1 level* and is a grouping of twenty-four voiceband channels by a D-type channel bank. Additional levels are designed to be compatible with high-capacity digital facilities. (Section 9.4.3 includes information on the D-type channel bank and the TDM hierarchy, and Figure 9-26 is a graphic representation of it.)

6.5.3 MULTIPLEX SYNCHRONIZATION

In the Bell System, the need for accurate frequency and stable timing signals to coordinate the numerous analog and digital transmission and switching systems now in service is filled by the Bell System Reference Frequency Standard (BSRFS) and a nationwide distribution network called the *Bell System Carrier Synchronization Network*. The center of this tree-like network is located in an underground repeater station in Hillsboro, Missouri. Here, the BSRFS (consisting of three interlocked cesium frequency standards accurate to one part in 10^{11}) generates two reference signals: one at 2.048 MHz and another at 20.48 MHz.

The FDM Synchronization Plan

As noted in Section 6.4.2, modulation using single-sideband (SSB) FDM terminals requires the insertion of a precise and stable carrier frequency at the receiving terminal for demodulation. If the frequencies of the transmitting and receiving terminals differ, an impairment called *carrier frequency shift* (see Section 6.6.1) that appears as a frequency offset in the information signal may result. To synchronize modulation and demodulation operations in SSB FDM terminals, reference frequency signals are transmitted from the BSRFS to regional synchronization centers deployed throughout the country. At each center, *regional frequency supplies* provide various control and operating signals (for example, carrier frequencies) to regional central offices needing FDM synchronization. These regional central offices are equipped with *primary frequency supplies,* whose signal frequencies are synchronized to the incoming signals from the regional frequency supplies. Multiplex synchronizing signals are then transmitted independently over coaxial or microwave facilities.

The PCM-TDM Synchronization Plan

As mentioned previously, a synchronization plan is a crucial component of time-division multiplexing because digital signals cannot simply be combined at the transmitting end and correctly identified at the receiving end unless transmitting and receiving multiplex terminals are locked to a common clock (that is, they are in synchronism). Two forms of synchronization are important at each level in the digital hierarchy. The first is synchronizing each bit stream to a nominal pulse rate. The second, and more fundamental, is synchronizing the relative timing of several bit streams so that they can be multiplexed together into one bit stream at a higher rate, and correctly identified and separated at the receiving end.

Since digital signals often originate in different locations that are separated by long distances and have independent timing clocks, synchronizing the relative timing of bit streams is a problem. The solution above the DS1 level is a process called *pulse stuffing,* in which bit streams to be multiplexed are each stuffed with additional dummy pulses to raise their rates to that of an independent, locally generated clock. The outgoing rate of the multiplexer[24] is therefore higher than the sum of the incoming rates. The dummy pulses carry no information and are coded so that they can be recognized and removed at the receiving terminal. The resulting gaps in the pulse stream are then closed to restore the original bit stream timing.

At bit rates above the DS1 level, delay variations caused by various transmission media can, in turn, cause timing discrepancies on the order of ±100 pulses in the received pulse stream. In principle, the departure from nominal timing could be handled by buffer storage,[25] but this becomes expensive at higher bit rates. With pulse stuffing, differences are corrected immediately, so that only a small amount of storage is needed to accommodate these variations. Another advantage of the pulse-stuffing method is high system reliability. The failure of a multiplexer or associated transmission facility will affect only those signals

---

[24] The term *multiplex* is often used interchangeably with or in place of *multiplexer* to mean the equipment that performs multiplexing.

[25] An intermediate storage medium between data input and active storage.

passing through the failed elements, since the timing at other multi-plexers is independently derived.

6.6 TRANSMISSION IMPAIRMENTS AND OBJECTIVES

Telecommunications signals are degraded by the practical limitations of channels (such as bandwidth), impairments arising from within the channel (for example, attenuation and echo), and impairments introduced from outside of the channel (for example, power-line hum and radio-frequency interference). The impact of many impairments depends on the type of signal carried over the channel and whether the channel is analog or digital. For example, impulse noise[26] from electromechanical switching systems has little effect on the reception of analog speech signals because of amplitude limiters in the circuit and the relative tolerance of the human ear. Digital data signals, on the other hand, can be seriously affected by impulse noise; blocks of data can be obliterated, resulting in high error rates in the received signal.

The planning, design, installation, operation, and maintenance of analog and digital transmission facilities are based on a thorough understanding of impairments and an appropriate set of transmission objectives which, when achieved, lead to customer satisfaction at a reasonable cost. Transmission objectives are dynamic: They respond to changing subjective reaction to impairments, new technology, new measuring techniques, and improved analysis tools. The steps involved in determining a new objective or adjusting an existing objective to achieve a balance between customer satisfaction and cost include:

- identifying the impairment that needs to be controlled, including defining the impairment in terms of a measurable parameter

- specifying a method to measure the impairment parameter

- evaluating the impact of the impairment on transmission performance

- finding the means to control the impairment

- formulating an end-to-end performance objective for the impairment that will provide a satisfactory quality of service

- allocating the end-to-end objective to individual parts of the network in terms of equipment design requirements, engineering application guidelines and rules, and maintenance requirements and limits for the use of craft forces

- establishing methods to monitor network performance to ensure that objectives are being met (see Chapter 16).

---

[26] Short bursts of high-level noise that sound like clicks over a telephone line. (See Section 6.6.2.)

For convenience, the impairments and objectives considered here are divided into those that primarily affect analog speech signals and those that primarily affect digital data signals.[27]

6.6.1 ANALOG SPEECH SIGNALS

Early in this century, the emphasis in communications was on the conquest of distance. The basic goal in telephone transmission was to provide a satisfactory signal volume at the receiver so that the talker could be heard. Once this goal was achieved, attention turned to improving the quality of transmission through technological advances.

Transmission objectives were initially established for the transmission of speech signals over the PSTN. As new types of signals and services evolved, these objectives were modified and new objectives were formulated. The following discussion begins with the problem of received volume and proceeds to other impairments and applicable objectives.

Subjective tests have been used to determine listener reactions to a range of received volumes and other impairments. Satisfactory transmission performance, in turn, is expressed in terms of a grade of service that is based on expected customer satisfaction with the quality of telephone connections provided by the network; that is, what percentage would rate the quality as excellent, good, fair, poor, or unsatisfactory. Objectives for impairments are generally set so that the majority of listeners would rate the transmission as excellent or good. Chapter 16 describes the setting of objectives in more detail.

Received Volume and Loss[28] Objectives

With many impairments, the more intense the impairment, the poorer the received signal. Received volume, in contrast, can impair transmission in two ways: The signal may be too weak or too loud. If it is too loud, the

---

[27] Bell Laboratories 1982, pp. 136-139, discusses impairments peculiar to video signals.

[28] In the Bell System, the term *loss* refers to *insertion loss*, a quantity that represents a specific relationship between the input and output of a network (for example, a customer connection or a circuit). The figures below illustrate a basic insertion loss calculation. In Figure 1, the generator is connected directly to the load, and the power delivered to the load is P_1. In Figure 2, a network exists between the generator and the load; with the network in place, power P_2 is delivered to the load. Insertion loss is the ratio of P_1 to P_2 and is expressed in decibels (dB):

$$\text{insertion loss (in dB)} = 10 \log_{10} \frac{P_1}{P_2}$$

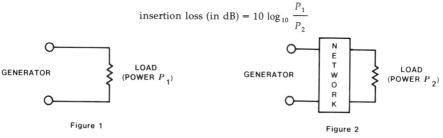

Figure 1   Figure 2

listener may be uncomfortable; if it is too faint, the listener may have difficulty understanding the received message. Received volume depends on several factors including telephone speaking habits, the acoustic-to-electric conversion efficiency of the station set, the amount of *sidetone*[29] in the telephone, and the attenuation in the built-up connection between speaker and listener.

Loss objectives (the amount of loss that provides a satisfactory grade of service) have been derived for individual portions (loops, trunks) of built-up, end-to-end, customer connections. Original analysis considered received volume only. This was followed by a model that combined received volume and idle circuit noise. Later the effects of talker echo were included to provide for performance evaluation in terms of a combined received volume-noise-echo grade of service.

Transmission objectives for loop loss have been derived based on achieving satisfactory performance in terms of received volume and noise impairments for a variety of built-up connections. Loop loss is controlled by specifying design methods to produce a satisfactory distribution of losses. (See Section 9.2.1.) When properly applied, the methods, together with trunk objectives, ensure that overall objectives on built-up connections will be met.[30]

Loss objectives for trunks are allocated on the basis of the need to provide satisfactory volume, the need to minimize the contrast as perceived by the customer between different types of calls (local, long distance, operator-assisted, etc.), the position of the trunk type in the switching hierarchy (direct, tandem, toll connecting, etc.),[31] and the need to control echo on trunks involved in longer connections through the network. On shorter trunks (less than about 10 miles for 2-wire voice-frequency [VF] trunks, 25 miles for 4-wire VF trunks, and 300 miles for carrier trunks) where echo is not a problem, the loss objectives are as listed in Table 6-1.

Talker Echo

Talker echo results when an electrical speech signal travels a long distance and some portion of it is reflected back toward the speaker. If the elapsed time (delay) is long and the echo path loss[32] is inadequate (that is, there is not enough loss to attenuate the echo), the echo can interfere

---

[29] The portion of the signal from a telephone transmitter that appears at the receiver of that telephone. Some sidetone appears to be desirable to assure a customer that the telephone is working and to help the talker adjust the level of speech.

[30] Based on measurements of 1-kHz loss made in 1960 and 1964, the Bell System average loop loss was 3.8 dB, and the standard deviation was 2.3 dB.

[31] Section 4.2.1 describes the types of trunks in the switching hierarchy.

[32] The return loss (a function of the hybrid balance) plus twice the circuit loss between the speaker and the point of reflection.

TABLE 6-1

RECEIVED VOLUME LOSS OBJECTIVES
FOR SHORTER TRUNKS IN
THE LOCAL NETWORK

Trunk Type	Loss (dB)	
	Nominal	Maximum
Direct	3	5
Tandem	3	4
Toll connecting	3	4

with the talker's speaking process. In extreme cases, echo may be so great that speaking is nearly impossible.

Figure 6-18 illustrates a frequently encountered situation in the PSTN that results in echo. The figure shows a 4-wire intertoll trunk terminating in a toll office. This trunk may be switched to any one of several class 5 end offices over 2-wire toll connecting trunks and, in turn, be connected to one of a number of loops with widely different impedances.[33] The transition between 4-wire and 2-wire transmission is provided by a hybrid circuit.[34]

If the hybrid termination is ideal (perfectly balanced; that is, impedances Z and Z_c are equal), the transmission loss across the hybrid (from B to C on the figure) is infinite, and no echo returns to the talker. This means that none of the signal in the transmission path A to B is returned in path C to D. Achieving perfect hybrid balance is difficult because the impedance Z is different for every connection through the switches, but the hybrid balance impedance Z_c is fixed. Nevertheless, the potential effects of highly variable loop impedances can be and are masked because the impedance of the 2-wire toll connecting trunks is carefully controlled. If, however, the 4-wire—to—2-wire interface is at the end office instead of at the toll office, there is no buffering by the toll connecting trunks and more serious echoes occur. Because of this lack of precise control over loop impedances and the insufficient degree of impedance matching possible at the class 5 end office, other means of controlling echo must be used.

[33] The opposition to the flow of alternating current by an element in a circuit.

[34] Hybrid circuits are used for various purposes. In this application (that is, as a transition between 2-wire and 4-wire circuits), it is also called a *4-wire terminating set*.

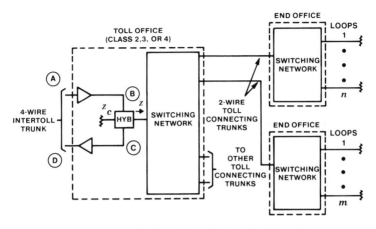

Figure 6-18. Talker echo path—4-wire intertoll
trunk to 2-wire toll connecting trunk.

The impact of talker echo depends on the echo's amplitude, the echo's delay in transmission (which is proportional to trunk length), and the speaker's tolerance of the echo. The echo becomes increasingly bothersome as echo amplitude and delay increase. Echo amplitude depends on the loss of the signal up to the point where reflection begins, the hybrid balance (return loss) at the reflection, and the loss in the path back to the talker. Figure 6-19 shows how the grade of service perceived by the talker varies with echo path loss and delay.

An inexpensive means of meeting grade-of-service objectives for echo is to increase the loss in the transmission path. (This loss appears twice—once in the path to the listener and again in the return path to the talker.) Additional loss, however, causes a reduction in the received talker volume. A loss administration plan called (rather arbitrarily) the *Via Net Loss* (VNL) *Plan* is a compromise between echo and received volume loss impairments. To control echo, the VNL Plan contains a loss component that depends on trunk length and trunk type. The trunks involved in longer connections where echo is a problem are in the toll portion of the network. They include intertoll trunks, toll connecting trunks, and intertandem trunks. Table 6-2 lists the loss objectives for these longer trunks, and Table 6-3 shows the VNL component versus length for high-usage intertoll trunks provided on carrier facilities.

Under the VNL Plan, whenever the echo path delay of a built-up connection exceeds 45 milliseconds, the 4-wire intertoll trunk that is part of this connection is operated at 0-dB loss and is equipped with an echo suppressor or echo canceler. This is also done when a high-usage intertoll carrier trunk is longer than 1850 miles or would require a VNL greater than 2.9 dB, as indicated in Table 6-3.

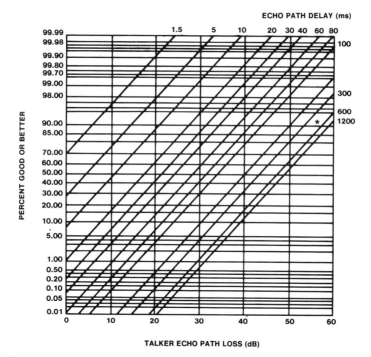

Figure 6-19. Grade of service versus talker echo path loss and delay.
Grade of service is determined by observing the percent good or better
(*horizontal lines*) at the intersection of the echo path delay (*oblique lines*)
and echo path loss (*vertical lines*) for a given connection. With satellite
communications, for example, a customer's signal travels over 45,000 miles
from talker to listener. The echo signal travels twice that distance, resulting
in an echo path delay of about 500 milliseconds. To achieve a grade of
service of 90 percent good or better with so great a delay would require a
talker echo path loss of about 57 dB (*asterisk* [*]).

An *echo suppressor* is an electronic device that compares the speech signals traveling in the two directions during a long-distance conversation.
The suppressor decides which person is talking at any given time and
inserts a high loss (35 dB or more) in the circuit in the opposite direction.
This prevents the echo from looping back to the talker over the listener's
speaking path. When the two persons talk at the same time, the suppressor inserts a loss in both directions, resulting in undesirable clipping of
speech signals.

An *echo canceler*, rather than inserting a loss in the return path, uses
the transmitted speech signal to generate a signal that is a replica of the
echo. Subtracting this signal from the actual echo signal cancels out the
echo, while allowing normal communications to continue undisturbed.
The echo canceler principle is illustrated by the simplified diagram in
Figure 6-20.

TABLE 6-2

LOSS OBJECTIVES FOR CONTROL
OF ECHO ON TRUNKS

Trunk Type	Loss (dB)	
	Nominal	Maximum
Intertoll	VNL*	2.9 (high usage) 1.4 (final groups)
Toll connecting	VNL + 2.5	4.0
Intertandem	VNL	1.5 (balanced offices) 0.5 (unbalanced offices)
Interregional direct	VNL + 6	8.9

*VNL = VNLF x 1-way length in miles + 0.4 dB. VNLF, the via net loss factor, depends on the type of facility used. For example, VNLF = 0.0015 for carrier facilities.

TABLE 6-3

VNL COMPONENT VERSUS LENGTH
(OPERATING ON ALL CARRIER FACILITIES)

Trunk Length (Miles)	VNL* (dB)
0-165	0.5
166-365	0.8
366-565	1.1
566-765	1.4
766-965	1.7
966-1165	2.0
1166-1365	2.3
1366-1565	2.6
1566-1850	2.9
Any length with echo suppressor or canceler	0.0

*VNL = 0.0015 x 1-way length in miles + 0.4 dB.

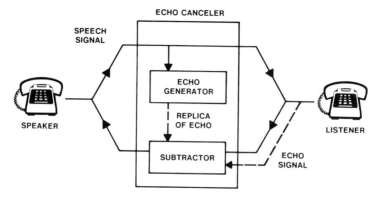

Figure 6-20. Echo canceler principle.

The current echo objective is for performance to be satisfactory on 99 percent of all telephone connections that encounter delay. The VNL Plan, hybrid balance, and echo suppressor/canceler administrative rules are currently being used in the analog portion of the Bell System network to meet this objective.

Message Circuit Noise

Message circuit noise is a complex signal, comes from a variety of sources, and has a great influence on the transmission quality of the network. Some of the components are induced power hum, or inductive interference (discussed below); thermal noise;[35] central office battery noise (see Section 12.2); and impulse noise that is generated by central office switch contacts.

For purposes of transmission planning and design, message circuit noise is defined as a weighted average of the noise level from a variety of sources within a voice channel as measured by a 3A noise-measuring set or equivalent, using a frequency weighting called *C-message weighting*. Measurements are expressed in dBrnC, decibels above reference noise (10^{-12} watts), using C-message weighting. The noise meter is designed so that a 1-kHz tone having a power of -90 dBm[36] will read 0 dBrnC.

The setting of noise objectives is based on the grade-of-service concept, the same concept applied to loudness loss (a measure of perceived speech volume) and echo impairments. Subjective tests were used in which speech was evaluated in the presence of noise. Observers were

---

[35] Random noise related to the operating temperature of electrical circuit elements.

[36] The unit dBm is a logarithmic measure of power with respect to a reference power of 1 milliwatt.

asked to rate a series of simulated telephone calls having different combinations of noise level and acoustic-to-acoustic loudness loss. The rating was in terms of five quality categories: excellent, good, fair, poor, and unsatisfactory. The results of these tests are shown in Figure 6-21.

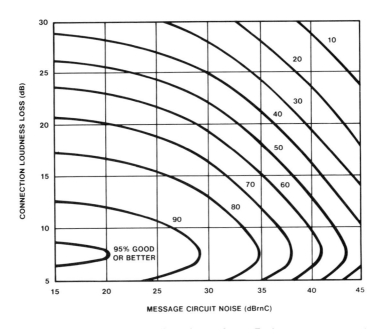

Figure 6-21. Loss-noise grade of service. Each curve represents combinations of loudness loss and message circuit noise for which a constant percentage of observers rated the transmission quality good or better.

Subjective reaction varies with both the level of circuit noise appearing at the terminals of the telephone set and the loudness loss. For example, when the message circuit noise level is less than about 25 dBrnC and the connection loudness loss is greater than 7 to 8 dB, the percentage of test subjects judging the performance as good or better depends more on loudness loss than on the noise level. For somewhat higher noise levels, the grade-of-service contours show more of a one-to-one relationship in dB between the effects of loudness loss and noise level (that is, both are equally responsible for the result). This indicates that over this range, subjective preference is roughly constant with a constant signal-to-noise ratio.[37] Finally, when the loudness loss decreases below 7 to 8 dB, an increasing percentage of test subjects finds the high received talker volume itself increasingly objectionable, regardless of the circuit noise level.

---

[37] On the figure, lines with a slope of −1 would represent constant signal-to-noise ratio.

To ensure high customer satisfaction in terms of a loss-noise grade of service at a reasonable cost, separate noise objectives have been established for trunks and loops. For trunks, the performance objectives recognize the fact that circuit noise on analog facilities accumulates with distance, as shown by the slanted solid lines in Figure 6-22. Trunk noise objectives are summarized in Table 6-4. Actual noise performance should be better than or equal to that shown. Consistent with these overall objectives, separate allocations have been made for trunks on short-haul and long-haul carrier systems (see Figure 6-22).

The message circuit noise objective for customer loops is stated in terms of a maximum or upper limit. Changes in noise level below about 20 dBrnC measured at the station set were found to have a negligible effect on the overall customer-to-customer connection performance grade of service, whereas the grade of service deteriorated appreciably when noise exceeded this value. A loop limit of 20 dBrnC, therefore, was chosen for all customer loops, including loop carrier systems.

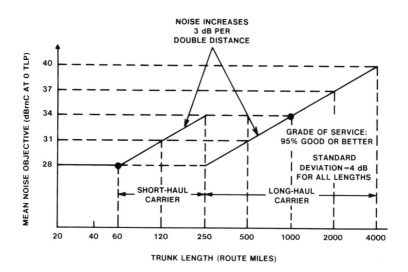

Figure 6-22. Message circuit noise objectives for trunks on analog carrier systems. Solid lines for short-haul and long-haul carriers reflect noise objectives for a grade of service of 95 percent good or better. *Transmission level points* (TLPs) define relative signal or noise levels at various points in a telephone connection. Section 8.6.1 contains more information on TLPs.

Intelligible Crosstalk

Intelligible crosstalk is a single, unwanted and understandable speech signal coupled from one message channel into another. Coupling of any type of signal between channels is a concern, but this interference is

TABLE 6-4

MESSAGE CIRCUIT NOISE OBJECTIVES FOR
CUSTOMER-TO-CUSTOMER TRUNK CONNECTIONS
(EXCLUDING LOOPS)

Type of Toll Connection	Distance Bands (Miles)	Percent Good or Better	Average Noise Power (dBrnC)	Maximum Standard Deviation (dB)
Short	0-180	99	19	6
Medium	180-720	97	23	5
Long	720-2900	95	27	4
Intercontinental	>2900	90	30	3.5

particularly objectionable because, in addition to impairing the quality of the channel, it creates a sense of loss of privacy. Crosstalk is caused by nonlinearities in circuits[38] within frequency-division multiplex (FDM) carrier systems and by electric and magnetic coupling between disturbing and disturbed circuits within various transmission media. Methods used to control crosstalk include shielding conductors, separating disturbing and disturbed circuits, impedance balancing, and designing and maintaining systems to suppress nonlinearities.

Stringent objectives have been established to minimize the probability that customers will encounter intelligible crosstalk. These objectives are expressed in terms of a *crosstalk index*, which is the actual percent probability of receiving an audible, intelligible speech signal on a call. This index depends on such factors as the number of disturbers, talker volumes, how much disturbing circuits are used, the coupling between disturbing and disturbed circuits, the listener's acuity in the presence of noise, and the noise level at the listener's telephone set. Table 6-5 lists typical objectives for trunks and loops. The objectives take into account the fact that exposure to crosstalk is generally greater on the longer inter-toll connections than on shorter local trunks.

Delay

Delay is the transmission time required for a talker's speech signal to reach the listener. Subjective tests have shown that delay has no major effect on speech transmission if customers are not anticipating it, and it is

---

[38] For circuits with nonlinearities, the output signal amplitude is not directly proportional to the input signal amplitude.

TABLE 6-5

INTELLIGIBLE CROSSTALK OBJECTIVES

Trunk/Loop Type	Objective Crosstalk Index (%)
Intertoll trunks	1.0
Toll connecting trunks	0.5
Direct trunks	0.5
Tandem trunks	0.5
Intertandem trunks	0.5
Loops	0.1

kept to 600 milliseconds or less one way. Once customers become aware of a long delay on a channel, however, they tolerate it less.

Terrestrial facilities within the United States may introduce about 20 milliseconds of 1-way delay depending on the length and the velocity of propagation of the transmission facility. Intercontinental connections using submarine cable may have 1-way delays of 100 milliseconds. Geosynchronous satellite circuits introduce considerably larger delays (on the order of 250 milliseconds one way); therefore, current delay objectives restrict the use of satellite trunks to no more than one up-down link per end-to-end connection.

Amplitude/Frequency Distortion

Amplitude/frequency distortion occurs when the relative magnitudes of the different frequency components of a signal are altered during transmission over a channel. The graph on the left in Figure 6-23 illustrates the amplitude response of a voiceband channel under preferred conditions. When conditions are significantly poorer than this, the amplitude response may be distorted as suggested by the graph on the right in Figure 6-23, where amplification is not the same for different frequencies across the channel bandwidth.

As mentioned in Section 6.1.1, an acceptable bandwidth for a voice signal is from 200 Hz to 3.5 kHz. Although a consistent set of bandwidth objectives for individual loops and trunks in the network has not been established, the factors that tend to increase the effective voiceband channel bandwidth also help make the amplitude response within the band more uniform. Amplitude distortion is controlled primarily by design rules for loops and trunks, office cabling limitations, and design requirements for carrier system channel units (see Section 9.4.3) and other multiplex arrangements.

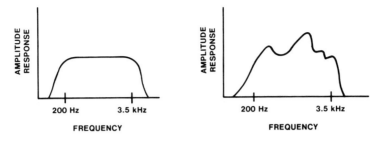

Figure 6-23. Amplitude/frequency responses. *Left,* preferred channel; *right,* poor channel.

A single inband amplitude distortion objective for speech signals is not available because of difficulties in expressing an objective that can be allocated optimally to trunks and loops.

Carrier Frequency Shift

Carrier frequency shift affects speech and program signals by reducing the naturalness of received signals. It results when carrier frequencies of transmitting and receiving analog transmission terminals using suppressed-carrier FDM (see Section 6.4.2) differ. Subjective tests have shown that customer-to-customer frequency shift should be held to a maximum of ±2 Hz to preserve a harmonic relationship that is satisfactory to discerning listeners. This objective has been selected as achievable in today's facilities, even for the longest connections through the network.

Inductive Interference

Since the power and telephone industries serve the same customers, they frequently share the same right-of-way (for example, the same utility pole). Because of this proximity, telephone cables are exposed to electromagnetic fields created by the power system currents. Under ideal conditions, this power influence results in equal voltages to ground on the two conductors of a wire pair. If the power influence is high enough and if the telephone circuit is not perfectly balanced, a voltage difference between the conductors will be produced that can result in audible noise levels. The standard telephone set is designed to have relatively low sensitivity at the fundamental frequency of power systems in the United States (60 Hz). Distortion of the power system current waveform, however, can generate harmonics of 60 Hz of sufficient amplitude to cause interference in the voice band.[39]

---

[39] As mentioned earlier in this section, the Bell System objective is for noise not to exceed 20 dBrnC on customer loops.

The Bell System employs various methods to minimize the possibility of inductive interference. These include using adequately balanced telephone cable with grounded sheaths, employing balanced telephone equipment, and separating power and telephone circuits. Cooperative efforts between the Bell System and the power industry, known as *inductive coordination,* may also be required to achieve satisfactory operation of telephone equipment. When total compatibility is impractical or impossible to attain using these measures, it may become necessary to install mitigation devices on telephone facilities. At present, these consist of a variety of magnetic core devices,[40] but they are used only as a last resort because they are expensive, bulky, and incompatible with certain systems.

Overload

When a signal is transmitted at an amplitude higher than that normally imposed on a telephone channel, it can produce intelligible crosstalk or interference. This impairment is considered overload when all channels in a transmission facility are affected. To control overload, a limitation is set on the amount of power that may be applied to an individual channel. The objective for both system load and customer load is that the per-channel long-term average power not exceed −16 dBm0.[41] This objective considers signal level variations with time and channel activity.

6.6.2 DIGITAL DATA SIGNALS

Transmission of significant amounts of digital data over the PSTN began about 1960 and is now a sizable portion of the traffic handled. Originally, data communications were carried only over analog transmission facilities using data sets. However, as digital transmission facilities and digital switching machines increase in number and connectivity, data signals can be transmitted in digital form more economically, generally at higher data rates and with potential advantages in transmission quality as well.

Digital data signals are generally transmitted in blocks containing error-detecting codes so that, when errors occur, the data may be retransmitted. Alternatively, the use of error-correcting redundant codes permits some or all errors to be corrected at the receiving end of transmission. Both techniques reduce or neutralize the effect of error-producing impairments. Because loss and noise are controlled by objectives for speech, specific objectives for data are confined to impairments

[40] Some of these devices are neutralizing transformers, drains, and chokes.

[41] The unit dBm0 is an expression of power level in decibels with reference to a power of 0 milliwatt.

such as impulse noise. In addition, there are service objectives that address the efficiency of data communications in terms of throughput.[42] These service objectives focus on transmission performance, availability, and maintainability.

For example, objectives for the Digital Data System (DDS)[43] are:

- The **transmission performance** objective is error-free transmission in at least 99.5 percent of all 1-second intervals. System performance is measured in *error-seconds*, where an error-second is a 1-second interval during which one or more bit errors occur.

- The **availability** objective is that data circuits remain connected station to station and error free 99.96 percent of the time as a long-term average. In terms of cumulative outage time, this objective equates to an average down time of less than 3.5 hours per year.

- The **maintainability** objective is that no single outage exceed 2 hours. This objective recognizes the impact of long outages on a customer's business and the perishable nature of some data.

The rest of this section discusses major transmission impairments that can affect data service over analog channels—**impulse noise, linear distortion, nonlinear distortion**, and **incidental modulation**.

Impulse Noise

Impulse noise consists of short-duration high-amplitude bursts (or spikes) of noise energy, which are much greater than the normal peaks of message circuit noise on a channel. An impulse is considered to have occurred if the noise voltage increases by at least 12 dB above message circuit noise for no more than 10 milliseconds. If impulse noise occurs often enough, it can obliterate large blocks of data and, in extreme cases, render a channel unsuitable for data transmission. Impulse noise sources include voltage transients generated by electromechanical switching systems, lightning, switching to protection channels,[44] and various maintenance activities.

Objectives for impulse noise are stated in terms of the maximum number of times noise above a particular threshold value occurs during a prescribed measurement interval. The present customer-to-customer

---

[42] The quantity of data a user can transfer during a given time interval.

[43] Section 11.6.1 discusses DDS.

[44] Broadband channels in carrier facilities used as spares. They can be switched into service if normal working channels fail.

objective for the switched network is for no more than fifteen impulse noise counts in 15 minutes at a threshold set 5 dB below the received signal level for 50 percent of all connections in the Bell System.

Linear Distortion

Linear distortion is a signal-independent impairment associated with a nonideal transmission channel characteristic. It is caused by transmission networks that produce a nonideal inband amplitude and phase response. To prevent the dispersion of pulses in a digital data stream, the channel characteristic ideally should have a bandwidth as wide as possible, a flat amplitude/frequency response, and a linear phase-versus-frequency[45] response within the channel. Otherwise, intersymbol interference may occur (the dispersion of pulses may cause a pulse to be detected where none exists), resulting in unacceptable error rate performance. Figure 6-24 shows an ideal pulse stream versus a pulse stream where dispersion has occurred.

PSTN objectives for linear distortion have been applied to loops and special-services circuits conditioned for data transmission (that is, specially amplitude and phase equalized), for example, *DATAPHONE* service[46] for data speeds above 300 bps. In this case, the amplitude/frequency distortion objective is that the loss at 2.8 kHz and at

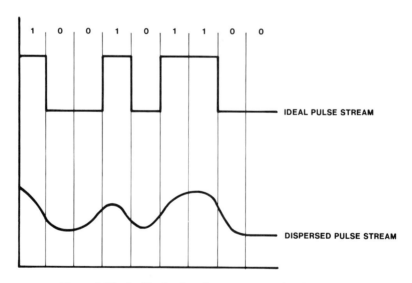

Figure 6-24. An ideal pulse stream versus pulse dispersion.

[45] With a linear phase-versus-frequency response, all frequency components of the transmitted signal experience the same delay.

[46] Section 2.5.4 discusses *DATAPHONE* service.

400 Hz not exceed the loss at 1 kHz by more than 3 dB. Moreover, the phase/frequency distortion objective permits no more than 100 microseconds delay difference between any two frequencies in the band from 1 to 2.4 kHz.

A simple method of indirectly measuring the linear distortion of a channel is the *peak-to-average ratio* (P/AR) meter method. To make a P/AR measurement, a representative pulse train is applied at the transmitting end of a channel, and the ratio of the pulse envelope peak to the envelope full-wave average at the receiving end is detected. This measures the pulse dispersion produced by the channel. Amplitude and phase distortion in the transmission channel tend to disperse the energy in each pulse, thereby reducing the peak-to-average ratio. The current objective for customer-to-customer connections requires a P/AR value of at least 48, on a scale of 0 to 100. A P/AR of 50 or more usually indicates that intersymbol interference will be acceptable for 2.4 kbps data. Although the P/AR method is simple to apply in the field, it has the property of responding to linear distortion and other impairments that produce pulse dispersion. It may be applied for assessing delay distortion when that is the predominant impairment on a channel.

Nonlinear Distortion

Nonlinear distortion (where the amplitude of the output signal from a channel does not have a linear relationship to the amplitude of the input signal) produces intermodulation products of the transmitted signal that fall within the channel bandwidth. (See Bell Laboratories 1982, pp. 389-401.) The amount of distortion may vary with time, even during a call. It is most often introduced by such transmission components as multiplex terminals, voice-frequency amplifiers, and station sets.

Nonlinear distortion within a channel is measured using a 4-tone test. Two pairs of equal-level tones (856/863 Hz and 1374/1385 Hz) with a composite power of −13 dBm0 are applied to the transmitting end of the channel, and second- and third-order products formed by nonlinearities within the channel are measured at the receiving end. The −13 dBm0 power level corresponds to the maximum power limit permitted for voiceband data averaged over a 3-second interval. This is roughly equivalent to a long-term average speech signal power of −16 dBm0. The overall customer-to-customer objective for nonlinear distortion is that second- and third-order modulation products should be at least 27 and 32 dB, respectively, below the received composite signal power.

Incidental Modulation

Incidental modulation is any unwanted amplitude or phase modulation appearing on a received data signal. Its effect on transmission is similar to that of impulse noise. When the variations in amplitude or phase are small, the impairment is called *jitter*. Larger changes in amplitude and phase are called *gain hits* and *phase hits*, respectively. Both jitter and hits involve rapid amplitude and phase changes, which may be either periodic or random.

Phase jitter, which perturbs the zero crossings of a data signal, tends to be more serious than amplitude jitter and can often be traced to an ac power hum on carrier frequencies and power supplies associated with FDM terminal equipment. Synchronization and timing signals as well as data signals are subject to phase jitter resulting in higher error rates. The current customer-to-customer objective restricts the maximum peak-to-peak phase modulation to 8 degrees for modulating frequencies between 4 and 300 Hz.

Gain and phase hits are produced in the network in several ways including automatic protection switching of radio channels during fading and manual switching of transmission facilities or multiplex equipment from working to standby status for maintenance work. Switching a transmission facility can cause a direct hit on the transmitted signal because of the phase or attenuation difference between the working and the standby facility. If the switch occurs within synchronizing equipment or on a facility carrying a synchronizing signal, a hit will affect the synchronizing signal and, in turn, several data signals. Gain hits of 3 dB or more begin to cause errors in data transmission. Similarly, phase hits of 20 degrees or more are likely to cause errors. The current end-to-end objectives are for no more than eight gain hits of 3 dB or more or eight phase hits of 20 degrees or more in a 15-minute interval.

AUTHORS

A. O. Casadevall
W. C. Roesel

7

Switching

7.1 INTRODUCTION

The primary function of switching in a telecommunications network is to interconnect the telephones or other telecommunications terminals of a large number of customers as economically as possible. As shown in Section 3.1, centralized switching requires only one telephone and one telephone line per customer; the interconnection of n customers without switching requires $n-1$ telephones per customer and a total of $[n(n-1)]/2$ wire pairs. Even with centralized switching, as n and the area served continue to grow, points are reached where using two or more switching systems interconnected by transmission facilities becomes the most economic way to provide telecommunications access for all n customers. The public switched telephone network (PSTN) extends the concept of a network of centralized switching systems to a very large geographic area through a hierarchical network of switching offices (see Section 4.2.1).

Although connection is the primary function of switching, other functions must also be provided, as discussed in Section 7.2. Some of these functions are associated with the operation of the telecommunications network and others are associated with customer services.

Interest in improved customer services has strongly influenced the design of modern switching systems that use *stored-program control,* a concept that enables new features to be implemented through changes in software rather than hardware. Section 7.4.3 describes stored-program control. Chapter 10 describes the various switching systems, including those that use stored-program control.

Traditionally, Bell Laboratories has been responsible for systems engineering of switching systems for both network and customer applications on behalf of AT&T and the Bell operating companies. These include systems developed by Bell Laboratories and manufactured by Western Electric, as well as systems from other suppliers.

The rest of this chapter provides a basic understanding of some of the concepts and considerations involved in the design of switching systems.

7.2 BASIC SWITCHING FUNCTIONS

Connection can best be understood by considering the two essential parts of a switching system: the **switching network** and the **control mechanism**. The *switching network* consists of the individual switching devices used to connect the communication paths. The *control mechanism* provides the intelligence to operate the appropriate switching devices at the proper time. (Sections 7.3 and 7.4 discuss switching networks and control mechanisms, respectively.)

In addition to the basic function of connecting communication paths, a switching system must be capable of receiving and sending network control signals between itself and customer terminals and other switching systems, performing functions required to administer and maintain the system, and providing customer services. Each of these switching functions is described below. These additional functions are many and complex and must interact properly, compounding the work of switching system designers.

7.2.1 CONTROL

Control is the technique by which a switching system interprets and responds to signals and directs the switching network. In the past, control was accomplished by logic circuits using relays and other electromechanical devices. Today, virtually all new switching systems employ stored-program computer control. By changing and adding to the stored program, the operation of the switching system can be modified and extended.

7.2.2 SIGNALING

All modern switching involves the transfer of information (for example, dialing and ringing) between users and the switching system and between two switching systems. This is known as *signaling* and can be thought of as a special form of data communication.

In early automatic switching systems, most signaling between systems involved dc electrical signals. Later, as signaling distances increased, single- and multiple-frequency tones were used. Recently, faster and more versatile digital signaling over dedicated networks separate from the voice network has been introduced. (Chapter 8 discusses the signals required in the network and the various signaling techniques.)

7.2.3 ADMINISTRATION AND MAINTENANCE

Today's switching systems provide separate features to ensure that the switch operates reliably and efficiently. These features monitor, test, record, and permit human control of service-affecting conditions of the switching system. Examples include:

- *network management*, which enables traffic to be rerouted to avoid congested portions of the PSTN (see Section 5.6)

- *traffic measurement*, which provides indications of the traffic loads being carried by various components of the switching system (see Section 5.7)

- *billing*, which allows recording of call-related information required to charge properly for service (see Section 10.5)

- *maintenance*, which involves features that automatically detect, isolate, and often locate system and component troubles to within several plug-in circuit packs (see Section 13.3.3).

Today, much of the operating data are reported by the switching system to operations systems that collect, analyze, filter, and summarize the data for human use. Chapter 14 describes some related operations systems.

7.2.4 CUSTOMER SERVICES

In addition to connecting communication paths, modern switching systems also provide a variety of customer services. As described in Chapter 2, customer services are enhancements to basic interconnection. They include operator services, coin services, Custom Calling Services, and special business features such as centrex service. Some examples of switching system functions related to these services are:

- routing a call for a nonworking number to an intercept operator

- returning deposited coins at a coin telephone when the called party does not answer

- routing a call to a line other than the one dialed (Call Forwarding)

- identifying the calling line for billing purposes on outgoing calls.

7.3 SWITCHING NETWORKS

The *network* is the portion of a switch that provides the connection between communication channels (lines or trunks) terminated on the system. Traditionally, switching networks are made up of connective devices or circuits arranged in a structure that allows the simultaneous connection of many pairs of communication channels. This mode of switching is known as *circuit switching*, denoting the dedication of circuits to each connection for the duration of the call. Other forms of switching are currently in use for data communications (see Section 11.6), but the telephone networks of today are virtually all circuit switching.

7.3.1 CIRCUIT-SWITCHING NETWORK TYPES, APPLICATIONS, AND TECHNOLOGIES

Two types of circuit-switching networks in use today are distinguished by the manner in which the information passes through the network. In **space-division networks**, the message paths are separated in space, as described in Section 7.3.2. In **time-division networks**, the message paths are separated in time, as described in Section 7.3.3.

As discussed in Chapter 4, switching systems are used in a variety of applications throughout the network. For **local service** (class 5 operation), 2-wire switching is sufficient. In 2-wire switching, one message path (the equivalent of two wires in a space-division switching network) simultaneously carries the transmitted and received information. For **toll operation** (classes 1, 2, 3, and 4), however, the need for more stringent control of transmission impairments (such as echo) due to the longer distances involved generally requires 4-wire switching. In space-division switching networks, 4-wire switching is provided over two message paths, one for each direction of transmission. As will be seen in Section 7.3.3, time-division switching systems are inherently 4-wire since the information flow in each direction is switched separately.

Three basic types of technology have been used to implement switching networks:

- The **manually-operated** switch where an operator places plug-ended wires (*cords*) in jacks is the oldest.

- **Electromechanical** switches are either motor-driven, gross-motion devices (such as rotary and panel switches), gross-motion devices driven by electrical impulses (such as step-by-step switches), or electromagnetically operated, fine-motion switches (such as crossbar and dry reed matrix switches). Gross-motion switches have inherent limitations in their operating speed; and they use common metal (for example, copper) contacts to withstand the wear of the sliding or wiping motions, which may, in time, result in noisy transmission paths. Also, the considerable wear on the contacts causes maintenance costs to be high. In fine-motion switches, contact motion is essentially perpendicular to the contact surfaces, resulting in much less wear. Therefore, precious metal (for example, gold) can be used to improve transmission quality. (Section 10.2 describes electromechanical switches in detail.)

- **Electronic** switching elements are usually made from semiconductor devices. The evolution of semiconductor technology has reduced the size, power consumption, and cost of electronic switching elements and, at the same time, increased their operating speeds, ruggedness, and reliability. This has made the use of electronic networks in switching systems practical. In particular, their operating speeds, which are orders of magnitude faster than electromechanical switching elements, are necessary for their use in time-division switching networks. These networks make the most efficient use of electronic switching elements in terms of the number of elements required. Because electronic switching devices tend to have less ideal transmission characteristics than metallic contacts, electronic switching networks that handle digitally encoded (for example, pulse-code modulation[1]) signals are also easier to design than those that handle analog signals. With the decreasing cost of digital coding circuits and the increasing use of digital transmission systems, most new circuit switching systems employ time-division networks to switch digitally encoded speech and data.

7.3.2 SPACE-DIVISION NETWORKS

The most common switching network in use today in the Bell operating companies is the space-division network. In space-division networks, a physical, electrical, spatial link is established through the network to connect terminated lines and trunks. Space-division networks are constructed of *stages* of metallic devices or semiconductor network elements. The electrical characteristics of the connecting links are generally suitable for carrying message signals requiring wide bandwidth or considerable power. The connecting path through the network is maintained for the duration of the message, and either analog or digital information (within the bandwidth of the network) may be passed.

Network Topology

The topology of a network describes the pattern in which the elements of the network are interconnected to allow input terminals to have access to output terminals. Most space-division networks involve **concentration, distribution**, and **expansion** as stages in switching. They are designed to make the required interconnections with little probability of blocking, while minimizing the number of switch elements in the network. The network must also be designed to be changed in size conveniently to adjust to changing demands in terminals served and traffic carried.

---

[1] Section 6.4.3 discusses pulse-code modulation.

Figure 7-1 illustrates a 4-point rotary switch. By directing the *wiper* (the moving rotary contact) to the proper position, any of the four customer lines can be connected to the rest of the network via the output link. While one of the lines is in use, however, the others are blocked. The usage of the four lines is *concentrated* on one output link. More paths and less blocking can be provided by multiple appearances of the same lines with access to two output links as shown in Figure 7-2. This allows two simultaneous conversations, but at the cost of an added switch and increased control complexity. (Some means must be provided to choose an idle concentrator switch for the second customer and to ensure that both switches do not inadvertently serve one customer line simultaneously.) With four inputs and two outputs, the concentration ratio is 2 to 1. An output link is in use, on the average, twice as much as an individual line.

Once the traffic has been concentrated, the switch must be able to *distribute* the traffic to the proper output terminals of the network. An example of distribution is shown in Figure 7-3 where each input has access to each output terminal. A network (or portion of one) that provides this function is known as a *distribution network*. In contrast to concentration networks, in which the ratio of inputs to outputs is greater than one, distribution networks have a ratio of one.

The ratio of inputs to outputs in one portion or stage of a network may also be less than one. This effect is the opposite of concentration and is known as *expansion* (see Figure 7-4). Expansion serves the same function for outputs that concentration serves for inputs. (It allows the distribution paths to operate at higher *occupancies* than that of the outputs.) Even when the outputs are trunks, they may operate at lower occupancies than the distribution paths. This may occur when the output terminals consist of many small groups of trunks connecting to different offices. It could also occur if groups of output terminals experience their busy hour of traffic at different times of the day.

The control of space-division networks falls into two general categories: **progressive control** and **common control**. In *progressively controlled* networks, a call progresses sequentially through the stages of switching. There is no capability to "look ahead," so the call may be blocked at a later stage. With *common control*, all stages of the network are examined simultaneously, and usually, a number of possible paths between the end-points are examined. This reduces the possibility of blocking in the network.

Progressively Controlled Networks

Setting up a path through a switching network with progressive control can be demonstrated using Figure 7-4, which represents a step-by-step network. First, the wiper of an idle switch in the concentration stage

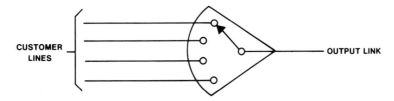

Figure 7-1. A 4-point rotary switch.

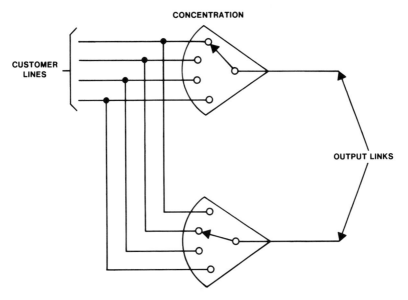

Figure 7-2. Concentration stage.

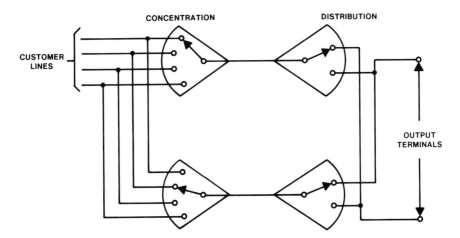

Figure 7-3. Concentration and distribution stages.

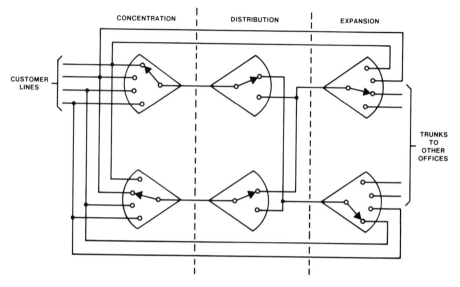

Figure 7-4. Concentration, distribution, and expansion stages.

must be moved to the input of the calling customer. Then, the connecting switch in the distribution stage must be moved to select the proper path to the expansion stage, where again a one-out-of-four choice must be made. Section 7.4.1 discusses control of such a network, in which the paths are established one stage at a time. Although a step-by-step network was used in this example, progressive control has also been employed in crossbar switching systems.

Progressively controlled networks have two inherent disadvantages:

• The blind progression of the connection through the network may not find an available complete path because all possible choices of connecting paths are not examined.

• They can be used to set up paths in one direction only. This implies that each customer must have two terminations, one for originating and one for terminating traffic.

Coordinate Networks

A *coordinate switch* consists of an array of contacts or crosspoints arranged in a matrix through which several inputs can be independently connected to several outputs. (As mentioned earlier, rotary switches are one-to-several or several-to-one devices.) Consequently, several communication paths can exist simultaneously in a coordinate switch.

Figure 7-5 represents a coordinate switch (typically used in crossbar and reed matrix switches). The Xs located at the intersections of the horizontal and vertical lines represent the crosspoints, which may be metallic

248

contacts, diodes, transistors, etc. In a coordinate network, generally only one crosspoint on any row or column can be in operation at any time. (Figure 7-6 shows the coordinate switch equivalents to rotary switch networks.)

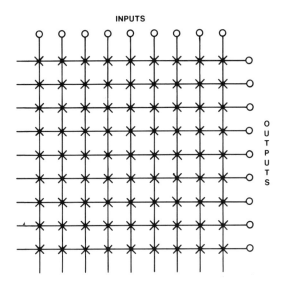

Figure 7-5. A 9-by-9 coordinate switch.

Figure 7-7 shows a *coordinate network* made up of several stages of coordinate switches. It provides full access between the nine inputs and the nine outputs and requires only fifty-four crosspoints as opposed to the eighty-one required in the single switch of Figure 7-5. In Figure 7-7, input A can be connected to output B through the single link connecting switch 1 to switch 6. However, once this connection is made, no further connections between inputs on switch 1 and outputs on switch 6 are possible until the connection between A and B is released.

By introducing three more switches (a third stage of switching), as shown in Figure 7-8, three paths (I, II, III) are available for interconnecting input A to output B. Each path consists of two links (for example, I_1 and I_2). This network requires eighty-one crosspoints, but it is still not nonblocking in contrast to the single 9-by-9 coordinate switch shown in Figure 7-5.

Although no crosspoints are saved in the example in Figure 7-8, in general, where the switching networks must handle between hundreds and tens of thousands of inputs and outputs, multiple stages of coordinate switches will produce a large savings in the total number of

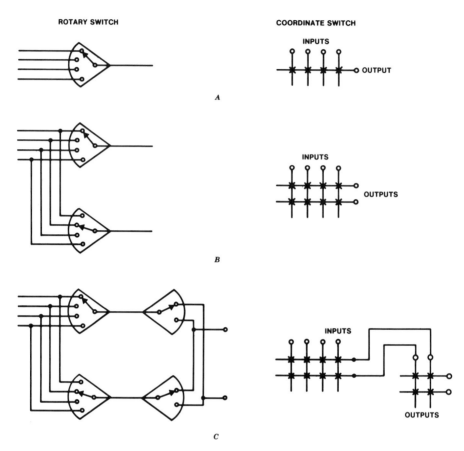

ROTARY SWITCH

COORDINATE SWITCH

INPUTS

OUTPUT

A

INPUTS

OUTPUTS

B

INPUTS

OUTPUTS

C

Figure 7-6. Rotary switches and equivalent coordinate switches. *A*, concentrator; *B*, multipled concentrator; *C*, concentration and distribution.

crosspoints (and, hence, the cost) compared with a single, large coordinate switch. For example, a 3-stage network of 10-by-10 coordinate switches would have relatively little blocking with a 70-percent savings in crosspoints over a single-stage 100-by-100 coordinate switch. The choice of switch size and the number of stages required depends on the expected range and the size of the network. Typically, 10-by-10 coordinate switches in network switching systems are found in a 4- to 8-stage coordinate network.

7.3.3 TIME-DIVISION NETWORKS

In space-division switching networks, each call has its own physical path through the network. In time-division networks, simultaneous messages are separated by assignment to separate time slots. Time-division networks can be divided into two types, depending on the form of the signals switched. **Pulse-amplitude-modulated switching networks** switch

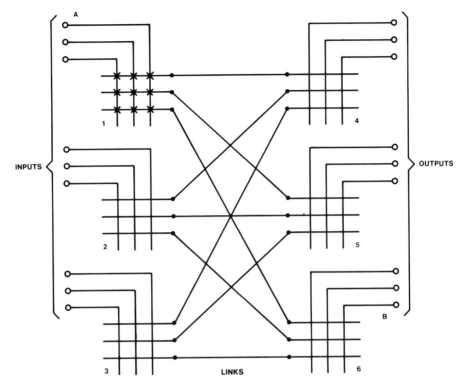

Figure 7-7. A 2-stage 9-by-9 switch using six 3-by-3 switches.

pulse-amplitude-modulated samples[2] of the signals; **digital time-division switching networks** switch digitally encoded samples of the signals.

The type and format of the digital encoding of voiceband signals is an important factor in the design of a digital time-division switching network. When the proper format is chosen, such networks can provide a direct interface with digital transmission facilities carrying digital time-division multiplex (TDM) signals.[3] Generally, with space-division networks, TDM signals must be demultiplexed (separated), and the separated pulse-code modulation (PCM) signals must be converted to analog form by digital-to-analog converters before passing through the switching network. Both conversions and converters can be eliminated when digital transmission facilities connect to digital time-division switching systems that have the proper format. The use of digital transmission is growing rapidly, and time-division switching networks offer significant economies in parts of the telecommunications network with a heavy concentration

---

[2] Section 6.4.3 discusses pulse-amplitude modulation.

[3] Section 6.5 discusses TDM.

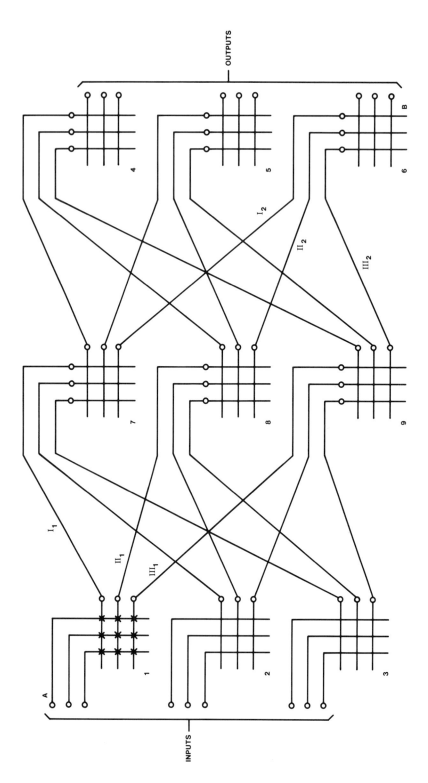

Figure 7-8. A 3-stage, 9-by-9 coordinate network using nine 3-by-3 switches.

of digital facilities. Even where analog facilities predominate, new time-division switching systems are more economical than the space-division systems they generally replace because of the new technology they employ.

Principles of Operation

In Figure 7-9, customers I_1 to I_n are talking to customers O_1 to O_n through a simple **time-division switch**. The talking paths are opened and closed by switching devices or *gates* labeled A_1 to A_n and B_1 to B_n. The switch connects the desired pairs of customers by controlling the operation of the selected gates. To connect I_1 to O_2, gates A_1 and B_2 are closed during the same "time slot," and a sample of the speech signal is carried over the common path (or *bus*). Customers I_3 and O_3 may be connected by closing gates A_3 and B_3 during another time slot. As discussed in Section 6.4.3, if the speech signals between I_1 and O_2 are sampled at least 8000 times per second, the resultant pulse-amplitude-modulated signal provides satisfactory speech transmission. The time slot for I_1 and O_2 must, therefore, be repeated at that rate, as must the rest of the n time slots.

In a digital time-division switch, the amplitude samples are encoded into PCM form, and the resultant sequence of binary digits is switched. Lines or trunks into a switching system from digital facilities present signals from multiple calls in a time-division-multiplexed format. The signals are separated and switched by a **time-slot interchange** (TSI).

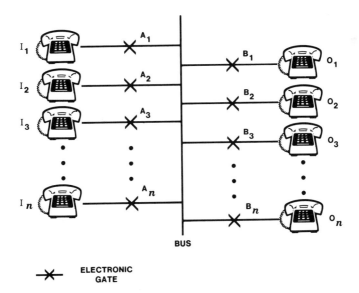

Figure 7-9. A simple time-division switch.

As shown in Figure 7-10, the TSI can be thought of as having storage locations (*stores* or *buffers*) associated with specific time slots in the incoming TDM bit stream and the outgoing multiplexed bit stream for each direction of transmission. A transfer (*read-write*) operation can change the order of the time slots by changing the order of the information from the input to the output stores. The transfer operation is determined by routing requirements and directed by the call processor. The transfer instruction for a given time slot is set for the duration of a call on that time slot. A mirror transfer is performed for the other direction of transmission through the TSI. It is worth noting that an n-channel TSI is equivalent to an n-by-n space-division switch since the n inputs have full nonblocking access to the n outputs.

The figure shows some samples from four calls that are in time slots 1, 2, 3, and n of the input multiplexed bit stream being routed through a TSI. The binary information from that line is directed to locations 1, 2, 3, and n of the input store. The routings for these calls have established that the calls should be moved to locations r, $r+1$, 2, and 1, respectively, of the output store and, thus, time slots $r+1$, r, 2, and 1 of the output bit stream.

In the simplest form, time-division switching can be accomplished by a switch consisting of one TSI, with n customers attached to the input side and the same n customers attached to the output side. A given customer's loop could be associated with a specified time slot (by means of

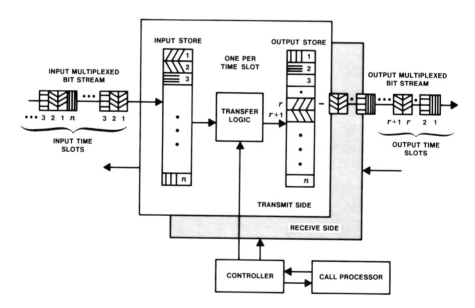

Figure 7-10. A time-slot interchange.

multiplex equipment such as a *channel bank*).[4] Then, by transferring samples from input time slot 3, for example, to output time slot 2, customer 3 could talk to customer 2. Likewise, on the receive side of the switch, time slot 2 would be transferred to time slot 3 to allow transmission in the other direction. In most applications, as described in the following section, the TSI is used as one of the primary building blocks of a time-division switching network.

The TSI shown in Figure 7-10 is a nonblocking network component, since any time slot has access to any other unused time slot, and both an input and output store are not necessary if there is flexibility in the order of reading from the input store directly to the output multiplexed bit stream.

Network Architecture

In many practical switching applications, a number of multiplexed input and output bit streams appear at the switching system, requiring the use of multiple TSIs.[5] Since the switching network must be able to switch any input time slot to any output time slot as determined by call routing, it must be capable of connecting the outputs of one TSI to the inputs of other TSIs. This leads to the other basic building block of time-division switching networks—the **time-multiplexed switch** (TMS). The TMS operates as a very high speed space-division coordinate switch whose input-to-output paths can be changed for every time slot to rearrange the interconnection of successive time slots of TSIs.

The TMS consists of high-speed electronic switching elements (gates) at the crosspoints of an m-by-m switching matrix as shown in Figure 7-11. Each of the m inputs and outputs is a multiplexed bit stream from a TSI. The shaded gates indicate the input/output connections during one time slot. When the clock signals the start of the next time slot, the controller changes the connections by sending control pulses to the appropriate gates. A TMS is essentially an m-by-m space switch with a third dimension—time. Instead of leaving the network paths up for the duration of a call as in a space-division switching system, a TMS is changed for each of the n time slots in one TDM frame. Conceptually, it can be viewed as n m-by-m space switches as shown in Figure 7-12, where each m-by-m matrix switches one of the n time slots from each of the m multiplexed bit streams.

---

[4] See Section 9.4.3.

[5] A TSI can be made to operate n times as fast as the incoming bit stream and thereby multiplex and switch n lines of incoming bit streams. However, a limit is reached in the speed with which TSIs can operate, and multiple TSIs are then required.

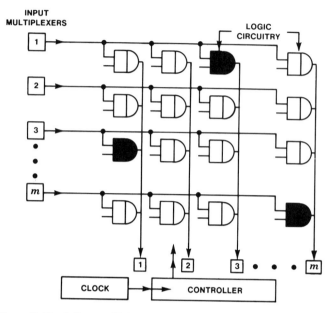

Figure 7-11. A time-multiplexed switch configuration shown during one time slot. Input lines 1, 3, and *m* are connected to output lines 3, 1, and *m*, respectively.

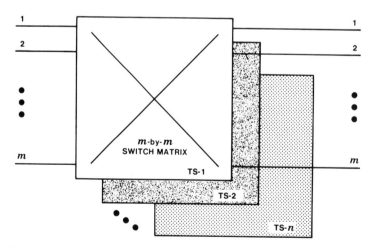

Figure 7-12. An analog equivalent of a time-multiplexed switch. The figure shows *n* different *m*-by-*m* switch matrices, one for each time slot of the input multiplexes.

Figure 7-13 illustrates one way of operating a TMS in conjunction
with TSIs in a time-division network. Since time-division networks are
designated according to the sequence of time and space switching stages,
the network shown is a time-space-time (TST) network, as are most digital time-division switches in the Bell System. The TMS network may
actually employ several stages of switching, such as in a time-space-
space-time (TSST) network. The figure shows the connection of two communications channels, each of which is represented by a time slot at one
of the TSIs. In the example shown, the directions of transmission have
been separated prior to arriving at the TSI. Samples from channel A
arrive in time slot 3 on TSI A. Samples from channel B arrive in time
slot 1 on TSI B. Each TSI has a mirror image (the shaded squares in the
figure) that handles the reverse direction of transmission. To connect
channels A and B, the central control establishes a path between TSI A
and TSI B during time slot 25 of the TMS. TSI A transfers channel A
samples to TMS time slot buffer 25. During time slot 25, the TMS connects the output buffer of TSI A to the input buffer of TSI B. TSI B then
transfers the samples to channel B's time slot, time slot 1. The other
direction of the transmission path progresses similarly and simultaneously on the other (shaded) halves of the TSIs, using the same time slot
of the TMS.

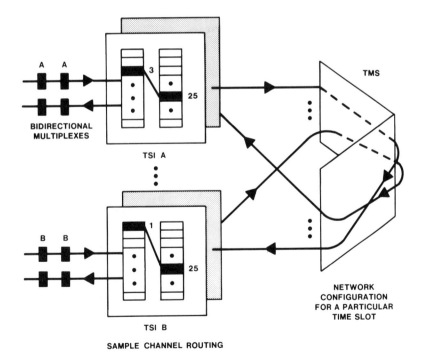

Figure 7-13. A time-space-time (TST) digital network.

Other arrangements of time (T) and space (S) stages are in current use in digital time-division networks. These include STS, TSTS, and SSTTSS.

As in space-division switching, time-division networks may use concentration and expansion to reduce the size of the switching network fabric. This concentration may be done prior to the TSI or may be done digitally as part of a TSI.

The number of time slots provided for in a time-division network depends on the speed of the technology employed. Typically, in a voiceband network, there may be from 32 to 1024 time slots. The coded information in digitized samples may be sent serially (typically, eight bits per sample), in parallel, or in combinations. Extra bits are sometimes added as they pass through the switch to check parity, to transmit supervision states, or to allow for timing adjustments.

Digital time-division networks are economically attractive because the network fabric may be constructed with large-scale integrated circuit components. Interfaces to digital transmission facilities are less costly and avoid the transmission impairments introduced in digital-to-analog and analog-to-digital converters that are required with analog space-division switches. Therefore, time-division networks promise better compatibility with the current trends in digital techniques being applied to data and speech transmission.

7.4 CONTROL MECHANISMS

The control mechanism in a switching system must interpret and respond to signals and operate the switching network. Techniques of control have evolved from early electromechanical switches (which were driven directly by dc pulses from telephones) to sophisticated computer systems.

7.4.1 DIRECT PROGRESSIVE CONTROL

In the earliest automatic switching systems, each switch has an associated controlling mechanism. Step-by-step systems, for example, have a stage of switching associated with each dialed digit. As the customer dials, successive stages of switching respond to the dial pulses that represent successive dialed digits and progressively select a path through the network; hence, the term *direct progressive control*. The dialed number is not stored in one place; rather, each dialed digit is represented by (in effect, stored in) the position of the related switching device.

Direct progressive control systems are advantageous in that the integration of switching network and control is economical for small- and medium-size offices. (Even in 1983, there are thousands of these smaller step-by-step offices in service in Bell and independent operating companies, and their replacement has only become significant in the last few years.) However, there are disadvantages:

- Customer lines must be connected to switch terminals in strict correspondence with directory numbers since no logic is provided for translation. Telephones with similar numbers that receive many calls will be subject to undue blocking. This may require a change in one of the numbers and may adversely affect that customer.

- Alternate routing is not possible since dialed digits are "used up" by the system instead of being stored for possible future reuse.

- The switching network is inefficient because it cannot look ahead to see if it will be blocked along the way.

- New services are usually impossible or expensive to incorporate in a direct progressively controlled switching system.

7.4.2 REGISTERS, TRANSLATORS, AND MARKERS — COMMON CONTROL

Switching designers alleviated some of the problems of direct control by adding a **register** to the control of a switching network. When customers take their telephones off-hook to originate a call, the control responds immediately (before dialing can start) to set up a connection through a first portion of the network from the originating line to an idle register among a group of registers. The portion of the network used may be the concentrator of the switching network or a special separate network called a *connector*. In this register (or indirect) progressive system, the dialed digits are collected in the register. Logic associated with the register then causes digits to be pulsed out to the rest of the switching network to make the proper connection. If blocking occurs, a second attempt to find a connecting path through the switching network can be made because the register still contains the dialed digits.

The registers are not dedicated to each customer on a per-line basis but instead are connected after the concentration stage and, thus, are common to many customers. Once the connection is established to the calling line, the register is cleared of its collected digits and returned to the idle state to serve another originating call. Thus, a few tens of registers can serve all the calls generated by thousands of customers. This sharing of control registers in a progressive network is a limited form of the *common-control* principle.

Translators convert one multidigit number to another. For example, once a register has collected the dialed digits, it can search for and connect itself to an idle translator circuit. The translator then converts the dialed number (or directory number) into another number that determines the location of the terminating connection (switch terminal) of the number being called. Thus, switching system design is freed from the constraints of the decimal numbers associated with the rotary dial. Also, by simple rewiring in the translator, a customer's terminating line connection can be moved without requiring a corresponding change in the directory number.

Since a register can obtain the information it needs from an electromechanical translator in less than a second, one translator can serve the calling traffic for a number of registers, further concentrating control and progressing toward full common control.

Common control takes advantage of the fact that control functions are not needed for the entire duration of a call but only for small portions of the time. With common control, many users share the control equipment, thus reducing the amount of control equipment that must be provided. The common-control principle is further exploited in **marker** systems, where path selection and switch control functions are concentrated into a wired logic unit called a *marker*.

The coordinate network in Figure 7-8 can be used to demonstrate marker control of a space-division network. The marker has access to the busy-idle condition of all the links in that network. If the control determines that terminal A is to be connected to terminal B, then it can examine the status of links I, II, and III for an idle path. Having found an idle path, it will operate the associated switch crosspoint to make the connection, and the links will be made busy, or "marked," for the duration of the call.

It should be noted that common-control techniques using a marker are not limited to coordinate networks. With link access, a marker can also be used to control the step-by-step network shown in Figure 7-4. In actual practice, all four combinations of progressive or common control with step-by-step or crossbar switches have been used in public telephone networks.

When markers are used to control switching networks, separate registers may still exist (see Figure 7-14), but instead of driving the switches directly, they pass the digits to a marker. The marker makes translations (for example, from directory number to terminal identification), tests many possible paths simultaneously to select one, and causes the proper switches to operate.

As shown in Figure 7-14, connectors efficiently access the functional elements of common control to communicate control information from a customer line to registers, markers, translators, and finally, to the network to set up the call. This functional approach allows more flexibility of action and ease of change. Modifications to such systems have resulted in many new services and wider applications of a single system to serve as local, tandem, and toll switches and as automatic call distributors.[6]

---

[6] Section 11.2.6 discusses automatic call distributors.

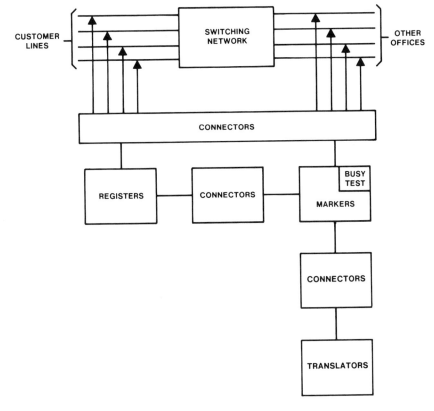

Figure 7-14. Marker-type common-control system.

Common-control functions originally used electromechanical switches and relays. With the advent of solid-state electronics, some or all of the common-control functions were implemented with electronic wired logic. This allowed a further reduction in the number of control elements, saving floor space and money. Whereas a large switching system carrying heavy traffic might require a dozen or more relay markers to control its network, one electronic marker may handle that same workload, although a second such marker might be included primarily to serve as a backup.

7.4.3 STORED-PROGRAM CONTROL

With stored-program control (SPC), specially designed processors that execute software programs replace the wired-logic common control. Since the addition of new customer services and special features, changes in translations, and other modifications can be implemented as changes in the program rather than as more complex changes in hardware, SPC greatly enhances flexibility. Figure 7-15 illustrates the basic components of a typical SPC system with centralized control. Besides the switching network, they include:

- one or more high-speed processors (central control) that interpret and execute the instructions of the program

- a memory (for example, program store) that stores the program

- a memory (for example, call store) that is used as an erasable scratch pad to record and accumulate data during call processing

- input signal devices (for example, scanners) through which the central control receives information such as customer on-hook, off-hook, and dialed digits

- output signal devices (for example, signal distributors) through which the central control causes network switches to operate.

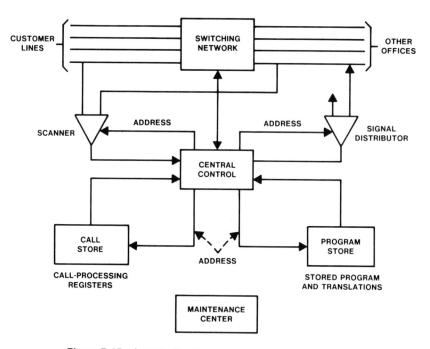

Figure 7-15. A centralized stored-program control system.

In the particular system shown in Figure 7-15, the principle of common control is fully applied. Processors perform all the functions necessary to switch calls. Through time sharing, processors simultaneously handle many calls in various stages of completion. The central control executes one function per call and then progresses to the same function on a different call or to another function on the same or on a different call. The processors are sufficiently faster than the network they control, so that they are able to interleave the control functions for many different calls without creating any delays noticeable by the customer.

Of the several types of SPC processor arrangements, the most common are a full **centralized control, independent multiprocessors** that perform load sharing, and **functional multiprocessing** arrangements in which different processing functions are allocated to different processors.

Centralized Processing

The application of electronics allows a large system to be fully controlled by a single high-speed processor. This has the advantage of greatly simplifying the interface circuits between the controller and the rest of the switching system. It also presents a single point (the processor's software) for introducing new features and services. It has the disadvantage that a complete control complex must be provided even for small office sizes, and complete redundancy is often required to allow the system to operate in the presence of a processor fault or while changes are made to the controller.

Independent Multiprocessing

By using more than one processor, the central controls may be designed to operate independently and share the workload. Load sharing allows two (or more) central controls to handle more calls per unit time than would be possible with a single (usually duplicated) unit. However, in the event of a failure of one unit, the remaining units must carry on with reduced system capacity. Under normal conditions, the two (or more) processors work independently, each performing the full range of functions required for call processing. At least part of the memory (for example, the busy-idle status of lines) must be accessible by all processors in order to avoid conflicting actions among the processors.

A major advantage of multiprocessing is that small offices require less than the maximum number of processors, lowering the cost of control. Increased capacity can be provided by adding processors. However, as more processors are added, the call capacity of each processor decreases. Conflicts on processor access to memory and peripheral equipment increase with the number of processors. Unless the amount of sharing of memory and peripheral equipment is carefully controlled, independent multiprocessing rapidly reaches a practical limit with respect to office size and would not be an economical approach for large switching offices.

Functional Multiprocessing

The advent of low-cost microprocessors has encouraged the distribution of the call-processing functions to a number of processors within a single switching system. Functional multiprocessing involves the allocation of different call-processing functions to different processors. As in load sharing, the processors frequently communicate with one another either directly or through common memory.

Functional multiprocessors may be arranged in a sequential, hierarchical, or hybrid structure. In sequential structures, as on an assembly line, each processor is responsible for a portion of a call and hands the next step to a succeeding processor. In hierarchical structures, as in a master-slave relationship, a master processor assigns tasks to subsidiary processors and maintains control of the system. Frequently, the subsidiary processors are dedicated to segments of the switching or signaling circuits. As network and peripheral equipment modules are added, this control capability is correspondingly enlarged. This enables processing capacity to be added as the system size increases and avoids incurring the cost of excess processor capacity. Most recent switching systems, such as the 5ESS switching equipment (discussed in Section 10.3.4), use functional multiprocessing.

AUTHORS

C. E. Betta
W. R. Byrne
W. S. Hayward
L. G. Raymer
F. F. Taylor

8

Signaling and Interfaces

8.1 INTRODUCTION

8.1.1 SIGNALING

Signaling is the process of transferring information between two parts of a communications network to control the establishment of connections and related operations. For example, a customer originating a simple call usually has three signals to send: call origination (receiver off-hook), call address (dialing), and call termination (receiver on-hook). The signaling considered here is normally carried on over a distance and is of two traditional application-related realms: **customer-line signaling** and **interoffice trunk signaling**. *Customer-line signaling* refers to the interaction between the customer and the switching system that serves the customer. *Interoffice trunk signaling* refers to the exchange of call-handling information between switching offices within the network. Certain applications, such as foreign exchange lines, may have some characteristics of each of these signaling realms and therefore are discussed as a third realm called **special-services signaling**. Alerting stations on a nonswitched private-line configuration is another signaling process in this realm. A fourth signaling realm has been created by the need to communicate directly among the computer-based systems that have been introduced to support telephone company operations. These signals are often not directly associated with a specific line or trunk but are carried by dedicated data links. The operations systems are linked by the operations system network (see Section 15.5), which also links some operations systems to switching and transmission systems.

The term *signaling*, as it is used in this chapter, does not include the many and varied call-handling interactions between components within a switching system or within the more complex forms of station equipment.

The amount of information sent in signaling is normally much less than in the associated communications. As a result, network control signals have generally been carried by the same transmission channel as message signals. Use of the voice channel when it is not occupied by conversation, as in *TOUCH-TONE* dialing, or use of frequencies not occupied by voice are typical examples of how this is done.

In considering signaling, then, it is necessary to describe interfaces between terminal equipment and transmission system, between transmission system and switching system, and between transmission system and transmission system.

8.1.2 INTERFACES

An *interface* can be thought of as the common boundary or set of points where two systems or pieces of equipment are joined. *Interface specifications* are the technical requirements for mating equipments. An *interface device* is any equipment used on one side of an interface to ensure that the interface specification will be met.

The set of boundary requirements at an interface, to a degree, separate responsibilities on the two sides of the interface. This allows each side the flexibility of rearrangement and evolutionary introduction of new equipment and services. The interface specification should be defined so as not to impede technological progress and to minimize the need for changes in the interface specification itself as new products and services evolve.

Interfaces also provide a demarcation point from which testing can be performed on the individual units being mated. If the interface is well defined, the units can be designed and tested independently, ensuring that they will work together as a total system. The interface specification also makes it possible to conduct tests at the interface to locate the sources of service impairment after equipment is put into service. Ideal interfaces are not always achievable, and, as will become apparent, the term *interface* sometimes is used in a broader sense than defined above.

The first five sections of this chapter discuss signaling as a principal component of a telephone interface specification, including descriptions of the **signaling functions, applications, interfaces,** and **techniques.** The rest of the chapter discusses other forms of interfaces as they apply to the telecommunications network.

8.2 SIGNALING FUNCTIONS

A typical telephone connection sequence from one line to another in the same central office illustrates the primary functions of signaling. These functions (and the corresponding actions by the customer where applicable) are listed below. Figure 8-1 is a schematic of this sequence including

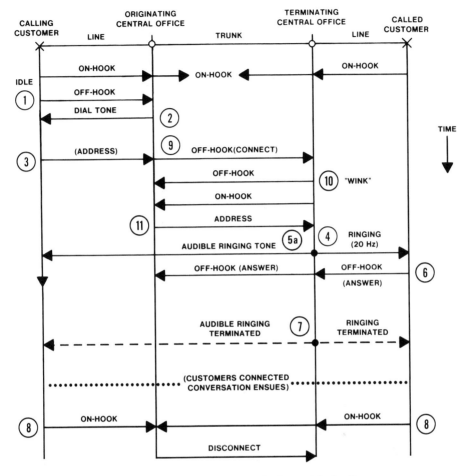

Figure 8-1. Signaling on a typical completed call.

the associated switchhook states.[1] The circled numbers in the figure correspond to the steps in the list.

1) A request for service is initiated when the caller lifts the telephone handset off the cradle or switchhook.

2) The central office sends a dial tone to the caller to indicate that the caller may begin dialing.

---

[1] When both customers are served by the same central office, that office handles both the originating and terminating central office functions shown in Figure 8-1 (steps 1 through 8 in the text).

3) The address of the called station is transmitted when the caller dials the called number.

4) If the called station is not busy, the central office alerts the called party by sending a *ringing signal* to the called station.

5) Feedback is provided to the originating station by the central office:

   a) if the called station is not busy, the central office returns an *audible ringing tone* to the caller; or,

   b) if the called station is busy, the central office sends a busy signal to the caller (not shown on Figure 8-1); or,

   c) if the call cannot be completed through the central office, the office sends a "reorder" message to the caller (not shown on Figure 8-1).

6) The called party indicates acceptance of the incoming call by lifting the telephone handset.

7) The central office recognizes the acceptance and terminates the ringing signal, removes the audible ringing signal, and establishes a connection between calling and called stations.

8) The connection is released when either party hangs up.

When the called customer is in a different central office, the following interoffice trunk signaling functions are required:

9) The originating office seizes an idle interoffice trunk, sends an off-hook indication on the trunk, and requests a digit register at the far end.

10) The terminating office sends a "wink" (an off-hook followed by an on-hook signal); this indicates a register-ready or start-dial status to the originating central office to initiate the output of address digits.

11) The originating office sends the address digits to the terminating office.

In Figure 8-1, the terms *on-hook* and *off-hook* have been extended to the interoffice trunk signaling junctions even though the switchhook does not exist as a part of the trunk. The "wink" signal is needed because the receiving register for the address signals is not permanently associated with a trunk; it is called for and switched to the trunk in response to the connect signal. Also, in interoffice calls, the originating office generates dial tone while the terminating office generates the ringing signal and the audible ringing tone.

The term *supervisory* is often used to refer to control functions that are basically on-off, such as request for service, answer, alerting, and return to idle. The term *address* refers to the telephone number of the called party. The term *information* refers to audible tones and announcements used to convey call-progress information to customers or operators.

The ways in which these signals are generated, transmitted, and detected differ for each realm of signaling (for example, customer line, interoffice trunk) and transmission facility used (for example, analog carrier, digital carrier).[2] Sections 8.4 and 8.5 describe, respectively, signaling interfaces and techniques used to implement signaling functions.

Most of the above signaling functions have been provided in some form since the initiation of telephone service and are apparent to the customer. Other signaling functions have been introduced more recently and are not apparent to the customer. Network management signals, for instance, are sent from one office to another to control the gross flow of traffic under unusual load situations, as described in Section 5.6. Other signals may interrogate a remote data base to access special billing or routing information for more flexible call handling. Also, a set of functional signals is provided for many electronic offices that are remotely monitored and administered by computer-based operations systems.[3] In general, with the widespread introduction of computer control and operation of the network, the list of functional signals can be expected to grow rapidly.

8.3 FUNDAMENTAL CONSIDERATIONS

A fundamental objective of the telephone industry is to make operation of the telephone as simple and universal as is practical. This has resulted in a relatively small number of arrangements for customer-line signaling, which are usually seen by the customer as highly standardized procedures. On the other hand, interoffice signaling is essentially a machine-to-machine interaction and is, therefore, less constrained by consideration of human factors. Rather, the emphasis is on overall efficiency and flexibility. Consequently, over the years, interoffice signaling has been extensively affected by new transmission techniques and advances in switching system design. This is reflected in the large variety of signaling arrangements in service.

To satisfy the objectives of uniformity for customer signaling and flexibility for interoffice signaling, a high degree of independence has been maintained between these two signaling realms. Facilitated by the

---

[2] Sections 9.3 and 9.4 describe analog and digital carrier systems.

[3] Section 14.3.2 describes one such operations system—the Switching Control Center System.

widespread use of common-control, switching systems in local central offices provide an effective signaling buffer between lines and trunks.

Another basic signaling consideration is promptness or its inverse, delay. The calling customer may experience three components of delay in completing a call: **dialing time, postdialing delay**, and **answer delay**; all involve signaling to some degree.

Dialing time is the time it takes for the calling customer to dial the desired number: from the time the customer lifts the handset (or goes off-hook) until the last digit is dialed. This delay is determined by the central office delay (called *dial-tone delay*) in recognizing the customer's request for service by providing the dial tone, the customer's reaction time, and the speed with which the customer operates the rotary dial or pushbuttons.

Postdialing delay is the elapsed time from the end of dialing to the start of ringing (or other feedback such as a busy signal) at the called end. Postdialing delay depends on many factors, including the number of switched links in the connection, the types of interoffice signaling used, switching system work times, and the traffic load.

Answer delay is the time from the beginning of ringing until the called station answers. It is determined primarily by the called customer's promptness in answering and, to a lesser degree, by the alerting method used.

The ability to interconnect with systems designed by a number of manufacturers has become more important recently. This requires greater attention and effort with respect to national and international signaling standards organizations, such as the Comité Consultatif International Télégraphique et Téléphonique (CCITT), so that interfaces are kept simple and few in number.

8.4 SIGNALING SYSTEM APPLICATIONS AND INTERFACES

For any signaling carried by a transmission facility, interfaces must be provided for the interchange of signaling information between the facility and the source and between the facility and the destination. In some cases, the interface is a well-defined demarcation point with specified impedances and voltages. (The E&M lead interface described in Section 8.4.2 is an example of this type.) In other cases, the interface may be so integrated into the circuit design that it is hard to define. In such cases, the signaling information interchange requires careful coordination in design and, to some degree, in engineering between the source, destination,[4] and intervening facility.

---

[4] *Source* and *destination* are used here to mean station equipment or a switching office.

This latter concept is best illustrated for a customer loop formed by a metallic pair. Supervision is provided between the central office and the station set by the presence or absence of direct current in the circuit formed by the two wires of the loop, the switchhook contacts in the station set, and the battery (or other source of dc voltage) to which the loop is connected at the central office as shown in Figure 8-2. (Other circuitry needed for message transmission and address signaling is omitted from the figure.)

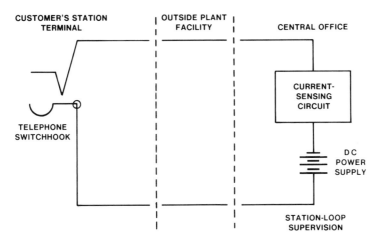

Figure 8-2. Schematic diagram of supervisory signaling interface between a customer's station terminal and the central office.

A current-sensing device is used at the central office to detect the station set's on-hook or off-hook signal. A single standard interface does not exist between the customer's line and the station set or between the line and the central office because of the wide variation in length and type of loop plant.[5] It would be prohibitively expensive to condition many millions of customer's lines to end in a standard impedance and signal level.

Until 1976, signaling in the Bell System was almost entirely on a per-line or per-trunk basis; that is, the signaling information for a particular channel was carried on the same channel as the voice or other message information. Of the various signaling interfaces, only two are described here: **loop signaling** and **E&M lead signaling**, as used for per-line and per-trunk signaling.

---

[5] AT&T 1983 and 1980a, respectively, contain definitions and some standards for the two interfaces.

Common-channel interoffice signaling, introduced in 1976, combines the signaling information for a number of channels and transmits the information over a data link derived from a dedicated facility. For common-channel signaling, the interface is primarily defined in terms of a set of binary-coded messages between digital processors. The rate and form of transmission, however, can vary. For example, the digital stream can be transmitted via data modem over an analog (frequency-division multiplex) channel or more directly over a digital (time-division multiplex) channel. The format used at this interface is often called a *protocol* (see Section 8.8). The following paragraphs discuss interfaces for the customer-line, interoffice trunk, and special-services realms (or areas of applications), which are summarized in Table 8-1.

8.4.1 CUSTOMER-LINE SIGNALING

Most customer loops consist of a pair of wires between the central office and the customer's station equipment. The percentage of loops implemented by carrier systems has, however, grown rapidly. Recent growth has primarily been in digital carrier systems that have become economically attractive as a result of advances in technology. When the loop is implemented on a carrier system, the customer's station equipment is connected by a pair of wires to a carrier remote terminal (near the customer's premises) rather than to the central office.

The signaling interface between a customer's terminal equipment and either the central office or the carrier remote terminal has a standard format known as *metallic loop signaling*. This arrangement provides for continuous application of a dc voltage from a dc power supply (see Section 12.2) toward the station, in conjunction with a current-sensing device to recognize the supervisory status of the station. The nominal value of the dc voltage is usually 48 volts and is commonly called *battery voltage*, or simply, *battery*.

Figure 8-2 shows the situation for metallic loops. Similarly, Figure 8-3 shows the situation for carrier-implemented loops. It should be noted that loop signaling is also shown between the central office switching system and the central office terminal. The carrier terminals convert the signal information to any signaling method appropriate for the carrier. (Sections 8.5.3 and 8.5.4, respectively, discuss signaling over analog and digital carrier systems.)

On-hook, or idle, is indicated by no current flow, whereas off-hook, or seizure, is indicated by the flow of current in the loop. For purposes of supervision, the battery polarity and, hence, current direction are not critical. Normally, however, negative[6] battery (−48 volts) is provided on the

---

[6] Positive voltages are generally avoided in outside plant cables because if there is any moisture present, copper may be lost through electrolysis.

TABLE 8-1

SUMMARY OF SIGNALING SYSTEM
APPLICATIONS AND INTERFACES

Signaling System Application	Interface	Characteristics
	• Loop signaling — basic station	Dc signaling
		Loop-start origination at station
		Ringing from central office
Customer line	• Loop signaling — coin station	Dc signaling
		Loop-start or ground-start origination at station
		Ground paths may be used in addition to the line for coin collection and return
	• Loop-reverse-battery	1-way call origination
		Directly applicable to metallic facilities
		Both current and polarity are sensed
Interoffice trunk		Available to carrier facilities with appropriate facility signaling system
	• E&M lead	2-way call origination
		Requires facility signaling system for all applications
	• Loop type	Standard station loop and trunk arrangement as above
		Ground-start format for private branch exchange (PBX)–central office trunks similar to that for coin stations
Special services		Automatic or ringdown for PBX nondial tie trunks
	• E&M lead	E&M for PBX dial tie trunks
		E&M for carrier system channels in special-services circuits

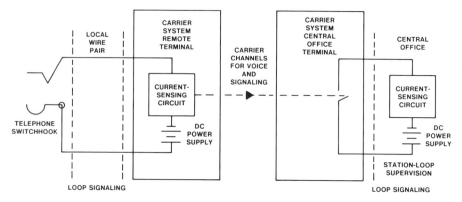

**Figure 8-3. Schematic diagram of signaling interface
for loop carrier application.**

ring conductor, with ground on the tip.[7] This format provides simple 2-state supervision from the customer toward the central office, with only minor variations introduced by the special features of different switching systems.

The customer is alerted from the central office by a ringing signal indicating the presence of an incoming call. Standard single-party ringing consists of 2-second intervals of 20-hertz (Hz) energy applied between the tip and ring conductors, followed by a 4-second quiet interval. Other ringing durations and intervals, including 1 second on, 2 seconds off, are used for multiparty lines. To provide for *ring tripping*, which turns off ringing when the customer answers, a superimposed dc voltage is generally used in association with a current detector. Following ring tripping, line potentials revert to the normal supervisory state. For loops implemented on carrier systems, ringing and other signals from the central office to the customer must be converted to a form suitable for carrier transmission. As in the case of on-hook, off-hook supervisory signaling (Figure 8-3), the carrier system is essentially transparent as viewed from either the station or central office end.

For 2-party and multiparty (4-party or more) service, the 20-Hz ringing voltage is applied to either the tip or ring conductor with respect to ground and is superimposed on either a positive or negative dc ring-tripping potential. In this way, up to four distinct signals are provided,

---

[7] In a 2-wire pair, the two leads are often called *tip* (T) and *ring* (R) after the parts of a standard telephone plug to which they connected in the days of manual switchboards. Similarly, a third wire (if present) is called *sleeve*. In 4-wire transmission, the four leads are called T, R, T1, and R1.

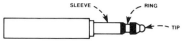

which, with appropriate connection of ringers in the station sets, provide full selective ringing of 2- or 4-party lines. With coded ringing durations, semiselective ringing of 8-party service can be provided. For 10-party service, additional coding is used beyond the standard one long and two short rings. By 1982, multiparty service was quite rare in the Bell System.

A special case of loop supervisory signaling exists for coin stations, in which additional control signals are required for coin collection and coin return. There are two basic types: **loop-start**, associated with dial-tone-first service, and **ground-start**, associated with coin-first service. *Loop-start* provides ordinary loop signaling, with the addition of battery polarity reversal used for answer supervision and a positive or negative 130-volt dc signal applied from ground to both the tip and ring conductors for coin collection or coin return. For *ground start*, seizure of the loop is initiated by applying ground to the ring conductor at the station in response to the insertion of the proper coin(s). A detector circuit in the central office recognizes this ground, applies dial tone, and establishes conventional loop supervision. Coin collection and return functions are then provided as described for loop supervision.

In addition to switchhook supervision, the customer must pass the address code of the called party to the switching office. This address information is communicated either as dial pulses or as tones from a pushbutton telephone. Dial pulses consist of short on-hook pulses occurring as interruptions in the normal off-hook loop supervision current at a 10-pulse-per-second rate. The number of dial pulses in a sequence equals the value of the digit, except for ten pulses, which equal the digit 0. The digits are separated by a relatively long off-hook interval, the length of which depends on how fast the customer can rewind and release the dial.

The ratio of break (on-hook) interval to total pulse cycle interval is frequently referred to as *percent break* and is typically between 58 and 62 percent. Twenty-pulse-per-second dialing is also used occasionally for special applications. Figure 8-4 shows dial pulsing for the digit 3.

Tones from pushbutton telephones consist of pairs of selected frequencies corresponding to the digits to be transmitted. The frequencies used are coded as shown in Figure 8-5.

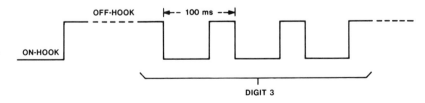

Figure 8-4. Dial pulsing.

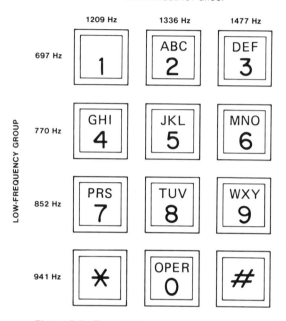

Figure 8-5. Tone-dialing frequency groups.

For *TOUCH-TONE* service, a pair of tones is generated when a button is pressed. If the number 7 is pressed, for example, the 1209- and 852-Hz frequencies are generated. Receivers at the central office recognize these tones as representing the number 7. The tones have been carefully selected to minimize harmonic interference and the probability that a pair of high and low tones will be simulated by the human voice, thus protecting network control signaling. For *TOUCH-TONE* service, signals may be transmitted over any voice circuit (providing loss and noise limitations are not exceeded). They are therefore facility independent and are frequently used for other applications. Upon receipt of address signals from *TOUCH-TONE* service, the tones are decoded to indicate the called address to the central office.

8.4.2 INTEROFFICE TRUNK SIGNALING

In the Bell System, network operation with automatic switching has historically required that signaling for a call begin at the originating station and follow the same path as the call itself. The obvious reason for this approach was to avoid the added costs of separate transmission channels for signaling. However, this mode of operation introduces the very real possibility of mutual interference between signaling and voice transmission. To minimize interference, the basic rule has been to keep signaling and talking from overlapping in time as much as possible on a given connection.

Because there was no separate signaling route or network, voice connections involving several trunks in a series were established by adding one trunk at a time, starting at the central office that served the calling customer and progressing toward the central office of the called station. On such calls, most of the signaling activity takes place before the called station is rung. During this time, the transmission path is available for signaling. This signaling technique is referred to as *per-trunk signaling*.

As shown in Figure 8-6, interoffice signaling may, in principle, be provided on a **per-trunk** or **common-channel** basis. Per-trunk signaling is widely used in the Bell System at present and is discussed first. Common-channel interoffice signaling (CCIS), whose use is growing rapidly, is discussed in subsequent sections.

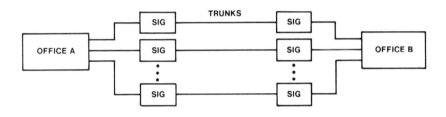

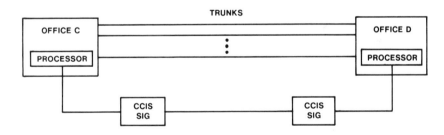

CCIS SIG   COMMON-CHANNEL INTEROFFICE SIGNALING EQUIPMENT

SIG   PER-TRUNK SIGNALING EQUIPMENT

Figure 8-6. Interoffice signaling techniques. *Top*, per-trunk signaling; *bottom*, common-channel interoffice signaling.

Per-Trunk Signaling

There are two major per-circuit trunk signaling interfaces: **loop-reverse-battery** (a form of loop signaling) and **E&M leads**. *Loop-reverse-battery* is applicable to those trunks that require call origination at only one end. These are called *1-way trunks*, although it should be recognized that it is only the call origination or trunk seizure that is one way. Basic signaling functions are still required in both directions. An *E&M lead* interface also may be used on 1-way trunks. On 2-way trunks, that is, trunks with call origination permissible at either end, an E&M lead interface is required.

Loop-reverse-battery, as used on metallic facility trunks, provides for application of nominal −48 volt battery on the ring conductor and ground on the tip at the terminating office end of the trunk. In addition, the terminating office has a current-sensing device and provision for polarity reversal (that is, −48 volt battery applied to the tip conductor). At the originating office end, means are provided for closing the loop, causing current to flow as an off-hook signal. A polarity-sensing device recognizes the state of the terminating end (for example, idle, disconnect, etc.). As described in Section 8.5, the loop-reverse-battery interface can also be used with an appropriate signaling system for applications on carrier systems.

There are numerous specialized variations of loop-reverse-battery, most of which have very limited applications and are not discussed here.

The E&M lead interface is a true interface as described in Section 8.1. The basic (Type I) E&M lead interface provides for signaling from the switching system toward the facility over the M-lead in the form of ground for on-hook and −48 volt battery for off-hook. From the transmission facility toward the central office, on-hook corresponds to an open E-lead, whereas off-hook is ground. Table 8-2 summarizes the signaling states. Figure 8-7 shows an example of E&M signaling from switching system A to switching system B with Type I interfaces. Switching system A indicates an off-hook status (−48 volts on the M-lead) by controlling the relay contact at point 1 on the figure. The M-lead detector at point 2 interprets the status, and the signaling circuit transmits it over the facility, closing the E-lead contact at point 3. The E-lead detector at point 4 in switching system B interprets the status as off-hook.

As an aid to remembering the functions of E&M leads, the "mouth" (M) and "ear" (E) analogy is useful because it is a reminder that the M-lead at one end of the facility drives the E-lead at the other end, although

TABLE 8-2

BASIC E&M LEAD SIGNALING STATES

State	From Switching System to Facility (M-Lead)	From Facility to Switching System (E-Lead)
On-hook	Ground	Open
Off-hook	−48 volts	Ground

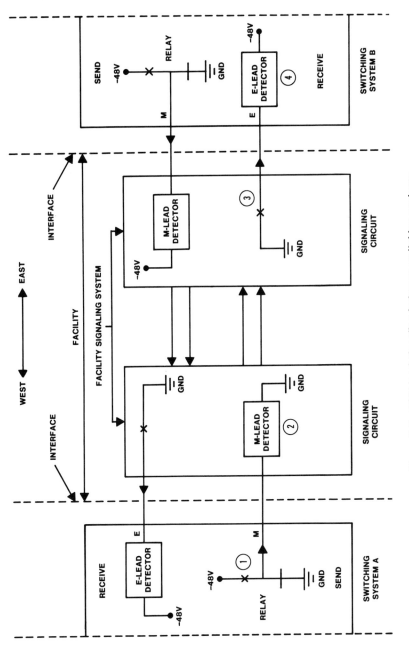

Figure 8-7. E&M lead signaling between switching systems with Type I interfaces.

that is not the source of the two terms. For example, when the switching system grounds the M-lead, signaling an on-hook state, the facility signaling conveys this to the signaling system at the other end of the facility using one of the techniques described in Section 8.5. The far-end signaling system will then provide an open condition on the E-lead to signal the next switching system.

With the introduction of electronic switching systems (see Section 10.3), the standard E&M lead interface was modified to avoid noise problems associated with ground return paths. Two varieties of looped (paired-lead) E&M interfaces exist. In some electronic switching system applications, full-looped E&M leads are used. These consist of loop closures on paired leads, both to and from the facility interface. For other electronic switching system applications, both ground and battery return for the M-lead are brought back to the facility interface as SG and SB leads. An example of one of these arrangements is the Type II interface shown in Figure 8-8.

Transmission facilities are often connected in tandem without an intervening switching system. If at least one of the facilities at a point of connection uses facility-dependent signaling, a through connection involving these dc signals must be made. This may not be simply a matter of connecting interface leads, because in the past, systems have not always been designed this way. Even if both of the facility signaling systems have E&M lead interfaces, it may be necessary to insert a signal conversion unit between the interfaces enabling the E-lead of one facility to drive the M-lead of the other (see Figure 8-9). If the Type II interface is used, direct interconnection is possible without a conversion unit because the contact closure that is the E-pair output is exactly what is needed to activate the M-lead pair detector (see Figure 8-10).

The loop-reverse-battery and E&M lead interfaces just described provide the supervisory interface for trunks. Destination codes are transmitted as either dial pulses applied through the same interfaces or multifrequency pulses[8] applied to the voice path. Section 8.5 discusses these and other signaling techniques.

Common-Channel Interoffice Signaling

New switching and transmission systems, as well as new features for existing systems, are continually being developed, and the complexities are such that providing intersystem compatibility is a major challenge.

Another challenge is to reduce postdialing delay, which is becoming more important because of new ways in which customers are using the network. For example, many calls involving computers at one or both

---

[8] The frequencies differ from those used for customer address signaling.

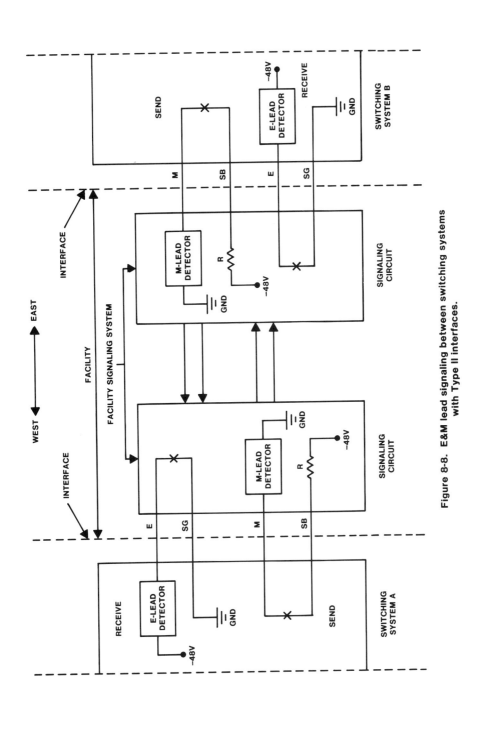

Figure 8-8. E&M lead signaling between switching systems with Type II interfaces.

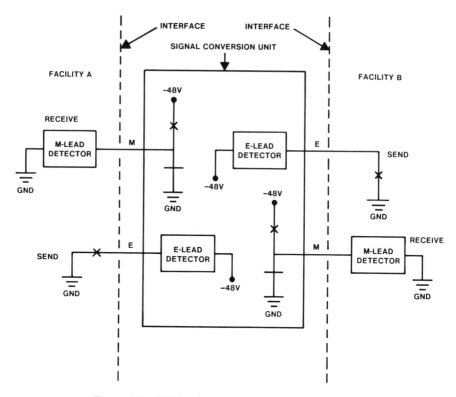

Figure 8-9. E&M lead signaling between facilities with
Type I interfaces using a signal conversion unit.

ends have relatively short messages, and therefore, the call setup time represents an appreciable part of the customer's total network time. Call origination, dialing, and answering functions are readily mechanized in these cases, so that these components of call setup time become small.

Still another concern is that traditional per-trunk signaling methods are not easily adaptable to certain evolving needs. Examples of such needs are (1) transfer of network management signals, (2) combining different types of traffic on one trunk group and yet retaining their identity at the far end, (3) far-end make-busy of trunks for maintenance purposes, (4) return of busy signal from originating rather than terminating office, so that the intermediate trunk(s) can immediately be made available to other calls, (5) increased transparency for the network (such as removal of constraints imposed on customer data transmissions to prevent harmful interactions with inband signaling equipment), (6) call tracing, (7) elimination or improved handling of simultaneous seizure of both ends of 2-way trunks, and (8) reduction of fraud.

282

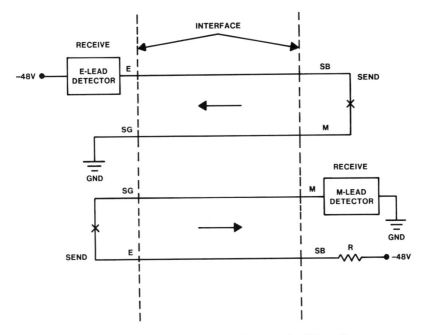

Figure 8-10. E&M lead signaling between facilities with Type II interfaces not using a signal conversion unit.

Since 1965, stored-program control switching systems have been used in the network in rapidly increasing numbers. With processor-controlled offices, a totally different signaling technique, which meets all the above needs, is possible for interoffice trunks. In such arrangements, there is no per-trunk signaling interface but rather a single data line (with backup) containing time-multiplexed signaling information. For economic reasons, this data link will usually go to a signal transfer point rather than provide one data link per trunk group. The data link will then serve multiple trunk groups from the office. Where such systems are used, neither loop-reverse-battery nor E&M leads are required, and similarly, neither dial-pulse nor multifrequency addressing is used.

Section 8.5.5 gives a functional description of common-channel interoffice signaling (CCIS). In 1976, CCIS became available for No. 4A Electronic Translator Systems and 4ESS systems. Later in the growth of CCIS, 1ESS systems and Traffic Service Position Systems were equipped.[9] It is expected that CCIS will be developed in other processor-controlled offices in order to permit the widespread availability of new services.

---

[9] Chapter 10 discusses these systems.

The initial system used a 2.4-kilobits-per-second (kbps) data link. In 1981, a 4.8-kbps terminal became available. In the future, data rates up to 64 kbps and an improved protocol may be justified.

8.4.3 SPECIAL-SERVICES SIGNALING

Special-services circuits frequently require special signaling arrangements because they have some characteristics of both customer lines and interoffice trunks. To provide these services, the standard customer-loop and trunk arrangements with various modifications have been used. It is important to recognize that the wide variability of implementation arrangements needed for special-services circuits represents a significant challenge in standardization and consolidation of signaling system equipment. It is beyond the scope of this text to consider all possible arrangements; however, a few major types are described.

A major category of special-services circuits includes special-access lines and trunks, such as foreign exchange lines, Wide Area Telecommunications Services lines, and private branch exchange (PBX)–central office trunks. Short special-access lines normally use standard single-party customer loop-start signaling. Long special-access lines, such as some foreign exchange lines, which use carrier transmission facilities, employ an appropriate form of signaling such as inband single-frequency signaling, described in Section 8.5.2.

Special-access trunks may use standard loop signaling but frequently employ a ground-start arrangement similar to that used in coin service. Special-services ground-start signaling is used to minimize the probability of simultaneous seizures on high-usage PBX trunks. This can occur on such circuits because a PBX attendant may attempt to initiate an outgoing call at the same time an incoming call is attempting to alert the attendant. If the incoming call connection occurs during the ringing silent interval, the attendant may not be alerted to this fact for up to 4 seconds, and a misdirected outgoing call could be placed. To prevent simultaneous seizures, ground-start provides for the application of ground on the tip lead toward the PBX as an initial contact or seizure signal even before the application of ringing. The attendant can test for this seizure and, if found, avoid placing an outgoing call on the busy trunk. Similarly, the trunk will appear busy to the PBX, so it cannot be selected for outgoing calls dialed from PBX extensions.

Tie trunks on PBX or key systems constitute another major class of special-services circuits. Tie trunks may be either **dial-repeating** trunks or **nondial** trunks.

Dial-repeating trunks are associated with those PBXs that are capable of routing tie trunk calls without attendant assistance and that, therefore, require dial-pulse address information. The E&M lead tie-trunk interface used is virtually identical to that used on interoffice trunks and, hence, passes dial-pulse address information. Although address signaling is normally dial-pulse, *TOUCH-TONE* dialing is also used. Multifrequency pulsing is normally not available at PBX systems.

Nondial trunks are associated with attendant-completed calls in which address information is transmitted verbally. Two modes of operation for nondial trunks are *automatic* and *ringdown*. Automatic and ringdown nondial trunks provide only a supervision alerting function. In the case of automatic operation, the signal to the interface at the originating end is a loop closure. For ringdown, the originating-end signal is the application of 20-Hz ringing by the operation of a ringing key by the attendant. When 20-Hz ringing is expected at the interface at the terminating end but there is an intervening facility that cannot pass the 20-Hz ringing signal, a combination of these types, known as *automatic ringdown*, is often used. The originating-end interface is a loop closure that results in the application of 20-Hz ringing at the receiving end. When the attendant answers the call, ringing is tripped.

Combinations of dial-repeating, automatic, and ringdown signaling are often used for different directions on a given tie trunk, depending on customer needs and the capability limitations of the terminating PBXs.

Private lines are another major class of special-services circuits that use many different types of signaling interfaces. For the simplest cases of private-line service, no signaling functions need be provided by the network. For example, in some private-line data services, the data sets provide the alerting function. For many private-line services, however, a separate alerting function is necessary. This is frequently provided by automatic or ringdown operation. The signals used are very similar to those just described for tie trunks, with automatic ringdown being the most common.

Many private lines are multipoint; that is, they have from several to many stations bridged together on a common private-line network. For such applications, selective alerting is frequently desirable. For simple multipoint networks with a minimum number of stations, code select may be associated with ringdown operation, or it may be combined with automatic operation in special arrangements to suit customer needs. In such an arrangement, coded ringing (one, two, three, etc., rings) may be used to alert a particular station. The ring count may be interpreted by an attendant or more typically by a code select ringing circuit in the terminating station that, recognizing its code, triggers an alerting signal at the station.

Services requiring more signaling features beyond simple alerting or multipoint networks with many stations may use a selective signaling system containing combinations of dial pulses or tones from *TOUCH-TONE* telephones to provide the selective ringing and control functions. Many other private-line signaling systems and interfaces are also in use.

Common-channel signaling has potential for switched special-services signaling, but at present, data link and signal transfer point costs are an obstacle because of the small number of circuits that are served in each application.

8.5 SIGNALING TECHNIQUES

To convey signaling information from one interface to another requires signaling transmission techniques that are compatible with the facility and the switching system involved. There are basically five major types of facility signaling systems: **direct current, inband tone, out-of-band tone, digital, and common-channel interoffice**. In addition, signaling systems are classified as **facility dependent** or **facility independent**.

Facility-dependent signaling techniques include dc signaling, out-of-band tone signaling, and digital carrier signaling. *Facility-independent* techniques are single-frequency (SF) inband tone signaling, multifrequency (MF) signaling, and dual-tone multifrequency (DTMF) signaling (for *TOUCH-TONE* service). For services built up from a combination of tandem facilities, there may be economic advantages in the use of facility-independent signaling systems. Common-channel interoffice signaling (CCIS) is also truly facility independent, as it is based on a separate data link carrying multiplexed signaling information for many different circuits. The CCIS data link may be over an entirely separate facility, often following a different geographic route. The interface with a switching system requires CCIS capability in that system.

The three modes of signal transmission are **continuous, spurt**, and **compelled**. *Continuous* supervision requires steady-state transmission and reception of status on a per-channel basis and has no memory associated with the signaling system. This mode is restricted to dedicated, non-time-shared facilities. *Spurt* signals are normally associated with short-duration address information but also may be used for supervision if the signaling system or connecting circuit provides the necessary state memory. Short-duration signals are particularly useful on time-shared facilities such as those used for CCIS. The *compelled* mode is a high-reliability system used for critical applications such as overseas signaling. With compelled signaling, every state change is transmitted, and its reception at the far end is acknowledged with an appropriate signal. Following acknowledgement, the signal is normally removed; hence, state memory is required.

The following sections describe each of the facility signaling systems mentioned here—direct current, inband, out-of-band, digital, and CCIS—in more detail, and Table 8-3 summarizes signaling implementation.

8.5.1 DC SIGNALING

For each of the major signaling system interfaces discussed in Section 8.4 (loop signaling, loop-reverse-battery, and E&M lead signaling), one or more facility signaling systems exist for use on dc (metallic) facilities.

The existing metallic facility signaling systems with their interfaces came into being through an interaction of desired performance, technical possibilities, and cost of both facility and switching system—an interaction too

TABLE 8-3

SIGNALING IMPLEMENTATION

SIGNALING FUNCTIONS
- Supervision
- Addressing

AREAS OF APPLICATION
- Customer lines
- Interoffice trunks
- Special services (includes PBX trunks and PBX tie trunks)

SIGNALING SYSTEMS INTERFACES
- Loop signaling (may be associated with station loops, trunks, or special services)
- E&M lead signaling (may be associated with trunks or special services, usually those requiring 2-way origination)
- CCIS

FACILITY SIGNALING SYSTEMS
Facility dependent (carry supervision, address, or both)
- Dc (design and engineering depend on resistance of signal path conductors) – station loops, short trunks, short special- services channels, and end segments on long channels
- Out-of-band (as on N1 carrier)
- Digital (as on T1 carrier)
Facility independent
- Inband tone
  SF (carries supervision, address, or both)
  MF (carries address) – interoffice
  DTMF (carries address) – primarily station loop
- CCIS (carries both supervision and address) – interoffice

SIGNALING MODES
- Continuous
- Spurt
- Compelled

complex to be treated here. In general, the main constraint is the dc resistance of the signaling path, including that of the facility.

At the simplest end of the spectrum is customer-line, or station-loop, signaling. This uses a wire pair connecting a pair of contacts at one interface with a battery and current sensor at the other (as shown in Figure 8-2). Constrained by design and engineering rules to ensure proper operation, this arrangement is a facility signaling system. With a typical central office and station, it will operate with external conductor loop resistances up to 1300 ohms, a value that permits satisfactory signaling and transmission over a majority of loops. Table 8-4 illustrates signaling range capability for different gauges of cable pair. In practice, the loop may consist of two or more segments of different gauge for best economy within the 1300-ohm limit.

When the station-loop resistance exceeds 1300 ohms, provision is made for *signaling range extension*. This is used only in exceptional cases; the great majority of loops do not require any such treatment. Range extension normally involves an extra circuit for detection and regeneration of signals, but it may involve no more than a higher-than-normal battery voltage at the central office. Range extension can be applied on trunks but is very seldom required. (Section 9.2.1 discusses loop design in more detail.)

For trunks, the loop-reverse-battery interface with dc signaling transmission, like basic station-loop signaling, uses only the metallic conductor, but it can work over a much longer circuit. In common with

TABLE 8-4

LOOP LENGTH ATTAINABLE WITH FIXED WIRE GAUGE

Gauge	Ohms Per Thousand Feet	Loop Length Limit* For 1300 Ohms (Miles)
26 — fine wire — least cost per foot	83	3.0
24 —	52	4.7
22 —	32	7.7
19 — coarse wire — greatest cost per foot	17	14.5

*On the longer circuits, load coils are required to improve speech quality. These are spaced at 6000-foot intervals and have a resistance of 9 ohms. This makes the actual loop length limit slightly smaller than shown in the table.

basic station-loop signaling, loop-reverse-battery uses a grounded voltage supply at only one end (the terminating end of the trunk), thus avoiding problems of earth potential differences between offices.

In station-loop signaling and loop-reverse-battery signaling over metallic conductors, the relays that receive signals are generally considered part of the switching system, and so the facility signaling system design for these applications naturally has fallen to people who design switching systems. The design of range extension equipment for metallic facilities is done primarily by transmission designers.

8.5.2 INBAND SIGNALING

Facilities not capable of passing dc signals must be provided with different types of facility signaling systems. This is true of all carrier systems. For analog carrier systems, a number of inband[10] tone-signaling techniques have evolved. Since these depend only on the existence of a voice channel, they are facility independent and may be used over tandem facilities including metallic pairs and digital carrier.

Included in inband systems are SF signals, MF signals, and signals from *TOUCH-TONE* service. SF signaling provides both supervision and, where appropriate, dial-pulse address signaling and is compatible with loop-reverse-battery and E&M lead signaling interfaces. All of these systems perform their functions without the need for special signaling range extension or conversion equipment.

SF signaling provides two basic states: on-hook (normally 2.6 kilohertz [kHz] tone-on) and off-hook (normally tone-off). Both the application and the removal of tone are controlled from the input dc interface; a received-tone detector provides an output signal to the dc interface. There is, of course, a mating SF unit at the distant end of the channel. A typical application is on a 4-wire carrier system with origination from either end.

The use of the two supervisory states is dictated by application requirements. A simple example is the use of SF to extend customer-line signals over carrier. Toward the central office, tone-off is off-hook, or seizure; and tone-on is on-hook, or idle. In the reverse direction, tone-off is both idle and busy, with tone-on corresponding to the application of ringing.

With SF signaling, time delays must be allocated for tone recognition, to allow for distortion caused by facility noise or transmission degradation and to prevent false signaling by voice simulation. The latter is referred to in signaling terminology as "talkoff" because a voice-produced

[10] Within the voiceband, 200 Hz to 3.5 kHz, as described in Section 6.1.

tone with the proper characteristics can simulate an on-hook condition, causing the circuit to be disconnected (the SF receiver functions as if the customer has hung up).

These factors normally necessitate sending a tone signal of minimum duration (typically over 50 milliseconds) independent of the input pulse duration. The detected signal must then be reconstructed to approximate that of the original source by a process referred to as *pulse correction*. The pulse-correction scheme used depends on the particular service requirements of the SF application.

Many signaling system interfaces can be used with SF signaling units. In toll applications, the interface is typically E&M lead. The interface also can be loop-reverse-battery, in which case, the SF unit appears transparent to a switching system. Where a customer line is quite long (as in the case of foreign exchange or some other special-services circuit), an SF unit can be used at the station, with a loop-signaling interface. On a foreign exchange line, for example, the off-hook signal from the station would remove the tone sent toward the switching office. Since special-services circuits often use various combinations of metallic and carrier facilities, optional signaling conversion circuitry with enough range capability for the majority of applications is provided in the SF unit for each of several signaling interfaces.

As in customer-line signaling, dial-pulse trunk address signals consist of interruptions of the off-hook supervisory state of the trunk; the rate of interruption may be ten or twenty pulses per second. Dial-pulse signals can be applied directly to both loop-reverse-battery and E&M lead interfaces. Multifrequency pulsing is very similar to *TOUCH-TONE* signaling in that pulses of pairs of selected tones are used to represent digits. In MF pulsing, the tones were not specifically chosen to protect against simulation by voice, since voice transmission from the calling customer is inhibited during MF pulsing. In addition to the actual address digits, a keypulse tone pair is sent as a start-of-address signal to unlock the terminating-end receiver, and a start signal is sent to indicate end-of-address (start call processing). These signals are sometimes faintly audible to the calling party.

8.5.3 OUT-OF-BAND SIGNALING

Analog carrier systems typically do not use the full 4-kHz voice-frequency channel slot because of channel separation filters in the channel *modems*.[11] These filters sharply attenuate the input voice signal above about 3.5 kHz. This has led to the development of out-of-band signaling

---

[11] A contraction of the words *modulator* and *demodulator* signifying an equipment unit that performs both of these functions.

systems using tones above this value but below 4 kHz. Such systems are not widely used, as they are normally facility dependent and are not applicable to tandem facilities without added hardware at the point of interconnection. However, one out-of-band system is often used with the N1 carrier system (see Section 9.3.3) that uses 3.7 kHz as the signaling frequency. While out-of-band tone signaling has been considered for other systems, including A6 channel banks, the universal applicability of SF signaling has outweighed any possible economic advantages of out-of-band signaling.

8.5.4 SIGNALING OVER DIGITAL FACILITIES

On most digital facilities, when transmission is provided by digital bit stream, it is very convenient to designate one or two bits periodically as signaling. The bit 1 or 0 then corresponds directly to a specific signaling state such as on- or off-hook. Signaling bits may be assigned to every encoded sample (8000 per second for pulse-code modulated voice)[12] or at a much lower rate consistent with the very low information transfer rate associated with per-channel signaling. Typical sample rates of about 1 kbps (sample interval of 1 millisecond) are used to simplify terminal equipment. Thus, for digital carrier, a signaling system not based on tones is possible and is used extensively. This leads to considerably simpler terminal equipment compared with SF units and, hence, to much more economical arrangements. This low signaling cost, combined with its inherent low transmission cost, has prompted the use of digital carrier as an alternative to analog carrier with SF and also to wire facilities.

The manner in which the signaling information is digitally encoded is now standard for facility terminal channel unit design. The channel unit detects the per-channel direct current signaling input (for example, the M-lead) and applies the necessary control to set the corresponding state of the digital signaling bits. The digitally encoded signaling information is then multiplexed (see Section 6.5) with the voice signals for transmission over the associated digital line. At the receiving terminal, the signaling information is recovered in the demultiplexing process. The appropriate dc conditions are then applied toward the connecting circuit (for example, the E-lead).

Since most digital signaling information is transmitted at a rate of at least 1000 samples per second, the timing of the signaling information is not appreciably disturbed. A state change at the source end is recognized at the destination end with essentially no increase in delay caused by sampling. For this reason, digital signaling is typically referred to as distortionless compared with SF signaling, which typically distorts or

---

[12] Section 6.4.3 discusses pulse-code modulation.

changes the timing of short-duration signals. Since it is essentially free from distortion, the digital system has the potential for providing superior performance, but it is, of course, facility dependent. Digital signals, therefore, cannot be extended to other facilities such as wire or analog carrier without conversion at the signaling interface.

Some stored-program control switching systems, including all time-division switching systems, can provide an internal interface to digital signaling without a digital-to-direct current conversion.

8.5.5 SIGNALING OVER SEPARATE FACILITIES —
COMMON-CHANNEL INTEROFFICE SIGNALING

The principle of the CCIS system is to transmit all of the signaling information pertaining to a group of trunks over a channel separate from the communication channels. As shown in Figure 8-11, one data link can be provided per trunk group, but it has been found to be much more economical to use a data link in common for many trunk groups and

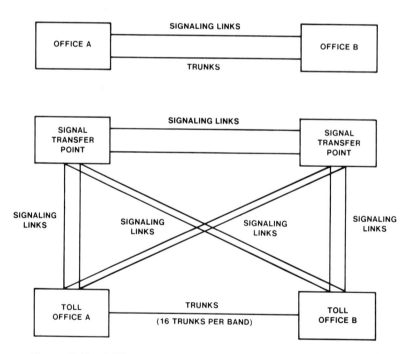

Figure 8-11. CCIS operating configurations. *Top,* associated mode; *bottom,* nonassociated mode. In the nonassociated mode, signaling for A-B trunks is by way of the signal transfer point. Both radio and cable paths may be provided for signaling links with only one path in operation at a time (the other is switched as needed) to provide redundancy with diversity.

"switch" the messages at a signal transfer point (STP). The STP is primarily concerned with the address of each message rather than with its content.

CCIS permits a great reduction in call setup time, not only by its inherent speed but also by signaling the next office in the route before an office has finished switching. Other inherent advantages are flexibility and low cost where the volume of signaling is substantial. These advantages stem from the fact that the signaling function is fully disassociated from the voice path and is handled in a highly concentrated form. By the same token, CCIS could introduce new problems unless special precautions are taken. For example, since signaling does not take place on the talking path, the signaling process does not inherently verify the transmission quality of the trunk. Therefore, it would be possible to route a call over a defective trunk. To avoid this problem, a test of the voice path is made part of the procedure for setting up each connection. Another concern is the possibility of simultaneously disabling all of the trunks (perhaps 4500) served by a CCIS link in the event of failure of that link. Duplication of STPs with diversity of facilities (for example, facilities connecting the same points over different geographical routes) provides the required reliability.

CCIS represents the first major Bell System application of *packet switching*.[13] Messages are transmitted as signal units of twenty-eight bits, including eight bits for error detection. Twelve signal units are linked into a block, or packet, for purposes of further error detection and retransmission. With the STP architecture, each group of sixteen trunks in toll office A (in Figure 8-11) having the same destination is given a unique numerical assignment (called a *band number*) on the CCIS link to the STP. The STP is arranged to transfer messages from the incoming band and link to an associated outgoing band and link in order to reach the distant office (toll office B) and the other end of the group of sixteen trunks. The present CCIS protocol (see Section 8.8) was designed to be optimal for medium-speed data links. This protocol, called *CCITT Signaling System No. 6*, or *System No. 6*, is used internationally and domestically.

The domestic version varies somewhat from the international version. The initial domestic application employed 2.4-kbps data links. This speed was later increased to 4.8 kbps for increased capacity, whereas the international application operates at 2.4 kbps. The domestic version of System No. 6 provides for generalized communications between computers not just the relaying of supervisory and address signals related to trunks. For example, in Automated Calling Card Service, customers at most coin telephones first dial 0 plus a destination number that is sent via the CCIS network to a data base. After receiving a prompt, customers dial their

---

[13] Section 5.8 discusses packet switching.

calling card number (a billing number, often the home telephone number plus a personal identification number), which is also sent via CCIS to the data base for validation. A favorable response then initiates the call and permits subsequent billing of the call to the billing number. No telephone operator is involved.

Another protocol, called CCITT Signaling System No. 7, has been specified to be optimum for high-speed digital data links (64 kbps). It uses signal units of variable lengths, a functional structure of levels, and a more straight-forward method of message addressing. System No. 7 will be introduced into parts of the domestic telecommunications network in the near future.

In Section 8.4.2 on **Per-Trunk Signaling**, there is little discussion of two topics that become more important with the introduction of CCIS: reliability and availability. *Reliability* is the degree of assurance that the received signal is the same as the transmitted signal; that is, it has not been altered by electrical dropout or interference. *Availability* is the measure of continuity of signaling service between two points. In per-trunk signaling, the signaling implementation is conservatively designed and uses the same path or channel as the call to which it relates. The dc and ac signals used with per-trunk signaling are relatively immune to transient electrical interference. As a result, if the path is adequate for voice, it will usually transmit the signaling information properly. With CCIS, however, the capabilities of the channel are exploited to achieve a higher information transmission rate. Thus, transient electrical interference can cause occasional bit errors. To minimize the effect of such errors, redundant information is added to the outgoing message in the form of error-checking codes. A parity code in which the parity bit is 0 for odd data and 1 for even data is an example of simple error-detection coding. The receiver can then detect most errors and call for retransmission of the message segments.

The status of CCIS in 1983 is that it is spreading rapidly in toll applications. Projections of load due to trunk-related signaling and for generalized communications indicate the need for higher capacity links and STPs in mid-1985. To meet this need, a new protocol optimized for 64-kbps digital links will be adopted. A new STP architecture is also planned for much higher capacity.

CCIS is of less value in the local network because of the signaling already provided with digital facilities (see Section 8.5.4). However, the value of features made possible by CCIS may motivate growth in local areas as well.

8.6 INTERNAL INTERFACES

An end-to-end connection between customers on a network such as the public switched telephone network (PSTN) requires a number of internal connections among pieces of transmission, switching, and signaling equipment. As indicated in Section 8.1.2, specification of interfaces for various equipment boundaries

- ensures the compatible interconnection of equipment on each side of the interface,

- separates design responsibilities for equipment on each side of the interface, and

- permits independent design and testing on each side of the interface.

Besides ensuring compatible operation initially, well-designed interfaces usually facilitate the introduction of new equipment and services as technology evolves. The interface specification permits new equipment to be designed and integrated into the network without requiring redesign of mating equipment. Stability is a desirable objective for an interface, and some internal interfaces have been stable for many years. However, achievement of cost savings and other benefits offered by new technology sometimes requires changing traditional interfaces or creating entirely new ones.

The next three sections describe three important internal interfaces:

- the **4-wire analog carrier** interface, which has provided a stable boundary between existing and new transmission and signaling equipment for many years

- the **DSX-1 digital system cross-connect**, which is a relatively new interface developed in response to the rapid growth of digital facilities in the network

- the **digital carrier trunk** (DCT), which is replacing a traditional interface between the 1ESS system and T-carrier facilities.

8.6.1 THE 4-WIRE ANALOG CARRIER INTERFACE

Figure 8-12 is a simplified representation of a connection between two 2-wire switching offices (offices A and C).[14] In the figure, an analog voice channel is carried from switching office A by a cable pair to building B, which contains only transmission equipment. From building B, the channel proceeds to switching office C over a 4-wire analog carrier system. (Office C illustrates another common situation—switching and transmission terminal equipment housed in the same building.) The terminating sets in building B and office C provide the hybrid function described in Section 6.6.1; the analog channel banks (see Section 9.3.5) perform the

---

[14] The connection shown in the figure is one of many possible configurations. Different equipment and interfaces would be involved if the transmission and/or switching systems were digital. The interfaces described in Sections 8.6.2 and 8.6.3 illustrate some of these differences.

required modulation and multiplexing functions; and T and R pads provide the desired loss between the 2-wire switching systems and the channel banks.

A classic example of an established, successful, pervasive interface is the 4-wire transmission interface between the single-frequency (SF) signaling unit (which performs the function of conversion between dc and SF signaling) and the A-type channel bank in the analog carrier transmission system. Two such internal interfaces are shown in Figure 8-12—one at point X and one at point Y. Each interface consists of four wires, designated T, R, T1, and R1, that carry message signals (for example, voice, data) and SF network control signals (for example, supervisory, address). The T and R leads carry signals from the SF signaling unit to the modulator input port of a channel unit in a channel bank. The T1 and R1 leads carry signals from the demodulator output port of a channel unit in a channel bank to the SF signaling unit.

The 4-wire interface is clean and simple, requiring only specification of the impedance in either direction (nominally 600 ohms), the power level in the transmit and receive directions (in terms of transmission level points, described below), and balanced pairs.[15]

Transmission Level Points

The *transmission level point* (TLP) concept is convenient for relating signal or noise levels at various points in a connection. As part of an interface specification, it facilitates the independent design and test of equipment by defining the signal amplitude or power level at the input and output terminals of the equipment.

A TLP does not define power in absolute terms but in relative terms. Specifically, the TLP at any point in a circuit is defined as the ratio, in decibels (dBs), of the power of a signal at that point to the power of the signal at a reference point. It is customary to consider the outgoing 2-wire class 5 switching system as the 0-dB TLP reference point.

The transmission losses and gains incurred during transmission from office A to office C are plotted on the upper scale in Figure 8-12; the losses and gains incurred during transmission from office C to office A are plotted on the lower scale. The portion of the upper scale beginning at office A and ending at building B plots the losses (in dB) from the 2-wire switching system in office A (0-dB TLP) to the 4-wire analog carrier interface at building B (−16 dB TLP). Analog carrier systems (which include

---

[15] A *balanced pair* consists of two wires (tip and ring) that are electrically alike and symmetrical with respect to ground. This circumstance has the advantage that noise and interference currents induced equally in both wires flow in the same longitudinal direction along the pair and are canceled out using a transformer-coupled load impedance.

the channel banks) are designed to provide 23 dB of gain; therefore, a
0-dBm[16] test tone applied at the 0-dB TLP will emerge from the carrier
system DEMOD OUT port in office C at a power level equal to +7 dBm.

It should be noted that the *absolute* power of a particular signal that
exists at any point along this circuit depends upon the service involved.
For example, the standard power for a data signal is −13 dBm0, which
implies a data signal power equal to −13 dBm at the 0-dB TLP. Such a
data signal would have an absolute power of −29 dBm at the MOD IN
(−16 dB TLP) port and −6 dBm at the DEMOD OUT (+7 dB TLP) port of
the 4-wire interface.

8.6.2 THE DSX-1 DIGITAL SYSTEM CROSS-CONNECT INTERFACE

The evolving Bell System digital transmission network is composed of a
great variety of digital signals, multiplexers, and transmission facilities
that operate at different levels or bit rates. Some of the digital signals
originate as voiceband analog signals, others as data. The initial multiplexing of the analog signals takes place in channel banks as one of
several functions that include the translation from analog to digital format. The result is the formation of the digital signal 1 (DS1) level signal
(1.544 megabits per second [Mbps]), which forms the basic building block
of the digital time-division multiplex hierarchy.[17] The DS1 signal may
result from the multiplexing of lower speed digital signals.

The DSX-1 cross-connect is a relatively new internal interface that
provides a convenient central point for cross-connecting, rearranging,
patching, and testing digital equipment and facilities at the DS1 level.
Figure 8-13 depicts the extensive variety of DS1 signal sources, multiplexers connecting to higher levels, and transmission facilities served by
the DSX-1 cross-connect. For example, at the top left of the figure, a signal at the DS1 level is formed by the time-division multiplexing of
twenty-four voiceband signals, each of which has been converted to digital form by pulse-code modulation. The modulation and multiplex functions are performed by the digital channel banks (D1, D1D, or D3) that
output the DS1 signal to the interface at the DSX-1 cross-connect. (Some
sources produce more than one DS1 signal. For example, the D2 channel
bank produces four DS1 signals as shown in the figure.) At the other
side of the DSX-1 interface, the DS1 signal may be connected directly to
another terminal or to a transmission system such as the T1 carrier system
(see Section 9.4.2) or combined through multiplex equipment to produce
a digital signal at a higher level (bit rate).

---

[16] A logarithmic measure of power with respect to a reference power of 1 milliwatt.

[17] Section 9.4.3 describes the digital time-division multiplex hierarchy. The various levels
or bit rates in the hierarchy are identified by a DSn (where n = 1, 2, 3, etc.) designation.

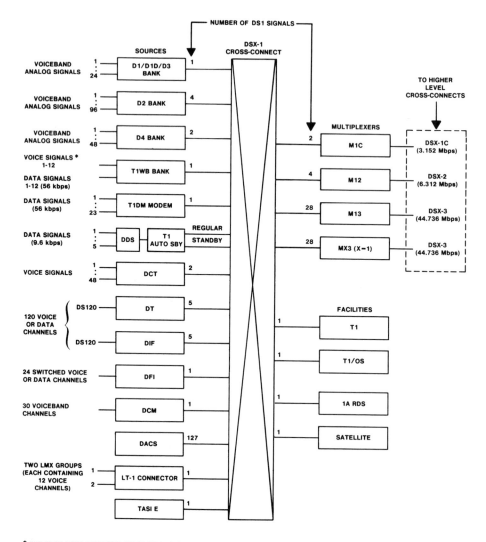

* THE T1WB BANK ACCEPTS A TOTAL OF 24 SIGNALS COMPRISING FROM 1 TO 12 VOICE SIGNALS AND FROM 1 TO 12 DATA SIGNALS.

DACS   DIGITAL ACCESS AND CROSS-CONNECT SYSTEM
DCM   DIGITAL CARRIER MODULE FOR DMS-10 SYSTEM
DCT   DIGITAL CARRIER TRUNK FOR 1 ESS SYSTEM
DDS   DIGITAL DATA SYSTEM
DFI   DIGITAL FACILITY INTERFACE FOR 5 ESS SYSTEM (15 DFIs—1 INTERFACE MODULE)
DIF   DIGITAL INTERFACE FRAME FOR 4 ESS SYSTEM
DT   DIGROUP TERMINAL FOR 4 ESS SYSTEM
OS   OUTSTATE AREA
TASI E TIME ASSIGNMENT SPEECH INTERPOLATION SYSTEM, VERSION E

Figure 8-13. DSX-1 (1.544-Mbps) digital system cross-connect. the figure shows a representative, but not exhaustive, collection of equipment and facilities served by the DSX-1 cross-connect at the end of 1982.

The DSX-1 interface specifications summarized in Table 8-5 describe the signal characteristics required of all DS1 signals appearing at this interface. As can be seen from the table, the DSX-1 interface requires a more complex specification than the 4-wire analog carrier interface described earlier. In addition to the parameters shown in Table 8-5, there is a requirement on the signal pulse shape at the interface[18] and on the maximum jitter impairment that connecting terminals or facilities can tolerate. Unlike the 4-wire carrier interface, the transmission level point is the same for both directions of transmission at the DSX-1 interface.

TABLE 8-5

DSX-1 INTERFACE SPECIFICATIONS*

Parameter	Requirement
Line rate	1.544 Mbps ±130 pulses per minute
Line code	Bipolar with at least 12.5% average 1's density and no more than 15 consecutive zeroes
Test load impedance	100 ohms resistive
Pulse amplitude	3 volts ± 20%
Pulse shape	Rectangular (50% duty cycle) fitting template for an isolated pulse
Signal power	12.4 to 18.0 dBm (771 to 773 kHz) <−16.6 to −11.0 dBm (1543 to 1545 kHz)
Pulse power balance	Difference in power between positive and negative pulses <0.5 dB

*Bell Laboratories 1982 contains a detailed discussion of the elements in this table.

8.6.3 THE DIGITAL CARRIER TRUNK INTERFACE

The use of electronic switching systems has created new, more efficient transmission and switching internal interfaces in terms of space, cost, and required maintenance. These interfaces have phased out several traditional internal interfaces. Examples of the new interfaces are listed in Table 8-6.

[18] AT&T 1978 contains the specification for the pulse shape.

TABLE 8-6

ELECTRONIC SWITCHING SYSTEM
TRANSMISSION/SWITCHING
INTERNAL INTERFACES

System	Internal Interface
1/1AESS	Digital carrier trunk
2/2BESS	Direct interface with T-carrier
3ESS	Direct interface with T-carrier
4ESS	Digroup terminal (DT) Digital interface frame (DIF)
5ESS	Interface module (IM) (composed of 15 digital facility interfaces [DFIs])
DMS-10	Digital carrier module (DCM) Subscriber carrier module (SCM) Office carrier module (OCM)

The following paragraphs illustrate how departure from a traditional design, which segregates transmission and switching functions, toward an integrated approach eliminates some previous boundaries between the two functions. They compare the interface associated with the new digital carrier trunk (DCT) channel bank and the traditional interface used with a D4 channel bank in a 1/1AESS system.

The D4 channel bank interface illustrated at the top of Figure 8-14 represents the traditional separation of the D4 channel bank and the trunk circuit (described below) associated with the switching system. The D4 channel bank is composed of a common, or control, equipment portion and a set of up to forty-eight individual channel units. Each channel unit handles a single trunk or special-services line. The D4 bank prepares analog signals from the switching system for transmission over digital facilities through the DSX-1 interface described in Section 8.6.2.

A trunk circuit connects the transmission facility to a specific terminal on the switching network at which the trunk terminates. Trunk circuits provide the interfaces that permit signaling over the trunk. Leads for each trunk are required to exchange signaling and supervisory information between the trunk circuit and the channel bank through the intermediate distributing frame (IDF). The 1/1AESS system central control

interfaces with the trunk circuit directly via a signal distributor and scanner.

The DCT combines certain T-carrier transmission functions and electronic switching system control functions to provide a more economical interface between the 1/1A*ESS* system and T-carrier facilities. As shown at the bottom of Figure 8-14, the DCT is a single, integrated channel bank frame that replaces the D4 channel bank, the trunk circuit, the signal distributor and scanner, the intermediate distributing frame, and the per-trunk leads. The DCT channel unit combines the D4 digital channel unit and the switching system trunk circuit functions. The exchange of signaling, supervisory, and maintenance information with the switching system central control is accomplished through the digroup[19] controller and the peripheral unit controller of the DCT frame. The peripheral unit controller includes a duplicated *WE*[20] 8000 microprocessor.

8.7 INTERFACES FOR INTERCONNECTION

8.7.1 INTERCONNECTION

Interconnection, as discussed here, is the direct electrical connection to the nationwide telephone network of (1) user-premises terminal equipment and communication systems or (2) the facilities of other common carriers (OCCs). It represents the relationship established for individual telecommunications services including not only the physical and electrical characteristics but also the maintenance and administrative procedures that have been agreed upon by the participants. Certain long-standing relationships, such as those between the Bell System and other telephone companies, may fit the basic concept of interconnection but are not included here.

A few definitions may be helpful in understanding the concept of interconnection.

- *Nationwide telephone network* refers to the combined telecommunications facilities networks of the various telephone companies throughout the United States, including the Bell System. *The Bell System network* refers to that part of the nationwide network owned and operated by Bell companies.

- *Facilities* is the telecommunications physical plant. The Bell System network as a whole is referred to as the *facilities network* (to distinguish it from traffic networks),[21] and a single channel provided by the facilities network is referred to as a *facility.*

---

[19] A *digroup* consists of twenty-four voice channels.

[20] Trademark of Western Electric Co.

[21] Chapter 4 discusses the facilities network and traffic networks.

- *Terminal equipment* is any separately housed equipment unit (a telephone set, for example) or group of equipment units treated as an entity (for example, a PBX and its on-premises stations). Such equipment is located on user premises and is interconnected with a Bell System network facility.

- *User-premises communications system* refers to those cases where equipment that is part of a separate communication system (for example, one end of a microwave facility) is interconnected with a Bell System facility.

- *Other common carrier* is a telecommunications common carrier, other than the Bell System, authorized to provide a variety of private-line services such as facsimile, measured voice, and wideband data as well as packet-switched digital data. The Federal Communications Commission (FCC) refers to these carriers as *domestic satellite* carriers, *miscellaneous* common carriers, and *specialized* common carriers. Many of them rely on Bell System local facilities to provide access to their facilities.

- *Separate ownership* is usually, but not necessarily, implied in the concept of interconnection. Separate ownership is involved where the terminal equipment interconnected with Bell System facilities is provided by the user. However, in those cases where the user leases the terminal equipment from the Bell System, the FCC has ruled that the Bell System must meet the same criteria for interconnection as is required of user-provided terminal equipment. Thus, the FCC rules governing network protection (see Section 8.7.5) apply equally in both cases. In any case, from the Bell System viewpoint, interconnection always involves the Bell System as at least one of the interconnection participants.

8.7.2 INTERFACES

In the interconnection environment, an *interface* is the boundary between two telecommunications capabilities. The physical connection may be either a jack and plug or a terminal strip for matching leads that are common to both telecommunications capabilities. Compatible electrical characteristics (such as impedances and signal power levels) are essential not only for proper operation but also to ensure that there is no interference with normal Bell System network functions. The interface may also be a boundary for maintenance and administrative functions.

Interfaces for interconnection are referred to as either *network interfaces* or *demarcation points*. Where user-provided terminal equipment is interconnected with the Bell System facilities network, the interface is referred to as a *network interface*. Interfaces for interconnecting OCC facilities to the Bell System facilities network are also network interfaces but

are commonly referred to as *demarcation points*. The two types of interfaces differ in two respects: (1) the physical and electrical characteristics
are not necessarily the same for a given service and (2) related functions
are handled differently.

Network interfaces tend to have standardized electrical characteristics,
pin arrangements, and protocols, particularly where FCC registration of
terminal equipment is involved (see Section 8.7.5). With such arrangements, the maintenance and administrative functions can be minimized
to ensure design and operating flexibility.

Demarcation points tend to be negotiable, since testing activities
extend across the interface and a more interactive approach is used for
circuit layout. But there is a trend toward standardized OCC interfaces.

8.7.3 INTERCONNECTION ENVIRONMENT

Bell System As Sole Provider of Service

The interconnection environment in those cases where a Bell System network facility is the sole provider of the telecommunications service is
shown in Figure 8-15.

In such cases, the network service is defined only between network
interfaces (NIs), and the user of the service is traditionally referred to as
the *customer*. Thus, the terminal equipment is located on the customer
premises.

Usually the NI is identified by some characteristic of the service with
which it is associated. For example, a 2-wire (tip and ring) interface with
loop-start signaling for interconnecting a telephone set to a PBX off-
premises station (OPS) line is referred to as an OPS interface (station
end). In those cases where digital transmission is involved, the NI may
be identified by the bit-stream characteristics of the digital signal, for
example, the DS1 interface described in Section 8.6.2. In some cases, an

EEI   EQUIPMENT-TO-EQUIPMENT INTERFACE
NCTE   NETWORK CHANNEL-TERMINATING EQUIPMENT (LIGHTNING
        PROTECTOR WITH OR WITHOUT OTHER CHANNEL-TERMINATING
        FUNCTIONS)
NI   NETWORK INTERFACE
TE   TERMINAL EQUIPMENT

Figure 8-15. Interconnection environment—Bell System services only.

NI does not have a service identifier. For example, the 2-wire loop-start NI mentioned above is referred to simply as a network interface where it is used for interconnecting a telephone set with a local central office line for access to any one of several Bell System PSTN services.

The equipment-to-equipment (EEI) interface shown in Figure 8-15 is an important part of the interconnection environment but is not an interface for interconnection since it does not involve a Bell System network facility as one of the interconnection participants. There is growing interest in developing standards for those interfaces that are applicable to both EEIs and NIs to facilitate portability of terminal equipment. Achieving such commonality among interfaces for interconnecting terminal equipment with public switched data networks and other digital services[22] is an objective of the current efforts in both domestic and international standards activities.

The network channel-terminating equipment (NCTE) shown in Figure 8-15 is part of the network facility. It may provide lightning protection for the terminal equipment in those cases where the facility involved is exposed to lightning. It may also include other functions such as loopback for remote testing on a 4-wire line, facility loss equalization, and digital signal regeneration.[23] The NCTE is not a network protection device; however, it may include inherent network protection such as hazardous voltage limiting (for example, diodes) and isolation from longitudinal imbalance (for example, transformers) and signal power control.[24]

OCC Service Using Bell System and OCC Facilities

The interconnection environment also includes those cases where an OCC service is provided by a combination of Bell System and OCC network facilities, as shown in Figure 8-16. In such cases, the Bell System provides the connecting facility between the user premises and the OCC facility. The user is referred to as a *patron* instead of a customer since the OCC has the overall service responsibility, and the interface between NCTE and OCC channel equipment (OCE) is referred to as a *demarcation point* (DP). The equipment configuration on the patron premises in Figure 8-16 is the same as previously described for those cases where a Bell System facility is the sole service provider (EEI is not shown), except that OCE may also be required in certain cases. Where such equipment is included, both a network interface and a demarcation point are required on the patron premises as indicated in Figure 8-16. Typically, OCE on

---

[22] Section 8.8 discusses data communications services.

[23] The FCC has not ruled on which functions may be included in the NCTE.

[24] Section 8.7.5 discusses inherent network protection.

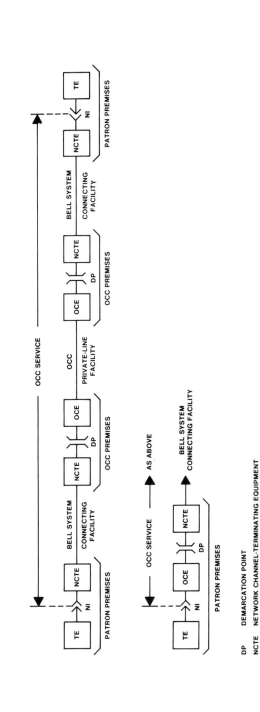

Figure 8-16. Interconnection environment—OCC service with Bell System connection facilities. *Top,* no OCE required; *bottom,* equipment configuration when OCE is required.

DP   DEMARCATION POINT
NCTE   NETWORK CHANNEL-TERMINATING EQUIPMENT
NI   NETWORK INTERFACE
OCC   OTHER COMMON CARRIER
OCE   OCC CHANNEL EQUIPMENT
TE   TERMINAL EQUIPMENT

patron premises might provide a single-frequency signaling function. Otherwise, OCE is located on OCC premises.

The Bell System may also be the sole facility provider between the following premises:

- OCC premises and OCC premises

- patron premises and patron premises

- OCC premises and patron premises

- patron premises and a Bell System central office building

- OCC premises and a Bell System central office building.

The configuration for such cases is the same as that shown in Figure 8-15 except that an NI is a DP, customer premises becomes patron premises, and in the latter cases, one of the premises is replaced by a central office building.

Typically, the services provided might include a 2-wire control facility between two OCC premises or between two patron premises for transmitting dc opens and closures for purposes such as status indication and telemetering, a 2-wire control facility between an OCC premises and a Bell System central office for echo suppressor disabling, or a video cable facility between two OCC premises or between two patron premises.

8.7.4 EVOLUTION OF INTERFACES FOR INTERCONNECTION

The Bell System has been an active participant in identifying and standardizing network interfaces for many years. Prior to the 1968 FCC Carterfone Decision (see Section 17.2.7), which permitted direct electrical connection of user-provided terminal equipment to Bell System and other telephone company facilities networks, much effort was directed toward standardizing interfaces for connecting customer-provided business machines to Bell System data equipment. Through such activities in the Electronics Industries Association (EIA), standards like the EIA RS-232-C and RS-366 were developed.

Direct electrical connection of customer-provided terminal equipment to the Bell System facilities network required a host of new interfaces to accommodate the existing Bell System services. At first, this need was filled by connecting arrangement service, an adjunct to existing services for those cases where network protection was required. A family of new interfaces was developed for connecting arrangement service, each interface being defined by a network protection device. Later, the FCC ruled that such protection devices could not be required and that, instead, terminal equipment was to provide the network protection. Such protection is required for all services except those classified as inherently protected (see Section 8.7.5).

Interconnection between OCC facilities and the Bell System facilities network evolved as a result of FCC decisions dating as far back as 1949. At that time, the FCC began licensing private microwave systems. These systems were not interconnected with the Bell System facilities network at first, but in 1969, the FCC permitted private microwave companies to compete with the Bell System in the sale of telecommunications services. In order for these companies (common carriers) to provide end-to-end service, they were permitted to interconnect with local Bell System network facilities. At first, the interfaces were specifically defined to isolate each carrier's area of responsibility with minimum maintenance, administrative, and design interaction. However, as a result of informal negotiations sponsored by the FCC, a more interactive means of supporting OCC interfaces was agreed upon (see FCC 1975). The new concept extended maintenance activities across the demarcation point and provided a more interactive design approach. Thus, even though standardization is becoming more prevalent, demarcation points tend to be negotiable.

8.7.5 NETWORK PROTECTION

Protection of the network is required by the FCC to ensure that terminal equipment on user premises does not cause physical damage to the Bell System and other facilities networks, injure employees, or impair service to other users (for example, usefulness to a third party). At first, when the Carterfone Decision went into effect, network protection requirements (for example, electrical limitations) were specified in appropriate tariffs, and other protection criteria were provided in various technical references published by AT&T. Network protection was ensured where possible by the network protection devices for connecting arrangement service. Initially, the FCC permitted these devices to be interposed between the terminal equipment and the facilities network.

The FCC also instituted proceedings to determine the conditions for interconnection and what rules, if any, should be adopted for a more suitable long-term solution to the network protection issue. As a result of these proceedings, the FCC identified specific network "harms" that could result from uncontrolled interconnection of terminal equipment and ruled that network protection must be provided in all cases. These proceedings led to the development of a comprehensive program, as embodied in the FCC Registration Program, for ensuring that such protection is always provided.

FCC Registration Program

The FCC Registration Program, instituted in 1975, replaced all previous programs for the direct electrical connection of terminal equipment to the facilities network and created a need to define and standardize a new class of network interfaces. Each new interface is rigidly defined at connection points normally provided between equipment units (or between an equipment unit and the network facility) on a customer's premises. Network protection is ensured by requiring manufacturers of terminal equipment to demonstrate compliance with FCC Part 68 Rules and Regulations. These rules establish a well-defined set of requirements for interconnection.

In this regard, terminal equipment manufacturers must submit documented proof of compliance with the Part 68 requirements for the particular service(s) to which the terminal equipment will be connected. The FCC then issues a registration number that must be affixed to each unit of terminal equipment connected. For each type of service, all terminal equipment connected to a network interface must be registered after a specific date, known as the "register-only" date. Terminal equipment connected prior to that date does not have to be registered unless modifications are made that affect compliance with the Part 68 requirements (see FCC 1977/1980).

Initially, the FCC Part 68 Rules and Regulations covered terminal equipment, with certain exceptions, for interconnection with the switched message network by a 2-wire interface with loop-start, ground-start, or reverse-battery signaling. The exceptions were PBXs, key telephone systems, main-station telephones, coin telephones, and terminal equipment connected to private and multiparty lines. Subsequently, standard interfaces were identified for 4-wire station-line service and certain private-line services including PBX message registration, PBX off-premises station lines, PBX tie trunks, and PBX automatic identified outward dialing. The FCC Part 68 Rules and Regulations were amended to include these services and main-station telephones. Standard interfaces have since been identified for many other existing services including 2-point private lines, multipoint private lines, local area data channels, and new data services such as those provided by DS1 facilities. However, to date, the FCC Part 68 Rules and Regulations have not been amended to include these services.

FCC Part 68 Rules and Regulations

FCC Part 68 Rules and Regulations specifies the requirements that equipment manufacturers must meet in order to be able to market their product for connection to the network (see FCC 1977/1980). Provisions are made for the means of connection, notification of the telephone company, verification requirements and procedures, labeling requirements, and specific physical and electrical requirements and limitations. The physical and electrical requirements and limitations include the following:

• environmental simulation: vibration, temperature, humidity, physical shock, metallic voltage surges, and longitudinal voltage surges

- leakage current and hazardous voltage limitations to control electrical hazards to telephone company equipment and personnel

- signal power and longitudinal imbalance limitations to control crosstalk and interference on other telecommunications channels

- on-hook impedance limitations to control pretrip (cessation of ringing before the called station actually rings) and to ensure the integrity of certain network maintenance procedures

- other requirements and limitations to ensure the integrity of network billing.

Inherent Network Protection

Inherent network protection refers to those cases where protection from network "harms" is an inherent part of certain private-line services. In such cases, hazardous voltage limiting and longitudinal imbalance protection are incidental to the NCTE functions. Signal power compliance may be ensured either by telephone company procedures normally associated with the service or by a limiting function otherwise provided by the NCTE. For services in this category, terminal equipment compliance with Part 68 Rules and Regulations is not required. Private-line services presently designated as inherently protected include a variety of types such as program audio, video, Washington Area Metropolitan Channels, 150-baud (code elements per second) telegraph, and others.

Certain digital services, for example, *DATAPHONE* digital service, are also considered inherently protected. However, in those cases where a digitally encoded signal level can be converted to an analog signal level and the service can be connected to an analog network service, inherent protection cannot be ensured. As currently proposed, signal power requirements for such services would be included in the Part 68 Rules and Regulations.

8.7.6 TYPICAL NETWORK INTERFACES FOR INTERCONNECTION

Two examples of existing network interfaces for interconnection are summarized in Table 8-7. One, referred to as a *Digital Data System* (DDS) interface, interconnects terminal equipment with a DDS facility. The other, referred to as a *network interface*, interconnects voice-frequency terminal equipment, such as a telephone set, with a switched message network facility.

The DDS interface is a 6-wire interface. Voltages used for status indication on two of the six leads conform to EIA RS-232-C. The physical

TABLE 8-7

TYPICAL NETWORK INTERFACES

	DDS Interface	Network Interface
Type of service	Private line	Switched
Signals transmitted	Digital	Analog, voiceband
Physical interface	6-wire	2-wire
Connector	Jack and plug	Jack and plug
Network protection	Inherent*	FCC registration
NCTE	CSU	Protector block
Control signaling	Bipolar violations and control code	Dc, loop-start
Transmit signal power limit	Not required	−9 dBm
Signal characteristics:		Not specified
Service rates	56 kbps; subrates of 2.4, 4.8, and 9.6 kbps	
Code	Bipolar return to zero	
Pulse shape	Rectangular, 50-percent duty factor	
Timing	Encoded in received bit stream	

*Hazardous voltage, longitudinal balance, and signal power protection in channel service unit (CSU).

connection is a 15-pin jack and plug (with only six pins used) with leads designated as follows:

DT1, DR1   -   Data transmit
DT, DR   -   Data receive
SI, GND   -   Status indication

These and other interface requirements are specified in the Bell System technical reference for the service. The NCTE, in this case, is a Bell System channel service unit (CSU) that is part of the DDS facility. The CSU provides several network functions. In the transmit direction, it restores the bipolar signals from the terminal equipment to a level suitable for application to the DDS facility. In the receive direction, it provides facility loss equalization, establishes a reference voltage level, and converts incoming signals to uniform bipolar pulses. In addition, the CSU provides loopback for the 4-wire line to facilitate the isolation of troubles from a remote test center.

Other DDS interface items listed in Table 8-7 are DDS characteristics that are part of the interface description required to distinguish it from other 6-wire interfaces. Control signaling, accomplished by control codes in the bit stream together with bipolar violations of the pulses, establishes the idle, out-of-service, and test (that is, loopback) conditions. A transmit signal power limit is not necessary since the signal power is limited by the CSU. This, together with the fact that the CSU incidentally provides hazardous voltage and longitudinal imbalance protection, qualifies the service as inherently protected. The last item in the table, signal characteristics, lists some of the other DDS characteristics that distinguish this interface from other network interfaces.

The network interface for interconnection with the switched message network is typically a 2-wire interface for voiceband analog terminal equipment. The physical connection is a modular jack and plug with leads designated T (tip) and R (ring). The NCTE for this case is typically a lightning protector and may, in some cases, include a maintenance terminating function. Terminal equipment connected to this interface must be registered.

Other items in Table 8-7 for the switched message network interface define the interface as unique to services with loop-start control signaling. Loop-start establishes the idle, off-hook, and dialing conditions. The −9 dBm signal power limit is also a part of the interface definition since it signifies a need for signal power control in the connected terminal equipment. It is the highest 3-second average signal power that can be applied to the network facility by the terminal equipment without potential for network "harm." In this case, potential network "harm" is crosstalk to a third party as a result of carrier overload. The limit does not apply to live voice signals or to network control signaling.

8.8 DATA COMMUNICATIONS INTERFACES AND PROTOCOLS

Data communications is concerned with the transfer of information in digital form between users. Information must be transferred in a manner that preserves its meaning; the communications mechanism must provide undistorted information transfer.

There is more to data communications than just the physical movement of bits between devices. The diverse nature of the information being transferred places varying requirements on the communications mechanism (for example, low transmission delay, high quality, low cost, etc.). In addition, transferring information between source and destination may involve traversing several different types of communications networks (for example, the direct distance dialing network, a packet-switching network, and a local area network). Thus, data communications generally involves a complex mechanism requiring precise rules to govern the transfer of information. These rules and their syntax (format) are called *protocols*.

A protocol may be thought of as a language used to interact with the communications mechanism to establish, carry out, and terminate communications. Since, in general, the data communications process could involve several systems and communications networks, the rules necessary to control the communications can become quite complicated. If only one protocol were defined to control all aspects of the communications, it would be extremely complex. Rather than define a single unwieldy protocol, the approach taken was to reduce the complexity to manageable proportions by structuring the information transfer process into smaller pieces. Extensive national and international effort has recently resulted in a model called the *Reference Model for Open Systems Interconnection* (OSI).[25] The model partitions the functions required for communications into seven hierarchical groups called *layers* (see Figure 8-17).

Each layer (or group of functions) provides capabilities, called *layer services*, to the next higher layer by building on (*enhancing*) the layer service provided by the layer below. A protocol associated with each layer controls the communications activities (*functions*) between *entities*[26] in the

---

[25] See ISO 1983 and CCITT 1983 for a discussion of this reference model.

[26] A logical element in a given layer, within a system, which executes the layer protocols. The logical element may be hardware, software, or a combination of both.

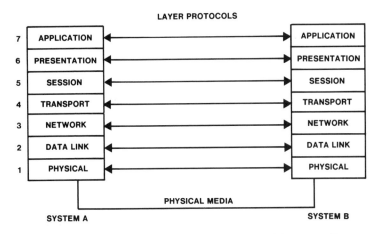

7	APPLICATION		APPLICATION
6	PRESENTATION		PRESENTATION
5	SESSION		SESSION
4	TRANSPORT		TRANSPORT
3	NETWORK		NETWORK
2	DATA LINK		DATA LINK
1	PHYSICAL		PHYSICAL

PHYSICAL MEDIA

SYSTEM A   SYSTEM B

Figure 8-17. Reference model for open systems interconnection.

layer residing in different systems (see Figure 8-17). Thus, the complex task of information transfer has been broken down into seven smaller tasks (layers) each of which has a protocol for controlling its activities. The protocols of each layer are independent of one another; that is, when a layer uses capabilities or services provided by the layer below, it does not know or care how the capability was generated.

One of the major advantages of defining the model as a hierarchical set of seven layers is that the end-users (users of the application layer, which may be application programs, human beings, etc.) do not need to be concerned with the details of the underlying communications mechanism. They need only be concerned with the information to be transferred and the application at hand.

Another advantage and primary goal of the model effort is that by standardizing the protocols for each layer, systems and networks built independently but in conformance with the protocols can operate with one another; this provides a potentially very large community of "open systems."

8.8.1 LAYERS OF THE REFERENCE MODEL

A useful way to view the model is to consider the distinction between the scope of the lower three layers and the scope of the upper four layers. The lower three layers provide all of the capabilities required for sending data over a network connection between end systems; that is, they provide the capabilities to establish, carry out, and terminate connections through one or more networks. The upper four layers of protocol provide all of the end system—to—end system signaling for data transfer. These protocols are carried transparently between end systems by the lower three layers and are only interpreted by the end systems. (See Figure 8-18.)

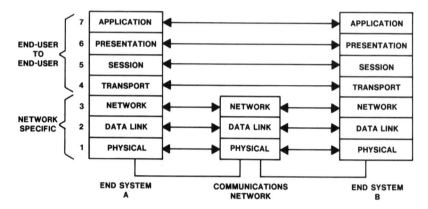

Figure 8-18. Protocol architecture.

The following paragraphs briefly summarize the functions of each layer. It should be noted that each layer contains the layer-specific procedures for establishing, carrying out, and terminating its own activities in addition to the specific functions mentioned below.

- The **application layer** provides protocols specific to particular applications (for example, airline reservations, credit checking, purchase orders) for the transfer of information meaningful to communications users. These protocols are realized by building application-specific protocols on top of the basic capabilities, called *common application capabilities*, of the application layer. All of the services available to users from the OSI environment are requested via the application layer.

- The **presentation layer** is responsible for representing the information to be transferred between applications in a manner that preserves its meaning (*semantics*) while resolving differences in its representation (*syntax*). Examples of different syntaxes include American Standard Code for Information Interexchange (ASCII) versus Extended Binary-Coded Decimal Interchange Code (EBCDIC), and screen-oriented data from a cathode-ray tube (CRT) terminal versus line-oriented data.

- The **session layer** is responsible for establishing, maintaining, and terminating an association between its users (presentation entities). It also negotiates the appropriate dialog discipline (for example, which user has the right to send at a given time) and has capabilities for synchronizing the activities of its users.

- The **transport layer** provides for the control of data transfer between end systems, independently of how the connection between end systems is established. The transport layer may provide for end system—to—end system acknowledgement, splitting of a transport connection onto many network connections, multiplexing many transport connections onto one network connection, and the capability to detect and, in some cases, correct errors in the data transferred between end systems.

- The **network layer** provides the capabilities and procedures required to control network connections (for example, set up, maintain, terminate) between end systems containing transport layer entities. The network layer isolates its users (transport entities) from the specifics of particular networks supporting the transfer of data.

- The **data link layer** enhances the basic capability of transferring bits so as to provide high-quality ("error free") transmission of data over a physical connection.

- The **physical layer** provides for the transmission of a bit stream over a transmission channel in some physical communications medium.

8.8.2 INTERFACES

Standard data communications interfaces have been defined to facilitate communications between pieces of equipment. At the point of demarcation between two pieces of equipment, two attributes of the interface are defined: the physical media aspects (for example, electrical and mechanical) and the protocols.

An example of an important data communications interface is that between data terminal equipment (DTE)—for example, a teletypewriter, CRT, or host computer—and data circuit-terminating equipment (DCE)—for example, a data set. This interface has been standardized in the United States by the EIA and internationally by the CCITT and the International Organization for Standardization (ISO). The most widely deployed DTE/DCE interface in the United States is EIA RS-232-C (see EIA 1969). Other important interfaces include EIA RS-449 (see EIA 1977/1980) and CCITT Recommendation V.35 (see CCITT 1981a). These interfaces pertain to the physical layer of the model and to the physical media (see Figure 8-19). Table 8-8 gives a listing of the interfaces provided by the Bell System data sets and Digital Data System (DDS) data service units (DSUs).

The advent of digital network services has produced a new set of interfaces at the network service boundary. Table 8-9 provides an illustrative listing of Bell System digital services and their respective interfaces. For packet-switching networks, CCITT Recommendation X.25 (see CCITT 1981b) covers the DTE/DCE interface for packet-mode operation, and Recommendation X.75 (see CCITT 1981c) covers the network-to-network interface. These interfaces consist of the physical media aspects plus the protocols at the lower three layers of the model (see Figure 8-20).[27] The physical layer and the physical media aspects can be, and typically are, those described in the previous paragraph.

Another example is Bell System Protocol Specification BX.25 (see AT&T 1980b) that was developed by the Bell System for use in the operations systems network (see Section 15.5). BX.25 fully specifies the actions of the DTE in a way that permits either connection to a network that supports CCITT X.25 or direct connection to another BX.25 DTE. BX.25 now

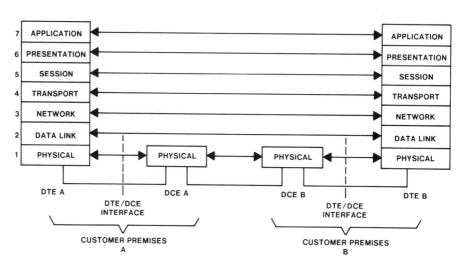

DCE   DATA CIRCUIT–TERMINATING EQUIPMENT
DTE   DATA TERMINAL EQUIPMENT

Figure 8-19. Application of OSI model to modem interfaces.

[27] As indicated in the description of the Basic Packet-Switching Service in Section 11.6.2, the protocols in X.25 are referred to as the physical (or bit), link, and packet levels. They correspond to the physical, data link, and network layers, respectively, of the OSI model.

TABLE 8-8

ILLUSTRATIVE LISTING OF
BELL SYSTEM DATA SETS AND DSUs*

Code	Technical Reference (PUB No.)	Interface
Voiceband Data Sets		
103J	41106	EIA RS-232-C
108F	41215	EIA RS-232-C
108G	41215	EIA RS-232-C
113C	41106	EIA RS-232-C
113D	41106	EIA RS-232-C
201C	41216	EIA RS-232-C
202S	41212	EIA RS-232-C
202T	41212	EIA RS-232-C
208A	41209	EIA RS-232-C
208B	41211	EIA RS-232-C
209A	41213	EIA RS-232-C
212A	41214	EIA RS-232-C
407C	41409	EIA RS-232-C
2024A	41910	EIA RS-449†
2048A	41910	EIA RS-449†
2048C	41910	EIA RS-449†
2096A	41910	EIA RS-449†
2096C	41910	EIA RS-449†
Wideband Data Sets		
303	41302	Coaxial cable
306	41304	CCITT V.35
DDS DSUs		
500B	41450	EIA RS-232-C
500B	41450	CCITT V.35

*Table 11-1 contains additional information.

†Adaptor available for RS-232-C.

TABLE 8-9

ILLUSTRATIVE LISTING OF BELL SYSTEM DIGITAL SERVICES

Service	Technical Reference (PUB No.)	Use	Interface	Speed (bps)	Operation	Type	Remarks
DDS	41021	Private line	6-wire	2.4, 4.8, 9.6, 56k	Full duplex	Sync	Point to point
DDS	41022	Private line	6-wire	2.4, 4.8, 9.6, 56k	Full duplex	Sync	Multipoint
DS1	41451	Private line	9-wire	1.544M 1.344M	Full duplex	Sync	Point to point
CSDC	61310	Circuit switched	8-wire	9.6, 56k	Full duplex	Sync	Alternate voice/data
BPSS	54101	Packet switched	6-wire	9.6, 56k	Full duplex	Sync	CCITT X.25

CSDC = Circuit-switched digital capability (see Sections 2.5.1 and 11.6).
BPSS = Basic Packet-Switching Service (see Sections 2.5 and 11.6.2).

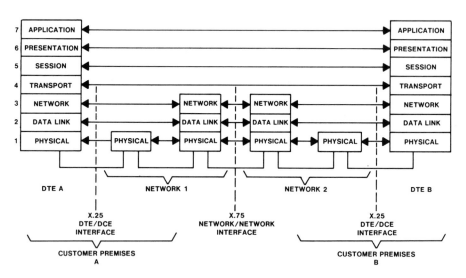

DCE DATA CIRCUIT—TERMINATING EQUIPMENT
DTE DATA TERMINAL EQUIPMENT

Figure 8-20. Relationship of X.25 and X.75 to the OSI reference model.

has four layers. The lower three layers correspond to the DTE side of the X.25 DTE/DCE interface. Currently, the uppermost layer of BX.25 combines the functions of layers 4, 5, and 6 of the model. This was done for efficiency and expediency; the operations systems network needs at these layers are relatively simple, and the standards committees had not completed the precise definition of these layers.

AUTHORS

P. D. Bartoli
H. V. Bertine
R. A. Fuller
W. S. Hayward
W. C. Roesel
W. R. Starrett
F. E. Weber

PART THREE

NETWORK AND
CUSTOMER-SERVICES SYSTEMS

Building on the broad introduction to the Bell System in Part One and the telecommunications fundamentals in Part Two, Part Three examines network and customer-services systems and associated equipment. The intent is not to catalog all systems, but to present advantages and disadvantages of selected systems relative to applications, describe system operation on a functional level, and discuss the evolution and current trends in each major area.

Chapters 9 and 10, respectively, describe many of the transmission and switching systems in use in the Bell System. Chapter 10 also covers billing systems, which are closely related to switching systems. Chapter 11 discusses the customer-services equipment and systems that provide many of the services described in Chapter 2. Chapter 12 highlights three types of common systems associated with the network and customer-services systems presented in Chapters 9 through 11. These types are power systems, distributing frames, and equipment-building systems. They are included here because, while they are not components of the network and customer-services systems, they provide essential supporting functions.

9

Transmission Systems

9.1 INTRODUCTION

This chapter describes many of the transmission systems now in use throughout the Bell System network.[1] Section 9.1 identifies the major applications for these transmission systems, namely, **loops** and the **interoffice networks** that include metropolitan, outstate, and long-haul intercity. Since the characteristics of each of these areas are unique, specific system features and capabilities are designed to match them.[2] There are three basic types of transmission systems: **voice frequency, analog carrier,** and **digital carrier**. Section 9.2 covers voice-frequency transmission systems, which are found extensively in the loop and metropolitan interoffice areas of the telephone plant. Sections 9.3 and 9.4 describe, respectively, analog and digital carrier transmission systems, which have found application in the metropolitan, outstate, and long-haul areas.

9.1.1 AREAS OF APPLICATION

The Loop Plant

The connection between the telephone customer and the serving central office is called a *loop*. Individual loops leave the serving central office building in a large main feeder cable via an underground conduit system that provides physical protection for the cable. Branching off from the main feeder cable are smaller branch feeder cables and then still smaller

---

[1] For a more thorough treatment of transmission system design, see Bell Laboratories 1982.

[2] Sections 4.2 and 4.4 contain related information on the structure of the traffic and facilities networks. It should be noted that Chapter 4 uses *rural* rather than *outstate*.

distribution cables that extend the loop eventually to a point near each individual customer's premises. The cables may continue in conduit to the distribution points, or they may be carried via telephone poles or buried directly (without conduit) in the ground. A *drop wire* or buried service wire extends the loop to the living unit or office building and terminates at the station protector. Inside wiring completes the loop connection to the customer's station set.

The loop area is generally characterized by changing population density and extensive rearrangements. The loop plant must be readily accessible to accommodate these changes. The major access points in the loop area are the *serving area interface* in the outside plant, which serves as a rearrangeable cross-connect point between the feeder and distribution cables, and the *main distributing frame* (see Section 12.3) in the central office building, where the loops are then cross-connected to the local switching system or to a dedicated interoffice circuit.

Loops must be capable of transmitting 2-way voice-frequency speech or data signals. In addition, they must handle signals from rotary and *TOUCH-TONE* dialing and supervisory signals peculiar to the switching machine in the serving central office. The equipment used on loop facilities must be able to transmit and to withstand the voltage used to ring the bell on the station set. The loop must also provide ample direct current to permit proper operation of the station set. Some of these design considerations are discussed later in this chapter.

As indicated by Figure 9-1, loops are generally quite short. The median length is about 1.7 miles, or 9 kilofeet (kft), and 95 percent of all

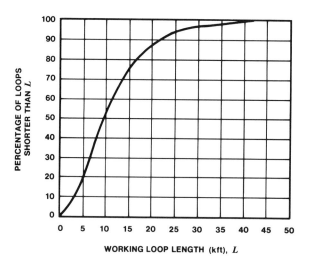

Figure 9-1. Distribution of loop length (1973).

loops are shorter than 5.2 miles (27.5 kft). As the distance from the central office increases, so does transmission loss. The primary design measure used to reduce the loss is the limitation of maximum loop resistance by the application of larger gauge wire pairs. Figure 9-2 shows the distribution of different gauge cables as a function of loop length. For loops longer than 18 kft (about 15 percent of all loops), loading and carrier techniques are used, as described in Section 9.2.1.

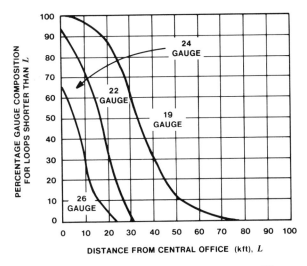

Figure 9-2. Average percentage gauge composition of loops versus distance from central office (1973).

Because of the need for flexibility to accommodate future assignments, a long-time telephone company practice has been to provide for multiple appearances of the same loop pair at several distribution points. These *bridged taps* make reassignments of loops easier but influence the transmission characteristics, so that rules have to be established to limit their length to 6 kft or less.

The Metropolitan Interoffice Plant

A metropolitan, or metro, area is served by a number of closely spaced central offices interconnected by interoffice transmission facilities. Two general types of circuits are provided by these interoffice facilities: **trunks**, which are shared by users of the public switched telephone network (PSTN), and **special-services circuits**, which are dedicated to those customers who have a long-term need to reach specified distant locations.

Special-services circuits utilize nearly half of the interoffice facilities. Both trunks and special-services circuits may be bundled together on the same cable or carrier facility as it leaves the central office (see Figure 3-2). These facilities provide paths to nearby offices, where the circuits terminate or pass through to their ultimate destination. A sequence of *spans* (subpaths), which carries a circuit from its origin to its destination, is designated a *route*. The average metro span is about 6 miles long. Route cross sections become thinner as the distance between offices increases, but economies of scale are achieved by concentrating circuits into spans and thus eliminating the number of geographic paths between pairs of offices. (Table 4-6 presents data on route cross sections and the distance between switching offices.)

Of some ten million interoffice circuits in the Bell System, about two-thirds are in the metro area. The very short metro area circuits use voice-frequency (VF) transmission, while the rest use digital carrier facilities. Transmission loss compensation is often required for the VF facilities in the form of inductive loading on many of the shorter VF circuits, and VF repeaters and loading on the longer ones. For digital transmission systems, the metro area requires low per-circuit terminal costs because the distances are so short (and line-haul costs correspondingly so low) that terminal costs dominate. The economic break-even length between voice-frequency transmission on paired cable and digital carrier has been dropping sharply in recent years, and in fact, for trunks connecting to 4ESS switching equipment, the break-even distance is 0 miles.

The Outstate Interoffice Plant

The outstate, or rural, area covers small cities and towns and remote suburban as well as sparsely settled areas. The networks are usually simple and tree-like with low connectivity between central offices and few alternate paths available for route diversity.

In the outstate area, trunk connections between class 5 and class 4 offices predominate. Since the central offices are widely spaced in the outstate area, transmission systems tend to be longer than in the metro area. Figure 9-3 plots the distribution of outstate area system lengths.

Interoffice cross sections (number of circuits) are also small by metropolitan standards: N-carrier routes average about fifty voice circuits. Annual circuit additions also tend to be small, which has a distinct impact on the type of facilities provided in outstate areas.

Currently, outstate networks consist predominantly of N-carrier and a small but growing number of T1 carrier systems (see Section 9.4.2) provided with protection switching to improve reliability. This protection is critical in an outstate area because alternate routes may not be provided,

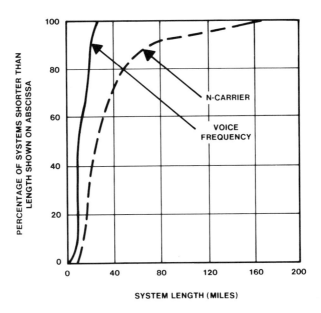

Figure 9-3. Distribution of system length
in the outstate area.

many small offices are unattended, and communities could easily become completely isolated.

The Long-Haul Interoffice Plant

The long-haul network overlays the metro and outstate networks and interconnects their toll offices. For this reason, the long-haul network is often referred to as the *intertoll network*. Traffic is highly concentrated and brought to a limited number of large switching offices. High-density intertoll trunks tie together class 4, 3, 2, and 1 offices throughout the United States. Cross sections in all but the extremities of the long-haul network tend to be large.

The discussion and accompanying facility map (Figure 4-11) in Section 4.4.2 indicate that multiple routes exist between metro areas. The heavy cross sections in the network generally have facility diversity and provide high reliability as well. (For example, parallel microwave radio and coaxial cable systems frequently are provided along the same route.)

Intertoll trunks are typically from 100 to 1000 miles long, with an average length of about 400 miles. Only 3 percent of the intertoll trunks are longer than 2000 miles, but because of their extreme length, they account for 25 percent of the total intertoll mileage.

Transmission loss objectives for the long-haul area are stringent. For example, an intertoll trunk of average length (400 miles) must meet a via net loss (see Section 6.6.1) objective of about 1 decibel (dB). Thus, costly high-quality transmission systems are generally required. On the other hand, the economics require low per-circuit-mile costs. This is a motivating force for wider system bandwidths so that large numbers of circuits may be combined on the system. This has resulted in a succession of systems that provide increasing amounts of capacity at decreasing per-circuit-mile costs. This is conceptually illustrated in Figure 9-4, where segments A through F represent cost versus circuit capacity for systems suitable for different network applications in terms of circuit capacity.

Only 21 percent of all circuits in the Bell System are long haul, but because they are so much longer, long-haul circuits represent the major percentage of the circuit mileage in the total trunk plant. At the end of 1980, the total number of circuit miles in the Bell System amounted to 1.03 billion. Transmission systems in the long-haul area account for over 75 percent of this total.

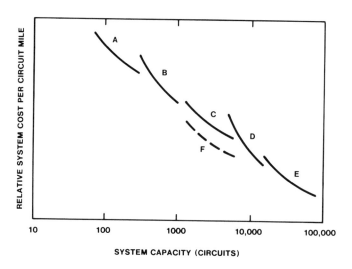

Figure 9-4. Economy of scale in the long-haul area.

9.1.2 INVESTMENT BY AREA

In 1980, the Bell System's total investment in transmission facilities amounted to nearly $60 billion. Almost 60 percent of this investment was in the loop plant. The remaining 40 percent was invested in the interoffice plant. The breakdown was approximately: metro area, 26 percent; outstate area, 5 percent; and long-haul area, 9 percent.

9.1.3 GENERAL TRANSMISSION SYSTEM TYPES

This section describes the basic characteristics of VF and carrier transmission systems. Details on specific systems are provided later in this chapter.

Voice-Frequency Transmission Systems

The medium most often used for VF transmission is a twisted wire pair in a multipair cable. The physical design of the cable (wire gauge, type of insulation, twist lengths) determines the transmission properties of the wire line. These properties include the attenuation, phase shift, and characteristic impedance of the transmission line. Relating these to the electrical properties of the pairs (resistance, inductance, conductance, and capacitance) is beyond the scope of this book. (Section 6.3 presents a general description of media.)

As Figure 9-5 indicates, VF transmission is limited to very short distances because of the high attenuation per mile of untreated cable pairs. To meet a 3-dB loss objective, for example, applications are limited to less than 3 miles even with 19-gauge cable.

To reduce the VF attenuation in cable pairs, lumped inductances called *load coils* are placed periodically along the cable. The most common loading system used in the VF plant is designated H88.[3] Figure 9-6

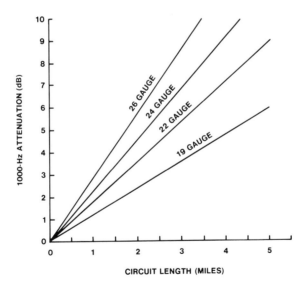

Figure 9-5. 1000-Hz attenuation versus length for nonloaded cables.

---

[3] This corresponds to load coils of 88 millihenries, each spaced 6000 feet apart.

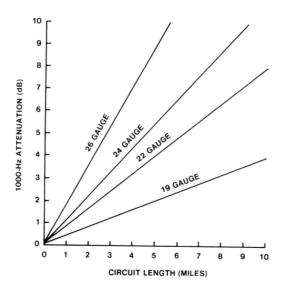

Figure 9-6. 1000-Hz attenuation versus length for H88-loaded cables.

shows the attenuation at 1000 hertz (Hz) for H88-loaded cables. Comparison with Figure 9-5 shows that loaded circuits one and one-half to three times the length (depending on wire gauge) of nonloaded circuits will have the same attenuation at 1000 Hz. Loading, however, produces attenuation that depends on frequency. As Figure 9-7 shows, a loaded cable behaves like a low-pass filter with a cutoff frequency of about 3000 Hz.

Loading introduces a number of other disadvantages. For example, compared to nonloaded pairs, it reduces the velocity of propagation by as much as a factor of 3. As a result, the absolute transmission delay[4] is increased. In addition, loading introduces considerable delay distortion and some attenuation distortion in band that requires equalization on longer VF lines.

To obtain additional range on nonloaded and loaded wire pairs, VF repeaters are necessary. In 2-wire operation, maximum repeater gains are limited to about 12 dB because cable characteristics change with temperature, terminations vary due to switching, and crosstalk becomes a problem. The maximum number of tandem repeaters is usually limited to two to avoid singing or other oscillations. For 22-gauge H88-loaded cables, this limits the range to under 35 miles, assuming a 3-dB loss requirement.

---

[4] Section 6.6 discusses transmission delay and other impairments mentioned throughout this chapter.

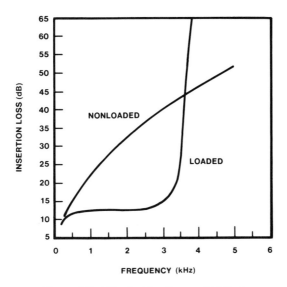

Figure 9-7. Effect of loading on 54-kft of
24-gauge cable.

To obtain additional range, 4-wire VF transmission may be employed. Four-wire transmission is inherently more stable than 2-wire, since there is only one singing path extending around the circuit from one end to the other. Repeater gains are then limited by nonlinear distortion, echo, and crosstalk considerations. In practical applications, 4-wire VF systems are limited to the length at which their cost per circuit is less than that of a competing carrier system.

Carrier Transmission Systems

Carrier systems, both analog and digital, are generally more economical than VF transmission where distances are longer and where cross sections are large. The lower line-haul cost of carrier transmission is the result of the increased utilization of the transmission media achieved by combining (multiplexing) a large number of message signals into a composite signal. This saving is at the expense of an increase in the cost of the required multiplex terminal equipment.

In addition to the economic advantage, carrier transmission also offers a number of performance advantages. The major one is a dramatic improvement in the velocity of propagation at carrier frequencies. This results in a very low absolute delay for carrier signals and a resultant advantage with respect to echo control. Combined with better control of

impedances, this permits long-distance circuits to operate at lower losses
while maintaining excellent stability.

Analog Carrier Systems. Analog carrier systems use repeaters that are
designed to compensate (equalize) for the loss characteristics of preceding
cable sections and reproduce at their outputs a linearly scaled version of
the cable section input signal. In practice, analog repeaters are not perfectly linear, nor are they noise free. Therefore, noise and nonlinear distortion accumulate with distance and determine to a large degree the performance achievable.

An analog carrier system is not limited in terms of the types of signals
that may be transmitted. Speech, data, video, and supervisory signals
may be combined, provided only that interference requirements are compatible with system performance and bandwidth. Speech signals are particularly suitable to analog carrier systems as compared to digital carrier
systems because system bandwidth is used more efficiently (see Section 6.4.3). However, techniques for placing a mix of different signals
(data, speech, and video) on an analog carrier system usually result in a
lower capacity (fewer channels) as compared to a digital system.

Over long distances, where line-haul costs are a greater factor than
terminal costs, analog transmission has traditionally proven more
economical than digital transmission, as illustrated in Figure 9-8. This
situation is changing, however (see the next section).

One important factor in the economic comparison between analog and
digital carrier transmission is whether the switching system associated
with the facility is analog or digital. When an analog facility terminates
on a digital switch, additional terminal equipment is required to convert

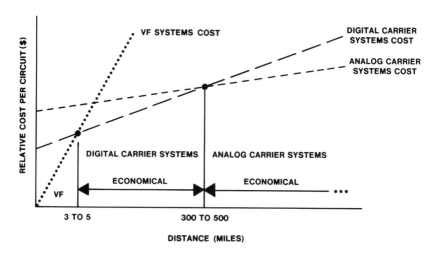

Figure 9-8. Relative economics of transmission systems.

the signal to the proper format. This increases the distance at which analog systems "prove-in" economically to a distance above that indicated in Figure 9-8.

Digital Carrier Systems. Digital carrier systems are characterized by discrete signals, regenerative repeaters, and time-division multiplex signals.

Converting signals, whether they be speech, data, or video, into digital form permits them to be intermixed and treated identically within a common transmission facility. Pulse-code modulation, described in Section 6.4.3, is used to convert speech signals into digital form. The major digital impairment is quantizing noise, which occurs only in the terminals and can be controlled by assigning a sufficient number of binary digits to represent the encoded samples.

The digital signal format is naturally well-suited for data signals. This "match" between the data signal and digital carrier transmission is very efficient. For example, T1-carrier can carry 24 times 64 kilobits per second (kbps), or 1536 kbps of information; while twenty-four analog voice channels can carry 24 times 9.6 kbps, or 230 kbps of information.

Digital carrier systems regenerate the signal at each repeater location. Line noise and interference in the media have very little effect on the signal since they do not accumulate. As a result, the required signal-to-noise ratio is lower for digital systems than for analog systems. Relatively noisy media, such as multipair cables where crosstalk is a limiting factor, are better suited to digital carrier than analog carrier. On the other hand, in high-quality media, such as coaxial cable, the ruggedness property of the digital signal is of no real advantage. In congested radio media, however, the digital signal is able to tolerate external interference to a much greater degree than analog transmission.

Digital multiplex terminal equipment is able to exploit the latest advances in low-cost, high-speed integrated circuit technology. Digital devices continue to become faster, smaller, longer lived, more reliable, and less power consuming.

As terminal costs decrease, the prove-in range for digital carrier systems will increase over that shown in Figure 9-8. The increase will result from both a shorter prove-in distance with respect to VF systems and a longer prove-in distance with respect to analog carrier systems on paired cable.

9.2 VOICE-FREQUENCY TRANSMISSION

9.2.1 LOOP AREA APPLICATIONS

Several VF system design methods have been used in the loop plant. They have evolved as a result of continuing efforts to reduce the cost of the loop cable plant while maintaining satisfactory transmission and signaling performance on an overall statistical basis. Table 9-1 summarizes these design methods, and they are described below.

The **resistance design** method controls transmission loss by limiting loop cable resistance and requiring loading on loops 18 kft or longer. The rules include limitations on the maximum allowable lengths of bridged taps and tolerances on load-coil spacing. By employing heavier gauge cable pairs where needed, loops designed with the resistance design procedure meet the needs of message telephone service generally without the need of electronic equipment for gain (amplification), equalization, or network control signaling.

Unigauge design has as its objective the use of finer 26-gauge cable rather than coarse gauge, thereby reducing the amount of copper required. This design uses 26-gauge pairs exclusively to a range of 30 kft, beyond which heavier gauge may be added to the loop. Loops longer than 24 kft are loaded, and range extenders, when needed for message traffic, are shared rather than permanently connected to each loop. These devices, which extend signaling as well as transmission range, are switched in by the local switching system. Longer loops for special-services circuits must, of course, be treated on a dedicated basis.

Long-route design is intended to serve the small fraction of telephone customers in rural areas who are located beyond the range covered by resistance or unigauge design. A specific combination of range extension and fixed gain is prescribed so that loop loss is limited. Loop lengths of as much as 210 kft (about 40 miles) can be achieved using 19-gauge, H88-loaded cable.

The **concentrated range extension with gain** (CREG) design method offers a uniform and flexible approach that is compatible with resistance design and long-route design methods. This method enables the increased utilization of finer gauge cables in the loop plant through the use of switched range extension shared by several customers. The application of CREG design rules results in lower loop losses than with other

TABLE 9-1

COMPARISON OF VF LOOP DESIGN METHODS

	Resistance	Unigauge	Long-Route	CREG
Resistance range (ohms)	0-1300	0-2500	1300-3600	0-2800
H88 loading (kft)	>18	>24	All	>15
Gauging	Mixed	26 gauge	Mixed (to 30 kft)	Mixed

designs over a considerable range of conditions. Because of the improved overall transmission performance of CREG loops as compared with unigauge loops, in addition to administrative and design advantages, CREG design has superseded unigauge design.

While these methods describe the design of the existing loop plant, new rules for loop design became effective in 1983. In almost all cases, digital loop carrier and **revised resistance design** will be used for future relief and extensions of the loop plant. The principal application characteristics for these approaches are

- Revised resistance design will be used for loops that are 1500 ohms or less and 24 kft or less.

  — Loops 18 kft or less[5] must be designed to 1300 ohms maximum and must be nonloaded.

  — Loops greater than 18 kft but less than or equal to 24 kft,[5] may be designed to 1500 ohms maximum and require H88 loading.

- Digital loop carrier will be used for loops longer than 24 kft.

9.2.2 METROPOLITAN INTEROFFICE APPLICATIONS

In the metro area, VF trunks and special-services circuits use multipair cables similar to those used for loop applications. With the average length of a metro circuit about 6.5 miles, the loss incurred requires between 2 and 3 dB of gain.

One approach to providing gain and cable loss equalization in 2-wire VF circuits is by use of the negative impedance E6 repeater. Some limitations of the E6 repeater, however, are:

- The maximum repeater gain must be reduced to provide adequate margin against singing, crosstalk, and overload.

- In many applications, additional signaling equipment is required to compensate for a reduction in dc signaling range and dial-pulse-delay distortion introduced by line build-out (LBO) networks,[6] transformer windings, and the capacitance of the amplifier.

- A fixed attenuation-versus-frequency characteristic, which cannot be adjusted to match specific cable layouts, limits the maximum number of repeaters on a circuit to one or two.

---

[5] Total length including bridged tap.

[6] Amplifiers (repeaters) in a cable transmission system may be designed to compensate for distortion of a specific length of cable. When the length of cable between amplifiers is less than that for which the amplifier is designed, one or more line build-out networks are used to bring the distortion to approximately the design level.

The more modern VF equipment concept is one of consolidation, using the metallic facility terminal (MFT). In the MFT, required transmission and signaling functions are provided by separate transmission and signaling plug-in units, which mount side by side in a factory-wired bay. When wired to the main distributing frame, the appropriate plug-in units automatically provide all functions needed, using a minimum number of cross-connections. The transmission unit provides gain, level adjustment, impedance matching, amplitude equalization, precision balancing, and access to the signaling unit. The signaling unit performs such functions as circuit range extension, dial-pulse correction, ringing regeneration, supervision regeneration, signal conversion, toll diversion, and dc control of the customer's station set. MFT installation is efficient since options are selected by switches on the plug-in units, and periodic testing is eliminated through the use of reliable, stable solid-state circuitry. The next section discusses 4-wire designs.

9.2.3 OUTSTATE INTEROFFICE APPLICATIONS

In the outstate area, the median length of a VF circuit is nearly twice that found in the metro area. The via net loss design, discussed in Section 6.6.1, controls the loss and echo of these longer VF circuits and requires 4-wire operation. Amplification is provided by separate gain units for each direction of transmission. (There are some 4-wire designs in the metro area, and this discussion applies there as well.)

The older equipment design is the V4, consisting of two major types: the 44V4, a 4-wire repeater, and the 24V4, used where a 4-wire circuit terminates at a 2-wire switching office or at a customer's premises. The MFT design for 4-wire facilities is the modern counterpart to the V4. Figure 9-9 illustrates the 44- and the 24-type MFT repeater configurations.

Two MFT shelf arrangements are provided, each having spaces for twelve plug-in units. When both signaling and transmission treatments are needed, one shelf assembly provides for six circuits using six double-module signaling and transmission plug-in units. When only transmission treatment is required, another shelf assembly is used, which mounts twelve single transmission plug-ins.

9.3 ANALOG CARRIER TRANSMISSION

9.3.1 LOOP AREA APPLICATIONS

In the loop area, continued growth in demand, the trend from multiparty to single-party lines, and the rising cost of copper pairs have provided impetus for development of pair-gain loop transmission systems that increase the utilization of loop cables by enabling more than one customer to share each physical wire pair. Analog carrier systems for loop applications fall into two basic categories: **single-channel** and **multichannel**.

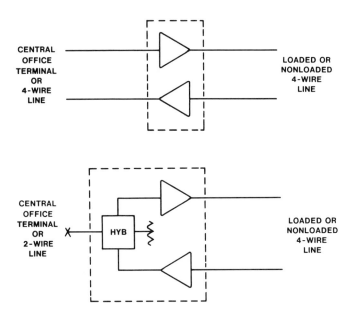

Figure 9-9. Metallic facility terminal 4-wire transmission configurations. *Top*, 44-type repeater; *bottom*, 24-type repeater.

Single-channel systems place an additional channel in the frequency spectrum above the baseband VF signal. These systems provide the additional voice channel by using double-sideband amplitude modulation (see Section 6.4.1). Bandpass filters separate the two carrier frequency bands associated with the two directions of transmission of the derived channel. Low-pass filters then separate the baseband channel from the derived channel. Since line repeaters are not used in the single-channel system design, applications are limited to short, nonloaded loops (less than about 18 kft). Single-channel systems are often used for the temporary deferral of new cable or conduit construction.

Figure 9-10 is a block diagram of a representative multichannel analog loop carrier system. No baseband channel is provided, and therefore, by allowing the use of line repeaters, these systems may serve longer routes than single-channel systems. The maximum range for the multichannel system shown is about 20 miles. These systems may be used on long rural routes experiencing low growth in order to defer costly new cable installations.

9.3.2 METROPOLITAN INTEROFFICE APPLICATIONS

Prior to the introduction of short-haul digital transmission systems (T1 carrier in 1962), short-haul analog wire line carrier systems (N1 carrier in 1950) were used in high-density metro areas to make more efficient use of

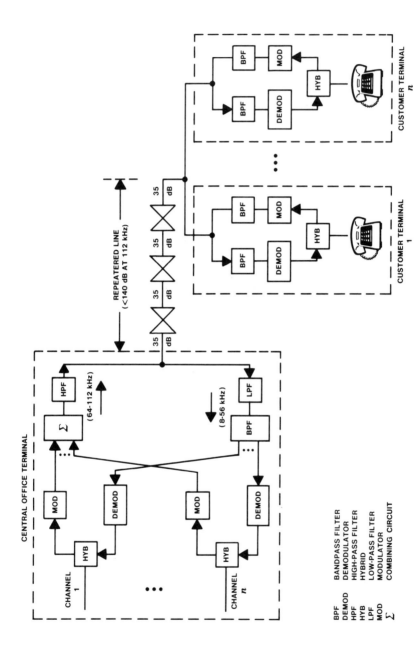

Figure 9-10. Multichannel analog subscriber loop carrier system.

BPF   BANDPASS FILTER
DEMOD   DEMODULATOR
HPF   HIGH-PASS FILTER
HYB   HYBRID
LPF   LOW-PASS FILTER
MOD   MODULATOR
Σ COMBINING CIRCUIT

cable pairs. Since 1962, the field of application for N-carrier analog systems has shifted to the outstate area. In a similar way, analog short-haul microwave systems, originally intended for the metro area, have been displaced by less expensive digital counterparts. In general, analog carrier is uncommon in large cities, although it still is an important part of the outstate and long-haul plant. The next section, which deals with outstate applications, discusses both of these former metro area families of analog carrier systems.

9.3.3 OUTSTATE INTEROFFICE APPLICATIONS

N-Carrier — Analog Carrier System

Since 1950, a number of analog carrier systems, designated N-carrier, have been developed to operate over multipair cables. Several versions of repeatered lines and terminal equipment have been introduced, the earliest based on vacuum-tube design and the more recent based on solid-state electronics. It is convenient to separate the discussion of the N-carrier repeatered line from that of its terminal since, for the most part, modernization of each proceeded independently.

N-Carrier Repeatered Line. The current N-carrier repeatered line design is called N2 and is based on a solid-state plug-in repeater. Earlier designs used the same transmission plan and have similar features. Repeater spacing varies between 3.5 and 5.0 miles depending on cable gauge, which may range from 16 to 24 gauge. Systems range in length from 15 miles to a maximum of 250 miles.

The N-carrier repeatered line employs a transmission plan designed to control circulating crosstalk paths around the repeaters and to compensate for frequency-dependent cable loss across the transmitted channels. As illustrated in Figure 9-11, channels are transmitted in a high-frequency group (172 to 268 kilohertz [kHz]) and in a low-frequency group (36 to 132 kHz). At successive repeaters, signals are translated from the low group to the high group, etc. As a result of this "frequency frogging," the loss-versus-frequency slope of the cable from across the 96-kHz band is self-equalized.

To control near-end crosstalk, opposite directions of transmission at each repeater are given different frequency assignments. As a result, there are two types of N-carrier repeaters: the "hi-lo" repeater and the "lo-hi" repeater. The signal from the output of any repeater is effectively blocked from coupling into the input of a repeater carrying signals in the opposite direction for the same N2 system (path 1 in Figure 9-11), for another system in the same cable (path 2), or in the same direction in a given system (path 3). Also, since the two directions of transmission

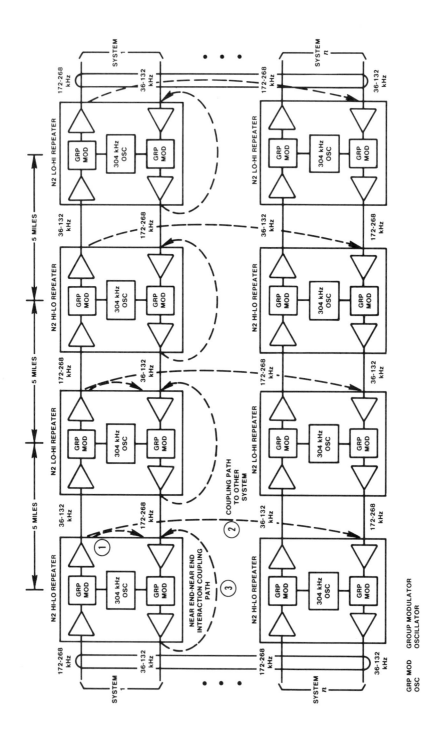

Figure 9-11. N2 carrier repeatered line.

GRP MOD GROUP MODULATOR
OSC OSCILLATOR

occupy different frequency bands, equivalent 4-wire transmission is possible using only two wires within a cable. In practice, full 4-wire transmission is generally used.

N-Carrier Terminals. The modern N-carrier terminal, N4, was introduced in 1979. It costs less, requires less space, consumes less power, has fewer individual parts, and provides more reliable data transmission than its predecessors. The N3 terminal was introduced in 1964 and, like the N4, provides twenty-four channels to the 4-wire line by using a single-sideband format.[7] The earlier N1 and N2 terminals provide only twelve voice channels using double-sideband transmitted-carrier transmission. Since the N4 terminal occupies only 14 percent of the space, uses only 15 percent of the power required for comparable N3 equipment, and costs less, it is the more attractive choice for outstate applications.

N-carrier terminals all use *compandors* to improve the signal-to-noise ratio of low-volume talkers. The compandor consists of a compressor at the transmitting terminal and an expandor at the receiving terminal. The compressor dynamically adjusts the amplitude of a speech signal so that a low-level signal, which would be most affected by noise, is raised and a high-level signal is unaffected. The expandor reverses the process by adding a loss equal to the compressor gain. The net effect is an improvement in the signal-to-noise ratio of about 20 dB, which permits a less costly repeatered line design.

Short-Haul Analog Microwave Radio Systems

6-GHz Short-Haul Radio Systems. The TM-2 and TM-2A are the modern short-haul analog microwave radio systems operating in the 6-gigahertz (GHz) common-carrier band. This band extends from 5925 to 6425 megahertz (MHz). Both systems use the same sixteen radio-frequency (RF) channel assignments, each 29.65 MHz wide, and provide message or video service for distances up to about 250 miles, or about ten repeater hops. Repeater spacing may vary between 5 and 35 miles. TM-2A is a more recent design that offers increased message (VF) channel capacity and improved performance.

Table 9-2 lists the major features of each system. It should be noted that both TM-2 and TM-2A derive a total of eight 2-way RF channels. One of these is assigned for protection. TM-2 provides 1200 VF channels in each 2-way RF channel, while TM-2A provides 1800 VF channels. This improved capacity is directly related to the increased transmitter power achieved by the newer system.

Two RF channel frequency assignment plans are used with these systems. The *regular* plan provides sixteen RF channels and corresponds to

---

[7] Section 6.4.2 discusses single-sideband amplitude modulation.

TABLE 9-2

6-GHZ SHORT-HAUL ANALOG MICROWAVE RADIO SYSTEMS
(5.925 TO 6.425 GHZ)

Characteristic	TM-2	TM-2A
VF channels per RF channel	1200	1800
Radio channel assignments	16	16
Total 2-way RF channels	8	8
Protection switching*	1 by 7	1 by 7
Working VF channels	8400	12,600
System length (miles)	250	250
Transmit power (watts)	1.0	1.6
Repeater type	Baseband	Baseband + intermediate frequency
Repeater spacing (miles)	5 to 35	5 to 35

*In protection switching, a spare RF channel is switched into service if a normal working RF channel fails. In TM systems, one spare channel provides protection for seven working RF channels (1 by 7).

the RF channel frequency used in long-haul TH radio systems (see Section 9.3.4). The *staggered* plan provides fourteen RF channels offset by 14.82 MHz (one-half the bandwidth of an RF channel) and is used to minimize interference where TM and TH long-haul routes intersect. Frequency frogging and dual polarization (see Section 6.3.4) are used to control interchannel interference, with a single antenna for both transmission and reception.

11-GHz Short-Haul Radio Systems. The TL-A2 and TN-1 are the modern 11-GHz short-haul radio systems. The 11-GHz RF band is 1000 MHz wide, extending from 10.7 to 11.7 GHz. Some of these systems have been used in combination with 6-GHz radio systems to provide crossband frequency diversity; however, the Federal Communications Commission (FCC) no longer permits further installations of this type because they constitute an inefficient use of the frequency spectrum.

The TN-1 system uses two RF channel allocation plans. The *regular* plan accommodates up to twelve 2-way RF channels (40 MHz apart) with just one antenna for each route direction. Adjacent RF channels are transmitted with alternate polarizations. With the *alternate* plan, two antennas are required in each direction, and the VF load per RF channel must be reduced from 1800 to 1200 VF channels, but up to twenty-three 2-way RF channels can be accommodated.

Table 9-3 compares the major feature of the TL-A2 and the TN-1 radio systems.

TABLE 9-3

11-GHZ SHORT-HAUL ANALOG MICROWAVE RADIO SYSTEM
(10.7 TO 11.7 GHZ)

Characteristic	TL-A2	TN-1
VF channels per RF channel	1200	1800/1200
Radio channel assignments	24	24/46
Total 2-way RF channels	6	12/23
Protection switching	1 by 5	1 by 11/2 by 21
Working VF channels	6000	19,800/25,200
System length (miles)	250	250
Transmit power (watts)	1.0	3.0
Repeater type	Baseband	Heterodyne or baseband
Repeater spacing (miles)	5 to 35	5 to 35

9.3.4 LONG-HAUL INTEROFFICE APPLICATIONS

L-Carrier Analog Carrier System

Over the past 40 years, a number of single-sideband amplitude modulation carrier systems, designated L-carrier, have been developed to exploit the characteristics of coaxial cable. Although coaxial cable is more expensive than wire pair cable, thousands of VF channels can be carried over a single pair of coaxial tubes. L-carrier systems have found primary application in the long-haul area where large cross sections of VF channels are transmitted over great distances. As shown in Figure 9-12, the attenuation of coaxial cable increases as the square root of frequency. This provides an exploitable bandwidth of about 100 MHz.

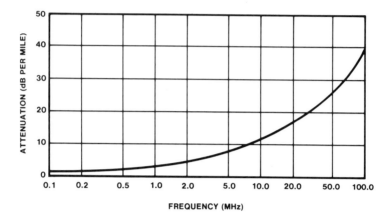

Figure 9-12. Attenuation of one mile of 3/8-inch coaxial cable.

The L-carrier family consists of the L1, L3, L4, L5, and L5E systems. In this progression, the capacity has increased from 600 to 13,200 VF channels per coaxial tube pair. In addition, cable size has increased from eight to twenty-two tubes per cable sheath. This was made possible through steady improvements in repeater performance (linearity, noise, bandwidth, and load-handling capability) for each successive generation. As a result, the total VF channel working capacity has increased from 1800 VF channels per cable (early L1) to 132,000 VF channels per cable (L5E).

Table 9-4 summarizes the main features of each member of the L-carrier family. The nominal repeater spacing for each successive system is exactly half that of its predecessor to achieve a comparable signal-to-noise ratio performance as the system bandwidths were expanded. The halving of repeater spacing simplifies the conversion from older, already installed systems whose repeater stations can easily be reused with this scheme.

By the end of 1980, there were over 239,000 tube-miles of L-carrier in the Bell System plant, over half of which is accounted for by the older L1 and L3 system designs that are no longer manufactured.

L5/L5E Coaxial Carrier Systems. Rather than describe all of the L-carrier systems, this discussion focuses on the latest L5 and L5E systems that use the same basic repeatered line. The repeatered line layout for L5 and L5E is illustrated in Figure 9-13. The coaxial cable is buried underground,

TABLE 9-4

L-CARRIER SYSTEMS

Characteristic	L1	L3	L4	L5	L5E
Service date	1946*	1953	1967	1974	1978
VF capacity/tube pair	600	1860	3600	10,800	13,200
Cable size (tubes)	8	12	20	22	22
Protection switching	1 by 3	1 by 5	1 by 9	1 by 10	1 by 10
Working VF channels	1800	9300	32,400	108,000	132,000
Top frequency (MHz)	2.8	8.3	17.5	60.6	64.8
Bandwidth (MHz)	2.7	8	17	57.5	61.5
Repeater spacing (miles)	8	4	2	1	1
Technology	Vacuum tubes	Vacuum tubes	Solid state	Solid state	Solid state

*Initial service on an L system was provided in 1941 using smaller-diameter coaxial tubes.

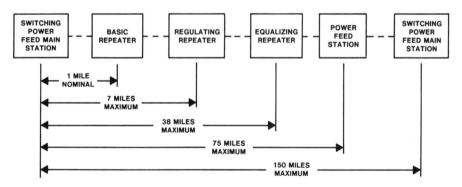

Figure 9-13. L5/L5E carrier system
transmission layout.

and all repeater stations are below ground in manholes designed to withstand earthquakes and nuclear attack.

Three types of line repeaters are powered remotely over the center conductors of the coaxial tubes from power feed stations spaced at maximum intervals of 75 miles. These repeaters are housed in pressurized watertight apparatus cases. The basic repeater compensates for the nominal loss of 1 mile of coaxial cable at 55°F. Regulating repeaters, spaced a maximum of every 7 miles, dynamically compensate for cumulative changes in cable loss caused by changes in ground temperature. A pilot tone is used for gain control in the regulating repeater. At the midpoint of each power feed span, an equalizing repeater corrects for residual deviations. Switching main stations are located up to 150 miles apart, and they may include two power feed spans.

One pair of coaxial tubes in the cable is reserved to protect up to ten working tube pairs. In the event of a working line failure, the fully powered protection line is automatically switched into service, based on excessive deviations in pilot tone or in the signal power itself.

Undersea Coaxial Cable Systems

In 1950, the first undersea coaxial cable system, designated SA, was installed between Key West, Florida, and Havana, Cuba. The first transatlantic system, the SB, followed in 1956. Both systems used two separate coaxial cables, one for each direction of transmission. Other similarities included repeater spacings approaching 40 nautical miles and the use of high-gain (greater than 60 dB) vacuum-tube repeaters. The SA system provided only twenty-four 2-way VF (4-kHz) channels. The SB originally provided thirty-six 2-way VF channels, but subsequent terminal improvements increased the capacity to forty-eight.

Figure 9-14 shows that, since 1956, there has been a succession of undersea coaxial cable systems installed every 6 or 7 years to meet the steady growth in transatlantic message traffic. These later generations use a single cable for both directions of transmission, employ 3-kHz VF channels,[8] and use single-sideband amplitude modulation multiplexing. Increases in capacity have been achieved by reducing repeater spacing, exploiting higher frequencies, and increasing the diameter of the coaxial cable.

The undersea coaxial cable differs markedly from the overland L-carrier coaxial cable shown in Figure 6-7. For example, the SG undersea cable consists of only one coaxial tube, with a solid dielectric to withstand the enormous sea pressure. The transmission design principles for an undersea coaxial system and an overland L-carrier are basically the

---

[8] In most other systems, the nominal voiceband or voice-frequency channel has a 4-kHz bandwidth.

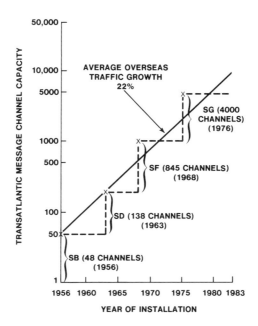

Figure 9-14. Growth in transatlantic traffic and the succession of undersea coaxial systems.

same. Repeater spacing is determined so as to optimize the received signal-to-noise ratio. Signal levels are set well above thermal noise and well below that which would produce excessive nonlinear distortion products. Differences between overland and undersea systems are a result of differences in the environment and the increased reliability required in undersea applications.

Since undersea repairs are very time-consuming and expensive, these systems must be extremely reliable. In fact, the reliability objective is 100 times more demanding than that of an overland system, requiring fewer than four service interruptions caused by component failures during a 20-year period. To achieve this, the number of components in an undersea system is held to an absolute minimum, and each component is specially manufactured and screened. Control of parameters is also undertaken as the cable is laid. For example, with SG, the 2-way repeaters are actually spliced into the cable (every 5.1 nautical miles) during the cable-laying process, and measurements are made continuously as the cable and repeaters are laid. Every 100 to 150 nautical miles, an equalizer (adjustable only when the cable is laid) is placed. At intervals of 700 nautical miles, equalizers that can be controlled from shore locations are installed. These are needed to compensate for changes in transmission characteristics due to cable aging.

Long-Haul Analog Microwave Radio Systems

During the last three decades, long-haul analog microwave radio systems operating in the 4- and 6-GHz common-carrier bands have provided an ever-increasing percentage of the total Bell System circuit mileage. By the end of 1980, TD and TH microwave systems accounted for over 57 percent of the total 1.03 billion circuit-miles; L-carrier coaxial systems provided only 19 percent of this total.

TD systems operate in the 3.7- to 4.2-GHz band, while TH and AR6A systems operate in the 5.925- to 6.425-GHz band. These systems have been designed to satisfy signal-to-noise and reliability objectives for long-distance communications paths up to 4000 miles long.

4-GHz Long-Haul Microwave Systems. The first TD microwave radio system was introduced in 1950. In its initial form, TD-2 carried 480 voiceband channels on each of five 2-way, 20-MHz RF channels providing a route capacity of 2400 VF channels. A sixth 2-way RF channel was used for protection. Over the years, numerous improvements have been introduced that have greatly increased the route capacity and, in turn, significantly lowered the cost per circuit-mile. There has also been a significant increase in system reliability. Some major improvements include: the introduction of the horn reflector antenna and its associated circular waveguide system (1955) which was introduced to permit the addition of a 6-GHz system on the same route and which also permitted both vertically and horizontally polarized signals to be transmitted (1959); the introduction of solid-state components starting in 1967, which improved system reliability; and the increase in the power output of the microwave transmitter from 0.5 to 5.0 watts (1973). The resulting route capacity is now 19,800 VF channels. Table 9-5 shows the steps in the evolution.

A portion of a typical system layout of a TD microwave system is illustrated in Figure 9-15. Repeater stations are usually placed between 20 and 30 miles apart, depending upon line-of-sight path clearance, tower, height, fade margin, interference, and economic considerations. Unlike analog coaxial systems, in which doubling the repeater spacing doubles the amount of loss, doubling the distance between radio repeater stations increases the loss by only 6 dB.

At each intermediate repeater station there are four antennas. One antenna is used for receiving and another for transmitting in each direction. Received RF channel signals are separated by waveguide networks, applied to intermediate frequency (IF) heterodyne radio receivers[9] and

---

[9] A heterodyne receiver "mixes" a single-frequency signal with the incoming RF channel signal to produce a difference frequency, in this case, the IF of 60 to 80 MHz. Amplification and certain signal processing can be done more efficiently at this frequency. In the transmitter, similar mixing of frequencies produces the RF channel signal to be transmitted.

TABLE 9-5

4-GHZ LONG-HAUL ANALOG MICROWAVE RADIO SYSTEMS
(3.7 TO 4.2 GHZ)

Characteristic	TD Radio*					
Service date	1950	1959	1967	1968	1973	1979
VF channels per RF channel	480	600	900	1200	1500	1800
Radio channel assignments	12	24	24	24	24	24
Total 2-way RF channels	6	12	12	12	12	12
Protection switching	1 by 5	Two 1 by 5	2 by 10	2 by 10	1 by 11	1 by 11
Working VF channels	2400	6000	9000	12,000	16,500	19,800
Transmit power (watts)	0.5	0.5	1.0	2.0	2.0/5.0	5.0

*Denotes TD-2 radio, introduced in 1950, TD-3 in 1966, TD-3A in 1970, and TD-3D in 1973.

transmitters, then recombined by waveguide networks, and transmitted to the next intermediate repeater station. Up to ten radio hops constitute a protection switching section. Within a switching section, a standby protection radio channel may be automatically substituted for any working radio channel in the event of a deep-frequency selective fade.

At a main terminal repeater station, frequency modulation (FM) transmitters modulate baseband signals, consisting of frequency-division multiplex (FDM) voice channels and data or video signals, to the IF band extending from 60 to 80 MHz. Frequency modulation is used in TD radio systems because FM signals are insensitive to the gain nonlinearity (compression) introduced by the typical IF and microwave amplifiers. FM receivers demodulate the 70-MHz IF signals to recover the baseband signals.

6-GHz Long-Haul Microwave Systems. The need for additional route capacity and the savings possible if 6- and 11-GHz routes could be added to existing 4-GHz (TD-2) routes led to the introduction of the horn

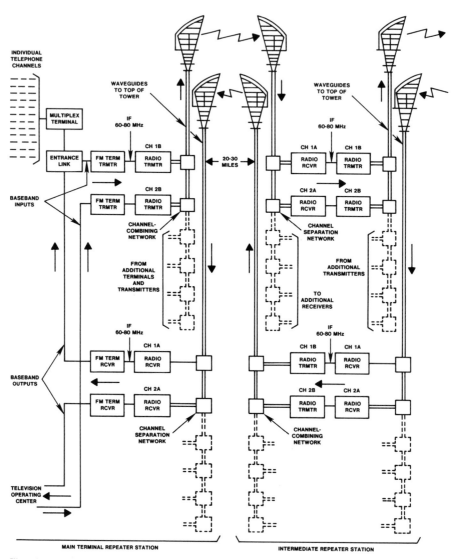

INDIVIDUAL
TELEPHONE
CHANNELS

MULTIPLEX TERMINAL

ENTRANCE LINK

BASEBAND INPUTS

FM TERM TRMTR

CH 1B
RADIO TRMTR

CH 2B
FM TERM TRMTR

RADIO TRMTR

IF 60-80 MHz

WAVEGUIDES TO TOP OF TOWER

CHANNEL-COMBINING NETWORK

FROM ADDITIONAL TERMINALS AND TRANSMITTERS

IF 60-80 MHz

FM TERM RCVR

CH 1A
RADIO RCVR

CH 2A
FM TERM RCVR

RADIO RCVR

BASEBAND OUTPUTS

CHANNEL SEPARATION NETWORK

TELEVISION OPERATING CENTER

MAIN TERMINAL REPEATER STATION

20-30 MILES

WAVEGUIDES TO TOP OF TOWER

IF 60-80 MHz

CH 1A
RADIO RCVR

CH 1B
RADIO TRMTR

CH 2A
RADIO RCVR

CH 2B
RADIO TRMTR

CHANNEL SEPARATION NETWORK

FROM ADDITIONAL TRANSMITTERS

TO ADDITIONAL RECEIVERS

IF 60-80 MHz

CH 1B
RADIO TRMTR

CH 1A
RADIO RCVR

CH 2B
RADIO TRMTR

CH 2A
RADIO RCVR

CHANNEL-COMBINING NETWORK

INTERMEDIATE REPEATER STATION

FM   FREQUENCY MODULATION
IF   INTERMEDIATE FREQUENCY
RCVR   RECEIVER
TERM   TERMINAL
TRMTR   TRANSMITTER

Figure 9-15. Typical TD microwave radio system showing
main and intermediate repeater stations.

reflector antenna and its associated circular waveguide in 1955. The 6-GHz TH-1 system was first placed in service in 1961 as such an overbuild. When TH-1 was first "overbuilt" on a TD-2 route, six additional working 2-way RF channels were provided, each having a VF capacity of 1800 VF channels. As a result, the route capacity was increased from 6000 VF channels using TD-2 alone to 16,800 VF channels using both systems. Table 9-6 shows the characteristics of TH-1 and TH-3 systems as well as the features of the newest 6-GHz microwave radio system, the AR6A (discussed in later paragraphs). All three radio systems use the 500-MHz common-carrier band extending from 5.925 to 6.425 GHz and provide sixteen radio channel assignments. TH-1 and TH-3 are both FM systems like TD-2 and TD-3. TH-3 is capable of a signal-to-noise ratio 4 dB better than that of TH-1.

A combined TH-3/TD-2D route has a working capacity of 32,400 VF channels. Even so, some of these routes may become filled by the early 1980s if the present growth in long-distance calling continues. Since

TABLE 9-6

6-GHZ LONG-HAUL ANALOG MICROWAVE RADIO SYSTEMS
(5.925 TO 6.425 GHZ)

Characteristic	TH-1	TH-3	AR6A
Service date	1961	1970	1981
VF channels per RF channel	1800	1800*	6000
Radio channel assignments	16	16	16
Total 2-way RF channels	8	8	8
Protection switching	2 by 6	1 by 7	1 by 7
Working VF channels	10,800	12,600	42,000
Baseband modulation	FM	FM	SSBAM†

*See footnote 10.

†Single-sideband amplitude modulation.

building new long-haul routes in parallel with existing ones is very expensive and often difficult because of RF interference, a new, larger capacity, "overbuild" microwave radio system has been developed. Introduced in 1981, the AR6A system puts 6000 VF channels on a single RF channel, compared to 1800 in TH-3.[10] As a result, the combined capacity of an AR6A/TD route is 61,800 VF channels. Because AR6A operates at 6 GHz, the system can use unoccupied TH radio channels on existing TH/TD routes. Also, existing TH radio channels may be replaced with AR6A. The capacity of the 6-GHz band is thus increased from 12,600 to 42,000 VF channels.

A 4000-mile microwave radio system has about 150 radio repeater stations spaced 25 to 30 miles apart. About 100 of these stations will simply be heterodyne repeaters, consisting of a transmitter and a receiver. The remaining 50 are called *main radio repeater stations*, or simply, *main stations*. In the case of AR6A, those main stations are equipped with a new one-for-seven microprocessor-controlled protection switching system. Typically, about ten of the main stations are also terminals where baseband signals enter and leave the system.

The AR6A is the first long-haul microwave radio system to use single-sideband amplitude modulation (SSBAM) rather than frequency modulation. In the past, SSB was not used in the long-haul radio network because available microwave power amplifier tubes produced unacceptable levels of nonlinear distortion. Two technological improvements have made SSB feasible: an ultralinear traveling-wave power amplifier tube (TWT) and a distortion correction circuit. The latter actually adds a controlled amount of distortion to the signal before it reaches the TWT so as to cancel out the distortion generated in the TWT and modulator circuits themselves.

Mastergroup[11] translators, designed for AR6A and L5E, assemble five 600-circuit mastergroups into a tightly packed 3000-circuit multimastergroup signal. Multimastergroup translators combine two of these signals plus five pilot tones to form a 6000-circuit IF modulating signal. Pilot signals control receiver gain and dynamic equalization as well as synchronize the frequency of the receiving multimastergroup translators along the radio route.

Satellite Communications Systems

A satellite communications system is a long-haul microwave radio system consisting of earth stations and orbiting satellite microwave repeaters. Between any two earth stations, regardless of route length, there are only

---

[10] A new TH-3 design, available in 1983, provides 2400 VF channels per RF channel or 16,800 total VF channels.

[11] Mastergroups and other levels of the analog FDM hierarchy are discussed in Section 9.3.5.

two radio hops as distinguished from terrestrial microwave radio systems that usually consist of several hops between radio terminals. By the end of 1982, about 6 percent of the long-haul radio circuit mileage in the Bell System was carried over domestic satellite systems.

In 1962, the Bell System's Telstar demonstrated the feasibility of satellite communications systems for telephony. The Telstar satellite had a low elliptical orbit, inclined 45 degrees to the earth's equator, with a rotation period of 158 minutes. Today most satellites are placed in high (22,300 miles) circular orbits with zero inclination. This gives them a period equal to that of the earth's rotation and has the advantage that the satellite is geostationary; that is, it appears to be stationary from a point on the earth.

Telstar provided a route capacity of only twelve voice channels. COMSTAR, a modern communications satellite designed and built to AT&T specifications for joint domestic service by AT&T and General Telephone and Electronics Satellite Corporation (GSAT), provides a route capacity of 18,000 VF channels. By 1986, the Telstar 3 satellite will provide a capacity of 21,600 VF channels.

Figure 9-16 shows the orbital positions of existing domestic satellites as of April 1981. Longitudes between 115 degrees west and 135 degrees west are highly desirable satellite locations since those farther east cannot serve Alaska, and those farther west cannot serve Puerto Rico.

In addition to domestic systems, there is also an international satellite system called INTELSAT, under the direction of the International

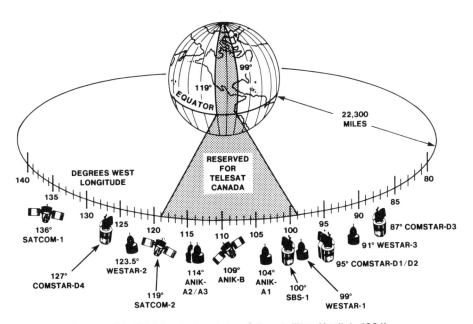

Figure 9-16. Orbit locations of domestic satellites (April 1, 1981).

Telecommunications Satellite Consortium. This system, which began in 1965 with INTELSAT I (Early Bird), currently involves over eighty-eight earth stations in sixty-four countries and satellites positioned over the Indian Ocean, the Pacific Ocean, and the Atlantic Ocean. The geostationary satellites view over 40 percent of the earth's surface. The Atlantic Ocean satellites carry over 60 percent of the global traffic in the system.

Satellite communications differ from terrestrial communications in several ways. Since there is only one intermediate repeater (the satellite), there is no accumulation of multiple-hop repeater impairments. Multipath selective fading, which affects terrestrial systems, is not a problem in satellite systems. However, atmospheric water vapor absorption is a problem. Each satellite hop is 1000 times longer than a terrestrial hop, and this results in a very great path loss and long transmission delay. Satellite systems require larger ground antennas than terrestrial systems to receive the considerably weaker signals. To minimize system outages due to component failures, extremely reliable components must be used in satellite repeaters, together with design redundancy, thorough testing routines, and protection-switching arrangements.

The frequency bands used by satellite communications systems are allocated by the Comité Consultatif International Télégraphique et Téléphonique (CCITT), the Comité Consultatif International des Radiocommunications (CCIR), and—for those systems operating in the United States—the FCC. Table 9-7 indicates that three frequency bands are available for satellite systems.

In the Telstar 3 system, six transponders provide coverage of the continental United States (CONUS) only. The down-link outputs of six transponders are switchable from CONUS to Alaska; six are switchable from CONUS to Hawaii; and six are switchable from CONUS to Puerto Rico.

Separate frequencies for the two directions of transmission reduce interference. By using the lower frequency band for down-link transmission, the path loss is somewhat reduced. This, in turn, reduces the

TABLE 9-7

FREQUENCY BANDS FOR SATELLITE
COMMUNICATIONS SYSTEMS

Frequency Bands (GHz)	Up-Link (GHz)	Down-Link (GHz)	Bandwidth (GHz)
6/4	5.925-6.425	3.7-4.2	500
14/12	14.0-14.5	11.7-12.2	500
29/19	27.5-31.0	17.7-21.2	3500

transmitted power required from the satellite, where power is limited. An additional consideration is that up-link 6-GHz signal leakage is less likely to interfere with 4-GHz terrestrial radio systems located near an earth station antenna. Most of today's satellite systems use the 6/4-GHz band pair, which is commonly used for long-haul terrestrial systems. The higher frequency bands permit more gain with the same size antennas and do not interfere with terrestrial systems; however, propagation loss due to rain is increased. The 14/12-GHz band pair is now being developed, while the 29/19-GHz band pair is more experimental.

The earth station antenna is very large (about 30 meters in diameter), thereby providing high signal gain and narrow-beam radio propagation. Although the satellite is in a geostationary orbit, automatic tracking is provided by means of a telemetry and control system to account for any small position deviations and to maximize the received signal. This system also performs a number of "station keeping" functions such as keeping the satellite at its assigned longitude and inclination by the use of small gas jets and adjusting the gain of the satellite receiver to balance up/down-link transmission.

The satellite is stabilized to maintain a fixed relation to the earth's axis and to eliminate tumbling. This permits a moderate-gain satellite antenna to be used. Solar cells and nickel-cadmium batteries are the primary power sources for the satellite equipment. The batteries are needed during solar eclipses.

Because satellite power is limited, FM/FDM modulation is used, since this provides a favorable tradeoff of power for bandwidth. In addition, an FM signal is not subject to amplitude nonlinearities. Therefore, the satellite amplifiers can be operated very close to saturation and thereby provide maximum power output.

Of the many design considerations that are unique to satellite systems, only six are mentioned below: (1) sun transit outage, (2) satellite eclipse, (3) rain effects, (4) transmission delay, (5) noise, and (6) interference coordination.

The geometry of the sun transit outage phenomenon is illustrated in Figure 9-17. As shown, the sun appears directly behind the satellite, and emissions from the sun fall upon the earth station antenna, causing a large increase in satellite circuit noise. This phenomenon takes place during the spring and fall equinoxes and persists a few minutes each day for a period of about six days. The resultant increase in noise is avoided by temporarily switching to a protection satellite located at a different longitude, using a separate earth station tracking antenna.

A second orbital phenomenon occurs when the satellite is opposite to that shown in Figure 9-17. During two 46-day periods each year, the earth's shadow causes the satellite to experience an eclipse of the sun. The duration of the daily eclipse varies, lasting up to about 1 hour. During this time, the satellite is deprived of solar energy and must rely or

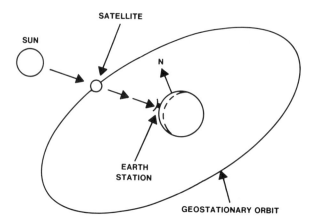

Figure 9-17. Geometry of sun transit outage.

battery power. In addition, this exposes the satellite and its circuitry to large temperature changes.

Rain causes three major impairments to a microwave satellite signal: attenuation, thermal noise, and depolarization. Raindrops scatter and absorb microwave energy, resulting in rain fades that, although small in the 6/4-GHz band pair, increase with frequency and are very serious in the higher 14/12 and 29/19-GHz band pairs. Rain is also a dissipative medium at microwave frequencies and radiates thermal noise. The combined effect is a decrease in system signal-to-noise ratio. With earth station site diversity protection, system outages due to rain fades are held to a minimum. Modern 6/4-GHz satellite systems, such as COMSTAR and Telstar 3, increase their channel capacity by using the same frequency with different polarizations for two channels. These systems are degraded by rain depolarization of the signal. The nonspherical shape of the raindrops converts the signal's linear polarization to elliptical polarization and thereby impairs the ability of the antenna to discriminate between the two channels. This impairment is more dominant than rain attenuation in the 6/4-GHz band pair, and unlike attenuation, its effect diminishes with higher frequency.

The 250-millisecond, 1-way transmission delay from one earth station up to the satellite and down to another earth station inhibits voice communications slightly, but it is a serious problem for data transmission. Although data sets are now available with modified protocols to operate over satellite circuits, many earlier data terminals would experience difficulties. The round-trip delay of 500 milliseconds on a satellite circuit would cause very annoying talker echo were it not for the use of echo cancelers. Prior to the invention of these devices, the practice was to use

360

a satellite facility in one direction and a terrestrial facility in the other direction of transmission.

Satellite transmission is impaired by noise from the earth itself. Thermal radiation from the warm earth extends across the entire beamwidth (main lobe) of the satellite antenna and is the dominant source of satellite antenna noise. The earth station is also significantly affected by noise from the earth. Since the earth station antenna is pointed at least several degrees above the horizontal, this noise is not in the main lobe of the earth station antenna but in the back and side lobes.

Interference sources include other communications satellites and terrestrial microwave systems. To control intersatellite interference, satellites currently operating in the 6/4-GHz band pair are placed no closer together than 4 to 5 degrees.[12] Figure 9-18 illustrates the basic interference paths between a satellite and a 4-GHz terrestrial microwave system. Since the 4-GHz down-link signal can interfere with a terrestrial system as well as other satellite systems, the FCC restricts the transmitted satellite signal power to no more than 5 watts. To control 4-GHz terrestrial interference into earth station antennas, site interference studies are conducted prior to locating an earth station. A large percentage of potential sites are rejected based on these ground exposure studies. There is also

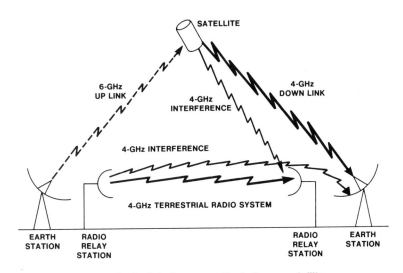

Figure 9-18. Interference paths between satellite communication systems and 4-GHz terrestrial radio systems.

[12] Minimum satellite spacings are determined by the FCC and international agreement. In 1983, the FCC proposed a minimum spacing of 2 degrees for satellites using this frequency band.

possible up-link interference into a 6-GHz terrestrial system. This is controlled by limiting earth station antenna elevation angles to no less than 5 degrees.

9.3.5 ANALOG FREQUENCY-DIVISION MULTIPLEX TERMINALS

Section 6.5 introduced the concept of multiplexing. As discussed there, analog carrier systems use FDM to combine large numbers of VF (0- to 4-kHz) channels for efficient use of wideband systems. Each VF channel occupies a specific 4-kHz portion of the broadband transmission media. FDM terminals modulate, filter, and combine voiceband signals to produce a stack of VF channels across the allowable broadband frequency spectrum.

The FDM plan and the various multiplex terminals in use in the Bell System today are illustrated in Figure 9-19. A large number of modulation steps are required to produce the twelve levels shown beyond voice frequency. The *basic group*, for example, is the first level in the multiplex hierarchy. Twelve VF channels are combined and occupy the frequency spectrum from 60 to 108 kHz, forming the output of the *channel bank* and serving as the input to the group bank equipment. Six of the remaining levels in the FDM hierarchy are system levels; that is, the frequency band at each of these levels corresponds to one or more broadband transmission systems. These are indicated by heavy horizontal lines in Figure 9-19 with the system name(s) at the left.

A total of five basic frequency translations accomplished by SSBAM are required to place VF channels in higher frequency levels and in larger groupings throughout the hierarchy. These translations are shown in the figure and are as follows:

- VF to basic group

- basic group to basic supergroup

- basic supergroup to basic mastergroup

- basic mastergroup to system level or basic jumbogroup or multimastergroup spectra

- basic jumbogroup or multimastergroup spectra to system level.

As new transmission systems have been developed, the FDM hierarchy has evolved, and certain features have become standard for the family of multiplex terminals required. These include:

- 4-wire transmission throughout

- SSBAM for maximum bandwidth utilization

- VF input and output levels of −16 dB and +7 dB

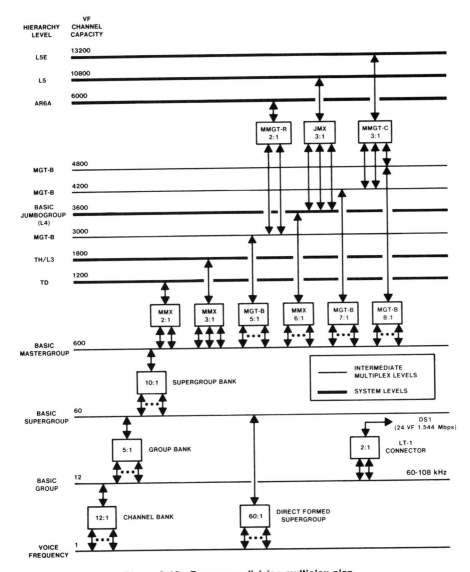

Figure 9-19. Frequency-division multiplex plan.

- modulating and demodulating circuits combined into 2-way units called *modems*[13]

- all carrier frequencies precise, stable, and multiples of 4 kHz

[13] These units should not be confused with the modems (data sets) used on customers' premises to convert digital information to analog form for transmission.

- pilot-controlled regulation used at the group, supergroup, and master-group levels.

A-Type Channel Bank

The first frequency translation performed in the FDM hierarchy places twelve VF channels into the *basic group* band from 60 to 108 kHz as SSBAM signals spaced 4 kHz apart. A block diagram illustrating the operation of the A5 channel is shown in Figure 9-20. Each of twelve voiceband signals modulates one of twelve carriers spaced 4 kHz apart, beginning at 64 kHz and extending to 108 kHz. The double-sideband

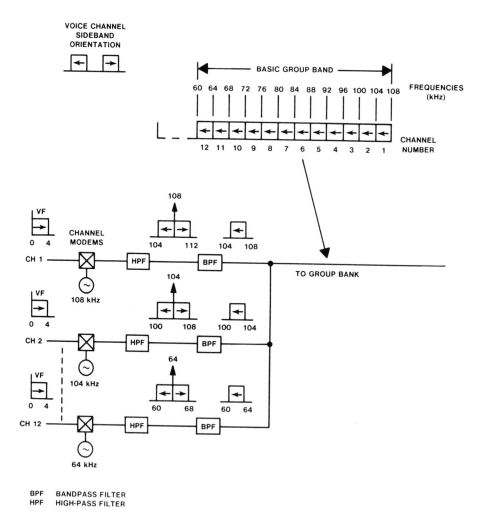

BPF   BANDPASS FILTER
HPF   HIGH-PASS FILTER

Figure 9-20. A5 channel bank frequencies.

signal at the output of the modulator is passed through a high-pass filter to suppress VF energy and then through a quartz crystal bandpass filter that passes only the lower sideband. Twelve translated channels are combined to form the composite basic group signal. In the reverse direction, identical filters select the channels that, in turn, are translated to VF by twelve demodulators.

Over the past 45 years, there have been six generations of A-type channel banks performing this function. The A1 through A4 channel banks used vacuum-tube technology and have been superseded by the A5 and A6 channel banks, which use transistors and hybrid integrated circuitry, respectively.

The A5 channel bank was a major improvement over previous vacuum-tube designs. It was smaller; consumed less power; and had improved frequency response, gain stability, reliability, and maintenance access. The A6 channel bank, the current design, uses monolithic quartz crystal filters and thin-film hybrid integrated circuitry for further reductions in cost and size.

A single-frequency alarm pilot may be transmitted along with the basic group as an option in either the A5 or A6 channel bank to actuate a carrier group alarm feature. The purpose is to minimize the effect of a carrier system failure on switching system load by making failed trunks appear busy. Otherwise, these trunks continue to be seized, the call attempt fails due to the faulty trunks, and the customer tries again.

LMX Group Bank

The second frequency translation in the FDM hierarchy places five groups (sixty voiceband channels) into the *basic supergroup* band. Each of the group signals modulates one of five carriers, which are spaced 48 kHz apart between 420 and 612 kHz. After bandpass filters select the lower sidebands of the translated group signals, they are combined and occupy the basic supergroup spectrum. The current vintage of group bank is the LMX-3.

LT-1 Connector

In the past, long-haul analog carrier signals, in order to be switched by the *4ESS* switching equipment or interconnected to digital transmission facilities, were required to be demodulated by A-type channel banks to voice frequency and then converted to a time-division format. This conversion from FDM to time-division multiplex (TDM) required considerable equipment floor space, power, and intraoffice VF cabling. The LT-1 connector, the Bell System's first transmultiplexer, converts signals from FDM at the basic group level to TDM at the digital signal level 1 (DS1). As shown in Figure 9-19, the LT-1 converter converts two analog group signals into a single, digital, 1.544-megabit-per-second (Mbps) DS1 signal. The DS1 signal, in turn, goes to a digital interface frame on the *4ESS* switch or to a digital system cross-connect DSX-1 frame.[14] A standard LT-1 frame accommodates 480 VF channels via twenty LT-1 connectors.

LMX Supergroup Bank

The third frequency translation in the FDM hierarchy following the formation of the 60-channel basic supergroup is performed by the LMX supergroup bank. It places ten supergroups (600 channels) into the *basic mastergroup* frequency band from 564 to 3084 kHz. The frequency format of this band, which is designated U600 (universal 600 channel), is the standard format used with most microwave radio and coaxial cable systems. It consists of six supergroups, a 56-kHz guard band, and four more supergroups. This wide guard band is used to place a line pilot for carrier system regulation. Between all other supergroups in the basic mastergroup band there are 8-kHz guard bands.

A standard basic mastergroup format common to several long-haul carrier systems greatly facilitates emergency broadband service restoration. In addition, the U600 format has the advantage that digital signals, such as 1A-RDS, with frequency components less than 500 kHz may be carried over TD/TH microwave radio systems.

MMX Mastergroup Multiplex and MGT-B Mastergroup Translator

The fourth frequency translation in the FDM hierarchy following the formation of the 600-channel basic mastergroup places between two and eight mastergroups (as shown in Figure 9-19) in frequency bands that are used directly as line-frequency signals for broadband facilities or as input signals to higher levels in the multiplex structure. The following paragraphs discuss two examples of how this translation is accomplished: the **MMX mastergroup multiplex** and the **MGT-B mastergroup translator**.

Figure 9-21 illustrates the MMX mastergroup multiplex, which was initially designed to provide the line signal for the L4 coaxial cable system. The MMX terminal places six mastergroups (3600 channels) into the *basic jumbogroup* band from 564 to 17,548 kHz. (It should be noted that MG1 is fed straight through the terminal without modulation.)

The mastergroups in the basic jumbogroup spectrum are separated by guard bands whose widths are proportional to their center frequencies. Wide guard bands are needed to permit individual mastergroups to be dropped or reinserted along a route without requiring the entire jumbogroup to be demodulated to basic mastergroup frequencies.

---

[14] Section 8.6.2 describes the DSX-1 cross-connect.

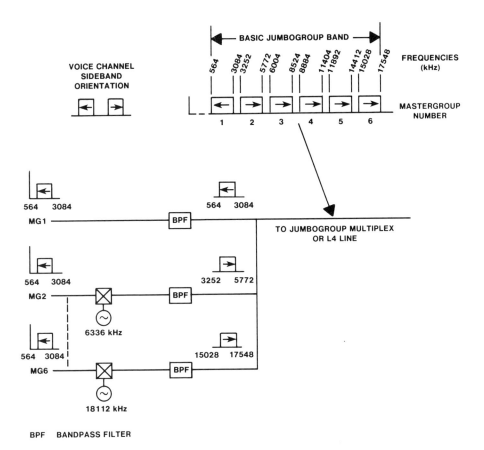

BPF   BANDPASS FILTER

Figure 9-21. MMX mastergroup multiplex frequencies.

As shown in Figure 9-19, the MMX terminal is also used to combine either two or three basic mastergroup signals to provide the line signals for TD systems and for TH and L3 systems.

The MGT-B mastergroup translator terminal illustrated in Figure 9-22 differs in many ways from the MMX terminal. The MGT-B terminal translates up to eight mastergroups (4800 channels) into the *multimaster-group* band from 3252 to 24,588 kHz. This wider bandwidth, combined with fixed 168-kHz guard bands between all mastergroups, results in additional VF channel capacity. More efficient use of the frequency spectrum is gained at the expense of mastergroup branching flexibility. The MGT-B terminal does not provide the mastergroup dropping and adding capability of the MMX terminal, although branching can be done at the basic mastergroup frequencies by demodulating the entire mastergroup signal.

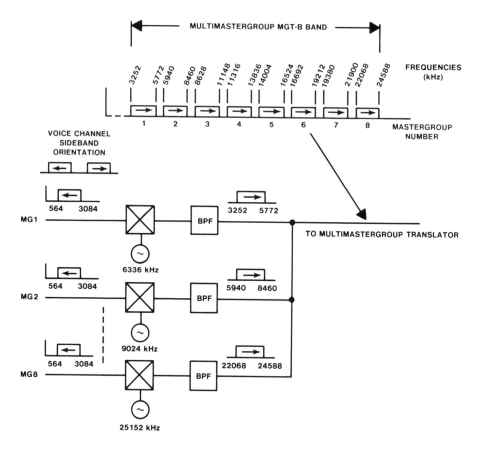

BPF BANDPASS FILTER

Figure 9-22. MGT-B mastergroup translator frequencies.

In addition to better bandwidth utilization, the MGT-B requires no costly protection switching equipment, as the result of improved reliability in design.

JMX Jumbogroup Multiplex and MMGT-C Multimastergroup Translator

Following the formation of the basic jumbogroup/multimastergroup signal shown in Figure 9-19, the fifth and final frequency translation in the FDM hierarchy places between two and three of these signals into frequency bands appropriate for the AR6A, L5, and L5E carrier systems. The **JMX jumbogroup multiplex** (developed specifically for the L5 carrier system) and the **MMGT-C multimastergroup translator** (developed for the expansion of L5 to L5E) are discussed below. The MMGT-R, developed for use with the AR6A microwave radio system, is similar in many respects to the MMGT-C.

The JMX terminal places three jumbogroups (10,800 channels) into the L5 line spectrum between 3124 and 60,556 kHz. A total of four steps of modulation and demodulation are used in the transmitting and receiving terminals to ease filter design and minimize interference problems. Protection switching is provided on a one-for-one basis to ensure high reliability in consideration of the large number of circuits carried by each working jumbogroup.

The MMGT-C multimastergroup translator terminal places three multi-mastergroups (13,200 channels)[15] into the L5E carrier system spectrum. Instead of two or three steps of modulation, the terminal uses a single step of modulation to translate MMG2 and MMG3; MMG1 is not modulated at all.

The guard band between multimastergroups in the MMGT-C terminal is about half that provided between jumbogroups in the JMX terminal. Even so, the guard band is sufficient to permit dropping and adding multimastergroups at L5E line frequencies, thus avoiding the need to demodulate the line signal to multimastergroup frequencies. The combined effect of (1) closer spacing between multimastergroups, (2) a broader L5E line frequency band, (3) closer spacing between mastergroups, and (4) a broader multimastergroup frequency spectrum permits twenty-two mastergroups to be loaded onto L5E as compared with eighteen mastergroups on L5.

A multimastergroup pilot is inserted between MG4 and MG5 at 13,920 kHz. It is used in conjunction with automatic multimastergroup protection switching and also as a continuity signal. Protection is provided on a basis of one spare modulator for every twenty working modulators in the MMGT-C terminal.

9.4 DIGITAL CARRIER TRANSMISSION

The digital transmission network consists of four major types of system components: **terminals, multiplexers, cross-connects**, and **transmission facilities** (or systems). Each of the system components has been designed to operate at one or more of the six bit rates, or levels, of the TDM hierarchy used in the Bell System digital network. The TDM hierarchy is described in Section 9.4.3.

Digital signals are created with the use of *digital terminals*. These terminals take a continuous-wave analog input and transform it, through the use of sampling and encoding, into a digital waveform.

---

[15] As shown in Figure 9-19, the MMGT-C combines two 4200-channel mastergroups and one 4800-channel mastergroup to produce the 13,200-channel L5E output.

Digital multiplexers provide interfaces between the different bit rates in the digital network. Many of the multiplexers include performance monitoring, failure detection, and alarm and automatic protection-switching features.

The *digital cross-connects* (DSXs) are the interconnection points for terminals, multiplexers, and transmission facilities. They are equipment frames where cabling between the system components is cross-connected to provide flexibility for restoration, rearrangements, and circuit order work. Each digital signal rate is handled by its own cross-connect. Hence, there are six cross-connects, identified as DSX-n, where n is one of the six TDM levels. The DSX-1 is discussed in Section 8.6.2 and illustrated in Figure 8-13. The other digital cross-connects are not discussed in this section.

Digital signals are transmitted from one location to another by *transmission facilities* or systems using a multitude of media (paired cable, coaxial cable, radio, optical fiber, and satellite) at the various bit rates. The rest of this section describes various transmission systems in relation to their application in the Bell System network and includes a discussion of the TDM hierarchy and several types of multiplex equipment.

9.4.1 LOOP AREA APPLICATIONS

Loop carrier systems bring electronic technology to the traditionally costly and active loop plant—the pairs of metallic conductors that connect subscribers to the central office. In suburban areas, the present loop plant is affected by rapid growth and movement, requiring costly cable installation to serve new customers. In rural areas, these conductors must extend over many miles to serve relatively few people. By substituting electronics for cable, subscriber loop carrier systems offer an economical way to serve suburban and rural areas. Because these systems increase the number of customers served by existing facilities (wire pairs), they are often called *pair-gain* systems.

Figure 9-23 is a block diagram of a digital carrier system representative of those used for loop transmission. Channels 1 through n on the left represent individual customer lines, which are multiplexed at a nearby customer terminal. Hybrid circuits separate the customer transmitted and received signals for processing. Following conditioning (for example, amplification, band limiting), transmitted signals are converted to a digital pulse-code-modulated (PCM) format as described in Section 6.4.3. The time-division multiplexer interleaves the digital signals from the n sources and transmits the combined pulse stream over a repeated line to the central office. At the receiver, a demultiplexer must be synchronized with the transmitting multiplexer so that the received pulses may be detected and routed to the appropriate channel.

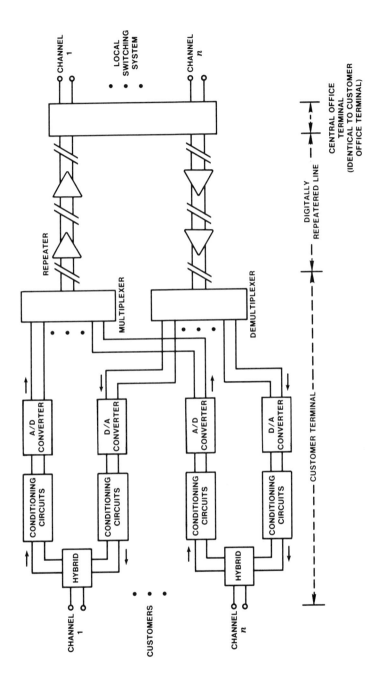

Figure 9-23. Digital loop carrier system.

The synchronization circuits are shared by all the channels in a digital system.

In contrast to analog loop systems, digital systems usually use two 2-wire pairs, each pair being dedicated to a particular direction of transmission. The signal transmitted over each pair occupies half the bandwidth of the signal that would have been necessary if a single pair had been employed for both directions of transmission. Generally, the choice of two pairs rather than one pair for transmission in digital systems results in considerable economic savings.

The SLC^{16}-40 system is a digital carrier providing forty full-time voiceband channels between a central office terminal and a single remote (customer) terminal. Using T1-type digital repeater line facilities, the remote terminal can be located up to about 20 miles from the central office terminal on 22-gauge buried cable. The remote terminal can connect customers on cable pairs up to 900 ohms or 5-dB loss beyond the remote terminal. This results in a customer—to—remote-terminal distance of 26 kft on 22-gauge loaded cable and almost 50 kft on 19-gauge cable. Radio systems that handle T1-type transmission can also be used between the remote and central office terminal. The modulation and signaling are performed in per-line channel units that can be added as customer demand develops, thus minimizing common equipment. This results in a low start-up cost and makes the system economical in long-route, low-growth areas as well as in shorter loop or higher growth areas where cable and structure relief are expensive. The system provides maintenance and alarm status information to the central office terminal and has a simple, straightforward maintenance plan.

The latest digital subscriber carrier system is the SLC-96 carrier system. It permits ninety-six customers to be served by as few as three digital transmission lines. The SLC-96 system is more economical, accommodates more customers, provides more services, is easier to maintain, and offers greater design flexibility than any of its predecessors. In addition, it has been designed to be compatible with the interoffice digital network.

Studies of the application of the SLC-96 system indicate that about one-third of the existing routes studied and roughly 10 percent of the growth expected on suburban routes could be economically provided by the SLC-96 system.

Along with these economic advantages, the SLC-96 carrier system provides a wide range of readily available benefits for rapidly growing suburban and rural areas. The specific benefits include:

- a full range of customer services, including single-party, multiparty, and coin service—both coin first and dial tone first

---

[16] Trademark of Western Electric Co.

- most special services such as Digital Data System service, foreign exchange lines, private branch exchange trunks, and private-line services

- remotely controlled testing of customer channels and wire pairs for single-party, multiparty, and coin service

- continuous transmission monitoring

- built-in aids for isolating system troubles

- a variety of remote terminals in different sizes and styles to accommodate conditions in the serving area.

The system consists of three basic components: a central office terminal, a remote terminal located in the area served, and T1 digital lines linking the two terminals. Each T1 line, composed of two wire pairs, can handle twenty-four channels. The T1 lines contain digital repeaters spaced about 1 mile apart.

The *SLC*-96 system employs time-division multiplexing and optional digital concentration to achieve pair gain. With optional digital concentration, a concentrator digitally switches active circuits to available channels. A two-to-one concentrator permits forty-eight customers to share the 24-channel capacity of a single T1 line. Only two T1 lines are needed to serve ninety-six subscribers, with a third line provided for protection. This is the *carrier/concentrator configuration*. When the *SLC*-96 system is used without concentration (in its *carrier-only configuration*), ninety-six subscribers are served by five T1 lines. Each of four lines carries twenty-four 2-way conversations, and a fifth line is available for protection in case a working line experiences transmission difficulties. If a working line fails, the protection line is automatically switched into service.

9.4.2 INTEROFFICE AREA APPLICATIONS

Metro Area Systems

This section describes the evolving technology that provides transmission services in local exchange, or metro, areas where interoffice distances are relatively short.

In short transmission systems, the cost of the terminals is considerably greater than the transmission line costs; simple baseband transmission on paired cable is preferred for these applications. For long routes, multiplex techniques prove economical since line costs now become significant, and multiplex techniques reduce the per-channel cost. The economic break-even point for baseband transmission on paired cable versus T1 digital carrier is about 5 miles, although this can vary considerably in any specific metro area. When baseband transmission on paired cable is combined with digital switching, the break-even point drops to 0 miles.

In recent years, almost all metro area trunks[17] have used paired cable as the transmission medium.[18] It is expected that, in the future, where traffic is sufficient, long trunks will be grouped into larger cross sections to achieve economies of scale by using higher speed digital carrier on **paired cable**, digital carrier on **coaxial cable**, digital microwave **radio**, and digital **lightwave** systems. The following paragraphs discuss these systems.

Paired Cable Systems (T1). T1 digital carrier was the initial Bell System short-haul digital transmission line; the first commercial application occurred in 1962. It has had wide application in metropolitan networks because of favorable cost and operational experience.

The signal transmitted in T1 is a 1.544 Mbps pulse stream that can be generated by a variety of terminals. Signals are applied directly to cable pairs in a bipolar format in which positive and negative pulses, always alternating, represent one state; and absence of pulse represents the other state. Figure 9-24 shows an example of a bipolar signal. The use of bipolar signals provides four significant advantages:

1) A bipolar signal spectrum has the significant portion of the signal power density below the frequency corresponding to the pulse-stream rate (that is, the frequency that is the reciprocal of the pulse period). A polar signal has twice the bandwidth.

2) The bipolar signal has a null in the power density spectrum at 0 Hz (dc) and at integer multiples of the pulse-stream rate. This avoids problems of dc (baseline) wander in ac-coupled input, output, and equalization circuits.

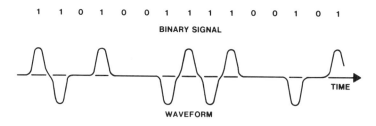

Figure 9-24. Bipolar signal.

---

[17] Traditionally, these have been called *exchange area trunks*.

[18] Section 9.3.2 discusses analog carrier systems used in the metro area.

3) The bipolar signal can be examined at any point along the transmission path, and any single-bit transmission error can be detected. If an error occurs in a 1, thereby converting it to a 0, adjacent 1's will be of identical polarity, violating the coding rule. If an error occurs in a 0, thereby converting it to a 1, there will be two successive 1's of identical polarity, again violating the rule.

4) A bipolar signal has good transition density for timing recovery.

Property (2) above is useful for remote maintenance margin testing, and property (3) above is useful for maintenance because working lines can be monitored automatically for excessive bipolar violation errors.

Error performance is a transmission performance quantity that must be controlled in digital system design. Line errors produce impulse noise in VF channels and can introduce serious errors in data transmission. A high occurrence of errors can cause terminals to lose synchronization, which is catastrophic (until reframing occurs)[19] in time-division multiplex systems. If line errors occur at rates less than, for example, 10^{-6} errors per bit transmitted, noise and signal distortion will be determined primarily by the terminal coding and decoding circuitry. The T1 objective is for 95 percent of all systems to have an error rate better than 10^{-6} errors per bit, under worst-case conditions.

Sources of error in T1 systems are similar to sources of noise and interference in analog multiplex carrier systems. Misequalization, interference from other services occupying the cable, and electronic circuit noise, for example, are sources of error. Examination of the components of a T1 repeater and its essential element, the T1 regenerator, yields a better understanding of how these sources of error are controlled when a system is engineered.

The T1 line repeater is a solid-state, plug-in unit, suitable for pole mounting or manhole placement and usually powered over the transmission pairs from the central office. Nominal spacing between repeaters is 1 mile. The repeater performs three functions: equalization of the input pulse stream to correct for linear distortion, extraction of the appropriate timing strobe for the regenerator, and determination of the input pulse value (that is, 0 or 1) with corresponding regeneration of a correct and properly shaped output pulse stream.

The optimum equalization does not correct for line distortion over the complete signal bandwidth but instead achieves a balance between noise and intersymbol interference. The amount of equalization required is determined by the type and length of cable used and is also influenced by temperature and manufacturing variations.

---

[19] A pulse-code modulation frame and location of the framing bit are shown in Figure 6-17.

T1 lines are self-timed, that is, timing information is extracted from the input waveform. Since a bipolar signal spectrum has a null at the pulse-stream rate, timing information cannot be directly extracted without conversion to a unipolar signal.

T1 carrier can be applied to a wide variety of cables. Although all cables exhibit an attenuation characteristic that increases with frequency, the loss-versus-frequency shape varies with different cable types and will change for a given cable with variations in temperature. Two approaches have been used in T1: selection of an appropriate equalizer from a series of fixed equalizers to match a specific regenerator requirement and use of a fixed amplifier gain characteristic together with a network to "build-out" the measured cable loss at a given regenerator to a predetermined value. Allowances are made in either case for temperature effects and variations between cable pairs.

T1 systems must be designed to cope with impulse noise generated by switching equipment in central offices. This type of impulse noise couples into carrier systems in the cables leaving the central office, where pairs used for VF trunks and carrier pairs occupy the same sheath. Usually, the repeater spacing in the first section out of an office is shortened so that the carrier signals are not allowed to be attenuated to too low a level, thereby achieving greater margin relative to impulse noise.

The crosstalk noise allocation determines the total number of T1 systems allowed in a given cable sheath. Near-end crosstalk is greatest where both directions of transmission are on pairs in the same sheath and in adjacent binder groups. In this case, as few as about 10 percent of the total pairs may be usable for T1 systems. If separate sheaths are used for each direction of transmission, near-end crosstalk is eliminated and only far-end crosstalk remains as a possible constraint. In practice, with existing regenerator designs, this does not limit full utilization of available cables.

T1 systems operate and are designed on a span basis. A *span* is defined as the aggregate of all span lines between two central office buildings. A *span line* is a regenerative repeatered line between the office repeater bays in two central office buildings. For longer routes, a number of span lines may be connected in tandem and together with channel banks on each end, constitute a 24-channel T1 system. By engineering and maintaining on a span basis, span lines can be connected ("hardwired") together as required to provide longer trunks without the need to engineer the system on a custom end-to-end basis. An average system contains four span lines and is 15 miles long. Where additional circuit capacity is needed on a route, additional span lines can be added to augment the span cross section by following a basic span design applicable to all span lines on that route.

Maintenance operations also are organized on a span basis. A span will have spare span lines provided to "make good" failed lines by patching when a working line fails. Individual repeaters are tested from the central offices by transmitting a digital signal that contains a superimposed audio component. The audio tone is returned to the testing office over a separate voiceband pair, and the tone's level and variation give significant information about the health of the individual repeater section.

Based on experience with T1 performance, there have been changes in operating parameters and improvements in design and engineering methods. One example has been new digital repeatered line designs that achieve an increased pulse-stream rate (and, therefore, a greater number of channels) by establishing a tighter balance between cable characteristics and regenerator design. These new lines, T1C and T1D, use techniques similar to the T1 design and provide for transmission at a 3.125-Mbps rate in 22-gauge plastic-insulated cable with the same regenerator spacings as T1. When D channel banks are combined with a new M1C multiplexer[20] for transmission over T1C or T1D, the system will provide forty-eight voiceband channels. Because the pulse-stream rate is twice that of T1, the equalization of the regenerator must accommodate a greater line loss at the higher frequencies required to support the 3.125-Mbps rate. *Automatic line buildout* (see footnote 6) is used to simplify engineering and installation. The engineering rules and design methods used provide greater control of crosstalk and interference required by the higher pulse-stream rate. Fault location principles for T1C and T1D are similar to those for T1.

Coaxial Cable Systems (T4M). The T4M digital repeatered line is a metropolitan area system but can be used for applications of up to 500 miles. It has manhole-located regenerators powered over the coaxial tubes and spaced up to 5700 feet apart. The line is divided into maintenance spans, with a maximum length of 111 miles between terminating offices where equipment for maintenance and restoration is located. Arrangements to add and drop circuits can be provided within the maintenance span without providing span-terminating equipment on through systems and without requiring additional protection lines over and above the common protection line.

T4M operates at a transmission speed of 274.176 Mbps and uses a polar rather than a bipolar signal on the line. Thus, only a single decision threshold is required instead of the two thresholds used for the 3-level (+, −, 0) bipolar signal. This provides increased margin, although certain advanced techniques are required to extract timing information and to compensate for cutting off the transmitted spectrum at low frequencies.

Design principles have involved consideration of many of the same factors as in the digital systems mentioned previously. However, T4M is

---

[20] Later sections of this chapter discuss D channel banks and multiplex equipment.

different in one important way: Interference between systems operating in a coaxial cable is not significant because of the high degree of shielding provided by the coaxial structure. As in all digital systems, intersymbol interference is an important factor. Allowance must be made in the design of the decision, timing, and cable equalization circuits so that even with normal environmental, production, and component-aging tolerances, the regenerator will be error-free. Automatic equalization without the need for installation adjustment is provided by automatic line buildouts. Four regenerator designs have been developed to cover the full cable loss range of from 0 to 56 dB.[21]

No routine testing of regenerators is contemplated, and a regenerator is considered to be operating satisfactorily with bit error rates better than 10^{-7} error per bit. An error rate of 10^{-6} error per bit will initiate alarms and a switch to a protection system.

The full system (500-mile) error objective of T4M is less than 10^{-6} error per bit for 99.9 percent of the time per line. The jitter[22] accumulation objective is no more than six time slots peak-to-peak (22 nanoseconds) over 500 miles. This value avoids a requirement for special "dejitter" circuits over and above those provided in M34 multiplexers (see Section 9.4.3).

Lightwave Systems (FT3). FT3 is the Bell System's first standard digital lightwave system. It transmits at the 44.7-Mbps rate, one of the Bell System's standard digital transmission rates.

A key element of the FT3 system is the lightwave regenerator. Mounted on a single plug-in circuit board, the regenerator contains a laser transmitter, an avalanche photodiode receiver, and electrical circuitry for timing recovery and pulse regeneration. This circuit pack contains both lightguide and electrical connectors, so both optical and electrical contacts are made when it is plugged into the equipment frame.

The FT3 system contains extensive maintenance features, including in-service performance monitoring of the error rate on digital transmission line, automatic protection switching to a spare line when the error rate exceeds one error in 10^6 bits, and test equipment to permit remote identification and isolation of faulty components.

The principal application of FT3 is for interoffice trunking in metropolitan areas. Digital transmission is already extensively used for metropolitan trunking; FT3 offers considerably greater duct efficiency and repeater spacing than metallic (wire pair) systems. The repeater spacing for FT3 is initially 4 miles—about four times that of metallic systems—

---

[21] At 137 MHz, the highest frequency important for the 274.176-Mbps transmission rate.

[22] Section 6.6.2 discusses jitter.

and as further reductions are made in fiber loss, increases in repeater spacing are anticipated.

In addition to standard FT3 applications, Western Electric is custom-engineering systems based on FT3 to meet special telephone company needs. One has been in service in Trumbull, Connecticut, since October 1979. This 4-mile system is in the feeder portion (the cables that leave the central office) of the loop plant and interfaces with the newly introduced *SLC*-96 system.

In the Connecticut system, the T1 signals coming from the *SLC*-96 system terminals in the central office are multiplexed by the MX3 and carried on fibers from the office to a remote terminal. MX3 has the capacity to multiplex twenty-eight T1 lines; thus, in the Connecticut configuration, a single MX3 can accommodate five *SLC*-96 system terminals (with five T1 lines each) serving 480 customers. For the initial service, two *SLC*-96 system terminals, serving 192 customers, have been installed. But additional capacity can readily be achieved by simply adding plug-ins to the MX3 multiplexer and more *SLC*-96 system terminals. Compared with metallic T1, the FT3 lightwave system employs a smaller cable, requires no intermediate repeaters, and has greater growth capability. The FT3C system, introduced in 1983, operates at 90 Mbps and provides twice the capacity of FT3 by transmitting two 44.7-Mbps signals over the same fiber.

Outstate Area Systems

To bring T1 quality and cost advantages to rural and suburban routes, T1 capabilities were expanded to adapt it to a telephone plant environment that includes long transmission distances, relatively few circuits, and unmanned switching exchanges (community dial offices). In September 1975, the first installation of this modified T1—known as T1 Outstate (T1/OS)—was placed into service between a toll office in Lamar, Colorado, and five of its outlying community dial offices. Three of the exchanges are independent telephone company offices.

The T1/OS system meets the needs of the rural environment with little modification to the basic T1 digital transmission equipment and without large developments. Improvements were added to T1 in three broad areas:

• New guidelines or engineering rules were developed to allow the operating company engineers to assemble long systems from the various T1/OS components.

• Features were provided to enhance reliability and maintenance. These include line automatic protection switching (APS), remote monitoring and APS control by means of built-in telemetry mated to the APS, and a new repeater fault-locating system.

• Modular and economical central office equipment arrangements were developed to provide the variety of combinations needed for small installations.

The outstate system shares basic components with T1—the D3 and D4 digital channel banks that encode twenty-four voiceband circuits into the 1.544-Mbps digital signal and the T1 repeatered transmission line for connecting between offices. The new maintenance subsystems are added to these components and, together with the set of engineering rules, make up a total system that gives T1/OS digital carrier a number of advantages over analog carrier for many rural routes.

Advantages obtained from using the D3 and D4 stem from the digital channel bank's low cost, low noise, and ability to handle a wide variety of switching system interfaces. In addition, because of the digital format, the quality of the line signal is monitored easily and directly. This makes maintenance and trouble location easier.

While conventional T1 is limited to about 50 miles, for T1/OS it was necessary to extend this range to over 150 miles. The new engineering rules achieve this goal while ensuring that overall performance objectives are met. The rules are based on connecting in tandem no more than 200 repeaters. The actual length of the system is a function of the quality of the individual repeater sections and the permissible repeater spacing.

The new fault-locating system is essentially a separate VF transmission facility and, as such, requires its own set of engineering rules. In T1/OS, fault-locating layouts can be complex; thus, the rules include a computer program for analyzing networks with branches, multiple terminations, and bridge taps. The program is called *T-Carrier Fault-Locating Applications Program* (TFLAP) and is a subprogram of the Universal Cable Circuit Analysis Program.

APS is central to the maintenance concept in T1/OS. Each T1 line is monitored at the ends of each protection-switching maintenance span; a failure triggers an alarm that identifies the troubled section.

When the APS is operating fully automatically (it can also be operated semiautomatically or manually) and a service line fails, through transmission is restored by transferring the digital signal to a waiting (good) protection line. The transfer is so fast that message traffic is not affected. The APS switches upon loss of signal, high error rates (10^{-4} or 10^{-6} error per bit), or wide pulses. When the problem is cleared, traffic is automatically returned to the original line, clearing the protection line for use by another failed system.

Typically, a number of maintenance spans are connected in tandem to make up a total T1/OS system. The APS is designed so that failures within one maintenance span do not affect APS operation in the others.

An important feature of the APS allows faulty repeaters to be located from one end of a maintenance span. On command, the two directions of T1 transmission at the far office are connected within the APS, thus "looping-back" the fault-locating signal to the testing office. The APS controls can be executed remotely.

Intercity Systems

Cable Systems. The T2 system provides transmission at 6.312 Mbps using paired cable. T2 provides intercity digital transmission for distances up to approximately 500 miles. Four DS1 signals, equivalent to four T1 lines, can be combined.

T2 uses a separate 22-gauge low-capacitance (LOCAP) cable for each direction to minimize crosstalk, permitting repeater spacings of 2.8 miles on repeatered sections of buried or underground cables not adjacent to office or powering stations. Aerial cable sections and sections adjacent to offices or powering stations require shorter spacings. The T2 system design is based upon a maximum of 250 repeater sections in tandem. Most systems are constructed in blocks of twenty-four T2 lines, including one protection line to which traffic from a working line can be switched automatically when errors on the former become excessive. In fully developed systems with D channel banks, this represents a cross section of 2208 voice channels. Using the largest size (104 pair) LOCAP cable being manufactured, a 2-cable route will provide a maximum cross section of 8832 voice channels. Because of the large cross sections available, T2 lines serve the intermediate-distance intertoll trunk market, connecting population centers between which traffic can be grouped to take advantage of economies of scale.

The T2 repeater consists of two separate 1-way regenerators powered over the transmission pairs. They may be mounted in manholes or above ground. The regenerator for each direction of transmission is mounted in a separate apparatus case. The T2 repeater performs the same functions as the T1 repeater: equalization, timing extraction, and regeneration.

Equalization is achieved with a combination of fixed equalization and automatic line buildout (ALBO). Five codes (designs) of equalizers feature different amounts of fixed equalization, selected on the basis of the cable route makeup. The ALBO consists of a variable equalizer controlled by a feedback signal developed from the fully equalized pulse stream at the ALBO output.

The pulse stream is in a modified bipolar format called *bipolar with six zero substitution* (B6ZS). In this format, 1's alternate in polarity as in T1, but a special code word is substituted when six 0's occur in a row in the original signal. This avoids loss of energy in the tuned timing extraction circuit. In addition, the restriction to no more than five 0's enables rapid detection of a loss of signal. It should be noted that B6ZS is defined as the format for the 6.312-Mbps signal (DS2 level in the time-division multiplex hierarchy) and is required from every terminal that will be connected to the T2 line.

T2 lines are divided into maintenance spans that can contain up to forty-four repeaters in manholes, offices, or intermediate power points. As an aid to sectionalization of troubles, the signal received at each end of the maintenance span is monitored on each T2 line. If errors in the B6ZS format exceed a threshold, appropriate alarms are activated. The signal is modified appropriately to eliminate format violations at any error rate, or in case of total signal interruption, a special all-1's signal is inserted. This is done to avoid erroneous error detection beyond the span where the fault occurred. It should be noted that violation-removal circuits do not correct errors; rather, they prevent the false indication of errors in succeeding spans.

Fault location is accomplished by a method similar to that used in T1, that is, with a special test signal containing a strong audio component.

T2 lines are being used for both intertoll trunks and private-line service between cities. The system error requirement for T2 is that 95 percent of all systems have an error rate of 10^{-7} or better under worst-case conditions. It is evident that T2 has been designed to more stringent requirements than T1, since the maximum length of T2 lines is about 500 miles, an order of magnitude greater than the usual maximum for T1, and the error objective is an order of magnitude better. With regard to jitter, it has been shown that 250 T2 regenerators (the maximum) in tandem have a root-mean-square (rms) jitter accumulation much better than the system objective of 10 nanoseconds.

Radio Systems (1A-RDS, 3A-RDS, DR6-30, DR11-40). The 1A-Radio Digital System (1A-RDS) is designed to transmit data signals over existing microwave facilities, including TD and TH. The primary application of 1A-RDS provides intercity connections for the Digital Data System. The 1A-RDS has 4000-mile, 2-way capability. As many long-haul systems as are necessary may be linked together to form the 1A-RDS channel. 1A-RDS provides radio facilities with the capability to transmit a DS1 signal in the radio baseband from 0 to 500 kHz, below the lowest frequency (564 kHz) used for transmission in the frequency-division multiplex basic mastergroup (described in Section 9.3.5). Since the basic mastergroup carries 600 voiceband channels, the 1A-RDS channel is referred to as *data under voice* (DUV).

The 3A-RDS is a high-capacity digital carrier system that operates at 44.736 Mbps. It provides for a total capacity of 560 DS1 signals (1.544 Mbps), or 13,440 voice channels,[23] in the 11-GHz radio band.

The 3A-RDS was intended primarily for applications on high-density intercity routes up to approximately 250 miles, as feeder for T4M or other systems, and to provide route diversity within the network. Repeater

---

[23] *Voice channel* or *voice circuit* refers to a 64-kbps channel.

spacings range from 6 to 25 miles depending on the intensity of local rainfall.

3A-RDS uses a combination of existing TN-1 analog radio equipment and a 3A-RDT digital terminal. The 3A-RDT provides for the digital processing modulation and demodulation steps necessary for transmitting the DS3 signal over a TN-1 radio channel. Regeneration is required at approximately every other repeater site; at these sites, a 3A-RDT regenerator is provided in addition to the TN-1 radio equipment.

DR6-30 is a high-capacity digital carrier system that operates at 6 GHz in channels of 30-MHz bandwidth and can carry a maximum of 9408 digital voice circuits, or the equivalent in data circuits, in seven channels.

It is most often used on hops of about 15 to 30 miles. The first DR6-30 system went into service early in 1981, between Eugene and Roseburg, Oregon—a 75-mile route over rugged mountain terrain in Pacific Northwest Bell territory.

DR11-40, operating at 11 GHz in 40-MHz channel bandwidths, has a capacity of 13,440 voice circuits, or the equivalent in data circuits, in ten channels. It is used for shorter hops, usually 15 miles or less. Signals at 11 GHz are more susceptible to rain-induced fading than 6-GHz signals, and so are usually restricted to the shorter distances. However, 11-GHz systems are used for the longer routes when the 6-GHz frequency spectrum is already fully occupied.

The first DR11-40 systems went into service in combination with DR6-30 on a Michigan Bell route between Flint and Kalamazoo, with branches to Saginaw and Big Rapids.

DR6-30 and DR11-40 are compatible with analog radio systems and can share the same towers and antennas. This joint use is possible because the design of the modulation scheme controls the spillover of digital signal power in adjacent channels. The DR6-30 and DR11-40 systems use 16-state quadrature-amplitude modulation. In this scheme, transmitted signal strength drops off sharply at the upper and lower frequency limits of a given channel, and for a given transmitted bit rate, the bandwidth is smaller than in some other modulation methods. The digital channel, therefore, presents a very small and acceptable amount of interference to any adjacent channels.

Lightwave Systems (FT3C). The FT3C lightwave digital transmission system transmits digitally encoded information between offices in the form of light pulses at the rate of 90.524 Mbps. One FT3C lightwave signal contains information for up to 1344 two-way voice channels over a pair of glass fibers within a lightguide cable. A cable may contain up to 144 individual fibers, giving a maximum capacity of over 80,000 voice circuits per cable, including protection.

The FT3C facility is intended for applications with medium-to-large service cross sections. The lower time-division multiplex transmission rates are multiplexed to the FT3C rate. An FT3C terminal can also be equipped to serve as an express terminal; that is, a terminal to interconnect back-to-back FT3C maintenance spans without appearing at a cross-connect bay. This permits violation monitoring, protection switching, and other maintenance functions to be performed in the express configuration.

Figure 9-25 is a simplified block diagram of a typical FT3C facility. The lightguide cable used between offices consists of glass fibers packaged in ribbons of twelve fibers each. One to twelve ribbons are enclosed in a protective sheath with an outside diameter of about one-half inch independent of the number of ribbons. A separate fiber is required for each direction of transmission.

Regenerators are required along the transmission path and are located in line repeater stations (LRSs). The maximum distance between LRSs is determined primarily by the grade of lightguide cable selected and the number of splices used to construct the cable path from regenerator to regenerator. At present, for example, a regenerator section length of 5.3 miles is attainable using the best grade of cable and an average distance of about 1600 feet between splices. Longer regenerator spacing is permissible when fewer splices are used. The grade of cable and permissible fiber path loss are system parameters subject to frequent modification due to rapid technological change.

The lightwave regenerator requires an environment similar to that found in central offices, but it may be located in a suitable hut, controlled environment vault (CEV), or leased space. In a hut, CEV, or leased space, power may be provided from a power plant that converts local ac power to dc power with a battery reserve.

The FT3C facility uses an MX3C lightwave terminating frame (LTF) to terminate up to ten 2-way lightwave service lines and up to two 2-way lightwave protection lines.

Lightguide cable interconnection equipment (LCIE) terminates the lightguide cable sheath and provides appearances of the individual fibers on an array of single-fiber connectors. Single-fiber interconnection cables connect the lightguide cable to terminal equipment or repeater station regenerators.

The FT3C lightwave line is designed to the following transmission objectives for system lengths up to 4000 miles:

• an average error rate of better than 10^{-8}

• a service outage caused by equipment failures of not more than 0.02 percent, that is, 1.7 hours per year.

Outage is that percentage of time during which service over a given line is interrupted because of an unprotected failure.

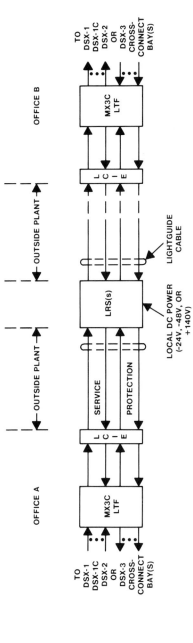

Figure 9-25. Block diagram of a typical FT3C lightwave system facility.

LCIE   LIGHTGUIDE CABLE INTERCONNECTION EQUIPMENT
LRS   LINE REPEATER STATION
LTF   LIGHTWAVE TERMINATING FRAME

9.4.3 DIGITAL MULTIPLEX EQUIPMENT

The digital carrier systems described in the previous sections operate at several different transmission or bit rates. As noted earlier, the bit rates are levels in the time-division multiplex (TDM) hierarchy. A variety of multiplex equipment has been designed to interface between the different bit rates. This section describes the TDM hierarchy and several kinds of multiplex equipment.

The TDM Hierarchy

Figure 9-26 shows the TDM hierarchy, generally described as DS levels 0 through 4. The 0- to 4-kHz nominal voiceband channel is shown below the DS0 level to illustrate the first step in combining voice or other analog signals into 64-kbps pulse-code-modulated signals. As described in Section 6.5.2, twenty-four voiceband analog channels are combined or multiplexed to form a DS1 signal (1.544 Mbps), also called a *digroup* (for digital group). Digital transmission systems are shown on the right at the corresponding transmission rates.

In the case of a short-haul T1 carrier system carrying a 24-channel circuit group, the channel terminal equipment bears a unique relationship to (and is normally thought of as part of) the transmission facility. For

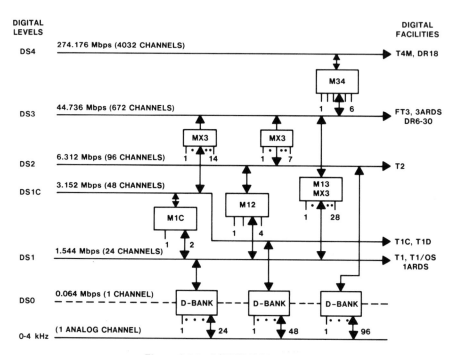

Figure 9-26. PCM-TDM hierarchy.

the higher bit rate systems (for example, T2, T4M, DR18, and FT3), the multiplex arrangements have evolved in a hierarchical manner to be consistent with the capabilities of the various media for high-speed digital transmission.

Channel Banks

Digital channel banks (or primary pulse-code modulation [PCM] multiplexers, as they are called in CCITT terminology) have two basic functions: They convert analog voice signals into digital form (and vice versa), and they combine the resulting digital signals into a single digital bit stream by time-division multiplexing (and vice versa). The D-type channel banks have a number of characteristics in common:

- They have equipment to provide the proper VF and signaling interfaces with the central office trunk circuits or other assigned circuits.

- They use filters to limit the transmitter input frequency range and to reconstruct voice signals at the receive output.

- They use *sampling gates* to sample transmitter input signals and to deliver receiver output samples to the correct channels.

- They have either a nonlinear encoder and decoder or a linear *codec* (coder-decoder) combined with a compressor and expandor to convert pulse-amplitude-modulated samples to PCM samples.

- They provide a means of controlling the timing of the sampling gates and the location of PCM and signaling bits in the bit stream format.

- They use the carrier group alarm to provide carrier failure alarms and trunk conditioning to minimize the effects of outages.

- They have power supplies to provide the dc-to-dc conversion needed to convert −48 volt office supply voltages to the voltages required in the bay.

D-type channel banks of the types D1A, B, C through D4 are 24-, 48-, or 96-channel carrier terminals that are composed of a common-equipment portion and a set of up to twenty-four, forty-eight, or ninety-six individual channel units. The common-equipment units have those functions that are common to all the channels and have characteristics that do not vary with the types of trunk circuits assigned to the channels. Channel units are individual to a particular channel, and their basic function is to provide the interface between the central office trunk circuits, or other assigned circuits, and the channel bank. They include hybrids (if needed) to convert between 2-wire and 4-wire transmission, transformers for 4-wire circuits, and maintenance access for each channel.

Plugging in the correct channel unit furnishes proper transmission and network signaling for a particular trunk.

The D1A channel bank was the first digital channel bank developed. It consists of twenty-four plug-in channel units and twenty-nine plug-in common-equipment units. The D1B channel bank was developed to allow 4-state signaling; it is identical to the D1A bank except in the format used to code signaling information. The D1C is a specialized bank designed to provide remote operator capability for Traffic Service Position Systems (TSPSs).[24] It permits transmission and reception of twenty-four voice channels and a 24-bit parallel data link between toll connecting trunks at a central office (TSPS base location) and a TSPS No. 1 at a remote office (remote operator group). The D1D channel bank provides direct, tandem, and toll connecting trunk facilities on an end-to-end basis with D2, D3, D4, other terminal equipment with a compatible DS1 interface, or another D1D bank.

All the D-type channel banks, except the D1 series, are suitable for use on intertoll trunks. A performance level superior to that of the D1 bank is required because toll calls may include several digital trunks interconnected by switching systems. For switching systems that can handle signals only in voice-frequency form, quantizing noise will accumulate because of analog-to-digital and digital-to-analog conversions at each switching system. For digital switching, a signal converted into digital form by one channel bank should be capable of being reconstructed by any other channel bank (except D1A, B, C). Thus, a high degree of standardization and uniformity is required of all such channel banks.

The D2 channel bank is a 96-channel carrier terminal delivering four independent 1.544-Mbps outputs, each one with the same output capacity as a D1 bank. It is effectively four independent channel banks of twenty-four channels each, even though some of its circuits are shared by all ninety-six channels. The D2 banks have 104 plug-in common units for ninety-six voice channels with circuits serving eight, twenty-four, or ninety-six channels on each unit. The number of channels, the sampling rate, and the output bit rate for D2 were chosen to match those of the D1 channel bank and the T1 line, but the D2 channel bank is not end-to-end compatible with the D1A or D1B because the amplitudes of voice samples are encoded differently.

The D3 channel bank is a carrier terminal used for processing twenty-four VF channels into a DS1 signal. It is suitable for use in direct, tandem, toll connecting, and intertoll trunks. The D3 channel bank is much smaller than D1; just five plug-in units are required for the common-equipment circuitry versus twenty-nine for D1. All the circuitry of a given channel is contained wholly within the channel plug-in unit and

---

[24] Section 10.4.1 discusses TSPSs.

not within common equipment, so that single-channel failures can be easily restored by simply replacing the unit.

The D4 channel bank is a 48-channel carrier terminal that is organized into two 24-channel digroups. It can be operated in any one of five different modes to produce DS1, DS1C, or DS2 outputs (2 times 48-channel terminals in the case of DS2) to match the line and far-end equipment. It has from eight to eleven plug-in common units, depending on the mode, for forty-eight channels with circuits serving twenty-four or forty-eight channels. D4 is end-to-end compatible with D1D, D2, D3, and other D4 channel banks. D4 channel banks provide a wide range of channel unit types to serve both message and special-services applications. In addition, the D4 bank can be used to provide the 64-kbps rate of the Digital Data System using dataport channel units.

ESS Switching Equipment Interfaces

Digital interfaces between *ESS* switching equipment and transmission facilities provide cost, space, and maintenance advantages over analog interfaces. That portion of the interface that directly terminates transmission lines is called the *transmission interface*. With this definition in mind, the *ESS* switching equipment interfaces can be thought of as comprising a number of transmission interface units (digital terminals), along with associated controllers that serve as control and maintenance interfaces with the particular *ESS* system used.

The first integrated digital terminal, the digroup terminal, was developed in conjunction with the 4*ESS* switching equipment. The digroup terminal unit (DTU) provides a bidirectional transmission interface between five DS1-rate lines and one time-slot interchange (TSI) port on the 4*ESS* switch. Eight working DTUs, protected by a switchable spare, and a digroup terminal controller (DTC) that serves as the control and maintenance interface with the 1A processor (which also controls operation of the 4*ESS* switching equipment) make up a digroup terminal frame. Each frame has the capability of terminating forty digroups or 960 working channels. Thus, a digroup terminal frame provides an interface between forty DS1-rate lines and the 4*ESS* switching equipment. In actual practice, a digroup terminal complex comprises as many digroup terminal frames as are required to accommodate digital trunks interfacing with a 4*ESS* switching office.

The digital interface frame (DIF) is a more recent 4*ESS* switch digital interface that performs functions previously performed by the digroup terminal and its associated signal processor (SP2). One DIF can replace four digroup terminal frames and one SP2. The design philosophy of the DIF is oriented towards a per-digroup approach rather than the common-equipment design approach used in the digroup terminal. The ability to do so economically stems from the incorporation of several custom large-scale integrated circuits in the per-digroup circuitry. The per-digroup arrangement also simplifies maintenance and isolates faults in per-digroup circuitry from common-control functions. The digital interface unit (DIU) provides a transmission interface between five DS1-rate lines and one TSI port of the 4ESS switch. Thirty-two working DIUs, protected by two spares, plus a digital interface controller (DIC), which provides the signaling and control interface with the 1A processor within the 4ESS switch, make up a DIF. Thus, each DIF has the capability of terminating 160 digroups or 3840 channels.

The digital carrier trunk (DCT) bank provides a transmission interface between digital trunks provided by the carrier facilities trunk link network (TLN) and the trunk distributing frame of the 1/1AESS switching equipment. It is a representative of flow-through technology: equipment based on the D4 bank designed to operate in the 1ESS switching environment. DCT banks differ from stand-alone D4 channel banks in two respects:

- DCT channel units combine both digital channel unit functions and switching system trunk circuit functions.

- DCT banks have a third port connection that permits the exchange of signaling, supervisory, and maintenance information with the 1/1AESS switch in digital format.

Ten DCT banks plus a peripheral unit controller—a redundant microprocessor system that serves as the interface with the processor of the 1/1AESS switching equipment—make up a digital carrier trunk frame. Since each DCT bank can accommodate 48 VF channels, each DCT frame has the capability of terminating 480 VF channels. On the carrier side, these channels are organized into digroups of twenty-four channels each and may be carried over T-carrier facilities at either the DS1 rate, DS1C rate, or the DS2 rate. So, the DCT frame may terminate twenty, ten, or five T-carrier lines depending upon the line rate.

The digital facility interface (DFI) is an interface unit that provides a transmission interface between a digital transmission facility and the 5ESS switching equipment. Up to fifteen DFIs compose one digital line/trunk unit (DLTU), which directly interfaces with DS1 terminations. Interfaces to DS1C and to the international 32-channel digital carrier are being planned.

Not covered within this section are the 2/2BESS and 3ESS switching equipment interfaces. The 2/2BESS switch direct interface with T-carrier permits direct control of T-carrier D3 and/or D4 channel units by the 2/2BESS switch processors. The 3ESS switch direct interface with T-carrier permits direct control of T-carrier D4 channel units by the 3ESS switch processor. These procedures entail the use of standard D3 and D4 channel banks with a new channel unit.

Digital Access and Cross-Connect System

The Digital Access and Cross-Connect System (DACS) provides per-channel DS0 (64 kbps) electronic cross-connection and test access to individual channels in analog or digital form from the DS1 signal. DACS terminates up to 127 DS1 signals (3048 DS0 channels), providing a maximum of 1524 DS0 cross-connections. DACS is a new element in the facility makeup of T1 systems. It is not a direct replacement for, nor an alternative to, any single existing hardware entity. The cross-connect capability of DACS permits the assignment and redistribution of 64-kbps channels among T1 systems at the digital level. This capability can be used to collect channels with common destinations and thus increase the fill of T1 lines and terminal equipment. It also reduces the need for back-to-back channel banks to convert signals to analog form and then back to digital form. The cross-connect capability of DACS can be used to segregate channels by type (for example, message/special services, 4-wire/2-wire), which will result in simplification of central office interconnection and testing arrangements. The test access capability of DACS may serve as an alternative to jack or Switched Maintenance Access System (SMAS)[25] arrangements in the digital environment. Benefits derived from DACS applications include capital savings resulting from reductions in the number of intermediate distribution frames and associated wiring, reductions in back-to-back D-type channel banks and SMAS connections, and more efficient use of T1 lines and digital terminals. Expense savings stemming from easier administration and improved facility maintenance are also expected.

Other Multiplex Equipment

The digital multiplexers provide the interfaces between the different bit rates in the digital network. A multiplexer contains one or several *muldems*, each consisting of a *mu*ltiplexer-*dem*ultiplexer combination. The multiplexer may also contain in-service performance monitoring and automatic protection-switching circuitry. All multiplexers except the M1C are available with time-shared, in-service performance monitoring and automatic protection switching. These two features are usually combined and are optional in some of the multiplexers. In multiplexers

---

[25] Through the use of relays, SMAS provides concentrated metallic access to individual circuits to permit remote access and testing by the Switched Access Remote Test System (SARTS).

equipped with monitoring and switching, the monitors sequentially check each muldem by comparing the inputs and outputs of the muldem with the inputs and output of a reference demultiplexer contained in the monitor. All input signals are also checked for integrity. Muldem failures are alarmed and result in an automatic switch to the standby muldem (if available). The failure detection and switch in the M13 and M34, but not the M12, are completed rapidly enough to avoid dropping calls in progress, although data service may be affected. Input signal failures cause an alarm but do not cause automatic switching. The time-shared monitor provides diagnostic indicators and issues the appropriate office alarms. These can also be sent to a remote location. For maintenance purposes, it should be noted that DS1 signals out of a multiplexer are always free of bipolar violations, even when bipolar violations are present at the entrance to the multiplex at the preceding end. For example, if a higher capacity transmission facility with the required multiplexers at each end constitutes one link within a cascade of T1 spans, bipolar violations occurring ahead of that facility cannot be measured downstream from that facility (as they could if that facility were just another T1 span). The multiplexer units described below and their position in the TDM hierarchy are illustrated in Figure 9-26.

M1C Multiplexer. The M1C multiplexer bay contains up to forty-eight operating muldems and one standby muldem. An M1C muldem accepts two digital signals at the DS1 rate and multiplexes them into a single DS1C signal that is transmitted over a T1C or a T1D digital line. The M1C muldem can also accept a DS1C signal from a T1C or T1D digital line and demultiplex it into two digital signals at the DS1 rate. The standby muldem is a hot spare that must be manually switched to replace any failed in-service muldem.

M12 Multiplexer. The M12 digital multiplexer combines four DS1 signals to form a DS2 signal. One standby muldem is used for the protection of up to twenty-three working muldems. An M12 multiplex frame can contain up to a maximum of twenty-four M12 muldems, depending upon the particular frame configuration desired.

M13 and MX3 Multiplexers. The M13 digital multiplexer provides a transmission interface between the DS1 and DS3 digital signal rates. Access to it is normally provided through the DSX-1 and DSX-3 crossconnects. An M13 multiplexer frame contains two M13 muldems. Multiplexing is accomplished in two steps. In the first step, up to four asynchronous incoming DS1 signals are multiplexed into a single internal DS2 signal. This is accomplished in the low-speed multiplexer portion of an M13 muldem. In the second step, up to seven of those DS2 signals are multiplexed into a single DS3 signal. This is accomplished in the high-speed multiplexer portion of the M13 muldem. The demultiplexer portion first recovers the seven DS2 signals from a received DS3 signal and then the four DS1 signals from each of the DS2 signals, thus arriving at the original twenty-eight DS1 signals. Protection for the M13 muldems is accomplished by providing one standby for each working high-speed muldem portion and one standby for up to fourteen working low-speed muldem portions.

The MX3 is a family of digital multiplexers that will be available in a number of configurations and options; the MX3 will serve as a stand-alone multiplexer, a line-terminating multiplexer for FT3 or DR6, or as a span terminal for FT3. The MX3 will interconnect a DSX-1, DSX-1C, or DSX-2 cross-connect with a DSX-3 cross-connect, the FT3 fiber optic line, or the DR6 digital radio. The MX3 multiplexer is expected to replace the M13 in multiplex applications and, as a line terminal for FT3, find use in up to 70 percent of the metropolitan digital trunk applications. The stand-alone version of the MX3, the MX3 multiplexer assembly, is a single MX3 muldem intended for small installations where cost, space, and power are major considerations.

M34 Multiplexer. The M34 digital multiplexer provides a transmission interface between the DS3 and DS4 digital signal rates. Access to it is normally provided through the DSX-3 and DSX-4 cross-connects. One standby muldem is provided for the protection of up to ten working muldems. A fully equipped M34 multiplexer arrangement contains up to eleven M34 muldems in a total of seven frames. The multiplexer portion of each muldem time-division multiplexes up to six asynchronous incoming DS3 signals into a single DS4 signal. The demultiplexer portion recovers the original six DS3 signals from a received DS4 signal.

9.5 TRANSMISSION APPLICATION OVERVIEW

Previous sections of this chapter describe various transmission systems. This section provides a breakdown by transmission facility category for the loop and interoffice application areas. Data are presented to give both a recent (1980) picture of the breakdown and an indication of the trends over the past 20 years.

9.5.1 LOOP APPLICATIONS

Nearly all customer loops are now carried at voice frequency on individual wire pairs. Only a few percent of all loops are on carrier facilities because most loops are short, making it difficult to compensate for the cost of carrier terminals. Where loops are long, the cross sections tend to be small, which in turn, tends to make carrier facilities uneconomical. Recent digital technology improvements, together with the increasing cost of wire, have resulted in some penetration of carrier systems into the loop plant. However, the increasing use of local digital switching systems, which permit lower termination costs for digital facilities, should permit digital carrier systems to become an important factor in the loop plant in the future.

9.5.2 INTEROFFICE APPLICATIONS

The application of different facilities to the interoffice area can be viewed in two important ways: in terms of the number of circuits (or voiceband channels) and in terms of circuit-miles. The first view is provided by Table 9-8, which shows the breakdown of circuits by carrier terminal type as of 1980. As indicated by the footnotes, each terminal type is used for two or more transmission systems. Because of the large number of transmission systems, a breakdown by system type would result in a large table and obscure an important point: that digital facilities terminated on D-type channel banks account for most (68 percent in 1980) of the interoffice circuits. Table 9-9 shows the trend in terminal types from 1960 to 1980. The evolution to digital facilities in the interoffice plant is clearly seen, with an increasing rate of replacement of analog terminals by digital terminals.

TABLE 9-8

CARRIER CIRCUITS BY TYPE (1980)

Type	Thousands of Circuits	Percent
A-type channel banks*	1164	11.8
N-carrier terminals†	1979	20.0
D-type channel banks‡	6752	68.2
Total	9895	

NOTE: Circuits shipped prior to N-carrier, with the exception of A-type channel banks, are not included since they have almost all been retired.

* Used for analog L-carrier and analog radio.

† Total includes O and ON terminals.

‡ Includes digital switching system interface terminals; used for all digital interoffice facilities.

TABLE 9-9

PERCENTAGE OF TERMINALS SHIPPED, 1960 TO 1980

Type	1960	1965	1970	1975	1980
A-type channel banks*	21.9	23.8	24.0	21.0	11.8
N-carrier terminals†	78.1	64.8	43.0	29.3	20.0
D-type channel banks‡	0.0	11.4	33.0	49.7	68.2

NOTE: Circuits shipped prior to N-carrier, with the exception of A-type channel banks, are not included since they have almost all been retired.

* Used for analog L-carrier and analog radio.

† Total includes O and ON terminals.

‡ Includes digital switching system interface terminals; used for all digital interoffice facilities.

Table 9-10 provides another view of interoffice applications. Here, the categories are generic facility types, and the data indicate that digital facilities account for only about 11 percent (in 1980) of the total interoffice circuit-miles. Analog systems dominate by this measure, with analog radio alone accounting for about 60 percent of all circuit-miles. As shown in Table 9-11, analog radio has essentially maintained a constant share of total circuit-miles in recent years, largely because of improvements discussed in Section 9.3.4 that increased system capacity.

TABLE 9-10

CARRIER CIRCUIT-MILES BY TYPE (1980)

Type	Millions of Circuit-Miles	Percent
Analog paired cable	38	3.7
Analog coaxial cable	197	19.3
Undersea cable	15	1.5
Analog radio	623	60.9
Satellite	36	3.5
Digital paired cable	102	10.0
Digital coaxial cable	1	0.1
Digital radio	11	1.1
Total	1023	

TABLE 9-11

PERCENTAGE OF CARRIER CIRCUIT-MILES
BY TYPE, 1960 TO 1980

Type	1960	1965	1970	1975	1980
Analog paired cable	29.8	19.6	10.3	6.6	3.7
Analog coaxial cable	31.5	25.7	23.6	22.8	19.3
Undersea cable	0.4	1.0	0.9	1.0	1.4
Analog radio	38.3	52.7	61.9	62.1	60.9
Satellite	0.0	0.0	0.0	0.0	3.6
Digital paired cable	0.0	1.0	3.3	7.5	10.0
Digital coaxial cable	0.0	0.0	0.0	0.004	0.14
Digital radio	0.0	0.0	0.0	0.00	1.0

Thus, while the use of digital carrier has been growing rapidly in the interoffice plant, the amount of penetration differs significantly according to the application. Short-haul circuits, such as those in metropolitan areas, have a high penetration of digital facilities. The long-haul intercity circuits are mostly on analog facilities, primarily analog radio. It should be noted, however, that there has been an active program to provide digital capability over long-haul analog facilities, thereby increasing the amount of end-to-end digital connectivity in the network. The 1A-RDS channel referred to as *data under voice* (see Section 9.4.2) is one example of this kind of adaptation.

AUTHORS

E. J. Anderson
R. A. Bruce
W. C. Roesel
H. R. Westerman

10

Network Switching Systems

10.1 INTRODUCTION

Switching systems used in the Bell System can be divided into two broad functional categories: those designed for **local** switching and those designed for **tandem** switching. Local offices connect customer lines to each other for local calls and connect lines to trunks for interoffice calls. Tandem switching has two applications. Offices that connect trunks to trunks within a metropolitan area are referred to as *local tandem offices*. Offices that connect trunks to trunks to form the toll network portion (class 1 to class 4) of the hierarchical public switched telephone network (PSTN) are called *toll offices* (see Chapter 4).

There are significant differences in requirements for these areas of application. Consequently, systems designed primarily for local switching applications (often called *local switching systems*) are different in architecture and function from those designed primarily for local tandem or toll switching. As discussed in Section 4.2 and shown in Figure 4-4, many switching systems serve more than one role in the PSTN, and in particular, they may provide both local and tandem switching functions.

Because lines and trunks connecting nearby offices usually generate a load per termination that is less than the link capacity within a switching network, local switching systems generally employ *concentration* of lines and *expansion* to trunks (see Section 7.3). In tandem and toll applications, however, trunks are more heavily used, so expansion of trunks to the network is frequently used.

Because lines and short-distance trunks generally use 2-wire transmission, the telephone connections within local and local tandem offices have generally been 2-wire. Time-division digital systems, however, usually switch on a 4-wire basis (that is, separate paths for each direction of transmission). Toll facilities are usually also 4-wire, so toll switching systems switch on a 4-wire basis.

Functions needed to provide exchange services[1] are built into the local switching system because of the convenience resulting from the direct interface with customer lines. The geographic centralization of the tandem office, however, offers efficiency in providing centralized billing and operator and network services.

The Bell System formally began installing automatic switching in 1919. Since that time, the market for switching systems has expanded continuously in terms of both the number of systems and their applications. The growth of cities, establishment of new population centers in the suburbs, increased use of the telephone, and demand for new services all contribute to the expanding market.

The evolution of switching systems has been marked by a flow of new technology. As new technology is developed, new capabilities become available at lower cost for the basic switching functions and new customer services. Switching networks have evolved from progressive switching to coordinate switching to the current technology of time-division switching. Switching control has evolved from direct progressive control through common control by electromagnetic registers and markers to the current technology of stored-program control.

Because of the large base of existing systems and the cost and effort associated with replacement, older systems that still provide satisfactory service may remain in service for a considerable time after introduction of a new system. As a result, several systems of various vintages exist in the PSTN.

10.2 ELECTROMECHANICAL SWITCHING SYSTEMS

10.2.1 EVOLUTION

Local Switching

The earliest automatic switching system was the **step-by-step** (SXS) system, known around the world as the Strowger system. Although it was invented in 1889 by A. B. Strowger and first installed in 1892, the Bell System did not begin using the step-by-step system extensively until 1919, and even then, the equipment was installed by the Automatic Electric Company. One reason for the delay in applying step-by-step systems was the high percentage of Bell System customers who were located in large cities where step-by-step systems were not economically attractive. However, by 1921, Western Electric did begin installing them in the smaller cities. Western Electric acquired licensing agreements with Automatic Electric in 1916 and began many design improvements leading

---

[1] Section 2.4 describes exchange services.

to its own system design, which appeared in 1926. Today, step-by-step systems are used in rural and suburban areas and even in some metropolitan areas that began small but grew extensively.

In the early 1900s, the Bell System began working on an automatic system to provide efficient service in large cities; the result was the **panel system**, first placed in service in 1921. Originally introduced as a local switching system, it was later adapted for local tandem operation. This system used a register, called a *sender*, to store dialed digits and to control progressive originating and terminating switching networks. At their peak, in the 1950s, panel offices served nearly 4 million customer lines in Bell operating companies in many large cities. Their replacement by more modern systems began in the 1960s, and the last panel system was retired on September 11, 1982, in Newark, New Jersey.

In 1938, following the invention of the crossbar switch and advances in relay technology, the Bell System introduced another metropolitan switching system, the **No. 1 Crossbar System**. The No. 1 Crossbar used separate originating and terminating coordinate switching networks. Each network was controlled by a group of markers that interpreted digits, selected a network path, and caused the proper network switches to operate. The No. 1 Crossbar was designed for large-size offices and used primarily to meet the substantial growth in demand for telephone service in cities.

After World War II, design was started on a crossbar system adapted to the needs of suburbs and small cities. This system, the **No. 5 Crossbar System**, met these needs through effective use of the common-control principle—using faster markers controlling a single coordinate switching network instead of separate originating and terminating networks. Local automatic message accounting (see Section 10.5.3) was incorporated into the design. Later, centrex service was added. In fact, this flexible crossbar system found economical applications beyond its original design range—in large city central offices and, as a special version, in AUTOVON.[2] The last new No. 5 Crossbar to be installed went into service in November 1977.

Toll and Tandem Switching

To a large degree, toll and tandem switching systems have evolved following the same technological advances as local systems. Automatic switching was first introduced in the toll environment during the early 1920s in Los Angeles, California, using step-by-step systems. Step-by-step toll systems were initially used to carry short-haul toll traffic. But, by the

---

[2] AUTOVON (automatic voice network) is a private voiceband network serving the Department of Defense. It employs automatic switching and handles both voice and data traffic.

early 1940s, cities were being tied together by step-by-step systems in
long-haul dialing networks. All step-by-step toll systems are 2-wire, and
most are directly controlled systems (no senders, decoders, translators,
etc.). About sixty have been modified for centralized automatic message
accounting (discussed in Section 10.5.4). These use common-control
features for the routing and charging associated with customer-dialed
traffic. Nearly all step-by-step switching systems that do tandem switching are in class 4 offices.

Many early panel systems were used within cities as local tandem
offices. The crossbar tandem switching system was designed as the suc-
cessor for these local tandem panel systems and went into service in 1941.
In cases where a local crossbar tandem office had access to all the local
offices in an area, it was naturally positioned to handle toll calls originating or terminating in that area. The ability to complete toll traffic was
added in 1947, and originating toll functions were added in 1953.

Improved transmission techniques made the introduction of crossbar
tandem (a 2-wire system) as a through toll (class 3, 2, or 1) switching system possible in 1955. Foreign-area translation (the ability to translate 6-
digit area and office codes to derive more extensive trunk group choices)
was added to crossbar tandem in 1958.

The first crossbar switching system designed exclusively for toll service was the **No. 4 Crossbar System** introduced in 1943 in Philadelphia,
Pennsylvania. In 1953, an improved version of the No. 4 Crossbar, called
the **No. 4A Crossbar System**, added foreign-area translation, automatic
alternate routing (the ability to route the call to other trunks groups if
the first route is busy), and address digit manipulation capabilities (converting the incoming address to a different address for route control in
subsequent offices, deleting digits, and prefixing new digits if needed).

The No. 4A Crossbar was intended for metropolitan areas and was the
largest of the Bell System's toll systems until 4ESS switching equipment
became available. The No. 4A Crossbar is a 4-wire switching system.
The 4-wire design was chosen to eliminate the echo problems associated
with converting 4-wire to 2-wire transmission after projections of both
intertoll and toll connecting equipment indicated increased carrier (4-
wire) operation in the future.

10.2.2 STEP-BY-STEP SYSTEMS

The term *step-by-step* describes both the manner in which the switching
network path is established and the way in which each of the switches in
the path operates. The basic step-by-step system is classified as a direct
progressive control system because the dial pulses generated by the customer's telephone directly control the stepping switches of the progressive networks.

The switches are functionally described as *line finders, selectors,* and *connectors.* Each of these switches combines vertical stepping and horizontal rotary stepping motions in a 2-stage selection process. A set of wiper brush contact fingers is moved, first in a vertical direction to select one of ten level positions, then in a horizontal direction until the selected position of ten at that level is reached. Figure 10-1 shows the terminal bank of a switch.

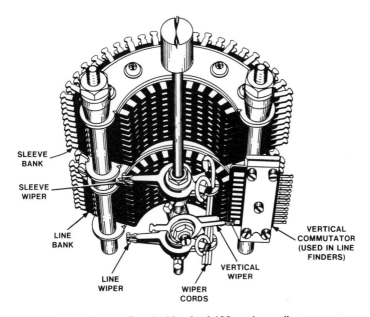

Figure 10-1. Terminal bank of 100 customer lines.

Figure 10-2 shows, in simple form, a step-by-step system interconnecting 1000 lines. For such a system, customers would be numbered from 111 to 000.[3] When a customer goes off-hook, an idle line finder locates the line requesting service through a vertical and horizontal hunt. The line finder is wired to a selector switch that returns dial tone to the caller. The selector switch moves up to the level determined by the first digit dialed (in this example, the hundreds digit). Next, the switch moves horizontally across the contacts in the selected level, hunting for a circuit to an idle connector switch in the called customer's hundreds group.

[3] Step-by-step switches respond to rotary-dial pulses from 1 to 10 (zero); in actual practice, the initial 0 is reserved as a single digit for connection to an operator.

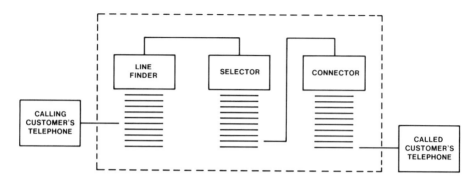

Figure 10-2. A 1000-line switching system requiring 3-digit selections.

When the second digit is dialed, the connector switch is stepped vertically to the level corresponding to the digit, thereby selecting a particular row of ten lines in that hundreds group. When the third digit is dialed, the connector switch moves horizontally across the row of terminals to the line dialed. If the line is idle, the called customer's telephone rings. If the line is busy, a busy signal is returned to the calling customer.

The step-by-step system has proven to be a popular system in the past because it is economical for basic functions and can be readily expanded as the need develops. On January 1, 1983, only 13 percent of all Bell System lines were served by this equipment, but more than 45 percent of the local switching systems (including community dial offices) were step-by-step. However, the progressive control nature of the system precludes the addition of new functions, such as *TOUCH-TONE* calling and alternate routing, without adding costly equipment to the office. Another disadvantage, or limitation, is that the customer's telephone number is determined by the physical termination (appearance) of the line or connector on the system. Customer lines cannot be moved to other terminations without changing the telephone number. Finally, the maintenance cost of electromechanical, large-motion switches is high. For these reasons, operating companies are replacing step-by-step equipment with flexible electronic switching systems (see Section 10.3).

10.2.3 NO. 1 CROSSBAR SYSTEM

The No. 1 Crossbar was developed for use in large metropolitan areas. It first went into service in 1938 in Brooklyn, New York, and over 300 systems were subsequently installed, serving more than 7 million lines. It is now being replaced by more modern systems. On January 1, 1983, 180 systems still remained, serving nearly 4 million lines.

As described in the previous section, step-by-step systems use devices that cause selector brushes to wipe over contacts in either rotary or linear motions to establish the network path progressively. The No. 1 Crossbar apparatus, in contrast, is a matrix of crosspoints associated with horizontal and vertical bars that are selected to operate an associated crosspoint within 60 milliseconds—much faster than the hundreds of milliseconds of the step-by-step switch. A common set of logic (a *marker*) makes network connections in a few tenths of a second. The marker finds a path through the network by finding a set of idle switch paths, one in each stage of the network, that can be connected serially. It then establishes the connection by direct orders to each switch. Other systems using crossbar switches preceded the No. 1 Crossbar, but this system was unique in that its markers could set connections for a call in less than 1 second and then move on to the next call. This also contrasts with step-by-step systems where the control logic—approximately six relays associated with each of four or more step-by-step switches—is retained for the duration of a call, however long it may take.

The No. 1 Crossbar derives its name from and is built around the crossbar switch shown in Figure 10-3. The name *crossbar* is derived from the use of horizontal and vertical bars to select the contacts. There are five selecting bars mounted horizontally across the front of each crossbar switch. Each selecting bar can choose either of two horizontal rows of contacts. The five horizontal selecting bars can therefore select ten horizontal rows of contacts. There are ten or twenty vertical units mounted on the switch; each vertical unit forms one vertical path. Each switch has either 100 or 200 sets of contacts, called *crosspoints*, depending on the number of vertical units.

When one or the other of the two horizontal select magnets controlling the selecting bar operates, the bar is rotated up or down. This action chooses one of the two horizontal paths available to this selecting bar by moving ten or twenty flexible selecting fingers either up or down to positions adjacent to a crosspoint on each vertical column of crosspoints. When rotated by a "hold magnet" and armature at one end, a vertical, or holding, bar along a column of contacts presses against deflected fingers in that column, closing the selected crosspoint.

After the operation of the hold magnet, the select magnet releases. This restores the horizontal bar and all selecting fingers (except that held at the closed crosspoint by an operated hold magnet) to normal. The other horizontal path can then be selected for connection to another vertical path by rotating the select bar in the other direction. In ten separate operations, up to 10 different crosspoints can be closed independently of each other in a 100-crosspoint switch. This allows ten calls to pass through a switch simultaneously, as opposed to only one call in a stepping switch.

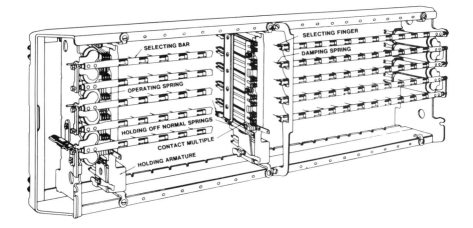

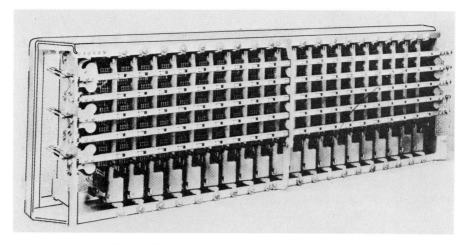

Figure 10-3. A typical 200-point crossbar switch.

Figure 10-4 is a simplified block diagram of a No. 1 Crossbar. The system uses separate originating and terminating networks. (The first stage of the line link frame is common to both networks.) A call is established as follows: On requesting service, the calling customer is connected to a *district junctor*[4] and subscriber sender. The sender provides dial tone and then receives the called number as it is dialed by the customer. It is then connected to an originating marker, which selects the trunk to be used for the call and sets up the originating-end connection (by way of additional stages of crossbar switches in the district and office link frames)

---

[4] The term *district junctor* is derived from a switching stage in panel systems that selected the district of a city to which a call was directed.

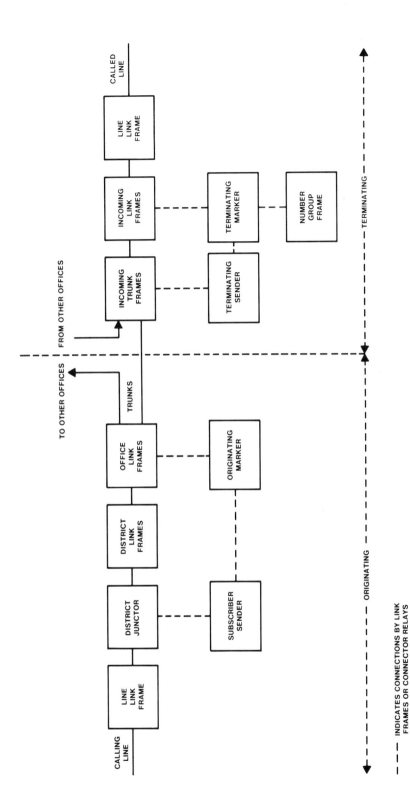

Figure 10-4. No. 1 Crossbar System block diagram.

from the district junctor to a trunk, to another office, or to the terminating network of the same office. When seized, the incoming trunk circuit at the terminating end is connected to a terminating sender; this sender registers the called number when it is sent by the subscriber sender. The terminating sender is then connected to a terminating marker, which sets up the connection of the trunk to the called customer's telephone.

Figure 10-4 also shows the *number group*. The terminating markers consult this circuit to determine the called line location. The number group also gives the marker the busy-idle status of the line being called before attempting to set up a network connection. This was an early design change that greatly increased the efficiency of the terminating markers, allowing them to handle more traffic.

10.2.4 NO. 5 CROSSBAR SYSTEM

The No. 5 Crossbar was developed to fill a need for a switching system suitable for suburban residential areas and smaller cities where it was expected that a high percentage of calls would be completed to customers in the same office. Another design consideration was the concept of direct distance dialing that required automatic recording of call details for billing purposes. The resulting system design proved to be suitable for applications and features that went well beyond original plans. First introduced in 1948, the No. 5 Crossbar went on to serve more than 28 million lines and more than 40 percent of the Bell System's telephones in the 1970s.

Figure 10-5 is a simplified block diagram of the basic No. 5 Crossbar equipment units. The equipment may be divided into two broad categories: the **switching network**, through which all talking paths are established, and the **common-control equipment**, which sets up the talking paths.

Customer lines appear on the *line link frames*, and trunks and originating registers on the *trunk link frames*. Each of these frames consists of a number of crossbar switches. Connections are established from lines to trunks or from lines to lines through an intraoffice trunk by the crossbar switches on the line link and trunk link frames. The common-control equipment used to establish the various connections includes registers, markers, senders, number groups, and connectors. A dial-tone marker sets up the connection to an originating register that provides dial tone and receives the digits dialed by the customer. When dialing is complete, a completing marker selects an appropriate idle trunk and sets up the network path to complete the call. For calls leaving the office, the completing marker connects a sender to the selected trunk so that the necessary signaling between offices can take place. For incoming calls, an

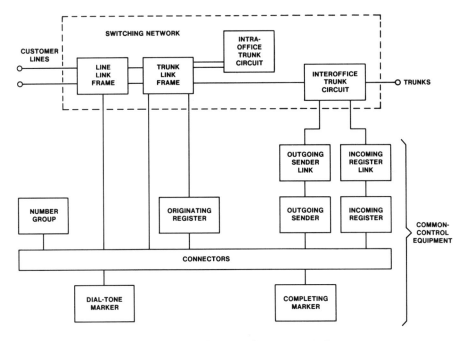

Figure 10-5. No. 5 Crossbar System block diagram.

incoming register receives the directory number of the called party from the distant office. A completing marker is required to obtain the line link frame location of the called customer from the number group. Once the connection is established, the common-control equipment is released; only the line link, trunk link, and trunk circuits (that form the transmission path) remain in the connection.

During the 30-year history of the No. 5 Crossbar, important improvements have been made and functions added:

- centralized automatic message accounting, making the No. 5 Crossbar more attractive as a small toll office.

- line link pulsing to facilitate direct inward dialing to stations served by a dial private branch exchange (PBX).

- international direct distance dialing, allowing customers to dial overseas calls with up to twelve digits.

- centrex service, including station-controlled dial transfer.

- automatic call distributor (ACD) capability. (In this capacity, a No. 5 Crossbar is used only as a large ACD for directory assistance and intercept service; it does not perform central office functions.)

- the **No. 5 Electronic Translator System**, which uses software instead of wired cross-connections to provide line, trunk, and routing translations. The system also stores billing information for transmission via data link to a centralized billing collection system.

10.2.5 NO. 4A CROSSBAR SYSTEM

No. 4A Crossbar is an electromechanical common-control system designed for toll service, with crossbar switches making up its switching network (see Figure 10-6). Trunk circuits provide signaling and transmission paths. Incoming and outgoing trunk link frames, using 4-wire crossbar switches, provide paths for connecting incoming trunks to outgoing trunks. Senders register destination codes pulsed from preceding offices and transmit the necessary information to the decoders. Decoders and translators determine the proper trunk group to the called office. Markers set up the connection from the incoming trunk to a selected outgoing trunk. A separate group of markers is associated with each specific train, as shown in Figure 10-6. A marker, sender, decoder, and translator are only associated with a call during part or all of the call setup time.

A number of improvements have been made to the No. 4A Crossbar. In 1960, centralized automatic message accounting (CAMA) equipment was added to record billing information automatically for local and toll calls at a central point. CAMA provides automatic billing of calls from customers served by local offices in which no AMA facilities are available.

In the No. 4A Crossbar, the translation function determines routing in terms of specific trunk groups from address digits.[5] It also provides other information, including the digits to outpulse and the type of outpulsing (for example, multifrequency or dial pulsing). Originally, translation was done by a device called a *card translator*; this device used phototransistors and represented the first use of transistors in Bell System equipment. In 1969, the card translator was replaced by the **No. 4A Electronic Translator System**. The No. 4A translator, a stored-program control processor, also allows the billing and route translation functions to be changed by teletypewriter input. This expedites the task of making routing changes to many switching offices in response to network changes, such as the addition of a new toll office or the creation of a new numbering plan area.

In 1973, the *peripheral bus computer* (PBC) was added. The PBC uses a minicomputer associated with the No. 4A translator to provide traffic and maintenance data. Summary reports are provided either by local cathode-ray tube (CRT) terminals or via data lines to centralized operations systems (see Chapter 14).

---

[5] The first three digits are used for calls within a numbering plan area (NPA); the first six digits are used for calls to another (foreign) NPA. Section 4.3 discusses NPAs.

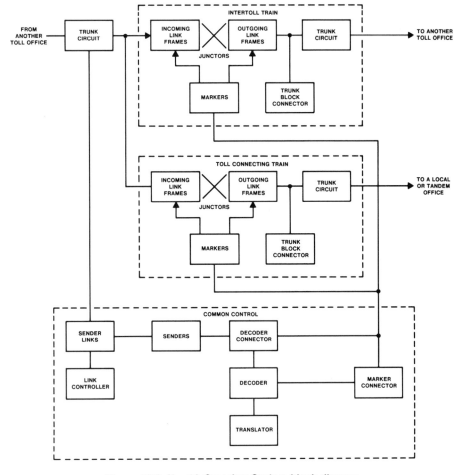

Figure 10-6. No. 4A Crossbar System block diagram.

In 1976, common-channel interoffice signaling (see Section 8.5.5) was added to the No. 4A translators to allow more efficient signaling between toll offices.

10.3 ELECTRONIC SWITCHING SYSTEMS

10.3.1 CONCEPTS

During the 1950s, Bell Laboratories began working on a new type of switching system to meet the growing demand for telephone services more economically. These *electronic switching systems*, made possible by the invention of the transistor in 1947, applied the basic concepts of an

electronic data processor, operating under the direction of a stored program, and high-speed switching networks. The stored-program control
concept allowed system designs to be far more flexible than before. With
stored-program control, it is still expensive to design new features but
relatively easy to install them.

Stored-Program Control

In most electronic switching systems, the logical steps involved in making telephone connections and providing services reside in a stored program (software). In wired logic systems, in contrast, the logic is designed
into electromechanical or electronic circuits (hardware). To change the
operation of an electromechanical system (for example, to provide a new
service), it was often necessary to redesign circuits and rewire them
extensively in the field. In electronic switching systems, new services
can frequently be implemented by making changes to the stored program. These services can be made available to customers soon after the
software has been centrally developed and distributed to the switching
systems.

The stored program controls the sequencing of operations required to
establish a telephone call, including such items as timing the duration of
signals. Thus, stored-program control can control a line or trunk circuit
according to its application, so that system cost can be reduced by minimizing the types of these circuits and the number of manufacturing wiring options needed.

The Processor

The switching network and peripheral units of electronic switching systems operate under orders received from an electronic processor.
Depending on the particular switching system, a central processor and/or
distributed processors are used. Each processor consists of general-
purpose registers and buffer circuits that perform information-processing
operations based on instructions in the stored program. These processors
are time shared by all calls they serve.

The processors used to control switching systems are similar to
general-purpose computers with some important differences. Processors
for switching systems are generally oriented toward logical rather than
arithmetic operations. Therefore, they do not require sophisticated logic
arithmetic units to multiply, divide, and perform high-order number
manipulation. Switching system processors are input/output driven and
are designed to process information in real time. That is, the system must
respond promptly to signals and data transmitted by customers and other
switching systems. Delays in processing could result in misrouted calls

or other incorrect treatment. Switching system processors must be extremely reliable. The system may be expected to provide continuous service 99.999 percent of the time—about 2 hours downtime in 40 years.

In general, processors have access to two types of memory. The first type, which contains the executable instructions, is usually protected against inadvertent program overwrites. The second type, a temporary or transient memory, is used by the system as a "scratch pad" and therefore must be fully readable and writable. The temporary memory contains information such as busy/idle status of lines and trunks, the digits being received on a particular call, billing information for an existing call, and the results of diagnostic tests.

Program Organization

The organization of the program is strongly influenced by the fact that it must operate in real time. It must also respond to trouble-detection circuits designed into the hardware to ensure dependable operation. For the program to meet all these requirements, it is necessary to establish a hierarchy of program tasks. Some tasks must be performed on a strict schedule; others may be delayed without significant adverse effects.

The central processor generally incorporates an interrupt mechanism that momentarily seizes control of the system when a demand (non-scheduled) or clock (scheduled) interrupt occurs. The interrupt circuit causes the system to stop its present program task, store the program address at which it was interrupted, and then transfer control to the appropriate program. When the interrupt programs are completed, control is returned to the program that was interrupted.

All systems of a particular type use one set of program instructions (called a *generic program*) that is the same for each office. New functions and features are provided in successive issues of the generic program, and different offices of one type may have different issues at any given time. This generic program includes all the functions necessary to cover the possible office sizes that the type of system serves and also includes means for handling changing traffic conditions and growth. The detailed differences for each individual system are listed in parameter information. This approach simplifies record keeping since only the parameter information that specifies the present office size and operating conditions is unique to each application of a switching system.

In 1975, feature loading was adopted to allow the selective inclusion of new program packages that, generally speaking, are large in size and limited in application. By this time, the designing and debugging tools and techniques had advanced to allow this capability. This made possible the addition of *ESS*-ACD (automatic call distributor) switching systems and cellular mobile telephone service with an electronic switching system as the serving switch.

Additional information that describes characteristics of each line and trunk within the office, as well as the routing and charging characteristics of the office, are found in the translations area of the program. This information is used to relate directory numbers to equipment numbers for lines and trunks, define classes of service and special treatments, and determine charging rates based on the called number.

A Typical Call

The call program analyzes a call in discrete steps. These steps start when a calling customer lifts the receiver or an incoming trunk is seized by a distant switching system. The program then proceeds through a chain of actions that ultimately establishes a talking connection. The system deals with each step of the chain in turn. The following paragraphs are a simplified description of how an electronic switching system processes telephone calls.

At a fixed periodic rate, the system checks (scans) each customer's line and each office trunk and records its state: on-hook (idle) or off-hook (dialing, busy in a talking connection, etc.). The state detected at each scan is compared with the state recorded at the previous scan. If there is no change, no action is taken. If a change is detected, the system consults the program for the action to be taken. A change of state in a line from receiver on-hook to receiver off-hook indicates that a customer wishes to place a call. The program, therefore, directs that dial tone be sent to the customer's line. Usually, this involves establishing a switching network path between the subscriber's line and a digit receiver. The digit receiver connects the dial-tone source to the calling line. Dial tone is removed upon the detection of the first received digit. A similar state change on a trunk indicates an incoming call. In either case, scanning of the digit receiver continues at regular intervals, and each digit is recorded in turn.

After the system registers the digits, it again consults the program. For an intraoffice call or an incoming call to a line, the next step is to determine, by scanning that line, the state of the called telephone. If the called telephone is busy, the program causes a busy tone to be sent to the calling party. If the called telephone is idle, the program causes audible ringing tone to be sent to the caller and the called telephone to begin ringing. For calls to stations on distant switching systems, the program directs the seizing of a trunk and the transmission of the called number.

If the call is answered, this state is registered, and the ringing signal and audible ringing tone are removed. A talking path is then established between the calling and called terminals. The lines, or line and trunk, involved in the call continue to be periodically scanned. No action is taken until one of the parties hangs up, at which time the program directs that the connection be taken down.

Networks

An electronic switching system may use either space-division or time-division switching networks (see Section 7.3 for general descriptions of these types of networks). The principles of stored-program control can be applied independently of the network selected. The type of network chosen for a system depends largely on the characteristics of the line and trunk transmission environment in which the system will be used. Space-division networks are economically compatible with analog line and trunk transmission interfaces. Time-division networks have significant advantages where transmission facilities employing digital multiplexing[6] are expected to dominate. Use of a space-division switching network in areas dominated by digital transmission facilities or use of a time-division network in an analog transmission environment requires extensive use of digital-to-analog and analog-to-digital converters at the interface between the transmission facilities and the switching system.

10.3.2 EVOLUTION OF ELECTRONIC SWITCHING

The Bell System's first trial of electronic switching took place in Morris, Illinois, in 1960. The Morris trial culminated a 6-year development and proved the viability of the stored-program control concept. The first application of electronic local switching in the Bell System occurred in May 1965 with the cutover of the first 1ESS switch in Succasunna, New Jersey.

The 1ESS switching system was designed for use in areas where large numbers of lines and lines with heavy traffic (primarily business customers) are served. The system has generally been used in areas serving between 10,000 and 65,000 lines and has been the primary replacement system for urban step-by-step and panel systems. The ease and flexibility of adding new services made 1ESS switching equipment a natural replacement vehicle in city applications where the demand for new, sophisticated business and residence services is high.

The need for economical electronic switching in the 2000- to 10,000-line offices was met with the introduction of the 2ESS switching system in 1970. While there are many similarities between the 1ESS and 2ESS switching systems (for example, stored-program control and switch elements), there are also important differences: The 2ESS switching equipment network architecture was designed to interface with customer lines

---

[6] Section 6.5 discusses multiplexing.

carrying lighter traffic, the features were oriented toward residential rather than business lines, and the processor was smaller and less expensive. The market for 2ESS switching equipment was suburban residential areas previously served by step-by-step systems and new suburban wire centers.

Electronic switching was extended to rural areas serving fewer than 4500 lines with the introduction of 3ESS switching equipment in 1976. Again, the 3ESS switching equipment design borrowed concepts from 1ESS and 2ESS switching equipment but implemented a less expensive processor and more economical network architecture.

In 1976, the first electronic toll switching system to operate a digital time-division switching network under stored-program control, the 4ESS system, was placed in service. It used a new control, the 1A processor, for the first time to gain a call-carrying capacity in excess of 550,000 busy-hour[7] calls. The toll transmission environment was experiencing an accelerated growth in digital time-multiplexed facilities. A toll stored-program control system with a digital time-division network provided large economic advantages over its space-division electromechanical predecessors.

The 1A processor was also designed for local switching application. It doubled the call-carrying capacity of the 1ESS switching system and was introduced in 1976 in the first 1AESS switch. The 1A processor was designed to be retrofitted into a working 1ESS switch. The network capacity of 1ESS switching equipment was also doubled to allow the 1AESS switch to serve 130,000 lines. Similarly, a new processor, added to 2ESS switching equipment in 1976, doubled its traffic capability.

In 1981, local digital time-division switching systems first began service in the Bell System. Northern Telecom's DMS[8]-10 switching system serves in rural and suburban offices and is more cost effective at these office sizes than comparable space-division switching systems. A second digital time-division switching system, the 5ESS system, began service in 1982 in local offices in Bell operating companies.

Electronic switching was also used to modernize toll operator functions. The stored-program control concepts and much of the 1ESS switching equipment hardware were used to implement the first Traffic Service Position System (TSPS) in 1969. Like other electronic switching systems, the TSPS has undergone many technological upgrades since 1969, including the application of a new processor. (Section 10.4.1 describes TSPS.) Modernization of operator intercept service with stored-program control concepts and automatically generated announcements began with the Automatic Intercept System (AIS) in 1970. (Section 10.4.2 describes AIS.)

---

[7] Section 5.2.3 discusses the busy-hour concept.

[8] Trademark of Northern Telecom, Ltd.

10.3.3 SPACE-DIVISION ELECTRONIC SWITCHING SYSTEMS

1ESS and 1AESS Switching Equipment

1ESS Switching Equipment Processor. Figure 10-7 shows the 1ESS switching equipment processor community. It includes a fully duplicated No. 1 Central Processor Unit (central control), program store bus, call store bus, program stores, and call stores. This duplication allows full interchangeability; for example, each processor has access to all the busses, and both call store busses and program store busses have access to both of their duplicated memory units.

The 1ESS switching equipment uses *permanent magnet twistor* program store modules as basic memory elements. These provide a memory that is fundamentally *read only* so that neither software nor most hardware malfunctions can alter the information content. Program stores contain 131,072 words, each forty-four bits long (thirty-seven bits of information and seven bits for error-correcting coding). The program store has a cycle time of 5.5 microseconds.

Call store provides "scratch pad" or temporary duplicated memory in the 1ESS switch for the storage of information related to the progress of calls and the status of equipment. Originally, 8192 words, each twenty-four bits long, were provided per call store in the form of ferrite sheets. Later, call stores containing 32,768 words were developed using core memory. Both types of memory are readable and writable.

The 1ESS switching equipment may optionally be equipped with a duplicated signal processor, a duplicated signal processor call store bus, and duplicated signal processor call stores. The signal processor performs

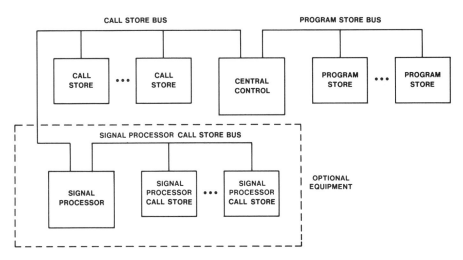

Figure 10-7. 1ESS switching equipment processor community with optional signal processor.

many of the input/output functions, thus relieving the central processor
of this work and, thereby, increasing the call-processing capacity of the
system. The signal processor call stores contain the signal processor program instructions.

1A Processor. The 1A processor community is, in some ways, similar to
the control processor of the 1ESS switching equipment. As with the 1ESS
switching equipment, the central processor, program store bus, and call
store bus are fully duplicated. However, the 1A processor uses readable
and writable memory for both program store and call store memory.
Originally, 65,536-word core-type stores were used, but 262,144-word
semiconductor stores are now available. Each word contains twenty-four
bits of information and two parity-checking bits. The cycle time of the
1A processor with the faster semiconductor stores is 700 nanoseconds,
compared to 5.5 microseconds in the 1ESS switching equipment central
control and 1.4 microseconds in the 1A processor operating with core
stores.

Unlike the 1ESS switch, program stores are not fully duplicated in the
1AESS switch, but two spare stores are provided for reliability. A portion
of the 1A processor call store memory is duplicated. However, only one
copy of certain fault recognition programs, parameter information, and
translation data is provided. Another copy of the unduplicated program
store and call store information is provided in file store. File store uses a
disk bulk storage device that can be used to load either the spare program
stores or call stores. In this way, additional reliability is provided. In
addition, magnetic tape units in the 1A processor provide for system reinitialization and detailed call billing functions.

Networks and Periphery. The 1/1AESS switches use the same peripheral
equipment. This allows for the transition from a 1ESS switch to a 1AESS
switch as the office capacity requirements increase, without the replacement of the entire switching system. The major peripheral equipment
items are: the **switching network and junctors**; the **scanners, signal distributors, and central pulse distributors**; and the **line, trunk, junctor,
and service circuits**. The following paragraphs discuss these items in
more detail. Peripheral equipment typically makes up 90 to 95 percent of
the 1/1AESS switching equipment.

Switching Networks and Junctors. The 1/1AESS switching equipment
networks perform two major functions: They provide a means for interconnecting lines, trunks and service circuits, and they match the relatively lightly used lines to the comparatively heavily used trunks and service circuits.

Figure 10-8 depicts the 1/1A*ESS* switching equipment *switching networks*. They are composed of two 4-stage switching arrays (line link networks [LLNs] and trunk link networks [TLNs]), interconnected by wires called *junctors*,[9] to provide a total of eight stages of switching.

Lines (generally associated with telephones) are connected to one side of a line link network. Because of the statistical characteristics of telephone traffic and the relatively small amount of traffic generated by a typical line, the LLN does not have to be a nonblocking network. That is, it does not have to provide paths to all customers simultaneously. Therefore, the LLN has a fixed concentration ratio between the number of lines terminated on one side and the number of junctors terminated on the other side. Standard 1/1A*ESS* switch concentration ratios are 2 to 1, 3 to 1, 4 to 1, and 6 to 1. All are based on the expected customer usage characteristics of an office. The smaller ratios are generally associated with urban offices. By comparison, a trunk link network, because of the heavier usage of trunk circuits and service circuits, uses a 1-to-1 or 3-to-2 concentration ratio between the number of trunks and service circuits and the number of junctors.

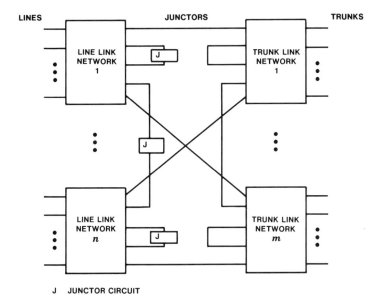

J   JUNCTOR CIRCUIT

**Figure 10-8. 1/1A*ESS* switching equipment
network structure.**

---

[9] Except that the connections from LLN to LLN contain *junctor circuits* to provide control access and power for line-to-line calls.

The 1/1AESS switching equipment switching networks employ a coordinate network (see Section 7.3.2) and use a ferreed or remreed switch as a crosspoint element. These crosspoint switches are compatible with both the existing outside plant (by providing metallic interconnection) and the high-speed electronic central control. The remreed crosspoint, the newer version of the two, is more compact and thus requires less floor space.

Scanners, Signal Distributors, and Central Pulse Distributors. Figure 10-9 depicts the communication links between the 1/1A processors and the peripheral circuits. Every telephone switching system embodies some mechanism for detecting service requests and for supervising calls in progress. Input information of this nature is furnished to the 1/1AESS switch by the operation of *scanners* that sample or scan lines and trunks at discrete intervals of time as directed by the system. The result of any given scan is generally compared to the previous scan to determine if any change in state (on-hook to off-hook, for example) has occurred. The sensing element used in all 1/1AESS switching equipment scanners is the ferrod sensor, a current-sensing device that operates on electromagnetic principles. It consists of a ferrite rod whose magnetic field changes depending on the state of the line or trunk being scanned.

Signal distributors translate orders received from central control into high-power, long-duration pulses that are distributed to the appropriate

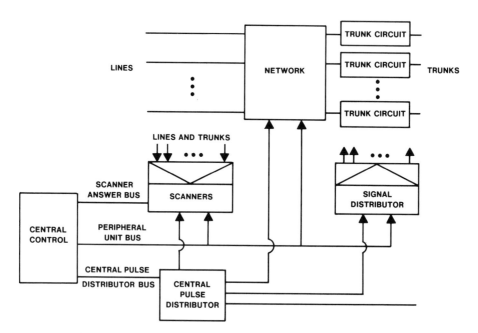

Figure 10-9. **1/1A**ESS **switching equipment peripheral systems.**

relays in trunk, service, and power control circuits in 1/1AESS switching equipment. These relays are controlled by polarized signals and are magnetically latched (held in an operated state), thus providing a memory function in the end device. The signal distributor has 1024 outputs. The decoding of the order from central control to provide access to one of these outputs is performed by relay contacts. Relay contacts were selected because, at the time of original design (around the early 1960s), no electronic device could economically compete with a contact on a large relay for such a decoding function when the required access cycle was about 20 milliseconds. Such was the case for most of the relays in 1/1AESS switching equipment trunk circuits as well. Since then, fully electronic signal distributors have been designed to replace the original signal distributor design. Western Electric began to manufacture those new distributors in 1975.

In electronic switching systems, some control functions must be carried out at electronic speeds or at speeds exceeding the capability of the magnetic latching relays controlled by the signal distributors. These control functions are provided by a *central pulse distributor* (CPD). In 1/1AESS switching equipment, a diode-transformer gate was chosen as the decoding element in the CPD. Two primary functions of the CPD are the enabling of other peripheral units to receive orders from the central processor and the control of outpulsing over trunks. The transformer provides a balanced output, so a pulse from such an output point can be transmitted over a twisted pair to remote locations without interference. In addition, bipolar pulses can be easily generated and transmitted. These pulses can control the operation and release of a relay over a single pair of wires by using a receiving device (called a *bipolar flip-flop*) that can recognize the two polarities.

Line, Trunk, Junctor, and Service Circuits. Individual circuits are required on a per-call and even per-line basis to match the widely variable outside world to the standardized "inside world" of the central processor. These individual circuits are the line, trunk, and service circuits. A *line circuit* is connected to each line. It provides initial battery power to the line as well as an indication of on-hook or off-hook state to the scanners and allows the removal of the scan elements during connection to the network to minimize transmission losses. A *trunk circuit* is connected to each trunk. It provides power to the associated line after the connection is established, provides trunk supervision signaling (an indication of the on-hook or off-hook state to the scanners), and isolates the network reed contacts during switching intervals. *Junctor circuits* serve in place of trunk circuits for the supervision of intraoffice calls. *Service circuits* include circuits that receive dial pulses or signals from *TOUCH-TONE* service, circuits that ring the called line, circuits that provide an audible ringing indication to the calling line, circuits that provide a busy tone, and circuits that transmit and receive multifrequency pulses. Each circuit performs only a few functions under program control, and different circuits are connected to the communication path as needed via the switching network.

Maintenance. The 1/1A*ESS* switching equipment reliability objective specifies that system downtime should total no more than 2 hours in 40 years. Because some failures of individual components are bound to occur over decades of system service, duplication is essential to meet this objective. Every system unit required to maintain service in 1/1A*ESS* switches is provided in duplicate, and troubles are found and corrected quickly in order to minimize the possibility of system failure caused by multiple troubles.

Somewhat less than half of all the stored-program instructions in 1/1A*ESS* switches are used for maintenance. Some of these programs, in conjunction with logic wired into the hardware, detect and report faults and troubles. Other programs control routine tests, diagnose troubles, and control emergency actions to ensure that the system operates satisfactorily, either by eliminating faulty subsystems or by reorganizing usable subsystems into a new operating combination. When trouble occurs, telephone switching actions are interrupted as briefly as possible to reestablish an operational system. Then, on a less urgent basis, the defective unit is diagnosed by the system itself, and the results are printed on the maintenance terminal—a teletypewriter for the 1*ESS* system and a CRT/keyboard terminal[10] for the 1A*ESS* switch.

Services. In addition to local telephone service, the 1/1A*ESS* switches offer a variety of special services. The Custom Calling Services described in Section 2.4.2 are available to all customers who are served by electronic switching systems. Business customers may select offerings such as centrex, *ESS*-ACD, Enhanced Private Switched Communications Service, or electronic tandem switching. (Chapter 2 discusses these services.)

2*ESS* and 2B*ESS* Switching Equipment

The 2*ESS* switching equipment was designed to extend electronic switching economically into the suburban market. Its first application was in 1970 in Oswego, Illinois. The 2*ESS* switch has a call capacity of 19,000 average busy season busy hour calls,[11] with a maximum of 24,000 terminals per system.

[10] Teletype Corporation's *DATASPEED* 40 terminal, described in Section 11.1.3, is used.

[11] Section 5.2.3 discusses the average busy season busy hour concept.

Network. The same design principles used in the development of 1*ESS* switching equipment were applied in 2*ESS* switching equipment. As in 1*ESS* switching equipment, the network in the 2*ESS* switching equipment uses eight stages of switching. The original network element was the ferreed switch, later replaced by the remreed switch. The 2*ESS* switching equipment network differs from the 1*ESS* switch in that lines and trunks terminate on the same side of the network. This is referred to as a *folded network*. There is no need for separate line and trunk link networks as in the 1*ESS* switch; however, a junctor circuit for power feed and supervision must be included in line-to-line connections. The 2*ESS* switching equipment allows for either a 2-to-1 or 4-to-1 concentration ratio.

Processor and Program. The 2*ESS* switching equipment processor is duplicated and constructed of transistor-resistor logic. Its memory is organized similarly to that of 1*ESS* switching equipment with ferrite sheet call stores and magnetic twistor card program stores. Unlike the 1*ESS* switching equipment, 2*ESS* memory units are associated with one of the dual processors and cannot be switched between them. To make stored-program control economically attractive for suburban local central office applications, the hardware and software design for the 2*ESS* system emphasized savings in call store and program store instead of real time.

The need for new and more complex features and the availability of improved technology led to the modernization of the 2*ESS* switching equipment. This was accomplished by the development of the 3A central control (3ACC), which replaces the 2*ESS* switching equipment processor. The 2B*ESS* switch is a 2*ESS* switch with a 3ACC as its processor. Older 2*ESS* switches can be updated by a processor retrofit procedure similar to that for 1A*ESS* switching equipment.

By combining integrated circuit design with semiconductor memory stores, the 3ACC doubles the call capacity originally available in the 2*ESS* switch. In addition to this advantage, the 3ACC requires one-fifth the floor space and one-sixth the power and air-conditioning that the 2*ESS* switching equipment central processor requires. Because of its state-of-the-art processor, newer features (such as automated data linking of automatic message accounting data) are available only on the 2B*ESS* and not on the older 2*ESS* switching equipment.

The 3ACC is a self-checking, microprogram-controlled processor capable of high-speed serial communication. Its average instruction execution time is 1.25 microseconds. Its memories are not separated by function as in 1*ESS*, 1A*ESS*, or 2*ESS* switching equipment. Instead, there is a single, insulated-gate, field-effect transistor main store, with portions allocated for call data, resident programs, and translation data. The resident programs are hardware write protected. Nonresident programs, such as maintenance and recent change, and backup for translations and resident programs are stored on a tape cartridge.

The first 2BESS switching equipment application was in Acworth, Georgia, in 1976. That was the same year that the first 1A and 3ESS switching equipment went into service. (The 3ESS switch also uses the 3ACC, as is explained below.)

3ESS Switching Equipment

The 3ESS switching equipment is the smallest Western Electric space-division, centralized electronic switching system, serving 2000 to 4500 lines. It was designed to meet the needs of a typical community dial office, with one busy hour call per line and approximately 65-percent intraoffice calling. Like the 2BESS switch, 3ESS switching equipment uses the 3ACC as its processor. As explained in the previous section, the memory configuration has microcoded instructions stored in read-only memory. The 3ESS switching equipment network uses remreed technology with a fixed 6-to-1 concentration ratio. As in 2/2BESS switching equipment, it is a folded network, with a maximum of 5760 terminals. The 3ESS switching equipment uses five switching stages on every connection.

The 3ESS switch was designed for unattended operation. To accomplish this, extensive maintenance programs are built into the software. Also, a sophisticated remote maintenance capability connects the Switching Control Center System (see Section 14.3.2) to the 3ESS switch.

One of the unique features of the 3ESS switch is the ability to "hot slide" at installation. Normally, when replacing an older system, the new system is assembled beside the old and brought into service while the old system continues to operate. This requires enough available building space for two complete switching systems. However, the 3ESS switching equipment is assembled at the factory, brought to the installation site on a platform, and put into service outside the building. The older system can then be taken out of service, removed from the building, and the new system "slid" into the building. A great savings in construction is realized. This type of installation is also available on a limited basis with the 2BESS switching equipment.

No. 10A Remote Switching System

A survey taken in the mid-1970s showed that approximately one-half of all switching systems in the Bell System served 2000 or fewer lines and that a large portion of these systems were No. 355A step-by-step systems, yet no electronic switching system had been developed to serve this market economically. The principal obstacle to electronic switching system penetration into small offices was the cost of a processor and its associated memory. Remote switching avoids this problem by sharing the processing capabilities of a nearby electronic switching system and using a microprocessor for certain control functions under direction of the host central processor.

Design Philosophy. The No. 10A Remote Switching System (RSS) is designed to act as an extension of a 1*ESS*, 1A*ESS*, or 2B*ESS* switching equipment host and is controlled remotely by the host over a pair of dedicated data links. Figure 10-10 illustrates the interconnection arrangement used between a No. 10A RSS and an electronic switching system host. Customer lines are terminated at the RSS in the conventional manner. The voice pairs are concentrated at the RSS and connected to the host system over T- or N-carrier facilities.[12] Outgoing and incoming calls are processed by the host system. Intra-RSS calls are initially set up through the host, with the final voice connection through the RSS only. However, if a data link between the host and RSS is severed, the RSS is capable of stand-alone operation for intraoffice calls. This is an important feature in the rural market, where as many as 65 percent of the calls placed are intraoffice calls.

Whether the call is outgoing or intraoffice, the RSS and host system take the same initial steps. Customer lines are scanned by the RSS for off-hook. When an off-hook occurs, the RSS asks the host system to set up a call path. The electronic switching system locates an idle channel in the T- or N-carrier, an idle digit receiver within itself, and a path from the receiver to the channel through its network. The RSS then sets up a path from the customer's line to the channel. The customer's call is now handled by the electronic switching system as if there were no RSS involved.

The RSS can handle a maximum of 2048 customer lines. It consists of termination and subscriber loop interface circuits for each line, signaling circuits, digit receivers, a data link to the host system, a switching network, carrier equipment including multiplexers, and a *WE* 8000 microprocessor and memories. The subscriber loop interface circuit feeds power to its customer line and monitors on-hook and off-hook signals. This circuit is scanned by the RSS's microprocessor, which is directed by the control programs contained in erasable, programmable, read-only memory. The control programs are activated by data link signals from the host electronic switching system. The RSS also has a short-term random-access memory to store data needed for call processing.

---

[12] Sections 9.3 and 9.4 discuss carrier systems.

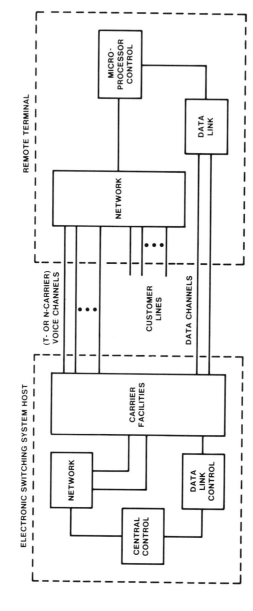

Figure 10-10. Block diagram of RSS host with T-carrier interconnection links.

When the duplicated data link between the RSS and its host fails, the RSS switches to a stand-alone mode. Only basic service on intra-RSS calls is allowed during stand-alone operation; custom calling, billing, traffic measurements, and similar services normally provided by the host are not available.

When operating normally, the host passes a message to the RSS every 10 seconds over the data link. Upon data-link failure, the RSS waits 30 seconds, then initiates its own call-processing procedures. The RSS has a copy of its line translations that is updated periodically. Established intra-RSS calls are maintained when the failure occurs, but calls in the process of being set up at the time of failure are lost.

The network in an RSS is composed of PNPN solid-state switching devices that draw little power and are small enough to fit on a standard circuit pack. However, because of the switch's low voltage tolerance, high-voltage signals such as ringing are switched through a separate high-voltage network.

10.3.4 TIME-DIVISION SWITCHING SYSTEMS

4ESS Switching Equipment

Design Objectives. The motivation for designing an electronic tandem switching system is basically the same as that for a local electronic switching system: to lower first-cost and operating expenses relative to an electromechanical design and to provide a flexible system that can be adapted to changing needs. An obvious approach to this challenge would have been to use the basic 1ESS switch control system with the substitution of 4-wire switching frames and appropriate trunk circuits. This approach was studied and abandoned because:

- Rapid growth of toll traffic and the penalties associated with multiple toll systems in a metropolitan area pointed to the long-range need for a system of very large capacity. The 1ESS switch would have required substantial increases in both processing and network capabilities.

- Since digital (pulse-code modulation[13]) transmission was predicted to become dominant in the toll network, it appeared economically attractive to use a digital time-division switching network in the toll machine.

Other important advantages of a time-division network in a large toll switching system arise from considerations of the installation costs of space-division networks and the cost of trunk rearrangements. One factor in both of these costs in space-division systems is the extensive

---

[13] Section 6.4.3 discusses pulse-code modulation.

cabling needed to connect a large number of network frames together and to many other functional units. With the 4*ESS* switching equipment, these costs are reduced through the integrated modular design of toll terminal equipment and switching equipment and by switching multiplexed signals to reduce the number of interframe conductors required. This also makes the rapid interconnection of frames using precut connectorized cables more economical. The choice of time-division switching for the network results in an economically practicable, nearly nonblocking network that obviates the need for equipment rearrangements to avoid network congestion.

Features. The 4*ESS* switching equipment is a large-capacity 4-wire tandem system for trunk-to-trunk interconnection. Initially, it handled 550,000 peak busy hour calls (for a typical call mix using the 1A processor equipped with core stores), 107,000 terminations, and 1.8 million hundred-call seconds (CCS) per hour. The 4*ESS* switching equipment forms the heart of the stored-program control network that uses common-channel interoffice signaling, while still supporting multifrequency and dial-pulse signaling. Stored-program control network features provided by 4*ESS* switching equipment include the Mass Announcement System that supports *DIAL-IT* network communications service (see Sections 2.5.1 and 11.3.2) and expanded inward Wide Area Telecommunications Services screening and routing. Further stored-program control network capabilities will be provided with future generics. The 4*ESS* switching equipment also provides international gateway functions.

Capacity. Rapid growth in toll traffic had created situations in metropolitan areas in which two or more of the available toll switching systems were needed to provide adequate capacity. When multiple toll offices serve the same area, trunking penalties result. In view of these situations and the projected continuing growth of toll traffic, an important objective in development of 4*ESS* switching equipment was to achieve a substantial increase in capacity.

The design of 4*ESS* switching equipment grew out of the 1*ESS* switching equipment concept in that it uses a high-speed processor as central control to handle the most complex aspects of call completion. Signal processors are provided for preprocessing the more elementary tasks, thus decreasing the per-call usage of the main processor. Through its use of core memories and higher speed logic, the 4*ESS* switch main processor, the 1A processor, is about five times as fast as the 1*ESS* switching equipment processor.

A large-scale integrated circuit semiconductor memory was developed to replace the 1A processor core memories and deployed in 1977. It provided a 3-to-1 size reduction and a 6-to-1 reduction in power consumption and was easier to maintain. Subsequent improvements in technology have resulted in further reductions in size and power consumption. Increased speed of operation has resulted in up to 30-percent increases in the call-carrying capacities of the 1AESS and 4ESS switches. These improvements began to appear in the field in 1979.

Software Structure. The 4ESS switching equipment software structure is based on a centralized development process using three languages: a low-level assembly language, an intermediate language called EPL,[14] and a high-level language called EPLX. The high call-processing capacity of 4ESS switching equipment is sustained through the judicious application of these languages. Real-time functions such as call processing are generally programmed in the assembler language, while measurements and administrative functions frequently are programmed in EPL. Some of the maintenance programs and audits that are not frequently run are programmed in EPLX.

The 4ESS switching equipment has an executive control loop operating system. The loop consists of short segments of low-priority call-processing tasks. High-priority call-processing tasks may be interjected at the end of any low-priority segment. This mode of operation contrasts with interrupt-driven schedulers found in the earlier systems. The 4ESS system executive control avoids the bookkeeping overhead and much of the memory-writing conflicts between high- and low-priority tasks.

In order to meet the stringent electronic switching system reliability objectives, highly defensive coding and an extensive audits structure have been used. Structured programming techniques have been and will continue to be used for development of the 4ESS switching equipment software generics.

Administration and Maintenance. Based on experience with the 1ESS switching equipment, greater emphasis was given to easing the operation and administration of the 4ESS switching equipment. For example, CRT input/output terminals permit an interactive mode of communication between human and machine. Low-cost, large-capacity magnetic disk memories (see Figure 10-11) store within the system data that formerly were kept as paper records. This permits activities such as assigning a trunk to a terminal to be done by the system rather than manually. The chosen network design maintains low blocking, even with undistributed loads; consequently, physical wiring changes traditionally associated with the periodic redistribution of trunks over terminals (required to intermix lightly and heavily loaded trunks) have been eliminated. Magnetic tapes controlled by the Auxiliary Data System provide for the output of traffic data, performance data, and billing records, as well as for the input of new program and translation information.

---

[14] Electronic Switching System Programming Language.

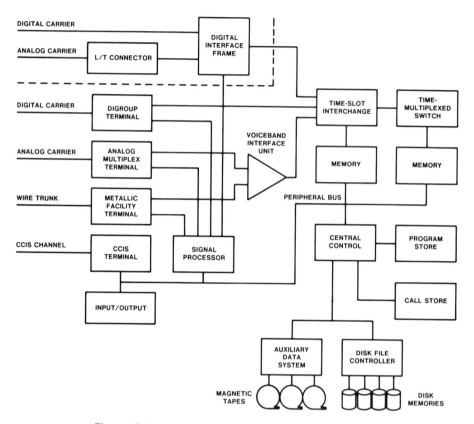

Figure 10-11. 4ESS **switching equipment block diagram.**

Up to six 4ESS switches can be remotely administered and maintained from centralized work centers. Remote trunk testing, data-base administration, and switching maintenance mean that very few functions must be performed on site.

Time-Division Switching. Because of the rapid growth of digital transmission systems in the network, the switching network in 4ESS switching equipment has been designed specifically to pass pulse-code-modulated (PCM) signals without conversion. This pays off on interfaces with the toll connecting facilities, where short-haul (up to 50-mile) PCM systems are already in extensive use. Short-haul intertoll trunks (up to 500 miles) using PCM are also in service and are expected to grow rapidly. Long-haul intertoll trunks are still predominantly analog, but long-haul digital transmission systems are now being enthusiastically introduced.

For analog transmission systems, the 4ESS switching equipment design provides for conversion of analog signals to digital form and vice versa.

Figure 10-11 shows these arrangements. The equipment below the dotted line indicates the original *4ESS* switching equipment configuration. The triangular shape of the converter, known as a *voiceband interface unit* (VIU), is intended to signify the multiplexing of 120 voice channels onto one digital path. Connections to trunks on PCM facilities do not need conversion and, therefore, use a much simpler interface, the *digroup terminal*. (*Digroup* is a contraction of *digital group*; in present systems, a digroup consists of 24 voice channels.) The digroup terminal (DT) multiplexes digital groups to put 120 voice channels onto one conductor and separates out supervisory signaling information. Trunks interface via the *metallic facility terminal* frame before conversion to PCM. Beyond these interface units, all signals are in PCM format and can be switched as required in the time-slot interchange (TSI)/time-multiplexed switch (TMS) complex under control of circulating solid-state memories.

Starting in 1981, the voice interface units have been replaced by the digital interface frame (DIF) shown above the dotted line in Figure 10-11. Digroups can terminate directly on the DIF, eliminating the need for digroup terminals, while analog carriers go through an L/T connector that converts from an analog (L3, L4, L5) to a digital carrier (T1). Metallic facilities go through a channel bank that converts to digital carrier. The DIF converts five digroups into one digital stream of 120 voice channels for input to the TSI.

Figure 10-12 shows the TSI-TMS complex in simplified form. The TSI associated with an incoming trunk stores the incoming coded PCM sample until the time slot selected for cross-office transmission of this call comes up. When it does, the TMS is configured by the circulating memory to provide a path to the TSI associated with the appropriate outgoing VIU, DT, or DIF. The coded PCM sample is thus sent through TSIs from the incoming VIU, DT, or DIF to a storage register in the TSI associated with the outgoing VIU, DT, or DIF, where it is held until the time slot corresponding to the appropriate outgoing trunk comes up. A coded sample is transferred from the outgoing trunk to the incoming trunk in the same way in the same time slot. The cross-office time slot is selected independently of the time slots corresponding to the incoming and outgoing trunks. The space-division portion of the switch (the TMS) is reconfigured for each time slot (about 10^6 times per second), in contrast to a conventional space-division switch that is reconfigured only in response to the arrival and departure of calls. (Section 7.3.3 discusses time-division switching further.)

Each TSI has 8 input and 8 output coaxial leads connecting to the TMS. Figure 10-12 shows only input leads from TSI 1 and output leads to TSI 128. The 1024 input leads (8 from each of 128 TSIs) and 1024 output leads are interconnected by the TMS, which is a 1024-input—to—1024-output matrix. Since each telephone connection is made up of two paths through the TMS, one for each direction of transmission, the maximum

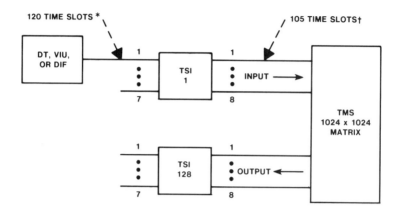

DIF	DIGITAL INTERFACE FRAME
DT	DIGROUP TERMINAL
TMS	TIME-MULTIPLEXED SWITCH
TSI	TIME-SLOT INTERCHANGE
VIU	VOICEBAND INTERFACE UNIT

* Actually 128 time slots. Eight reserved for maintenance, leaving 120 for telephone connections.

† Actually 128 time slots. On the average, only 105 can be occupied because of the 7-to-8 expansion in the TSI.

Figure 10-12.   Diagram of TSI-TMS complex.

number of simultaneous telephone connections that can be accommodated may be found by multiplying the number of time slots on each TSI lead by the number of possible connections through the TMS and dividing the answer by 2:

$$105 \times \frac{1024}{2} = 53,760 \text{ connections.}$$

Floor-Space Savings.   An important part of an operating telephone company's initial cost for switching is the building to house the equipment.   The 4*ESS* switching equipment is very compact compared to a No. 4A Crossbar; the initial 4*ESS* system occupies about one-fifth the floor space for equivalent systems serving 10,000 to 20,000 trunks.   This comparison includes all equipment and work space and assumes small crossbar switches in the No. 4 Crossbar.   Subsequent designs of 4*ESS* systems accomplished significant further reductions in size; recent designs occupy about one-fourth the floor space of the initial 4*ESS* system.

DMS-10 Switching System

The *DMS*-10 system is a digital time-division switching system for local offices.   It was designed for use in community dial office applications. The system has a 13,000-per-hour peak call capacity; line size ranges from

200 to 6000 subscriber lines. Its first standard application in the Bell System was in March 1981. Flexibility is emphasized through modularity in software and hardware architecture. Hardware equipment uses printed circuit packs that plug into printed circuit backplanes for easy and rapid replacement and growth. The *DMS*-10 system is designed for service in an unattended rural office. Figure 10-13 is a functional block diagram of the *DMS*-10 system.

Processor Complex. The system architecture is organized into system control, the switching network, and peripheral equipment. System control is provided by a fully duplicated central processing unit (CPU). The two processors operate in active/standby mode with automatic switchover control. Two cartridge tape units also provide system backup. The primary tape unit stores complete system software and office data; the secondary tape contains overlay software consisting of infrequently used programs. System memory for program store, call store, and data store is subdivided into 32K-word blocks, each block packaged as one printed circuit pack. Memory for the Custom Calling Services resides on a separate 32K-word module. For reliability, $n+1$ *redundancy* is used, whereby one spare memory block is provided for every n ($n \leqslant 8$) blocks of memory. A duplicated high-speed control bus interconnects the CPU with memories, input/output devices, and network modules.

Network Complex. The switching network is composed of one to four network groups, each serving approximately 2500 lines and associated trunks and service circuits. Intergroup traffic is carried by digital junctors. The network also provides service circuits for tone generation, dial-pulse and multifrequency transmission, conference calls, terminal input/output, various test functions, and scanning of the peripheral equipment (PE). The PE modules provide the hardware interface to subscriber lines, trunks, and service and maintenance circuits. The analog-to-digital and digital-to-analog conversions are provided in the PE by one channel codec (coder-decoder) for each individual subscriber-line interface. Custom large-scale integration logic performs the time multiplexing of speech signals from any 1 of 112 analog terminals of two PE modules to any one of thirty channels of a digital multiplex (MUX) loop. MUX loops provide thirty PCM voice channels and two signaling channels to carry the digitized speech and signaling information from the PE to the network. Sparing of MUX loops occurs at the PE end; one MUX loop serves two PE shelves, with the capability to handle the load on four shelves should its "mate" fail.

Digital trunks terminate on a digital carrier module (DCM) in the *DMS*-10. DCMs convert the internal 2.048-megabits-per-second (Mbps) transmission format of the MUX loops to the 1.544-Mbps format of standard digital signal level 1 (DS1) interfaces, along with the time-slot switching from thirty to twenty-four channels. There is one dedicated

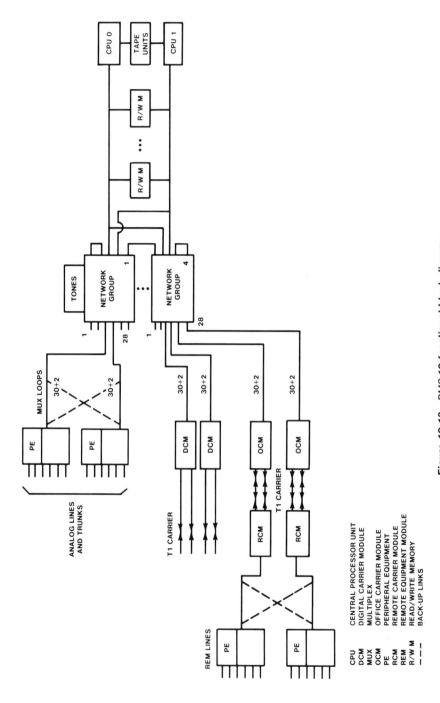

Figure 10-13. *DMS-10 functional block diagram.*

CPU — CENTRAL PROCESSOR UNIT
DCM — DIGITAL CARRIER MODULE
MUX — MULTIPLEX
OCM — OFFICE CARRIER MODULE
PE — PERIPHERAL EQUIPMENT
RCM — REMOTE CARRIER MODULE
REM — REMOTE EQUIPMENT MODULE
R/W M — READ/WRITE MEMORY
– – – — BACK-UP LINKS

MUX loop per DCM, each DCM accommodating up to twenty-four digital trunks.

Use of a remote equipment module (REM) provides the capability to extend the internal MUX loop functions to PE modules located up to 70 miles away from the central office. Two MUX loops, terminated on an office carrier module (OCM) at the central office, are interconnected by a dedicated DS1-type carrier to the two MUX loops from the remote carrier module (RCM) at the remote location. Thus, the REM handles traffic from four PE modules corresponding to two paired MUX loops, or 224 subscriber lines and analog trunks. Since the same peripheral equipment is being used, the same central office physical environment is needed at the REM location.

The *DMS*-1 digital remote concentrator serves as a pair-gain multiplexer when operating with the *DMS*-10 switching system. Traffic for up to 252 customer lines can be concentrated in two T1 lines over a distance of up to 70 miles to an integrated interface in the *DMS*-10 switching network. That interface eliminates the need to reconvert and re-expand the signals and terminal appearances at the central office.

Software Structure. The *DMS*-10 software and firmware can be viewed as hierarchical layers surrounding the CPU, with inner layer functions "hidden" from outer layers. The inner layer closest to the CPU includes firmware for handling CPU interrupt messages, the bootstrap loader for starting up the system from scratch, run-time-sensitive routines, and automatic fault detection and recovery software. The next layer, stored in random-access memory, contains the scheduler for arranging the machine's working schedule. The outer layer has the software modules for terminal input/output, overlay program handling, and call processing. The software for *DMS*-10 systems is written in *SL-1*,[15] a high-level language, similar to ALGOL,[16] developed at Bell Northern Research, Ltd.

Scanning for state changes is done in two levels by hardware logic circuits. Subscriber-line state changes are noted by peripheral control packs that are scanned by network signaling packs before sending interrupt messages to the CPU. Input messages to the CPU are "time-stamped" and queued in call store for processing. The CPU processes the state change interrupts and is not involved with a call during the stable, or talking, state. Call processing in the *DMS*-10 system is said to be an *event-driven system*, similar to that used in the 2B*ESS*.

The *DMS*-10 is flexible in allowing the operating telephone company to set up translations and routing as desired. A full range of residential

---

[15] Trademark of Northern Telecom, Ltd.

[16] Developed by a committee of the Association of Computing Machinery in conjunction with their counterparts in Great Britain, France, Germany, and the Netherlands.

and small business features can be provided, and billing capability can be provided either locally or remotely. Interfaces for operations systems are being developed for traffic, maintenance, and memory administration.

5ESS Switching Equipment

The 5ESS system is a digital time-division electronic switching system designed for modular growth to accommodate local offices ranging from 1000 to 100,000 lines. It was designed to replace remaining electromechanical switching systems in rural, suburban, and urban areas economically. A major design goal was to use equipment modularity to achieve an economically competitive system over this wide range of line sizes. The 5ESS switching equipment uses distributed processing and modular software and hardware to provide a flexible architecture and simplify the addition of new features.

The 5ESS switching equipment architecture is shown in Figure 10-14 and consists of a number of interface modules (IMs) connected via a time-multiplexed switch. A duplex administrative module processor provides centralized routing control and administrative maintenance features. It is connected to the TMS through a message switch for communication with the interface modules. The initial application, put into service in March 1982, in Seneca, Illinois, offered features similar to 3ESS switching equipment and was limited to a single interface module. Succeeding issues of the generic program, cut over in 1983, offer the multimodule configuration and local/toll features for combined class 4 and class 5 operation. A remote switching module (the No. 5A Remote Switching Module) is scheduled for 1984. A fully integrated interface, also scheduled for 1984, will enable a SLC-96 carrier system to terminate economically and directly on electronic switching equipment.

The 5ESS system's time-space-time network consists of time-slot interchanges in each interface module that are connected via two optical fiber network control and timing links to the solid-state TMS. These links operate at the rate of 32 Mbps and are used for messages between the administrative module and interface modules and among interface modules as well as for voice transmission.

Administrative Module. The duplex administrative module processor of the 5ESS switching equipment consists of two $3B^{17}20$ computers, each equipped with disk storage and input/output processors (IOPs). The duplex processors operate in active/standby modes and rely on hardware and software sanity checks to switch modes. The IOPs provide interfaces (such as teletypewriters) to technicians, data links to operations systems,

---

[17] Trademark of Western Electric Co.

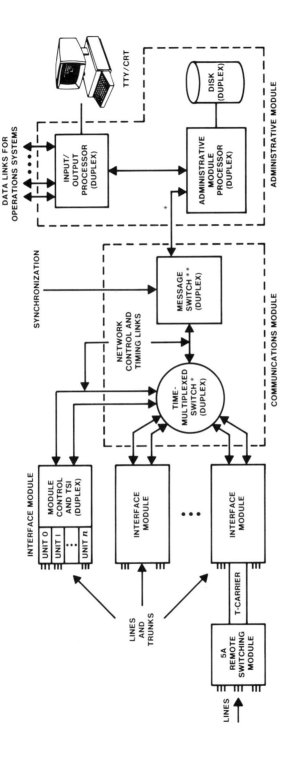

Figure 10-14. 5ESS switching equipment system architecture.

*OMITTED IN SINGLE-MODULE CONFIGURATION
**SUBSTANTIALLY SIMPLIFIED IN SINGLE-MODULE CONFIGURATION

and administrative module processor scan-and-distribute circuits via a number of peripheral controllers. Control messages and data pass between the administrative processor and the interface modules via the communications module. The administrative module processor also communicates with operations systems for traffic, billing, and maintenance data[18] through the IOPs.

Communications Module. The communications module consists of a message switch and a time-multiplexed switch. In addition to interfacing the administrative processors to the TMS, the message switch terminates special data links such as the one for the synchronization carrier (normally from a 4*ESS* switch) used to maintain digital synchronism. The TMS is used to connect voice channels in one interface module to voice channels in other interface modules as well as for data messages between the administrative modules and interface modules and for data messages between interface modules.

Interface Modules. Each interface module has a duplex microprocessor-based module controller; a module processor, used for call processing; and a duplex 512-slot TSI, used for time-division switching. In addition, there are several interface units. One, called a *digital services unit*, is used to generate call-progress tones and interfaces with the TSI. Other interface units terminate lines, trunks, and T1 carrier facilities. Once digitized at the interface unit, each 8-bit PCM voice sample, either received from T1 or converted by the interface module from an analog input, has eight bits of control information added to it. The resulting 16-bit word can be switched via the TSI (for intramodule calls) or sent via the network control and timing links to the TMS (for intermodule calls). Data interfaces at each TSI and control interfaces at each module controller terminate serial data and control busses from the interface units. A duplex digital signal processor in the interface module receives dial pulses and processes busy/idle bits to relieve the module controller of that real-time-intensive task.

Each interface module can host analog line/trunk units, digital line/trunk units, digital carrier line units, digital service circuit units, or metallic service units in addition to miscellaneous test and access units. These units are tied via busses to the module controller and (with the exception of the metallic service unit) to the TSI. The exact mixture of units in a module depends on the traffic and customers being served and can be different in each interface module. The interface module can also

---

[18] Respectively, the Engineering and Administrative Data Acquisition System (EADAS) discussed in Section 14.3.1, the Automatic Message Accounting Recording Center (AMARC) discussed in Section 10.5.5, and the Switching Control Center System (SCCS) discussed in section 14.3.2.

be used to terminate the T-carrier connections between a remote switching module and its host office. The digital line/trunk unit terminates the digital T1 carrier. The digital carrier line unit terminates *SLC*-96 carrier systems directly.

Both space division and time division are used for economical termination of subscriber lines. The analog line unit includes a concentrator available in 4-to-1, 6-to-1, and 8-to-1 ratios. Its solid-state gated diode crosspoints can pass high-level signals such as ringing and battery so that separate crosspoints are not necessary. *SLC*-96 carrier systems using the integrated T-carrier feature and terminating on a digital carrier line unit can be concentrated digitally in the digital carrier line unit. Trunks are unconcentrated. The metallic service unit contains a metallic test access network, some high-level service circuit functions, and scan-and-distribute points.

No. 5A Remote Switching Module. Remote switching is made available via the No. 5A Remote Switching Module (RSM). The RSM can be located up to 100 miles from the 5*ESS* switching equipment and can terminate a maximum of 4000 lines with a single interface module. Several No. 5A RSMs can be interconnected to serve remote offices as large as 16,000 lines.

The No. 5A RSM is a standard 5*ESS* system interface module with software augmented to provide stand-alone switching capability in the event of a failure in the host-remote link. A No. 5A RSM is linked to a host 5*ESS* system via T1 carrier or other facilties such as lightguide. The T1 carrier facilities terminate on a digital line/trunk unit on the 5*ESS* system host. Using the same basic hardware and software modules in the 5A RSM and 5*ESS* switching equipment ensures compatibility and reduces development effort. It also provides for a structural modularity that allows a smooth transition from a remote to a full 5*ESS* system or vice versa.

An important feature of the RSM is direct trunking. Most remote switches (for example, the No. 10A Remote Switching System) require that all interoffice calls pass through the host switch. The 5A RSM can be equipped with trunk units that directly interface facilities to other offices. This avoids costly trunk rearrangements when remote switches are deployed.

Software Structure. The 5*ESS* switching equipment software is divided into two segments. The portion in the administrative module processor is responsible for officewide functions such as the human interfaces, routing, charging, feature translations, switch maintenance, and data storage and backup. The portion in the interface module is responsible for the standard call-processing functions associated with the lines and trunks terminating on that interface module. These functions represent about 80 percent of the processing on calls and include interpreting status and address digits; controlling the interface unit, analog concentrator, and TSI path hunts; and signaling. Most software is written in C, a programming language developed by Bell Laboratories, and has a modular structure to afford easy expansion and maintenance. The administrative module processor and the interface module processors share a common software environment composed of a common data-base management system and an operating system for distributed switching. This facilitates software portability.

The 5ESS switching equipment was designed for flexibility. Processing power is added with the addition of each interface module. Modular software and modern data-base management concepts allow the software to be easily extended and modified to accommodate new services. Likewise, the modular hardware allows units to be upgraded to take advantage of new technology.

10.4 OPERATOR SYSTEMS

The same technology used in switching systems has been applied to operator-services equipment. The goal of this modernization is to reduce the number of operators required by automating the more routine tasks and by using operators more efficiently. Two examples of such modernized systems are described below—one for **toll services** and one for **number services.**

10.4.1 TOLL SERVICE

Before the introduction of mechanization, toll operators worked at 3CL switchboard positions, pictured in Figure 10-15. Many tasks involved in operator-handled calls were done manually, including switching (connecting circuits with cords), timing the call duration, and billing. Mechanization of these tasks leaves the operator free to perform the personal functions of speaking with customers, gathering needed information, and entering it into the system by pressing the appropriate keys.

The first major mechanization of the toll operator's functions was the development of the *traffic service position* (TSP) in 1965 as part of the crossbar tandem switching system. The operator's position was a cordless console with a numerical display and pushbuttons. In order to make the mechanization of operator-services capabilities independent of the designs of present and future toll and tandem switching systems, a new electronic *Traffic Service Position System* (TSPS) was developed and first introduced in 1969. The TSPS is an autonomous system—it stands apart

Figure 10-15. Manual toll switchboards.

from both the local and toll offices. Because the signaling and transmission interfaces for TSPS are standard, it functions with all the various designs of local and tandem switching systems.

The system is able to handle the types of calls shown in Table 10-1. It can also handle guest-originated calls from hotel rooms and provide the hotel with an automatic, immediate teletypewriter printout or a timely operator voice report of the charges for these calls.

For customer-dialed station-to-station calls, TSPS can serve as a centralized automatic message accounting (CAMA) point to record billing details without operator intervention. TSPS operators also provide assistance on customer-dialed international calls, and when local offices are not modified for international dialing, customers can place calls through TSPS on a dial-zero basis (that is, the customer dials only zero; the TSPS operator keys in the international number).

The TSPS uses both the basic hardware components and the system structure of the 1ESS switch. At first, the real-time capacity of the stored-program control (SPC) processor used in TSPS was approximately 16,000 initial position seizures per busy hour. The maximum number of major system elements are:

- three thousand trunks

- 310 operator positions, accessible as a single team

- eight chief operator groups (local, remote, or both)

- 62 operator positions per chief operator group.

TABLE 10-1

TSPS OPERATOR FUNCTIONS

TSPS Operator Functions	Type of Call	
	From Coin Stations	From Noncoin Stations
Obtaining billing information for calling card or third-number calls	1+, 0+	0+
Identifying called customer on person-to-person calls	0+	0+
Obtaining acceptance of charges on collect calls	0+	0+
Identifying calling number*	1+, 0+	1+, 0+
Monitoring coin deposits	1+, 0+	
Handling operator assistance calls	0−	0−
Type of call (as it appears on TSPS console)	1+ = Customer-dialed station-to-station calls 0+ = Customer-dialed special calls 0 = Operator assistance calls	

+ Indicates that more digits are to be dialed, that is, the called telephone number.

− Indicates that no more digits are to be dialed.

* Needed only when calling number is not automatically identified and forwarded from the local office.

As a result of automating parts of the operator's job and centralizing several small teams into one larger, more efficient team, TSPS operation requires many fewer operators than the cord switchboard.

Figure 10-16 shows a typical TSPS operating room. Each console contains two positions in a desk-like arrangement. A position becomes available to receive calls when an operator plugs the headset into its jack. Calls are automatically distributed to all attended positions, so all operators receive an equal share of the load. When a position is given a call, the operator hears a distinct tone and is given a lamp display to indicate

Figure 10-16. A typical TSPS operating room.

whether the originating station is coin or noncoin and whether the customer has dialed zero alone or zero followed by seven or ten digits or if the call requires only operator number identification for billing purposes. With these indications, the operator is able to respond appropriately. In particular, on calls received from coin stations, the initial deposit and duration of the initial period are indicated in the numerical display.

Whenever a call is connected to a TSPS position, all call details are available from the system memory. The details are directly equivalent to those that would be written on a ticket if the call were handled at a cord switchboard. Under key control, the calling number, the number that is being called, a calling card number if keyed into the system, the number of a third telephone if one is being billed, or the charging rate on coin calls can be displayed. Other operator controls allow the operator to release connections forward or backward, to ring the stations forward or backward, to collect or return coin deposits, and to connect to special-services operators over outgoing trunks.

As shown in Figure 10-17, all TSPS trunks have two 2-wire appearances on the TSPS link network. The network connects to various service circuits: digit receivers, outpulsers, coin control circuits, tone circuits, and operator positions. The basic logic instructions for handling calls are in the memory and are executed by the SPC processor. Changes in the state of trunk, service, and other peripheral circuits, including the positions themselves, are detected by scanners together with programs and memory

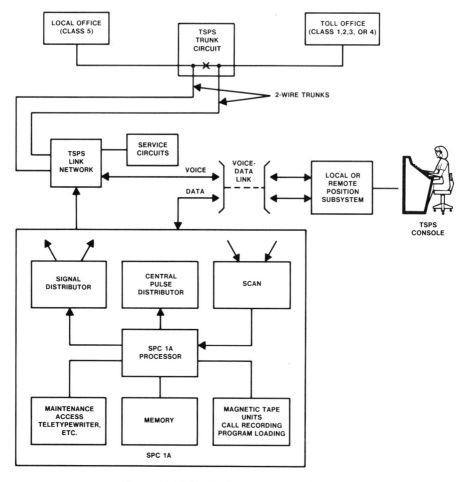

Figure 10-17. Traffic Service Position System.

indicating previous states. Output instructions via signal distributors and central pulse distributors control these circuits and the position lamps.

The processor-memory complex—including such support units as the control and display panel, a signal distributor, a central pulse distributor, a master scanner, the maintenance teletypewriter, and the program tape unit for loading and unloading memory—constitutes a subsystem called the *SPC No. 1A*. The switching network (link network) used in TSPS to connect the trunks to the service circuits and positions is a 4-stage, 2-wire, space-division network using ferreeds.

The program structure for TSPS closely follows the 1*ESS* switch program structure. An executive control program, interrupt levels, priority work lists, fault-recognition programs, and diagnostic programs are used to provide the real-time characteristics of the system.

Since its initial cutover in 1969, many enhancements have been made to TSPS. For example:

- The remote trunk arrangement extended TSPS service to remote areas that could otherwise not be served economically by TSPS.

- The more recent Automated Coin Toll Service (ACTS) automates operator functions on coin toll calls. The Station Signaling and Announcement Subsystem required for ACTS synthesizes requests for deposits, counts the deposits, and sets up and times the call, thereby eliminating operator intervention from these types of calls in most cases. By taking over these routine tasks, ACTS frees operators to concentrate on more complex calls, such as collect or third-party billing, and to help any customers requiring assistance.

- Automated Calling Card Service eliminates operator intervention on calling card, collect, and bill-to-third-number calls by allowing customers to key in billing information on these calls.

- The SPC No. 1A processor is being replaced with the *SPC No. 1B* processor, which uses the *3B*20 computer control unit (the *3B*20 computer is also used in *5ESS* switching equipment); this has increased the real-time capacity of TSPS by 60 percent. By the mid-1980s, close to 100 percent of the telephones served by Bell operating companies will be served by TSPS.

10.4.2 NUMBER SERVICES

The *Automatic Intercept System* (AIS), which became available in 1970, was developed to automate and centralize the processing of calls to nonworking numbers. The various types of calls that may be routed to AIS include calls to vacant or unassigned numbers, calls to changed numbers, calls to disconnected numbers, and trouble intercepts.

Incoming intercept trunks from local offices are connected to equipment that synthesizes recorded announcements specifically tailored to each intercept call. For those relatively few nonroutine cases requiring operator assistance, AIS provides an improved method of operator handling. As such, AIS totally eliminates operator intervention on most intercept calls. This, together with the centralization of many small intercept bureaus into one AIS, has resulted in operator savings of about 75 percent.

As shown in Figure 10-18, an AIS contains one or more Automatic Intercept Centers (AICs). There is one centralized intercept bureau associated with a *home AIC* in an AIS. An AIC contains a time-division

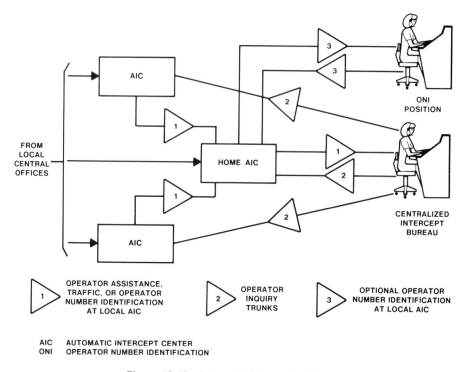

FROM
LOCAL
CENTRAL
OFFICES

AIC

HOME AIC

AIC

ONI
POSITION

CENTRALIZED
INTERCEPT
BUREAU

1 ▷ OPERATOR ASSISTANCE,
TRAFFIC, OR OPERATOR
NUMBER IDENTIFICATION
AT LOCAL AIC

2 ▷ OPERATOR
INQUIRY
TRUNKS

3 ▷ OPTIONAL OPERATOR
NUMBER IDENTIFICATION
AT LOCAL AIC

AIC AUTOMATIC INTERCEPT CENTER
ONI OPERATOR NUMBER IDENTIFICATION

Figure 10-18. Automatic Intercept System.

switching network, a stored-program processor, magnetic disk memories for filing unassigned numbers, service circuits, announcement machines, and the centralized intercept bureau. The network connects to a maximum of 512 intercept trunks, announcement circuits, and service circuits and can accommodate as many as sixty-four connections simultaneously.

When a call comes into the AIC from the local office, it is connected to a multifrequency receiver that receives the called digits and passes them to the processor. The processor decides whether to connect the call to a centralized intercept bureau operator or to look up the called number in the files.

For situations in which local offices are not equipped to identify the called number on intercept calls, operator number identification is provided at the AIC.

The files of unassigned numbers are stored in duplicated magnetic disk memories for reliability. The processor uses this information to compose an announcement. Recorded words and phrases, including numbers, are supplied by a duplicated 96-track announcement machine. The output of each track can be connected to customer lines through the switching network.

444

The locations of recorded numbers, words, and phrases are stored in the administrative processor, which selects the items required for an announcement in the proper sequence for the particular call. The administrative processor is the same as that developed for 2*ESS* switching equipment; the program structure and maintenance strategy closely follow those of the 2*ESS* switch.

The intercept file represents a large data base that requires frequent updating. A typical installation serving a metropolitan area (New York has several) or an entire state, such as North Carolina, may have as many as 500,000 records and may require as many as 18,000 changes per day. A minicomputer-based Data Base Administration System provides for efficient updating via teletypewriter or data-link channels, as well as an off-premises backup of the entire file.

Recent enhancements to AIS provide for automation of parts of some of the calls that previously required operator handling. The network capability has been expanded to a total of 1024 terminations and 256 simultaneous connections. By the mid-1980s, approximately 85 percent of the operating company telephones will be served by AIS.

10.5 BILLING EQUIPMENT AND SYSTEMS

10.5.1 INTRODUCTION

In 1980, the Bell operating companies handled, on the average, over 500 million calls a day. Under current tariffs, billing data is not required for every call placed since approximately 70 percent are covered under the provision of a monthly flat-rate charge. Even on these calls, however, the switching system must make a decision *not* to record call details for billing.

The billing information required for the remaining 30 percent depends on the type of call. About 10 percent of the total calls originated are classified as toll. These require the most billing information, including time of answer, call duration, and calling and called numbers. The remaining 20 percent are measured local calls. These require a range of billing information from full details to an abbreviated form with no called number or call duration, depending on local tariffs.

Current trends indicate that the percentage of calls requiring some form of per-call measurement will continue to increase, reaching close to 90 percent by 1990. At the same time, the volume of total calls placed is rising at about 3 percent per year. Under these conditions, automatic message accounting (AMA) equipment will be called on to record substantially increasing amounts of data in the next decade. (See Figure 10-19.) In addition, since most of the additional calls to be measured are local calls, the average revenue per call recorded will drop. The need to

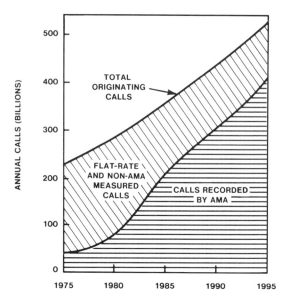

Figure 10-19. AMA recorded volume projection.

reduce the per-call cost of billing to maintain a low cost-to-revenue ratio
has been a primary motivation in the recent evolution of billing systems
and will continue to motivate future evolution.

Before discussing the evolution of billing systems, a brief description
is in order of the final product of automatic message accounting—the customer's bill. Three types of formats are used for customer billing. Each
has implications for billing-data collection as noted throughout the text.
The definitions of the three types given below should be considered typical rather than an absolute standard.

- The **full-detail** format provides all details on each call, including date,
  answer time, calling number, called number, duration, and dollar
  charges.

- The **bulk-bill** format provides three entries on the bill, showing (in
  dollars) the usage amount, the allowance amount, and the total usage
  charges over the allowance amount. All applicable calls in the billing
  period are totaled in one line on the bill.

- The **summary-bill** format provides the bulk-bill information and a
  summarization by zone called and rate period (day, evening, night),
  showing total calls, overtime minutes, and the associated dollars.

The full-detail format is typically used for all toll calls and increasingly for nontoll, measured calls. The bulk- and summary-bill formats

are used for nontoll measured calls for which local tariff is based on message unit charging.

10.5.2 EARLY BILLING-DATA COLLECTION SYSTEMS

Initially, records were required only for toll calls. These were tickets manually prepared by the operators. When the need for measuring local calls appeared, the manual ticketing method was supplemented by the use of a small electromagnetic counter, called a *message register*, that was associated with each line that required measurement and was operated once for each call to indicate a unit charge.

The operator-prepared tickets for toll calls and periodic photographs of the message registers for measured local calls were sent to a telephone company center known as the *Revenue Accounting Office* (RAO), where customers' bills were prepared.

Subsequently, capability was provided for continuous scoring of message registers throughout the conversation to indicate duration. As the metropolitan areas grew larger and customers began to call regularly beyond their local areas, a method known as *zone registration* was adopted. With zone registration, a call placed to a distant zone might cause the message register to be scored two or more times per time period rather than once (like calls to the local zone).

Although zone registration provides an economical method of charging for interzone (short-haul toll) calls, it does not provide a record of the called number or time of day for each call, which was desirable for calls requiring more than five message register operations per time period. To secure such a record and to obtain the benefits of economy and increased speed possible from automatic operation, automatic ticketing arrangements were developed for use in the step-by-step switching system environment. With this system, a ticket was automatically printed for each chargeable multimessage unit and short-haul toll call, with the destination and time of answer included. As a result, all essential information pertaining to the call was available on a printed record.

The use of operator-prepared and automatically generated tickets has been essentially eliminated in the Bell System. The use of message registers, although substantially reduced, still exists in many older switching systems. These changes have been accomplished by the use of the **local automatic message accounting** (LAMA) and **centralized automatic message accounting** (CAMA) systems discussed in the next two sections and by the 2900B Local Message Metering System (LMMS). The LMMS, manufactured by the Vitel Division of Vidar, was developed to replace message registers in No. 1 Crossbar and panel offices. The system uses duplicated hardwired controllers to process information on the M-lead of each customer's line and write the assembled call data on local magnetic tape. The first installation of the 2900B LMMS to generate billing data occurred in New York in 1972.

Before the availability of modern computers, the work in the RAO of sorting per-customer information, computing the charge per call, totaling the charge per customer, and preparing the bill was labor intensive and represented an appreciable expense item. A brief discussion of message processing in modern RAOs is given in Section 13.2.2.

10.5.3 LOCAL AUTOMATIC MESSAGE ACCOUNTING

The advent of direct distance dialing in 1951 provided a natural stimulus to increase the volume of toll calls. Because of the large processing costs and limited capacity associated with operator and automatic ticketing arrangements as well as the lack of details provided with message registers, a completely automatic method of providing a machine-readable record of call details was needed. The first implementation of this concept, now commonly known as AMA, was the paper-tape system installed in the No. 1 Crossbar and extensively in the No. 5 Crossbar. This system is now designated Local Automatic Message Accounting-A (LAMA-A) and was first used to collect billing data on a No. 5 Crossbar office in 1948 in Pennsylvania. By 1980, LAMA-A still recorded about 25 percent of all AMA data. Today, however, its use is on the decline in favor of the more modern methods discussed in Sections 10.5.5 and 10.5.6.

The principle used with LAMA-A is illustrated in Figure 10-20. With this system, the recording of call details takes place in three distinct parts or entries as explained below and uses the existing switch common-control and supervision-detection equipment to avoid per-line recording devices.

A typical call proceeds through the system as follows: After the customer completes dialing, the dialed digits, class of service, calling-line equipment number, and call type are passed from the switching common-control apparatus to the AMA common-control apparatus. Here, translations, such as calling-line equipment number to billing number, are made. The resultant information is presented to one of up to twenty electromechanical paper-tape perforators. The specific perforator used is determined by the trunk or district junctor selected to switch the call, resulting in a natural load balance among the perforators. This record containing the information gained from the customer's dialed digits and the common-control circuits is known as the *initial entry*. The last line of the initial entry contains a 2-digit code, known as the *call identity index*, that identifies the trunk or district junctor used for the call (100 trunks or district junctors per perforator). This index is the same for each of the entries recorded for the same call and is used by the computer in the RAO for call assembly, a process that brings together all information pertaining to a single call.

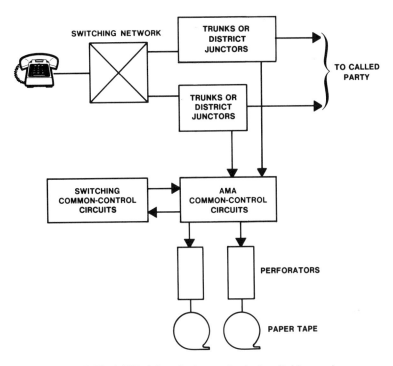

Figure 10-20. LAMA-A for electromechanical switching systems.

When the call is answered, a second entry, called the *answer entry*, is made. This entry consists of a single line on the paper tape and contains the call identity index and the minutes and either tenths of minutes or seconds of the hour in which the call was answered. (An *hour entry*, not part of the initial or answer entry, is recorded on the tape at the beginning of each hour and applies to all subsequent calls until a new hour entry is recorded.)

When the call is terminated, a third entry, called the *disconnect entry*, is perforated (usually on the same tape unless the previous tape was removed and replaced while the call was in progress). This entry is also a single-line entry and contains the call identity index and the minutes and either tenths of minutes or seconds of the hour in which the call was terminated.

In order to ensure correct RAO processing of the call data, numerous other entries are recorded. Information that is common to all data from a particular central office is perforated only at the beginning and end of the tape. Typical of these are the entries denoting the month, day, and recorder number. As noted earlier, an hour entry is perforated every hour on the hour. Day entries are recorded once a day at three o'clock in the morning along with a splice entry. The latter offers a convenient point on the tape for starting a new record or for cutting and removing a tape. Cancel entries are perforated to allow the RAO to continue processing at a point on the tape when the switch equipment encounters an irregularity such as loss of time synchronization.

At the end of each day, or several days if a weekend is involved, the tapes from all perforators in the system are cut and forwarded to the RAO. Collectively, these paper tapes provide all the information required for billing the AMA-recorded, customer-dialed, and chargeable messages.

The same principle discussed above of avoiding per-line recording equipment in electromechanical offices was also applied to electronic offices. However, in contrast to electromechanical offices where the AMA function is performed by equipment auxiliary to the switching equipment, in electronic switching systems (such as 1/1A*ESS*, 4*ESS*, and *DMS*-10 switching equipment), the AMA function is an integral part of the system. In these systems, the AMA function is implemented in software that gathers the billing data while the call is in progress and stores them in a single software register. Figure 10-21 outlines the 1/1A*ESS* switching equipment AMA program in block diagram form. All data associated

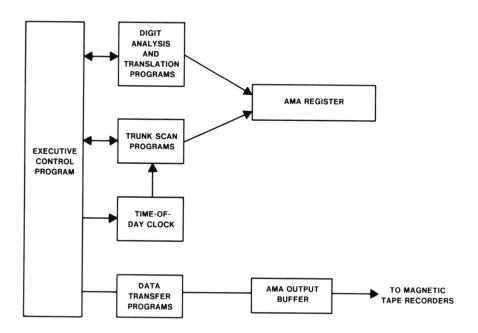

AMA   AUTOMATIC MESSAGE ACCOUNTING

Figure 10-21. LAMA program block diagram
for 1*ESS* and 1A*ESS* switching equipment.

with a single call are written to magnetic tape in a single entry at call termination. The recording equipment typically consists of two tape recorders and associated circuitry. The tape units are provided on an active/standby basis to ensure reliability and to augment maintenance operations. With this arrangement, the system has access to either AMA unit at all times. Software tape buffers avoid loss of data when the active tape is momentarily busy recording previous data. Use of the single-entry format eliminates the call-assembly process at the RAO.

10.5.4 CENTRALIZED AUTOMATIC MESSAGE ACCOUNTING

Before local AMA can be introduced, certain practical limitations must be considered. For instance, many offices are in areas where the dialable traffic requiring AMA treatment is relatively light. In these cases, installation of AMA equipment is not economically feasible. Furthermore, local AMA can only handle traffic originated by individual and 2-party customers; multiparty customers must place their charge calls through an operator. Also, the design of LAMA equipment for panel and most step-by-step systems is not practical because of the absence of common-control equipment. To apply AMA to toll calls for customers not served by LAMA, centralized automatic message accounting (CAMA) was developed in 1953. This system locates the recording equipment in a central switching location where it serves a number of central offices, as shown in Figure 10-22. Details of toll calls normally switched by the central location are recorded by the CAMA equipment as the calls are switched.

Initially, CAMA arrangements were developed for use in crossbar tandem offices, using paper-tape equipment similar to LAMA-A. Subsequently, CAMA features were developed for the No. 4A Crossbar; No. 5 Crossbar; intertoll step-by-step; and, using magnetic tape equipment, for the 1/1AESS and 4ESS switching equipment offices. In 1973, a minicomputer-based system that replaced paper-tape perforators was introduced in existing crossbar tandem and No. 4A Crossbar CAMA offices. This arrangement, referred to as CAMA-C, uses International Business Machines minicomputers and magnetic rather than paper tape to record AMA data.

In addition to AMA recording, CAMA must provide additional capabilities. First, whereas LAMA equipment is arranged to translate up to approximately ten originating office codes associated with the local system, CAMA must translate originating office codes associated with many local systems.

A second capability provided in the CAMA design is the ability to obtain the calling customer's telephone number from the local switching system. Two methods are used, operator number identification (ONI) and automatic number identification (ANI). When a local system is not equipped with ANI, the CAMA office switches the connection to an

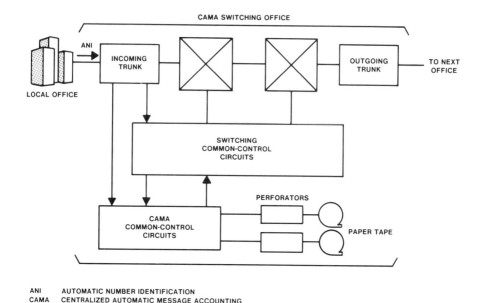

CAMA SWITCHING OFFICE

INCOMING TRUNK

OUTGOING TRUNK

TO NEXT OFFICE

LOCAL OFFICE

ANI

SWITCHING COMMON-CONTROL CIRCUITS

PERFORATORS

CAMA COMMON-CONTROL CIRCUITS

PAPER TAPE

ANI   AUTOMATIC NUMBER IDENTIFICATION
CAMA CENTRALIZED AUTOMATIC MESSAGE ACCOUNTING

Figure 10-22. CAMA recording arrangements (LAMA-A example).

operator who requests the calling customer's telephone number before allowing the call to proceed. The operator keys the number into the CAMA equipment for transfer to the AMA tape. Where a local system is equipped with ANI, operator intervention is only required to handle calls from multiparty lines and on ANI failures. For all other calls, the ANI equipment, which is located in the local switch, responds to signals from the CAMA equipment and transmits the identified number to the CAMA system via multifrequency pulsing over the talking path.

10.5.5 REMOTE RECORDING

The use of LAMA and CAMA for recording call details provided the Bell System with a vehicle well suited to the needs of AMA through the early 1970s. At that time, however, it became clear that changes in local tariffs would soon require complete call details on many local calls as well as toll calls. Furthermore, it was important to offer this capability in the older switching systems serving a majority of Bell System lines as well as the newer electronic switching systems. Some switches, such as step-by-step, required access to CAMA for recording toll calls. Switching local calls to the CAMA point for recording and back to the local switch for completion was shown to be very costly in trunking. Other switches, such as No. 5 Crossbar, had the LAMA system that could be expanded for local-call recording, but this meant a significant investment in the older, electromechanical AMA equipment. Fortunately, availability of integrated circuit logic devices, reliable data transmission facilities, and flexible microprocessors and minicomputers offered an economical solution to the problem of providing AMA on a large percentage of local calls. This new technology allowed low-cost data-acquisition devices, known as *sensors*, to be interfaced with the local switch in order to gather AMA data as they were generated; the data were then transmitted separately from the talking path to a central, shared minicomputer. At the central computer, processing and subsequent recording in a single-entry format on magnetic tape could be accomplished, all at a relatively low cost per office. The first such minicomputer, introduced in 1975, was known as the Automatic Message Accounting Recording Center (AMARC).

- increased revenue due to more accurate call timing than LAMA-A

- reduced Revenue Accounting Office processing costs through use of a single-entry output recording format

- elimination of per-office tape pickup and delivery

- elimination of loss from electromechanical AMA equipment blockage and paper-tape mutilations

- recorded data on 1600 bits per inch (bpi), industry-compatible magnetic tape.

Nos. 1 and 1A AMARC

The No. 1 AMARC, the first in a series of two major AMARC developments, receives billing data in real time over dedicated voice-grade data links. It is based on a pair of *DEC PDP*[19]-11/40 minicomputers operating in a duplex configuration.

The No. 1 AMARC controls and receives the data from a maximum of thirty dedicated channels. Each channel consists of a dedicated voice-grade line equipped with 202T (or equivalent) data sets, operating asynchronously at 1.2 kilobits per second (kbps). In case of channel failure, the AMARC automatically dials a back-up channel to the affected sensor over the direct distance dialing or local message network. A maximum of two such back-up channels is provided to cover the thirty input channels.

Studies indicated that the No. 1 AMARC would exhaust its memory capacity (needed to add new sensors) before it exhausted its data channel and real-time capacity. In addition, operating companies wanted more detailed maintenance reporting and features such as remote alarms to allow easier monitoring of the AMARC. This meant an increased processing load for AMARC. As the No. 1 AMARCs grew, they were being

[19] *DEC* and *PDP* are trademarks of Digital Equipment Corporation.

clustered in administrative centers by the operating companies. Considering these factors, it was determined that an AMARC with larger capacity could lower costs by offering greater economy of scale.

It was against this backdrop that the No. 1A AMARC was developed. This AMARC uses higher-capacity minicomputers (*PDP*-11/70s) and peripheral equipment with Western Electric–manufactured communications interfaces and alarm circuits to provide ninety input channels, improved maintenance capabilities, and sufficient capacity for additional sensor types. To ensure reliability, all major central units (processor, memory, tape) are duplicated as with the No. 1 AMARC. The first full-capability No. 1A AMARC began operation in 1981 in Illinois. The No. 1 AMARC is no longer being deployed.

The following paragraphs discuss several equipment configurations designed to interface different switching systems with the AMARC.

Call Data Accumulator

The call data accumulator (CDA) gathers message billing data on local calls in step-by-step offices and transmits the data to the AMARC. It consists of a hard-wired logic device that gathers data (calling and called number, answer and disconnect times) from the system. Toll calls in step-by-step offices equipped with a CDA continue to be recorded via CAMA. The first CDA was cut into service in New York in 1975.

Billing Data Transmitter

The billing data transmitter (BDT) is a sensor that provides a low-cost modernization vehicle in offices where LAMA-A costs become unattractive because of maintenance of the older equipment and tape pickup and delivery. It consists of duplicated microprocessors that collect the data from the existing AMA paper tape recorder circuitry and send them over a data link to the AMARC. In addition to the economics of improved technology, the BDT provides additional revenue by introducing more precise call timing.

The BDT replaces LAMA-A perforators and associated timing circuits in the following electromechanical switching systems:

- No. 1 Crossbar

- No. 5 Crossbar

- crossbar tandem (excluding TSP)

- step-by-step CAMA

- No. 4 Toll Crossbar.

The first BDT was cut over to billing service in New York in 1976.

LAMA-C

LAMA-C is a duplex minicomputer system based on *PDP*-11/40s and designed to improve billing economics in No. 5 Crossbar offices. Although no longer being deployed, many LAMA-C systems are still in use. The system was designed for large office applications where substantial growth in message billing data collection was expected, either through rapid growth in customers or through implementation of a new tariff. Because of the variety of services available on the No. 5 Crossbar and the desirability of eliminating the wired cross-connections used in LAMA-A equipment, LAMA-C was developed using an on-site minicomputer rather than wired logic or microprocessors. The system is capable of collecting billing data on all calls, assembling, formatting, and then transmitting the call records via data link to No. 1 AMARC. Software translation of line equipment number to directory (billing) number and automatic number identification outpulsing to TSPS enhance the economic value of LAMA-C.

The first LAMA-C installation was cut into service in New York in 1975, working with the same No. 1 AMARC as the first CDA.

No. 5 Electronic Translator System

The No. 5 Electronic Translator System, which collects billing data and transmits them to the AMARC, was developed primarily to provide line, trunk, and routing translations in No. 5 Crossbar offices via software rather than wired cross-connections. The No. 5 translator consists of duplicated Western Electric 3A auxiliary processors with associated scanners and distributors. It is bridged onto the No. 5 Crossbar System and receives call details from markers, trunks, and certain lines via scan leads. Trunk scanning on outgoing, intraoffice, and certain miscellaneous trunks provides information for trunk selection, routing, and billing. Line scanning on inward Wide Area Telecommunications Services lines also provides data for AMA.

The No. 5 translator reduces the cost of AMA expansion required for measured service in many cases. Benefits accrue from reduced floor-space requirements and operating expense savings. The incremental cost per line of extending the size of the No. 5 Crossbar is reduced when the No. 5 translator replaces the existing translation and billing equipment.

The first No. 5 Electronic Translator System was cut over to billing service in Ohio in 1977.

3ESS Switching Equipment

The 3ESS switching equipment was the first to be connected to AMARC without the need for custom-designed interface hardware. Internal software provides all the required logic, and standard data sets are used for data transmission. Data are sent to AMARC in three entries similar to the LAMA-A system's initial, answer, and disconnect entries. Just as in the case of step-by-step switches, connection to AMARC was motivated by the need for economical recording of local call data.

The first 3ESS switch was connected to AMARC in 1976 in the New York Telephone Company.

Call Data Transmitter

The call data transmitter (CDT) is the most recent addition to the sensors that provide an AMARC connection from the No. 5 Crossbar. It is a microprocessor-based system designed to fill the need of those No. 5 Crossbar Systems that do not have LAMA-A or would require a major expansion of LAMA-A and for which the cross-connection activity is not sufficiently high to make them candidates for the No. 5 translator. The CDT allows connection directly to markers and trunks and can record local calls, toll calls, or both. The CDT and BDT may be used simultaneously in the same office, for instance, if a BDT was installed for LAMA-A modernization and then a significant expansion in local call detail recording was required as the result of tariff changes.

The first CDT was cut over to billing service in Illinois in 1980.

2BESS and 5ESS Switching Equipment

The latest additions to the No. 1A AMARC system of sensors are the 2BESS and 5ESS switching equipment. The 2BESS system was modified and the 5ESS system originally designed to connect to the AMARC using a 2-entry system to transmit the data for a single call. Both systems perform most of the billing function in software and require only data sets in addition to the common switching equipment to connect to the AMARC.

The first 2BESS and 5ESS switches were cut over to billing service using the AMARC in Kentucky and Illinois, respectively, both in 1982. Generics of the 5ESS system scheduled for 1983 will use a new billing architecture known as *store and forward*, which is described below.

10.5.6 REMOTE RECORDING USING STORE AND FORWARD

With the AMARC system, data were transmitted to the AMARC as they became available (that is, in real time). During the late 1970s and early 1980s, the reliability and cost per bit of small- to medium-size disks (10 to 200 megabytes) improved dramatically. This new technology provided the capability to place enough storage at the switch to hold several days' worth of AMA data and led to the development of a new concept in AMA data collection known as *store and forward*. AMA data are stored on determined by a data base at the collector. This system can take advantage of dial-up connections and eliminates the need for dedicated data determined by a data base at the collector. This system can take advantage of dial-up connections and eliminates the need for dedicated data links. The storing and transmitting hardware and software at the switch are known as an *AMA transmitter* (AMAT) and at the receiving end as the *collector*. The first application of the AMAT/collector system in the Bell System is the *DMS*-10 AMAT and No. 1A Collector that was cut over to billing service in Pennsylvania in 1983. Based on the flexibility of the store-and-forward architecture and continued improvement in disk technology, it appears this concept will continue to influence AMA design for some time.

A diagram of current billing-data collection equipment available in the Bell System is shown in Figure 10-23. Using this equipment, it is possible for the telephone companies to record all call details on magnetic tape and to record full details on local calls for any system type economically.

10.5.7 AMA SUMMARY AND FUTURE TRENDS

AMA systems will continue to evolve to meet the changing needs of the Bell operating companies. One trend influencing data collection systems is the increasing volume of lower priced calls due to wider application of measured-service tariffs. The reduction in average revenue per call creates a continuing need to reduce the per-call cost of billing. The application of low-cost disks and microprocessors appears certain to continue to influence the design of future billing systems, as does the provision of new services that require specialized billing information. The desire to eliminate completely the manual handling of data is a long-term goal that emphasizes the need for teleprocessing from the data source (the switching system) to the data user (the RAO). Future developments in packet-switching networks and connection to large batch-processing computers may allow highly flexible, reliable, and cost-effective methods to accomplish this goal.

The introduction of new switching systems, as the result of continually improving technology and the need to bill for innovative new services, requires AMA design standards that are both flexible enough for future needs and specific enough to stimulate economic developments. Finally, divestiture will create major changes in requirements for AMA systems and billing.

10.6 SWITCHING APPLICATION OVERVIEW

The previous sections of this chapter discuss in some detail the evolution and characteristics of selected Bell System switching systems for both local and tandem applications. This section summarizes the current and

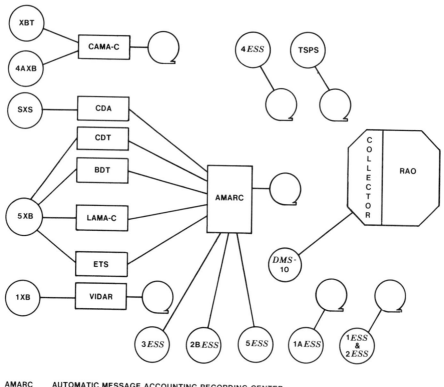

AMARC   AUTOMATIC MESSAGE ACCOUNTING RECORDING CENTER
BDT   BILLING DATA TRANSMITTER
CAMA-C   CENTRALIZED AUTOMATIC MESSAGE ACCOUNTING
CDA   CALL DATA ACCUMULATOR
CDT   CALL DATA TRANSMITTER
ETS   ELECTRONIC TRANSLATOR SYSTEM
LAMA-C   LOCAL AUTOMATIC MESSAGE ACCOUNTING
RAO   REVENUE ACCOUNTING OFFICE
SXS   STEP-BY-STEP
TSPS   TRAFFIC SERVICE POSITION SYSTEM
XB   CROSSBAR
XBT   CROSSBAR TANDEM

Figure 10-23. Current billing-data collection equipment.

projected view of the use of these switching systems. In addition, the future trend in the Bell System towards an integrated services digital network is discussed.

The evolution of stored-program control (SPC) switching systems and their characteristics is shown in Table 10-2 in chronological order for local, local tandem, and toll applications. Improvements in software and the introduction of newer processors, processor memories, and network designs have provided a variety of switching systems to cover the applications over the full range of office sizes. The 5ESS system introduced in 1982 provides this same application coverage with one system.

10.6.1 NUMBER OF LINES AND LOCAL SWITCHING SYSTEMS

Table 10-3 shows the January 1, 1983, census of Bell System local switching systems, including the number of systems and total number of lines
served by equipment type. Step-by-step equipment (including community dial office [CDO] applications) is still a large part of the local network:
about 46 percent of existing systems are step-by-step and about 13 percent
of all Bell System working lines are served by step-by-step systems.
No. 1 Crossbar Systems and 1/1A*ESS* switching equipment have the
highest average number of lines per system, reflecting their design for
application in large metropolitan areas. Space-division electronic switching systems, as a whole, serve about 55 percent of the total lines but do
so with about 29 percent of the switching systems. Time-division systems, the newest entry in the switched network, currently account for
only a very small percentage of total lines and systems.

    The projected view of local switching systems indicates the continued
phasing out of electromechanical systems, with replacement of non-CDO
step-by-step systems through the early 1980s and replacement of crossbar
systems and modernization of the many new CDOs through the mid-
1990s. Both space-division and digital time-division SPC systems are
expected to contribute to the replacement of older technology for some
years to come.

    Continued growth in the use of SPC electronic switching will result
in more flexible routing and office code usage, savings in floor space, and

TABLE 10-3

CENSUS OF LOCAL SWITCHING SYSTEMS (APPROXIMATE) (JANUARY 1, 1983)

Type of System	Number of Systems	Percent of Total Systems	Total Lines (Millions)	Percent of Total Lines	Average Lines Per System
Step-by-step	800	8.2	6.1	7.1	7,600
CDO (step-by-step)	3,700	37.8	5.1	5.9	1,400
No. 1 Crossbar	180	1.8	4.1	4.8	22,800
No. 5 Crossbar	2,120	21.7	23.5	27.3	11,100
1/1A*ESS*	1,750	17.9	42.1	48.9	24,100
2/2B*ESS*	700	7.2	4.4	5.1	6,300
3*ESS* and 10A RSS	420	4.3	0.6	0.7	1,400
Time-division	110	1.1	0.2	0.2	1,800
Total	9,780	100.0	86.1	100.0	8,800

less maintenance expense. Large systems such as the 1A*ESS* switching equipment will be used primarily for metropolitan modernization applications. Time-division digital switching systems such as the 5*ESS* system will be the primary replacement vehicles for the No. 5 Crossbar suburban modernization. Remote systems, such as the No. 10A Remote Switching System, No. 5A Remote Switching Module, and the *DMS*-10 remote equipment module will be used mostly for small-office rural modernization. By the early 1990s the Bell operating companies' local networks may be entirely under stored-program control.

10.6.2 NUMBER OF TOLL TERMINATIONS AND TOLL SWITCHING SYSTEMS

Table 10-4 shows the January 1, 1983, census of Bell System toll switching systems and terminations. Currently, the No. 4A Crossbar and 4*ESS* toll switching equipment serve the majority of toll terminations, although the relative number of these systems is very small (146 out of a total of 718 toll systems).[20] These switching systems, together with the crossbar tandem and some 1/1A*ESS* switches, are located in major metropolitan areas. Other systems serve the more sparsely populated rural areas where the toll demand is too small to warrant larger switching systems. In these rural areas, many of the systems serve both local and toll traffic. It is estimated that about 86 percent of toll terminations are in major metropolitan areas.

As shown in Table 10-4 (as of January 1983), space-division electronic switching systems serve about 21 percent of total terminations and represent about 23 percent of total (both electronic and electromechanical) systems. The 4*ESS* time-division switches represent about 52 percent of total terminations and about 12 percent of total systems. The 4*ESS* switches have been installed at a rate of about one per month since their introduction in 1976.

As a result of their capabilities (for example, large capacity, 4-wire network, electronic translation), 1*ESS*, 1A*ESS*, and 4*ESS* switches are expected to handle most tandem system terminations. Replacement of electromechanical systems will continue and should be complete by the early 1990s. Growth is expected to be handled mostly by time-division systems, which will account for perhaps 75 percent or more of total terminations by that time.

---

[20] In March 1983, a *DMS*-200 digital time-division switching system was placed in service in the Bell System toll network.

TABLE 10-4

CENSUS OF BELL SYSTEM TOLL SWITCHING SYSTEMS
(JANUARY 1, 1983)

Type of System	Class				Number of Systems	Percent of Total Systems	Total Terminations (Millions)	Percent of Total Terminations	Average Terminations Per System
	1	2	3	4					
Step-by-step				86	86	12.0 ⎫			
Crossbar tandem		2	28	96	126	17.6 ⎬	1.50	27	3,200
No. 5 Crossbar			18	176	194	27.0 ⎭			
No. 4A Crossbar	2	8	43	9	62	8.6 ⎫	1.25	21	7,500
1/1AESS		2	35	129	166	23.1 ⎭	3.00	52	36,000
4ESS	8	40	24	12	84	11.7			
Total	10	52	148	508	718	100.0	5.75	100	—

10.6.3 FUTURE TRENDS

The projected view of local and toll switching systems for the Bell operating companies and AT&T was described above. This section is more speculative and postulates the further evolution of nationwide communications to a fully integrated services digital network (ISDN). The ISDN is conceived of as an end-to-end public digital network that will economically satisfy a broad range of customer needs. Customer needs for services such as telemetry, security monitoring, electronic mail, electronic funds transfer, computer inquiry and response, facsimile, computer bulk data transfer, and video need cost-effective network solutions if some or all are to be provided on a large-scale public basis. These solutions are expected to evolve from the digital connectivity offered by an ISDN.

The spectrum of customer needs is expanding rapidly. Voice communication has traditionally been the predominant force in telecommunications. However, this is changing. In addition to human-to-human communication, human-to-machine and machine-to-machine digital communications are becoming significant aspects of future customer needs.

The need for digital communications is expected to expand in the 1980s and 1990s as the result of the decreasing costs of digital circuitry and the maturing of large-scale integration. An end-to-end digital transport network, including switching of digital voice and data channels, presents opportunities for many new services, such as those mentioned above. At present, these services cannot all be anticipated nor quantified. However, the concepts of stored-program control and electronic switching promise to offer the capabilities required to provide these services.

AUTHORS

S. D. Aspie
C. E. Betta
W. R. Byrne
W. S. Hayward
C. E. Johnson

E. M. Johnson
C. A. Kienbaum
D. J. Marutiak
W. B. Perkinson
F. F. Taylor

11

Customer-Services
Equipment and Systems

11.1 STATION EQUIPMENT

Station equipment, which provides the user of telecommunications services access to the nationwide network, includes **telephone sets**, (including **public telephones**) **data sets**, and **data and graphics terminals**. This section briefly discusses the operation and major components of these products and some design considerations. It describes a number of the features present products provide and includes some discussion of current trends in station equipment.

11.1.1 TELEPHONE SETS

The 500-Type Telephone Set

An electromechanical telephone has two primary functions: **transmission** and **signaling**. The familiar 500-type set, several varieties of which are pictured in Figure 11-1, was first produced in 1950. Its functional elements consist of a carbon **transmitter**, electromagnetic **receiver, network, switchhook, dial**, and electromagnetic **ringer** (see Figure 11-2).

Transmission. For transmission, the electroacoustic transmitter and receiver units and network ensure the desired levels of transmission, reception, and sidetone. The carbon *transmitter*, which provides high output power at low cost, transforms an acoustic signal into an electrical signal. The pressure of sound waves impinging on the diaphragm of the transmitter varies its resistance. This modulates (varies) the direct current provided by the central office battery and generates an alternating current on the loop to the central office. This alternating current modulates the

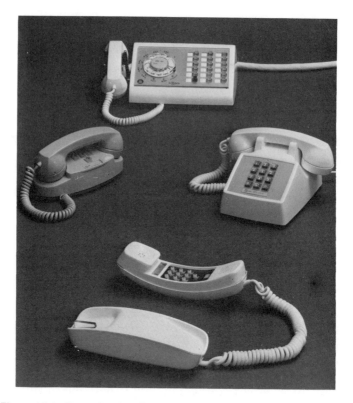

Figure 11-1. Several sets with 500-type circuitry. *From the top, CALL DIRECTOR,* PRINCESS,* traditional *TOUCH-TONE,* and *TRIMLINE* telephones.

---

* Registered trademark of AT&T Co.

magnetic field of the electromagnetic *receiver* in the listener's telephone, moving a diaphragm in the receiver. This motion generates sound waves that correspond to those impinging on the transmitter of the speaker's telephone.

One function of the *network* in the telephone set is loop equalization. The set must be able to operate with various switching systems (electronic switching equipment, crossbar systems, step-by-step systems, for example)[1] over different loop lengths (from less than a mile to several miles) and still provide the desired signal levels for transmission and reception. The network also controls the level of sidetone (that portion

---

[1] Chapter 10 describes switching systems.

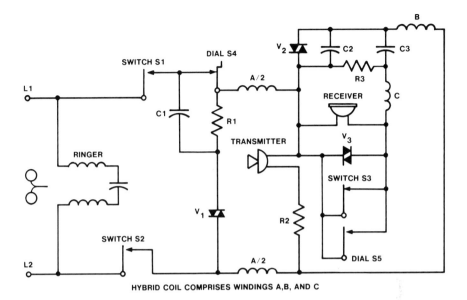

HYBRID COIL COMPRISES WINDINGS A,B, AND C

Figure 11-2. Schematic diagram of a standard 500-type telephone set.
The functional elements are carbon transmitter, electromagnetic receiver,
transformer-coupled passive network, dial, and electromagnetic ringer.
Resistor R1 and varistor V1 make up the loop equalizer circuit.* On short
loops, relatively high levels of current lower the resistance of the current-
sensitive varistor, thus shunting the rest of the network and reducing the
transmit and receive levels. On long loops, current levels are lower; the
varistor remains at a relatively high resistance, and the transmit and receive
levels are not significantly changed. Capacitors† C2 and C3 and resistor
R3 make up the antisidetone balance network impedance (see Figure 11-3).
Varistor V2 compensates for changes in varistor V1 and the impedance of
the loop. A third varistor (V3) is placed across the receiver to limit
received signals and loud sounds that might damage the ear.

---

* A *resistor* is a device with electrical resistance that is used in a circuit
  for protection, operation, or current control. A *varistor* is a device whose
  resistance is not constant but depends on the applied voltage.

† A *capacitor* is a device, consisting of both conducting and nonconducting
  (dielectric) materials, that can store electrically separated charges.
  Impedance is (or offers) electrical opposition to the flow of alternating
  current. It consists of both resistance and reactance and is measured in
  ohms.

of the signal from a telephone transmitter that is returned to the receiver
of that telephone). Sidetone conveys liveliness and affects the loudness
of the telephone conversation. If sidetone is not controlled, a speaker's
voice sounds much louder than the other party's voice. Consequently, a
person tends to talk more quietly. When the level of sidetone is too low,

a person tends to speak louder. Since the controlling circuit reduces the level of sidetone, it is commonly called an *antisidetone circuit* (see Figure 11-3).

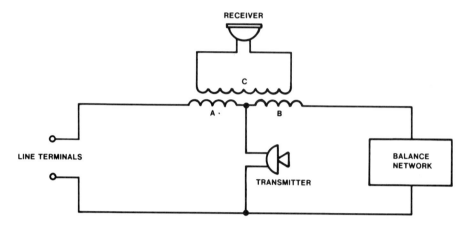

Figure 11-3. Antisidetone network of a standard 500-type telephone. This network consists of a 3-winding hybrid transformer and an impedance balancing network that efficiently couple the transmitter and receiver to the telephone loop. The alternating current generated by the transmitter induces voltages in the three windings so that only a little of the signal power goes to the receiver; most of it is divided between the balance network and loop impedances. Signals from the loop induce voltages so that most of the signal power is delivered to the receiver and relatively little goes to the transmitter and the balance network. Both the turns ratio of the transformer and the impedances have been selected to produce a sidetone signal of about the same amplitude as a nominal received signal. Because it influences return loss and thus affects the quality of transmission, the impedance of the set must also be matched to that of the loop.

Signaling. In the 500-type set, signaling involves use of the switchhook, dial, and electromagnetic ringer. The *switchhook* actuates central office supervision equipment to indicate origination, answer, or termination of a call.

A rotary *dial* signals the central office by means of dial pulses, controlled interruptions (nominally ten pulses per second) of the loop current. TOUCH-TONE dialing may also be provided. A pad with twelve pushbuttons (digits one to zero, plus an asterisk [*] and a number, or pound, sign [#]) and associated oscillator circuitry generates tones that are in the voiceband.[2] Each digit and symbol is uniquely represented by two tones, one from each of two mutually exclusive groups of four frequencies. This type of dialing or signaling is often referred to as *dual-tone multifrequency* (DTMF).

---

[2] Some manufacturers produce sets with pushbutton dials that generate dial pulses.

The *ringer* is connected across the loop ahead of the switchhook contacts and operates when a nominal ringing voltage of 86 volts root-mean-square (rms) at 20 hertz (Hz) is superimposed on the telephone loop at the central office.

The Electronic Telephone

In newer design electronic telephones, active electronic devices replace several components of the familiar 500-type set. For example, the functions of the transformer-coupled speech network, dial, and electro-mechanical ringer can be incorporated into large-scale integrated (LSI) circuits, the use of which offers advantages to the designer, the manufacturer, the telephone company, and the customer. Reductions in component weight and volume give the designer much more freedom. More automated assembly techniques make manufacturing savings possible. Overall transmission performance is improved, and with fewer mechanical parts to wear out, field performance reliability is improved, benefiting both the telephone company and the customer.

A typical electronic *TOUCH-TONE* telephone set contains a **tone ringer, surge protector, polarity guard, electronic dial, active network,** and **linear transmitter** (see Figure 11-4).

- The *tone ringer,* which replaces the conventional electromechanical bell ringer, contains a ringing detector. When a valid ringing signal is detected, a ringing-tone generator produces an output signal that drives an electroacoustic transducer, which emits the alerting tone.

- The *surge protector* is connected across the telephone line to protect the circuitry from damage by high-voltage surges such as those induced by lightning.

- The *polarity guard* guarantees proper tip and ring (see Section 8.4.1) polarity.

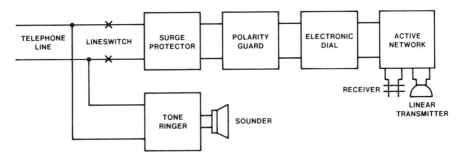

Figure 11-4. Block diagram of an electronic telephone.

- The *dial* on the electronic *TOUCH-TONE* telephone has a pushbutton keypad to generate DTMF signals. Electronic logic accomplishes switching functions, like muting out the receiver, that are normally performed by a series of mechanical switches in 500-type telephone sets. The DTMF circuitry may also be replaced by circuits generating dial-pulse signals.

- An integrated *active network* replaces the transformer-coupled passive network used in the 500-type set. The gain of the transmit and receive amplifiers is automatically adjusted by the loop current to compensate for transmission loss, which varies with loop length.

- A more stable *linear transmitter,* or microphone, can replace the carbon transmitter (whose efficiency is affected by age) because of the amplification provided by the active network.

Sets with Additional Features

In addition to sets that provide basic telephone service, other equipment is available to provide special features and services. These sets include apparatus for the handicapped (see Section 2.2.5), repertory dialers, and speakerphones.

A *repertory dialer* permits a user to dial telephone numbers automatically from a preprogrammed directory. In early designs like the card dialer, the directory consisted of prepunched cards, which were read by mechanical feelers. Currently, the *TOUCH-A-MATIC* repertory dialer, which incorporates solid-state memory, can automatically dial from a repertory of stored numbers at the touch of one button (see Figure 2-4). Low-cost integrated circuits and microprocessors permit the addition of special features such as dial-tone detectors, which sense dial tone and initiate automatic dialing, and last-number dialing, where the last number dialed by the user is automatically redialed at the push of a button.

There are several types of hands-free telephones and loudspeaking telephones manufactured by Western Electric and other companies. Some models are used as adjuncts to a telephone set and others are used in place of a telephone set. The Western Electric 4A *speakerphone,* which is used with a telephone set, offers hands-free voice telephone service between a group of people and one or more other groups or individuals. The 4A speakerphone is a simple audio terminal that consists of a transmitter unit (which contains an on-off switch, volume control, and an indicator light) and a loudspeaker unit (see Figure 2-5).

The transmitter should be placed centrally among the conferees and the loudspeaker at least 18 inches from the transmitter. The distances of the transmitter and receiver from the users introduce a loss of 20 decibels (dBs) or more in each path. More gain must be added to both units than that normally provided in a conventional telephone, where the transmitter and receiver are held close to the user's mouth and ear. When this gain is added, however, singing[3] and distant-talker echo (see Section 6.6.1) can occur. These problems are solved by *voice switching*. Voice switching activates transmission in only one direction at any one time. Logic added to the speakerphone compares the speech energy in the transmit and receive paths at each location to determine which path should be activated.

Public Telephones

Two types of public telephones are available: **coin** and **Charge-a-Call**. A *coin* telephone can handle both sent-paid (coins deposited) and non-sent-paid (billed to another telephone or calling card) calls, while a *Charge-a-Call* station can only complete non-sent-paid calls (see Section 2.6).

The coin telephone differs from a Charge-a-Call station primarily in the functions required to detect, identify, totalize, collect, and return coins. From a functional standpoint, a standard 500-type set could be used for Charge-a-Call service. The central office, however, must be able to identify a Charge-a-Call station and deny completion of local calls and toll calls dialed without an initial zero.

This section discusses the evolution of coin telephones, the operation of coin systems, coin-handling functions, design considerations, and current trends.

Evolution of Coin Telephones. The present standard single-slot coin telephone was introduced in the mid-1960s. The first version, the A-type set, was produced until 1968. When the need for dial-tone-first service became apparent, the C-type set, which can be equipped with either rotary or *TOUCH-TONE* dialing, was introduced to allow conversion to dial-tone-first operation.[4] Both the A- and C-type stations were electromechanical—they registered and totalized coins mechanically. They generated coin signals as spurts of a single-frequency tone to indicate the denomination of deposited coins: one spurt for a nickel, two for a dime, and five for a quarter. These spurts were recognized by the operator and manually recorded on the toll message ticket.

By the early 1970s, it had become apparent that Automated Coin Toll Service (ACTS) equipment could readily automate routine operator functions on sent-paid toll calls. Moreover, when added to the widely

---

[3] An undesirable, self-sustained oscillation in a channel caused by excessive positive feedback. In a telephone connection, singing manifests itself as a continuous whistle or howl.

[4] A large proportion (about one-third) of Bell System stations have not yet been provided with dial-tone-first service. Because of the high cost of converting some older electromechanical central offices, operating companies may delay conversion until electronic switches replace older offices.

deployed Traffic Service Position System (TSPS),[5] ACTS would also provide substantial savings by reducing the number of operators required. Automated detection of coin deposits made certain station changes necessary to prevent simulation of coin signals by noise and voice energy on the coin line. All coin sets manufactured after 1977 have a chassis equipped with a dual-frequency oscillator. In addition, a systemwide program was started to replace the single-frequency units in the older A- and C-type sets before the introduction of TSPS/ACTS.

In 1978, with the production of the D-type station, electronic apparatus (for self-testing before coins are accepted and for coin detection) was introduced into the coin telephone. Compared with the C-type set, the D-type station has a different chassis, and the totalizer is physically replaced by electronic apparatus (a 47A signal unit). Because of the power requirements of its electronic apparatus, this station can only be used in a dial-tone-first office. It generates dual-frequency coin tones and has other features that permit its use in locations equipped with TSPS/ACTS and Automated Calling Card Service.

In the late 1970s, studies indicated that a new electronic set with design characteristics substantially different from those of the D-type set would provide considerable maintenance savings and improve coin collection. These studies led to the development of the 3A2 electronic coin public telephone (ECPT), which is scheduled for introduction in late 1983 or early 1984. This set, shown in Figure 11-5, contains a new coin chute and a new chassis with a WE 4000 microcomputer to eliminate problems like coin jams and to prevent acceptance of coins when in an unpowered (inoperative) state. Whenever the set goes off-hook, the microcomputer tests itself, the other chips in the ECPT, and the coin-processing mechanism. The microcomputer allows considerable flexibility and still operates on only loop power. A self-test button inside the ECPT helps maintenance personnel identify malfunctioning components.

Coin System Operation. Coin telephones require access to special equipment in the central office to determine the presence of coins and handle their collection or return. In coin-first offices, a specific coin deposit is required to initiate the call. For calls that do not require them, the coins are returned either during call setup (collect and other calls served by TSPS) or at the end of the call (repair and other service calls). In dial-tone-first offices, the presence of an initial rate deposit is detected automatically to permit completion of a call that requires the initial rate.

In most coin telephones, including the new ECPT set, the station functions on a prepay basis. That is, the coins must be deposited before call

[5] See Section 10.4.1.

Figure 11-5. An electronic coin public telephone.

completion is permitted. A few of the older central offices with step-by-step equipment, usually found in sparsely populated rural areas with only a few public telephones, are arranged to accept deposits after completion of the call. This method of operation, called *postpay coin service*, was usually chosen because of the long distance between the local community dial office and the serving toll switchboard and the resultant large cost of returning coins on uncompleted calls. TSPS remote trunk arrangement (RTA)[6] is reducing the number of such locations.

Local prepay coin service can be either timed or untimed, depending on telephone company preference. Most service in the United States is untimed (without overtime charging). A few states, mostly in the Northeast, have local coin overtime charging. At the end of an initial

---

[6] See Section 10.4.1.

interval, an additional deposit is required. If the additional deposit is not made, the local call is terminated.

Coin-Handling Functions. As indicated earlier, a coin telephone must be able to detect, identify, totalize, collect, and return coins.

- *Detection and identification* — All types of coin telephone sets must size deposited coins as they travel down the coin chute to determine whether they have the appropriate diameter and thickness. Depending on the set, this is done either mechanically or electrically. The sets must also determine whether the coins are the proper weight and material. If these three criteria (size, weight, and material) are not met, the deposited coin is not accepted. If they are met, the coin is identified as a nickel, dime, or quarter, and when appropriate, this information is sent to the central office.

- *Totalizing* — All coin sets totalize deposited coins, that is, add them up until the initial rate has been met. C-type sets in use for coin-first service totalize coins before sending a signal to the central office to begin dial tone. In dial-tone-first offices, the station sets must totalize to ensure that a coin-present test will be successful and the call appropriately processed once the initial rate has been correctly deposited.

- *Collect/return coins* — Coins that are accepted after testing for size, weight, and material are totalized and fall onto a coin hopper. The coin hopper trap is controlled by a single-coil coin relay, to which is applied either a negative or a positive dc voltage, depending on whether the coins are to be returned or collected. After activating coin collection or return and before releasing the equipment used on the coin call, most central offices make coin disposal tests to see if the function was successfully completed. If the first attempt was not successful, the coin relay is reactivated and the test repeated. The coin disposal test is usually made at the end of a call when the receiver of the telephone is on-hook so that the user is not disturbed while talking. When coins are collected during a conversation (as for a TSPS call or local coin overtime call), tests are not made then so as to minimize the interruption. Coin disposal failures are detected when subsequent deposits are collected at the end of the call or when a coin-cleanup cycle is generated at the central office.

The amount of line current (23 milliamperes) required for proper coin telephone operation limits the distance of the set from the central office. In most central offices, the limit is 1300 ohms, typically a distance of about 3 miles. Coin range extenders can be used to permit proper operation of the coin telephone over ranges from 1300 to 2800 ohms, typically up to a distance of about 8 miles. A number of Bell System products such as the 8A REG, as well as products from outside manufacturers, can be used as range extenders.

Design Considerations. Reliability and ease of maintenance are particularly important considerations in the design of public telephones. Sets must be designed to function in a variety of environments, including remote outdoor locations where temperatures may range from −40°F to +130°F. They must also withstand abuse by customers, fraud attempts, vandalism, and extremely high use. They must be easily maintained and quickly repairable. The instructions on public telephones for both local and toll calls must be simple and easy to follow.

Current Trends. The trend toward complete dial-tone-first service is expected to continue, although more slowly since the urban centers have largely been converted. ACTS will probably become universal because of savings to be derived from its use. The introduction of Automated Calling Card Service, which is superimposed on ACTS, will also continue.

The Transaction Telephone

Transaction telephones (see Figure 11-6) are business telephone sets that not only provide standard telephone services but also automate business transactions such as credit verification, authorization on credit-card purchases, inventory control, and electronic funds transfer. These sets are

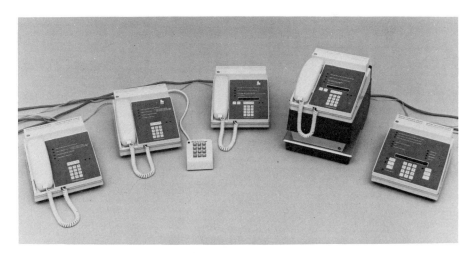

Figure 11-6. Transaction telephone sets. *Left to right,* Transaction I, Transaction I with auxiliary manual entry pad, Transaction II, Transaction II with printer, Transaction III terminal.

early examples of a type of telephone that may see widespread use in some of the voice and data services now emerging. The Transaction telephone relies on miniature integrated circuits for its memory and logic and for its card-reader components. It reads the information encoded on a magnetic strip on the back of credit cards. Additional information can be entered with a 12-button manual entry pad.

A sequence of lights leads the user through a series of input procedures. In a routine transaction (depicted in Figure 2-7), a clerk slides two magnetically encoded cards (the merchant's identification card and the shopper's credit card) through a magnetic tape reader and then enters the amount of the transaction with the pushbutton pad. To ensure against fraud, the shopper can also use an optional remote pad to enter a personal identification number. The Transaction terminal automatically generates the proper *TOUCH-TONE* dialing sequence to call a computer at a credit agency or bank, transmits information about the transaction, and obtains a purchase authorization in the form of a computer-synthesized voice or a light signal. Details of the transaction are recorded in computer memory for billing and accounting purposes.

The Transaction set can also be used as a repertory dialer to call frequently needed numbers automatically. Cards, specially encoded for this purpose, can be stored in a slot at the back of the set.

The Transaction II telephone set is equipped to collect, store, and exchange information with a remote data base. Data responses may be configured to provide a light-emitting diode display, to light either of two tone-to-light response lamps, or to activate the Transaction printer. A built-in speaker permits the set to be used with or without the handset.

The Transaction III data terminal can receive transaction approval, denial, or referral from a data center in a frequency-shift-keyed mode (see Section 6.4.4). This information appears on an 8-character display, on one of four response lamps, or on a Transaction printer.

11.1.2 DATA SETS

Data Set Functions

The need for data sets became clear in the mid-1950s when the development of digital computers stimulated an expanding commercial market for data communications. Until that time, the telephone network was almost entirely a voice network. The analog transmission systems in use, while suitable for voice communications, were not designed to handle high-speed digital signals.

The primary function of a data set is to convert a data signal from a computer or data terminal, usually in digital form, into an analog signal suitable for transmission over conventional telephone channels. As shown in Figure 11-7, data sets are used at both ends of a typical connection between data terminals and a computer. The digital signal is converted to an analog signal by one of several modulation techniques (see

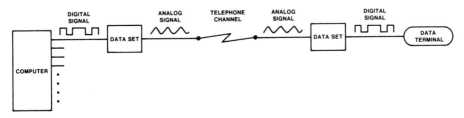

Figure 11-7. Typical data-set application. The input to the data set is typically a binary stream (that is, a sequence of bits). The data sets convert the customer's conventional signals into a form that is suitable for transmission over telephone channels.

below). Since many data sets both transmit and receive, they must both modulate the signal to be transmitted and demodulate the received signal. The term *modem* (from *modulation-demodulation*) is often used as a synonym for *data set*, although a data set usually provides additional control functions such as those involved with call setup and call termination.

Data Set Types and Applications

Data sets can be described in terms of several operational characteristics.

- *Speed* — This refers to the rate of data transmission (in bits per second, bps) provided by a data set. The usual categories are:

  — low: below 300 bps

  — medium: 300 bps to 9600 bps, or 9.6 kilobits per second (kbps)

  — high: above 9.6 kbps.

- *Voiceband/broadband* — A *voiceband* channel has a nominal bandwidth of 4000 Hz, or 4 kilohertz (kHz), and is suitable for voice transmission. A *broadband* channel has a greater bandwidth and is suitable for data transmission. (See Section 6.2.)

- *Use* — Data sets are used on the public switched telephone network (PSTN), on private lines, or on both.

- *Synchronous/asynchronous* — In synchronous transmission, the time interval between characters is fixed and based on a controlled clock rate. In asynchronous transmission, the interval between characters is variable.

- *Half duplex/full duplex* — In half-duplex transmission, data are alternately sent and received. In full-duplex transmission, data may be sent in both directions simultaneously.

- *Serial/parallel* — In serial transmission, the elements of a signal occur at successive intervals. Only a single transmission channel is used. In parallel transmission, the signal elements are sent simultaneously, either over separate channels or on different carrier frequencies on the same channel.

• *Modulation technique* — Modulation is the process by which a characteristic (that is, the frequency, phase, or amplitude) of a wave (such as a carrier) is varied in accordance with another wave (such as a signal). The techniques commonly used by data sets are:

— *Frequency-shift keying (FSK).* In this method, the modulating signal shifts the instantaneous frequency of a carrier between predetermined, discrete values. FSK is preferred when hardware economy and simplicity are more important than the efficient use of channel bandwidth.

— *Phase-shift keying (PSK).* In this method, the modulating signal shifts the instantaneous phase of the carrier between predetermined, discrete values.

— *Digital amplitude modulation.* Amplitude is the wave characteristic modulated in this technique. Two types of digital amplitude modulation that have been used in the Bell System are *vestigial sideband modulation* (*VSB*) and *quadrature-amplitude modulation* (*QAM*).[7]

Evolution of Bell System Data Sets

The first Bell System commercial data set was introduced in 1957. It used frequency-shift keying and transmitted data at a rate of 1 kbps. By the early 1960s, a whole line of *DATAPHONE* data sets had been developed and categorized as follows:

100 Series	-	Narrowband (low-speed) data sets
200 Series	-	Voiceband serial data sets
300 Series	-	Broadband serial data sets
400 Series	-	Voiceband parallel data sets
600 Series	-	Voiceband analog data sets

Throughout the 1960s, a rapid progression of new developments led to ever higher bit rates for a given channel bandwidth. The most dramatic increase came as a result of the development of the *adaptive transversal equalizer* (see Lucky, Salz, and Weldon 1968). This adaptive filter automatically compensates for the linear distortion (see Section

---

[7] Section 6.4.4 discusses FSK and PSK. See Lucky, Salz, and Weldon 1968, pp. 159-164, for information on VSB and QAM.

6.6.2) introduced by a telephone channel. Without automatic equalization, voiceband data rates had been limited to about 3 kbps; with automatic equalization, transmission at rates up to approximately 10 kbps became feasible.

The use of customer-provided modems on voice bandwidth private lines had been permitted for many years, but following the Carterfone Decision in 1968 (see Section 17.2.7), many new data-set vendors entered the marketplace. Then, in 1975, the Federal Communications Commission (FCC) permitted direct connection of customer-provided data sets to the PSTN, further stimulating competition. Bell System data sets also evolved during this period and less popular lines, such as the 600 series, were dropped.

The most recent developments in data sets have been more the result of advances in technology and in customer features than improvements in modulation techniques. First, beginning in 1972, the Bell System introduced a new family of data sets in the 100, 200, and 400 series. The dramatic reduction in size and power of these sets was the result of the exploitation of medium- and large-scale integrated circuit technology and microprocessors (see Kretzmer 1973).

In 1980, a family of more sophisticated microprocessor-based, synchronous data sets for private-line applications was introduced. Included in these *DATAPHONE* II data sets are *WE* 8000 general-purpose microprocessors and up to nine custom-designed large-scale integrated circuits. The microprocessors control the overall operation of the data sets; the large-scale integrated circuits perform signal-processing functions such as encoding, decoding, modulating, demodulating, and equalizing.

This built-in intelligence provides a new capability for data sets: They can continuously monitor themselves and the quality of the received signal and, when necessary, identify and report faults to a central location where diagnostic control devices display the faults for the customer's operations personnel (see Figure 2-2). Reporting is done over a narrow band of frequencies at the lower end of the voiceband channel using a low-speed FSK transmitter-receiver integrated into each data set.

In addition, the data sets provide customers with a menu of tests and commands that they can use to identify, isolate, and diagnose faults within the system without technical assistance. Maintenance personnel can then make rapid repairs, thus reducing system downtime. Table 11-1 lists the available Bell System data sets and their characteristics.

Design Considerations

In considering data set performance, the basic criterion is error rate. Both the *bit* and *block* error rates are of interest. Because data are usually transmitted in blocks, the block error rate is perhaps the better measure of performance for higher-speed voiceband transmission (above 2 kbps).

For example, a survey[8] of data transmission performance on the PSTN indicates that, for a 202-type data set operating at 1.2 kbps, 82 percent of the calls will achieve a bit error rate of 10^{-5} or better (that is, at most, only one wrong bit in one hundred thousand). On 85 percent of the calls, the block error rate for 1000-bit blocks is 10^{-3} or better (that is, one erroneous block per thousand). A 208-type data set operating at 4.8 kbps will achieve a block error rate for 1000-bit blocks of 10^{-2} or better on about 88 percent of its calls.

11.1.3 DATA AND GRAPHICS TERMINALS

Data Terminals

Data terminals are the devices through which information enters or leaves a communications system. They permit people to communicate with computers and with each other. They convert information from alphanumeric form (the words and numbers of ordinary language) to electrical signals for entry, storage, and transmission and then reverse the process to deliver information. Terminals may also facilitate data communications by performing such functions as code conversion, error control, formatting, and acting on control information. Typically, keyboards are used to enter information through a terminal, and printers or cathode-ray tubes (CRTs) deliver information received by the terminal.

Many different kinds of data terminals are now being offered by various manufacturers. For example, the electronic cash register is becoming increasingly popular. In most cases, these cash registers have built-in modems with which they communicate with a computer over private lines.

The following discussion concentrates on Bell System terminals, especially those made by the Teletype Corporation, a subsidiary of Western Electric. These include teletypewriters, CRT (or video display) terminals, cluster controllers, and intelligent data entry and retrieval terminals.

The *DATASPEED* terminal product line is a family of modular, interactive data terminals that can be used to enter, store, display, edit, print, send, and receive data over a dial-up switched network or private-line facilities. Four versions of terminals, *DATASPEED* 40/1 through 40/4, operate either synchronously or asynchronously at speeds up to 9.6 kbps. The *DATASPEED* 40/4 terminals enable users to cluster several terminals on a common controller. Several arrangements are available:

- single-display — one single-display controller, a keyboard-display terminal, and an optional printer

---

[8] See Balkovic, Klancer, Klare, and McGruther 1971.

- minicluster — one minicluster controller and up to three devices, two of which can be keyboard-display terminals (the other may be a printer)
- maxicluster — a station cluster controller, up to six device cluster controllers, and as many as thirty-six devices (a maximum of twenty-four keyboard-displays with the rest printers).

The Teletype Corporation's 43 teleprinter operates on either the PSTN or private lines at speeds of 10 or 30 characters per second. It incorporates a data set and is available in either keyboard send-receive (KSR) or receive-only (RO) configurations. The pin-feed version shown in Figure 11-8 provides 132 characters per line on fanfold, 12-inch-wide paper. The friction-feed version prints a 72- or 80-character line on continuous-roll, 8-½ inch wide paper. The 43 teleprinter offers a full 94-character ASCII[9] character set and uses a 7-by-9, dot-matrix, impact printhead.

Figure 11-8. The 43 teleprinter.

The recently introduced line of *DATASPEED* 4540 terminals is a series of intelligent (microprocessor-based), keyboard-display, interactive input/output devices (see Figure 2-3). These terminals are designed for applications such as inquiry/response, data entry, and data retrieval and for use with synchronous private-line data sets or *DATAPHONE* digital service channels at speeds of from 2.4 to 9.6 kbps. Station arrangements

[9] American Standard Code for Information Interexchange.

from one to thirty-two devices, including both keyboard-display units and/or printers, are possible under the control of a single, microprocessor-based cluster controller. Figure 11-9 shows a typical station arrangement.

Finally, in the area of disk storage, the Bell System offers the *Comm-Stor*[10] II communications storage unit. This is a single- or dual-drive magnetic diskette system for storage of messages received on-line or prepared locally for transmission to a remote computer. It uses floppy disks to improve the message preparation and communication functions of *DATASPEED* 40 terminals, 43 teleprinters, and other asynchronous terminals.

Figure 11-9. A *DATASPEED* 4540 terminal station arrangement.

Graphics Terminals

GEMINI 100 electronic blackboard system (see Figure 2-6) can send handwriting and voice over conventional telephone lines and is used for remote teaching and teleconferencing. It instantly transmits characters written on a special pressure-sensitive blackboard to one or more remote locations over the PSTN or private lines. At the remote location(s), the

[10] Registered trademark of Sykes Datatronics, Inc.

writing appears on a monitor and can be viewed comfortably in ordinary room light. A tape recorder can be connected at either end to tape both voice and graphics signals for later replay.

A basic *GEMINI* graphic communications system for 2-way transmission consists of the following equipment at each end of the installation: a blackboard, a remote monitor with video memory circuitry, and electronic equipment such as a control unit, transmitter, receiver, and a data set (see Figure 11-10). In addition, a commercial hard-copy printer can be connected to the monitor to provide paper copies of the graphics.

Figure 11-11 shows the multilayered structure of an electronic blackboard. Writing generates two voltages: one representing the horizontal position of the chalk, the other its vertical position. These voltages are sampled forty times per second, often enough to keep up with even the fastest handwriting. The voltages are encoded in digital form and then transmitted over a telephone line at a speed of 1.3 kbps with a 202-type data set. At the remote location(s), the received signals are converted into analog form, stored in an electronic scan converter memory, and then sent to the television monitor for display. A portable conference telephone transmits the audio portion of the lecture or conference over another line.

The blackboard is erased with an ordinary felt eraser, and as the eraser wipes a spot off the blackboard, a corresponding alteration is made on the remote monitor(s). A CLEAR button on the control unit may also be used to clear the video memory at the remote location(s).

Chalk switching allows the person at either end to write, and lockout circuits preclude the transmission of writing in both directions simultaneously. This allows the use of less-expensive half-duplex data sets (see Section 11.1.2). Half-duplex transmission is necessary for multipoint conferencing so that any location may send to all the others. Experience has shown that half-duplex transmission is sufficient from a human factors point of view. People do not compete for the chalk as they may for the microphone in a voice-switching conference system.

A significant feature of the *GEMINI* 100 electronic blackboard system is its provision for recording the graphics transmission on a conventional audio tape recorder. If a stereo recorder is used, both voice and graphics can be recorded simultaneously.

The *GEMINI* 100 electronic blackboard system can be a significant aid in education, extending classrooms into areas that are either too remote or too small to justify the expense of special courses.

11.1.4 GENERAL DESIGN CONSIDERATIONS

Station equipment must be designed to be compatible with the Bell System telecommunications network and to provide service consistent with Bell System objectives. Other factors that must be considered are safety,

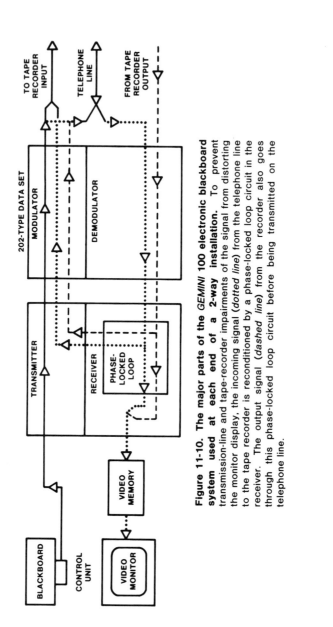

Figure 11-10. The major parts of the *GEMINI* **100 electronic blackboard system used at each end of a 2-way installation.** To prevent transmission-line and tape-recorder impairments of the signal from distorting the monitor display, the incoming signal (*dotted line*) from the telephone line to the tape recorder is reconditioned by a phase-locked loop circuit in the receiver. The output signal (*dashed line*) from the recorder also goes through this phase-locked loop circuit before being transmitted on the telephone line.

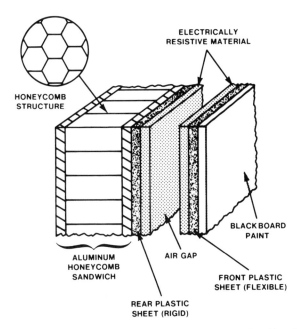

Figure 11-11. Structure of the electronic blackboard. The blackboard is about 1-¼ inches thick and consists of two plastic sheets in front of an aluminum-honeycomb / aluminum "sandwich." Writing causes the front plastic to touch the rear plastic, which is cemented to the sandwich. Contact between the plastic sheets generates voltages representing the position of the chalk. (*Drawing not to scale.*)

human factors, and telephone company concerns about installation, maintenance, and logistics.

Compatibility requirements relate to the characteristics of network components. For example, rotary dials must operate with central offices equipped with all types of switches (for example, the No. 5 Crossbar, the 1*ESS* electronic switch, etc.). From a transmission standpoint, both variations in customer loops and limitations of the central office are factors designers must consider.

The equipment must not introduce noise, which would degrade service. It must not interfere with standard tests such as line insulation tests nor with the performance of billing equipment. All new telephone products, both Bell and non-Bell, must be "registered" and must pass tests to ensure compliance with FCC Part 68 Rules and Regulations (see Section 8.7.5).

To protect the service provided by the network and the people working on it, functional requirements must be satisfied, both before and after the application of specified environmental stresses. For example, the level of signal power generated by terminal equipment to the telephone line must be limited to prevent crosstalk on other subscriber lines.

The safety of users, installers, and repair personnel is also an important design consideration, both during ordinary use and when the equipment is subjected to high voltages like lightning. For example, insulation is required to prevent the appearance of unsafe voltages or currents.

From the standpoint of physical design, the telephone must operate over a wide range of temperatures and humidity. It must also continue to function after being subjected to shocks such as falling from a table onto an uncarpeted floor. Materials used in its manufacture must be fire resistant and not cause allergic reactions.

Since station equipment is the most important interface with customers, it must be designed with potential users in mind; that is, it must not only work well but be both comfortable and easy to use. A telephone handset, for example, must have proper acoustic coupling with the user's mouth and ear. The ringer has to have a pleasant sound, yet be loud enough to be heard in a variety of environments. The dial must minimize dialing errors.

Designs for telephone equipment must also take into account the installation, maintenance, and logistic problems confronting telephone companies. A modular telephone set design, which uses plug-ended cords and connectors in the station set and house wiring interface, allows installation and maintenance personnel to assemble or repair a telephone set quickly. The availability of plug-ended cords reduces the number of sets that an installer must carry to be sure of having the proper colors and lengths of cords. Modular design also facilitates customer participation through Bell PhoneCenters[11] and telephone company service centers (see Section 2.7).[12] The higher manufacturing cost of connectors in the station set is more than offset by reduced operating costs.

Design changes are constantly being incorporated into station sets to reduce the amount and cost of maintenance and repair. For example, the newer models of *TRIMLINE* telephones use light-emitting diodes for dial lighting, eliminating the need for an installation visit to add a dial light transformer as well as reducing the number of service calls associated with lamp failures. These sets also incorporate a chassis design that allows the internal mechanism of the handset to be removed easily from the plastic shell, substantially reducing the cost of refurbishing.

---

[11] As a result of the FCC's Computer Inquiry II decision (see Section 17.4.3), many Bell PhoneCenters were taken over by American Bell Inc. on January 1, 1983.

[12] For non-Bell station equipment, customers will observe the arrangements provided by the manufacturer or merchandiser of the equipment.

11.1.5 CURRENT TRENDS

Until very recently, the use of electronics in residential telephones has generally been limited to low-volume, special-purpose station equipment that provides special services for which customers are willing to pay (for example, automatic dialing). Now, however, the availability of low-cost electronic devices, like microprocessors, is substantially reducing the cost of adding features and services. Microprocessor control and custom large-scale integrated circuitry will make possible a wide variety of feature modules that customers can select and install themselves. These devices, which incorporate timers, digital displays, and memory for data storage, will make the telephone an intelligent terminal with data-handling capability that customers can customize to suit their needs.

11.2 CUSTOMER SWITCHING SYSTEMS

This section describes the equipment that provides the customer switching services described in Section 2.3. Both older equipment still in service and new products are included. With few exceptions, new installations use only the most modern equipment. Figure 11-12 shows the basic relationships between a central office, private branch exchanges (PBXs), and key telephone systems (KTSs).

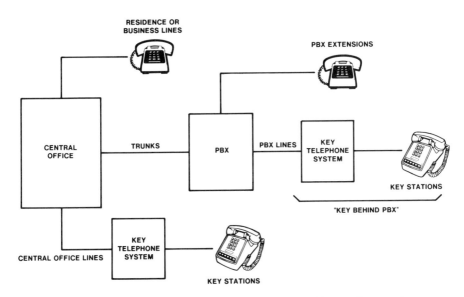

Figure 11-12. Relationship between customer switching systems and a central office. *Note:* Trunks may be PSTN ("message") trunks, Wide Area Telecommunications Services (WATS) trunks, foreign exchange (FX) trunks, or tie trunks to other PBXs.

11.2.1 KEY TELEPHONE SYSTEMS

A key telephone system is basically an arrangement that permits the station user to operate switches (*keys*) to access more than one central office line or other line (typically up to five) and to perform other desired functions. Typical functions include answering or originating a call on a selected line, putting a call on hold, or operating an intercom feature for phones at the same location, a buzzer, or a message-waiting light at another (nearby) telephone.

The earliest KTSs, known as *wiring plans*, were simply collections of lamps, keys, and wiring. These evolved into special hardware arrangements, 1A KTSs, that reduced the amount of field labor required for installation. For example, the assemblies of the 1A1 KTS, introduced in 1953, could be installed with a smaller amount of special wiring. The 1A2 KTS, first offered in 1963, was even easier to install. In addition, its solid-state components, printed wiring boards, and miniature relays resulted in smaller equipment arrangements on customer premises.[13]

In 1973, in order to reduce the amount of labor associated with the installation of key telephones even further, preassembled KTS packages were introduced. These COM KEY[14] key telephone systems come from the factory prewired and equipped for a specific number of lines and stations.

In addition to the packaging, several new system features and a new line of station instruments were introduced with the COM KEY key telephone system. While similar to other key sets of the time, the new station sets incorporate some new features. For example, an amplifier and speaker have replaced the electromechanical bell. This change serves two purposes. First, it replaces the ringing bell with a tone signal to indicate the arrival of a call. Second, the amplifier and speaker are accessible to the intercom calling feature so station users who are using their handsets for other calls may also receive voice messages. A new system feature of the COM KEY key telephone system is a simple conference arrangement. A user first establishes two independent calls on two central office or PBX lines and then completes the 3-way connection by depressing buttons for both lines simultaneously. Conference calls can also be established with two intercom calls.

Both the 1A2 and COM KEY key telephone systems are still being installed. They are characterized by heavy cables (twenty-five pairs) from each station instrument to central KTS hardware, which is typically

[13] Numerous companies (including Stromberg-Carlson, ITT, and Automatic Electric) manufacture key telephone switching equipment and station sets. The equipment built by these companies is essentially the same as Western Electric products. In the past, it was sold primarily to independent telephone companies.

[14] Registered trademark of AT&T Co.

located in a nearby "closet." Figure 11-13 shows the configuration of these key telephone systems.

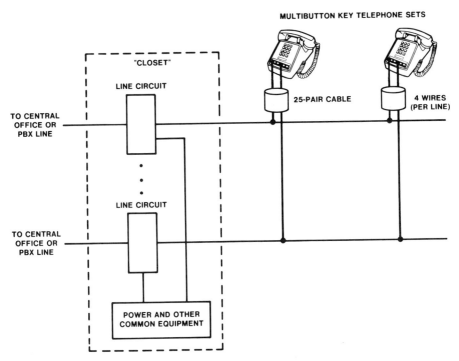

Figure 11-13. Basic 1A1, 1A2, or *COM KEY* **key telephone system.** The buttons on each telephone set are pushed to use particular lines. The line circuit operates a lamp under each button to indicate the status of the line: ringing, busy, on hold, or idle. The hold function on the line circuit permits a user to answer a call on another line on the set without releasing the first call.

11.2.2 MODERN SMALL COMMUNICATIONS SYSTEMS

The availability of low-cost, solid-state integrated circuit technology has made it possible to design new, small communications systems that are more versatile than earlier key telephone systems. These communications systems are characterized by a new line of multibutton electronic telephone sets equipped with light-emitting diodes (LEDs) in place of lamps and speakers rather than bells. A "skinny" cable of approximately 3- or 4-wire pairs (in place of the earlier 25-pair cable) is used to connect the instrument to the common equipment.

The *HORIZON*[15] communications system, illustrated in Figure 11-14, is an example of a modern, small communications system. The *HORIZON*

---

[15] Registered trademark of AT&T Co.

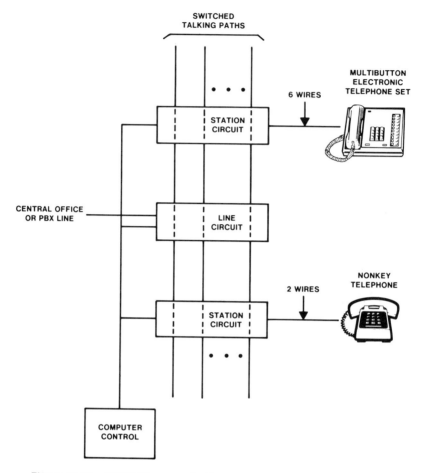

Figure 11-14. *HORIZON* **communications system.** Signals from telephone set buttons (or rotary dials) are interpreted by the computer control. The computer then directs the appropriate line and station circuits to engage the talking paths to establish the desired connection. The maximum number of lines provided is thirty-two. These can be PBX lines or central office lines or some combination of the two. A maximum of seventy-nine telephone stations can be served by a *HORIZON* communications system.

communications system uses both multibutton electronic telephone sets and conventional single-line instruments. The common equipment in this system is controlled by a microcomputer. In addition to conventional key telephone features, this system can provide built-in paging arrangements, call-transfer features, special intercom capabilities, and numerous other features similar to those found in a modern PBX.

Many features are made possible by the programs running on the control computer. For example, the call-coverage features allow calls to be directed to other telephones when the dialed number is either busy or does not answer. Such features are controlled in the *HORIZON* communications system by the software, which can be changed with a special piece of hardware known as an *access unit*. Telephone company personnel use a *service access unit* to make software changes in features like call coverage as well as to troubleshoot (that is, isolate faults) or modify the system. The customer can also change some of the feature-related software with a *customer access unit*, which provides a subset of the capabilities of the service access unit.

Many other small communications systems manufactured in both the United States and Japan have features and characteristics similar to those of the *HORIZON* communications system. Because these systems provide so much more than the key telephone equipment they replace, the term "communications system" is often used to describe them as well. By providing many of the same features as a PBX, they are also blurring the distinction between a traditional key telephone system and a traditional PBX.

11.2.3 VINTAGE PBX'S

The term *vintage PBX* as used here includes all of the PBX switching systems built prior to the introduction of computer control. The following paragraphs discuss a wide variety of these—both manual and dial (sometimes called *automatic*) arrangements. Section 11.2.5 discusses modern computer-controlled systems.

Manual PBXs

The earliest PBXs were essentially scaled-down central office systems. Thus, the first PBXs were manual cord switchboards. As shown in Figure 11-15, all PBX lines and trunks terminate in switchboard jacks. An attendant connects two stations or a trunk and a station manually with a cord circuit. A cord circuit includes, among other things, two cords with plugs that are inserted into the appropriate jacks. The attendant must monitor the lamp signals on all jacks and cord circuits to recognize requests for service and the completion of calls. Although eventually replaced by dial PBXs, the attendant and cord switchboard remained popular for many years because it was a convenient way to answer and extend incoming calls.

Dial PBXs

The first dial PBXs introduced in the late 1920s automated only the intercom calling and outgoing call functions. This was done with step-by-step switching equipment of the same type used at that time in central offices.

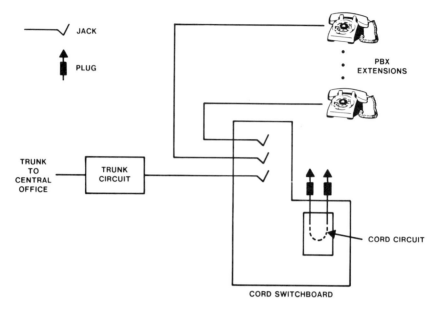

JACK

PLUG

TRUNK TO CENTRAL OFFICE

TRUNK CIRCUIT

PBX EXTENSIONS

CORD CIRCUIT

CORD SWITCHBOARD

Figure 11-15. Manual PBX. To answer incoming calls, the attendant inserts one of the cord plugs into the trunk jack. A call can be extended to a PBX station by inserting the other plug into the appropriate station jack. For outgoing calls, the first plug is inserted into the jack of the station requesting service; the other plug goes into a trunk jack. To set up intercom calls, the attendant uses a cord circuit to connect the calling extension with the called extension.

Figure 11-16 shows the relationships among the switches, the lines, and the cord switchboard. The most popular version of this system was the 701-type PBX. Small systems with some special switch arrangements were known as 740-type PBXs. During the 1950s and 1960s, the step-by-step systems were modernized and improved. Later versions had attendant consoles in place of switchboards and provided such features as direct inward dialing, *TOUCH-TONE* calling, and automatic identified outward dialing.

Automatic Electric, Stromberg-Carlson, and others also manufactured step-by-step PBX systems, which were essentially the same as those made by Western Electric for the Bell System. These were sold to independent telephone companies to provide PBX service to their business customers.

Crossbar PBXs

At about the same time that the 701-type PBX was being modernized, work began on new crossbar PBXs. Development and manufacture continued from the late 1950s, through the sixties, and into the early

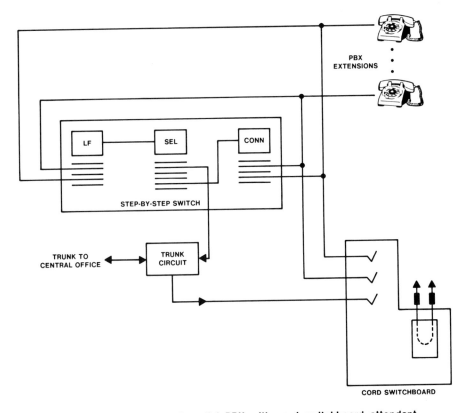

PBX EXTENSIONS

LF

SEL

CONN

STEP-BY-STEP SWITCH

TRUNK TO
CENTRAL OFFICE

TRUNK
CIRCUIT

CORD SWITCHBOARD

Figure 11-16. Step-by-step dial PBX with cord switchboard attendant position. Incoming calls from the central office ring into the cord switchboard. The attendant answers the call and uses a cord circuit to extend the call to the desired location. Outgoing calls are dialed by station users. When a station set goes off-hook, dial tone is provided by the combination of line finder (LF) and selector (SEL). A single digit, 9 for example, drives the selector to the ninth level, where an idle trunk is selected. For intercom calls, the first digit operates the selector and the next two operate a connector (CONN). The connector controls ringing to the called extension.

seventies. This technology was also borrowed from central office systems. Figure 11-17 shows a basic crossbar PBX.

Crossbar systems had several advantages over step-by-step systems. First, they used consoles rather than cord switchboards. Second, the equipment was much smaller and quieter. As a result, the switching system could be mounted in cabinets suitable for general office space, eliminating the need for an equipment room. Third, common control replaced progressive control switches. Common control permitted the

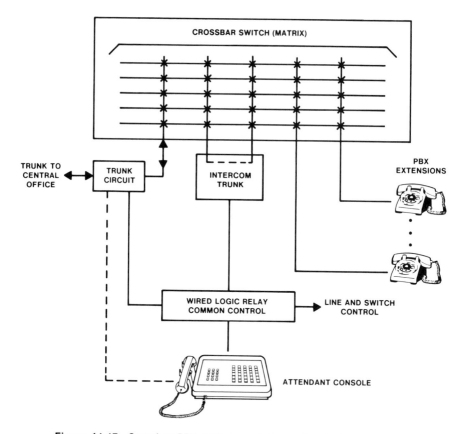

Figure 11-17. Crossbar PBX with console-type attendant position. All switched connections are made by the crossbar switches under the direction of the common control. Incoming calls are answered by the attendant but completed to the desired station over a crossbar switch path. Outgoing calls are dialed by the station user. The trunk and the crossbar switch path are selected by the common control. Intercom calls involve two switched paths: one from the calling party to the intercom trunk, the other from that intercom trunk to the called party.

design of improved line- and trunk-hunting algorithms and lowered the cost of providing *TOUCH-TONE* service. For added reliability, the common-control hardware was sometimes duplicated.

The major crossbar PBXs manufactured by Western Electric during this time were the 756 (for up to 60 lines), the 757 (for up to 210 lines), and the 770 (for up to 400 lines). Figure 11-18 shows the station capacities of the principal dial PBXs. It should also be noted that, in the early 1960s, the Bell System pioneered an advanced customer switching system. The

496

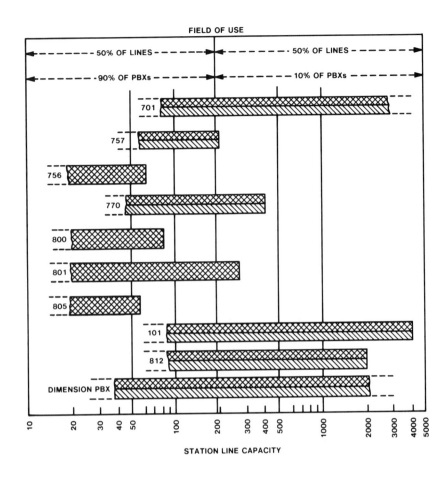

Figure 11-18. Fields of use for principal dial PBX systems.

101*ESS* electronic switch used a centralized digital computer to control time-division switching modules on the customer's premises.[16] Many of the techniques used in the 101*ESS* switch were used again in the *DIMENSION*[17] PBX described below.

---

[16] Section 7.3.3 describes time-division switching.

[17] Registered trademark of AT&T Co.

During the 1970s, numerous crossbar PBXs were imported from Japan. This equipment, manufactured by Nippon Electric, Fujitsu, and OKI Electric, used crossbar switches and wire spring relay hardware that looked the same as Western Electric products. The Japanese crossbar PBXs were purchased by both Bell and independent telephone companies as well as directly by customers.

Wired-Logic Electronic PBXs

The last of the vintage PBXs was a series of wired-logic electronic systems whose development and use overlapped that of the crossbar PBXs. These PBXs used solid-state (transistor) devices for the logic elements in the common control and a variety of devices for the switching matrix. For example, the 800 and 801 PBXs used a reed relay matrix (see Figure 11-19); the 805 PBX used full-size crossbar switches, and the 812 PBX used miniature crossbar switches. These systems covered various market segments from a few lines up to 2000 lines. As with earlier PBXs, the

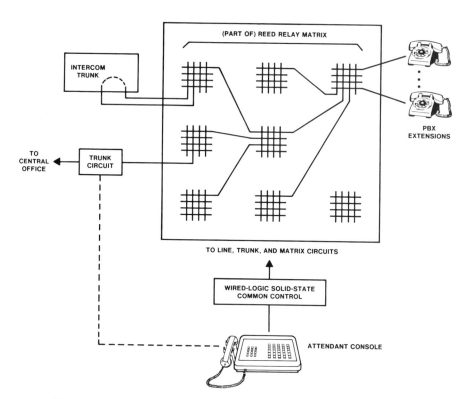

Figure 11-19. 800A electronic PBX with reed relay switch matrix. This system operates essentially like the crossbar PBX shown in Figure 11-17.

switching components were borrowed from central office designs. The remainder of the hardware and the wired-logic solid-state common control were unique to the PBX.

During this era (primarily the 1970s), other companies were also manufacturing PBXs with wired-logic solid-state common controls. While some used the same switching-matrix technology as the Bell System, others such as WESCOM and ITT used solid-state devices for switching matrixes.

11.2.4 RECENT DEVELOPMENTS IN THE DESIGN OF PBX'S

Recently, the evolution of PBX design has accelerated. This is partly the result of technological developments, such as the availability of small, low-cost microprocessors and integrated-circuit memory devices. These devices are well suited to the design of switching system control arrangements. The application of digital switching techniques has also been a factor. In this case, the analog voice signals are converted to a digital format,[18] switched by digital circuits, and then converted back into analog form. The combination of digital switching and digital microcomputer control has blurred the distinction between data- and voice-switching systems.

In addition to technological developments, the evolution of PBX designs has been affected by changes in the marketplace. In 1968, the FCC decided to allow business customers to purchase their own customer switching systems and connect them to the network. Prior to this ruling, switching services were only available from the serving telephone company. Following the ruling, several new companies began to manufacture and sell PBX equipment, while established firms began to offer existing, enhanced, or new products directly to users. Since the 1968 decision, the market for customer switching services and equipment has become highly competitive.

11.2.5 MODERN PBX EQUIPMENT

In 1974, the Bell System introduced a new electronic PBX. This system uses a special digital computer to control a time-division switch. The computer was designed by Bell Laboratories to provide both high reliability and an instruction set that is well suited to the requirements of telephone switching. Figure 11-20 shows a diagram of this system, the *DIMENSION* PBX.

---

[18] Section 6.4.3 discusses pulse-code modulation of analog signals.

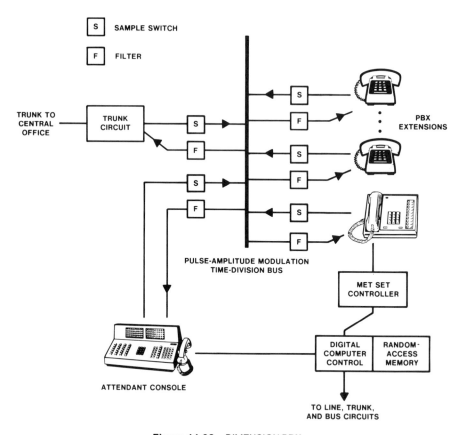

Figure 11-20. *DIMENSION* **PBX.**

The switching medium of the *DIMENSION* PBX is a time-division bus.[19] The time-division bus simultaneously carries pulse-amplitude-modulated (PAM) samples of voice signals from sixty-four different conversations. In order to reconstruct an original voice signal, it is necessary to take samples at least 8000 times per second (see Section 6.4.3). The *DIMENSION* PBX samples voice signals 16,000 times per second.

The digital computer controls the connections on the PAM bus and most other functions. This control is established by the system software. Since this software manifests itself as a series of PBX system features, the Bell System calls its software *feature packages.*[20] By the end of 1981, Bell Laboratories had written twelve feature packages for the *DIMENSION*

---

[19] Section 7.3 discusses the concept of a common path, or *bus*, as a switching medium.

[20] Other companies with digital control systems call their software *releases* or *generic programs.*

PBX. One of the most advanced programs is feature package 8 (see Figure 11-21).

In addition to time-division switching and digital computer control, the *DIMENSION* PBX also uses digital signals to control the attendant consoles and special multibutton electronic telephone (MET) sets. In the case of the attendant console, a data link is provided from the computer to the console. Messages sent to the console operate LEDs and alphanumeric displays. When the attendant pushes a console button, a data message is sent to the PBX computer. Data links from MET sets do

---

DIMENSION PBX FEATURE PACKAGE

SYSTEM PBX FEATURES

Automatic Identified Outward Dialing
(AIOD)
Automatic Route Selection Code
Restriction
Direct Department Calling (DDC)
Direct Inward Dialing (DID)
Flexible Numbering of Stations
Inter-PBX Call Transfer
Inter-PBX Coordinated Station
Numbering
Loudspeaker Paging—Deluxe
Main/Satellite
Music-on-Hold Access
Power Failure Transfer
Radio Paging Access
Remote Access to PBX Services
• With Voice-Switched Gain
• Via Authorization Codes
Station Message Detail Recording
(SMDR)
Tandem Tie Trunk Switching
TOUCH-TONE to Dial Pulse
Conversion
Uniform Call Distribution (UCD)
Wide-Frequency-Tolerant Power
Supply

STATION FEATURES

Automatic Callback—Calling
Call Forwarding
• All Calls
• Busy and Don't Answer

Call Waiting Services
• Originating Call Waiting
• Terminating Call Waiting
Controlled Restrictions
Custom Intercom
Data Privacy
Executive Override
Night Station Service
—Full Service
—Fixed Service
Outgoing Trunk Queuing
Speed Calling
Station Hunting
• Circular
• Terminal
Threeway Conference Transfer
Trunk Answer from any
Station

ATTENDANT FEATURES

Alphanumeric Display for Attendant
Position
Attendant Conference
Attendant Direct Station Selection
with Busy Lamp Field
Attendant Transfer—All Calls
Busy Verification of Station Lines
Calling Number Display to Attendant
Centralized Attendant Service (CAS)
• Branch Identification to Attendant
• Combined PBX/Attendant
Concentrator
• W/Separate Attendant Concentrator

Class of Service Display to Attendant
Incoming Call Identification
• Alphanumeric Display
Multiple Line Directory Numbers
(DID and Non-DID)
Serial Call
Switched Loop Operation
Timed Reminders
Trunk Group Busy Indicators on
Attendant Position
Trunk Verification by Customer

**ELECTRONIC TANDEM SWITCHING
FEATURES**

Authorization Codes
Automatic Route Selection—Deluxe
• With Time-of-Day Routing
• Controlled Alternate Facility
Restriction Levels
Deluxe Queuing
Traveling Class Marks
Uniform Numbering
• Automatic Alternate Routing
• Automatic Overflow to DDD

**CUSTOMER MANAGEMENT AND
CONTROL FEATURES**

Customer Administration Center (CAC)
Facilities Administration and Control
Station Rearrangement and Change
Traffic Data to Customer

DIMENSION **CUSTOM TELEPHONE SERVICE (DCTS) ELECTRONIC KEY TELEPHONE FEATURES**

STATION FEATURES

Idle Line Preference
Last Extension Called
Manual Signaling
Preselection
Prime Line Selection
Recall Button
Station Busy Indication
Station Direct Station Selection
Station Power Failure Transfer
Station Message Waiting

LINE FEATURES

Automatic Intercom
Bridged Call
Common Audible Ringing
Dial Intercom
Manual Intercom
Music-on-Hold Access
Personal Central Office Line

CONTROL FEATURES

Exclusive Hold
Manual Exclusion
Priority Hold
Ringing Transfer

Figure 11-21. Some of the features offered as
part of feature package 8 on the *DIMENSION* PBX.

not connect directly to the PBX computer but to a MET set controller (another computer-operated device), which, in turn, connects to the system computer. The advantage of data link control is that all functions can be accomplished with very few wires (usually three pairs). The *DIMENSION* PBX also makes several kinds of management information available, including data on the call load being carried by the PBX, details on specific calls, and special measurements on groups that answer many incoming calls.

Other manufacturers, of course, also make computer-controlled PBXs, some of which use digital switching. The difference between the PAM technique used in the *DIMENSION* PBX and the digital technique can be seen by comparing Figures 11-20 and 11-22. Both types of PBX are based on the principle of digital computer control and time-division switching. The difference in techniques is a matter of how the voice is represented on the time-division bus. The digital technique, usually pulse-code

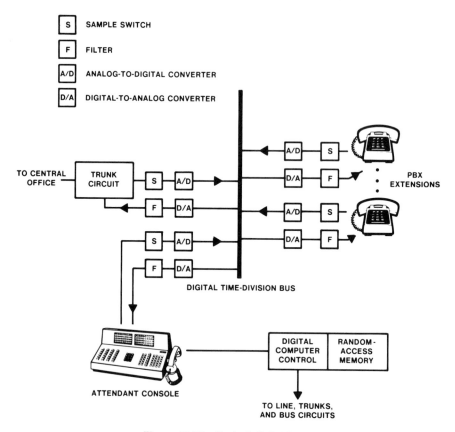

Figure 11-22. Typical digital PBX.

modulation (PCM), represents the signal as a coded set of binary pulses, whereas, in the *DIMENSION* PBX, they are amplitude-modulated pulses.

11.2.6 AUTOMATIC CALL DISTRIBUTORS

Automatic call distributors (ACDs) are used to switch large volumes of incoming calls to agent (answering) positions automatically. ACDs are used both for telephone company services such as directory assistance and for commercial applications such as airline reservation bureaus.

ACDs have the following characteristics:

- When all agents are busy, incoming calls are queued and answered in approximately their order of arrival.

- Calls are distributed among the agents to maximize the efficiency of the group.

- Management information data are collected to administer both the switching facilities and the size of the agent group.

ACDs range in size from those with fewer than 10 agent positions to those with 500 or more. Approximately 70 percent of all ACD installations have fewer than 50 positions, and 90 percent have fewer than 100 positions.

While PBX service is usually thought of as facilitating a firm's communications, an ACD is directly involved in the business operation itself. Generally, all of the personnel associated with an ACD work full time handling incoming calls, and the agents who answer calls usually service them completely. PBXs have many more lines than trunks; an ACD may have more trunks than agents (lines). This surplus of trunks permits the queuing of incoming calls until an agent becomes available. Queuing reduces the probability that an agent will have to wait for the next call.

Should a customer need both an ACD and a PBX, either of two options may be selected: two separate switching systems or one system that can do both jobs. Many modern PBX systems can provide ACD-type features for a subset of lines and trunks. In the *DIMENSION* PBX, for example, these features are known as Direct Department Calling and Uniform Call Distribution. Some modern systems can even provide a complete set of ACD features and PBX features simultaneously.

The ACD business has generated some jargon that is widely accepted. For example, an airlines reservations center may divide its agents into two specialized groups or "splits": one to handle calls from travel agents, the other to take all other calls. Calls would arrive at the ACD on two different trunk groups, but one ACD switching system would support both trunk groups and both splits. Should the call load on one split become too heavy, calls can be offered to agents on another split automatically. The shifting of some calls from one split to another on the same ACD is known as *intraflow*. Where there is only one agent group or all splits are fully loaded, calls can be passed on to another ACD. This feature is called *interflow*. Interflow is also used to pass calls to another ACD after business hours.

Since an ACD is essentially a specialized type of PBX, its evolution has paralleled that of the PBX. In fact, since there is no such thing as a manual automatic call distributor, the modern ACD cannot predate the automatic PBX. Early call distributors were arrangements known as *order turrets*, the first of which were manual answering positions. The 6A order turret, introduced in the late 1940s, was actually an ACD. Improved ACDs using both step-by-step and crossbar switching technology were introduced in the early 1960s. These could be arranged to provide up to 600 agent positions. In 1975, a central office–based ACD became available. This is implemented in a central office equipped with a 1ESS switch (see Section 10.3.3), which can provide up to one thousand agent positions in up to thirty splits. Supplementary computer facilities create management information reports.

The latest ACD offering by the Bell System as of this writing, the 80/5 Call Management System (CMS), is manufactured by WESCOM, Inc. This is a computer-controlled digital switch, normally used for PBX service, that has been programmed for ACD operation. The 80/5 CMS provides up to 144 agent positions and can operate as both an ACD and a PBX simultaneously.

Several manufacturers produce computer-based management information systems for older ACDs. Typically, these connect to the trunks and agent positions of electromechanical systems. Upgrading with modern management information system features can postpone the replacement of an older ACD.

A small manual call distributor, which is based on key telephone hardware, was introduced in 1973. The 4A call distributor is intended for customers with fifteen or fewer agent positions.

11.2.7 TELEPHONE ANSWERING SYSTEMS

As described in Section 2.3.5, a telephone answering bureau provides a centralized location where the telephones of clients are answered. This is a 2-part job. The telephone company provides equipment to do the necessary switching, while the bureau management provides sufficient operators to answer calls and record messages accurately.

As with other customer switching systems, older arrangements are still in service, and newer arrangements are available for new installations. The oldest, and simplest, telephone answering system (shown in

Figure 11-23) is a direct *bridge*[21] of the client's telephone, which appears at the answering bureau. Cord switchboards at the answering bureau contain a jack for each line to be answered. The average telephone answering bureau has three such positions. While this arrangement can provide basic service, it requires a large amount of cable between the switchboard and the central office. The bridged connection also compromises the privacy of the client's telephone service, and installation of a new or temporary connection is often time consuming.

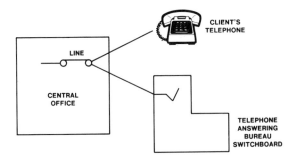

Figure 11-23. Telephone answering systems: simple bridged connection.

In order to reduce the number of cable pairs between the central office and the answering bureau, a special switching arrangement was developed. As shown in Figure 11-24, as calls are received, a unit (called

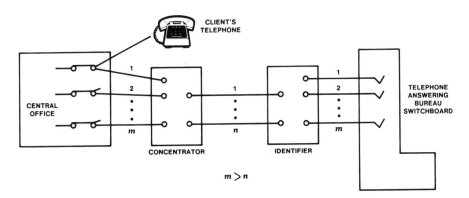

Figure 11-24. Telephone answering systems: use of a concentrator/identifier to reduce cable-pair requirement between the central office and the telephone answering bureau.

---

[21] A bridging connection is a parallel connection that draws some of the signal energy from a circuit, often with imperceptible effect on the circuit's normal operation.

a *concentrator*) at the central office connects the lines of *m* clients to one of a relatively small number (*n*) of trunks to the bureau location. Another special unit (called an *identifier*) then connects the trunk to a switchboard and, based on signals from the concentrator, indicates exactly which client's telephone has been connected by the concentrator. This arrangement results in considerable savings in copper and improves privacy. However, like the simple bridge arrangement, this setup is still awkward for clients who only need an answering service occasionally, for example, during an emergency or while on vacation.

The Call Forwarding feature now found in electronic central offices has made new telephone answering arrangements possible (see Figure 11-25). A telephone answering bureau simply installs a group of business lines. When clients require the service, they forward calls to one of these lines. Regular clients can have dedicated lines; occasional users can forward calls to a spare line provided for this type of service. While this arrangement does not necessarily reduce the number of copper pairs needed, it does solve the privacy problem and serve occasional users efficiently.

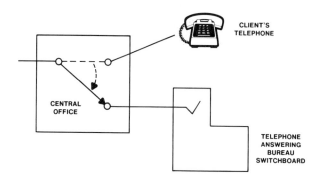

**Figure 11-25. Telephone answering systems: use of simple
Call Forwarding to telephone answering bureau.**

The latest step in the evolution of telephone answering equipment combines the Call Forwarding feature with the direct inward dialing (DID) feature normally used for PBX service. This arrangement, shown in Figure 11-26, also requires a special switching system and attendant consoles at the answering bureau. (This special system looks like a PBX with many attendants and no telephones.) A relatively small number of DID trunks connects that switching system to the central office. To use this service, a client forwards calls to what looks like a business line. The central office then connects forwarded calls to one of the DID trunks and

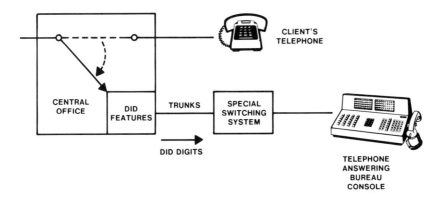

Figure 11-26. Telephone answering systems: use of Call Forwarding and direct inward dialing to telephone answering bureaus. *Note:* The number of clients' telephones is greater than the number of trunks.

sends DID digits over that trunk to the switching system at the answering bureau. Complete call identification can therefore be provided at the answering console.

This arrangement provides complete privacy, requires few cable pairs, and is suitable for occasional service. A lingering disadvantage, however, is the inability of a client to join a conversation between a calling party and the answering bureau. This is because Call Forwarding actually switches the call from the client's line and redirects it to the answering bureau.

11.3 PUBLIC SWITCHED NETWORK SERVICE SYSTEMS

The services available from the nationwide PSTN were described from the customer's point of view in Section 2.5. The following sections examine the systems and equipment that provide these network services. The discussion emphasizes the configuration, functional operation, and customer service features provided by systems associated with direct services dialing capability and *DIAL-IT* network communications service.

11.3.1 THE STORED-PROGRAM CONTROL NETWORK

The PSTN has thousands of switching nodes that employ the variety of switching systems described in Chapter 10. As indicated in Chapter 10, older systems are rapidly being replaced by more advanced stored-program control (SPC) systems. The collection of SPC systems in the PSTN that are interconnected by the common-channel interoffice signaling (CCIS) network (see Sections 8.4 and 8.5) is called the *SPC network*.

The SPC network enables the Bell System to offer the network services described in Section 2.5.1. This section describes the elements of the SPC network and its architecture as it pertains to providing these services.

Figure 11-27 is a schematic of the SPC network. The solid lines represent communication channels (trunks) that interconnect the major SPC network nodes: toll switching systems and operator-services systems such as the Traffic Service Position System (TSPS). The broken lines represent the CCIS channels (data links). A simplified version of the nationwide signaling network is also included. For convenience, nodes called *network control points* (NCPs) and their associated data bases are shown inside the signaling network, but the architecture of the SPC network does not actually restrict them to the CCIS network. They may also be integrated into existing SPC network nodes or associated directly with them. The network control points and data bases are used to store customer service data and are accessed by signals routed over the CCIS network. Figure 11-27 can be viewed as two networks: one for basic call transport and the other for signaling and information exchange among the SPC network nodes. The integration of these two networks forms the basis of the SPC network.

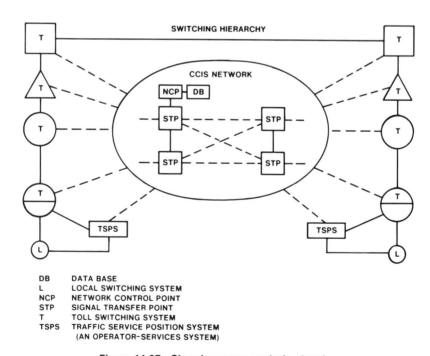

Figure 11-27. Stored-program control network.

By 1981, the CCIS network had been extended nationwide, with twelve pairs of signal transfer points (STPs)[22] handling about 38 million signaling messages per average busy hour over 2700 signaling links. As new SPC switching nodes with CCIS capability are added to the network, CCIS links are connected to their two regional signal transfer points and become members of the SPC network. There are 197 toll switching systems in the SPC network, and CCIS is routinely used on 400,000 trunks (about 35 percent of the intertoll trunks). It is projected that, by 1985, 80 percent of the intertoll trunks will be using CCIS. TSPSs are now being equipped with CCIS capability, and 40 percent of these operator-services systems are expected to belong to the SPC network by the end of 1983.

The architecture of the SPC network has been designed to meet two network service needs: the need to customize services and the need for ubiquitous access.

- The need to customize services to fit the particular requirements of individual customers results in three service objectives:

  — Customer control over service options should be quick, easy, and direct. Customers prefer to invoke, use, and cancel them as they wish.

  — A variety of services should be available. While many of their basic service needs are similar, different customers may want significant variations in how their overall service is put together.

  — As new services are defined, they should be introduced quickly, without major changes in the entire network.

- Many of the new service applications have great value when they can be invoked from any telephone. These services are obvious candidates for network solutions rather than specialized arrangements outside the network. On the other hand, the universal deployment of new service capabilities in every node of the network is clearly impractical, too slow, and too costly. The solution is a carefully planned network architecture.

The architecture of the stored-program control network has been designed with these service needs in mind (see Figure 11-28). Calls from customers requesting special SPC services are routed to certain nodes called *action points* (ACPs). Two observations can be made here. First, an extended, open-ended dialing plan has been worked out for service requests from customers. Second, the routing of calls requiring special

---

[22] A pair of signal transfer points is associated with each region of the PSTN in the United States and Canada.

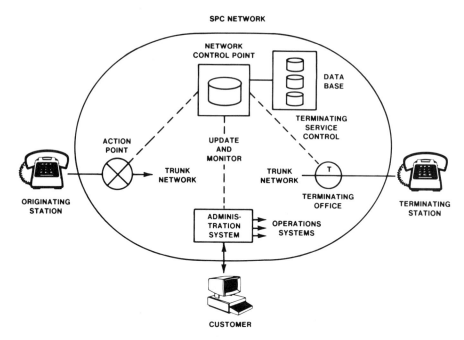

Figure 11-28. Generic architecture of the stored-program control network.

actions from any local switch to a special ACP node provides a mechanism for ubiquitous access to new service capabilities. Ordinary local offices need not have any new service capabilities themselves. They need only apply their regular routing functions to send special calls to "more intelligent" nodes in the network for handling. They must also pass along the automatic number identification of the calling line. (This function is already widely available in Bell System local offices.)

The action point has the ability to recognize special calls by the dialed digits and to send messages requesting instructions to an appropriate network control point. The ACP itself has no knowledge of the customer service but responds as directed by the NCP. The NCP is the repository for all customer network service information. The ACP is equipped with basic capabilities, called *switching primitives,* that it applies as instructed by the NCP. This functional split is an important characteristic of the SPC architecture.

The NCP can query any SPC network node for information or action. When lines at one business location are busy, a customer may prefer to have calls routed to another location. "Hubbing" all service logic in NCPs also assists in making changes in customer services since changes can be made rapidly in data stored there. If a rich set of ACP capabilities has been deployed, nothing new may be needed in the network control nodes themselves.

A necessary part of the architecture is the methods and administration system for entering customer service information into the data bases associated with the network control points, for monitoring the service to ensure performance, and for interacting as needed with operations systems to maintain service quality.[23] Ultimately, the customer will want to interact directly with the service management process.

The architecture of the stored-program control network has provided a set of direct services dialing capabilities (DSDC) that are used both to improve current services and to offer new services. Automated Calling Card Service and Expanded 800 Service are the first two services to use DSDC.

Automated Calling Card Service permits customers to charge calls to a number other than that of the originating station without the assistance of an operator. As shown in Figure 11-29, the network control point is

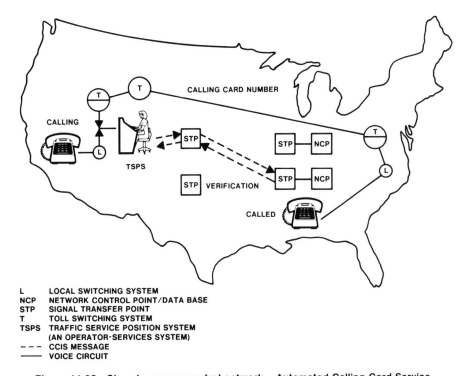

L   LOCAL SWITCHING SYSTEM
NCP   NETWORK CONTROL POINT/DATA BASE
STP   SIGNAL TRANSFER POINT
T   TOLL SWITCHING SYSTEM
TSPS   TRAFFIC SERVICE POSITION SYSTEM
         (AN OPERATOR-SERVICES SYSTEM)
– – –   CCIS MESSAGE
———   VOICE CIRCUIT

Figure 11-29. Stored-program control network—Automated Calling Card Service.

[23] Chapter 14 discusses operations systems.

accessed by a TSPS serving as an action point. The calling customer, with appropriate prompting from the ACP, keys in the number being called (including the numbering plan area,[24] if needed), the number to be billed, and a private personal identification number (PIN):

$$0 + NPA \text{ - } NXX\text{-}XXXX \text{ - } NPA \text{ - } NXX\text{-}XXXX \text{ - } YYYY$$

called number   billing number   PIN

calling card number

After validating this information, the NCP either authorizes the call, allowing it to be completed, or denies the call, requesting that the customer take appropriate action.

Expanded 800 Service, shown in Figure 11-30, is an improved method of handling inward Wide Area Telecommunications Services (INWATS) traffic. When a call is originated toll free to a customer of Expanded 800 Service, the 800 plus the 7-digit dialing sequence is received by the local

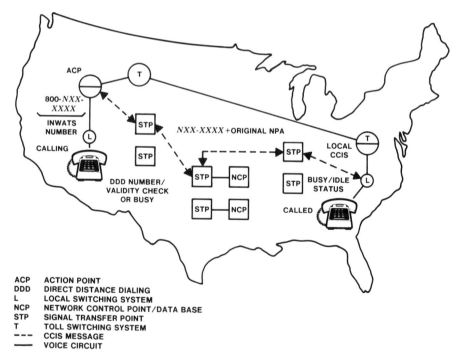

ACP   ACTION POINT
DDD   DIRECT DISTANCE DIALING
L   LOCAL SWITCHING SYSTEM
NCP   NETWORK CONTROL POINT/DATA BASE
STP   SIGNAL TRANSFER POINT
T   TOLL SWITCHING SYSTEM
— — —   CCIS MESSAGE
———   VOICE CIRCUIT

Figure 11-30. Stored-program control network—Expanded 800 Service.

[24] Also called the *area code*.

office, and the call is routed to the nearest action point. Initially, the ACPs for Expanded 800 Service will be large SPC switching systems such as the 4ESS electronic switch. In order to determine how to handle the call, the ACP sends a message over the common-channel interoffice signaling network to an appropriate network control point. The NCP data contain the routing instructions as requested by the customer of Expanded 800 Service. When the NCP extracts the proper call-handling instructions from the data base, it sends an answering message to the ACP telling it how to route the call. The ACP then forwards the call. When the Expanded 800 Service customer lines terminate on a local office in the SPC network, busy/idle status messages keep the NCP informed when all the lines are busy. Consequently, when all lines are busy, the NCP tells the ACP either to turn back the call or to route it to a different location (which has been predetermined by the customer of the service). This increases completions when multiple terminating locations are available and also reduces ineffective network attempts to busy lines.

The NCP is one of the first applications of the 3B20 computer. The computer is configured with a CCIS network interface to communicate with action points, an administration and maintenance interface, and a disk storage system to hold the customer data. For reliability, NCPs are provided in geographically separated pairs. Each member of the pair has the processing capacity and information necessary to provide uninterrupted service to all users of the pair should the other member not be in service.

Other components of the stored-program control network such as electronic switching systems, TSPSs, and the CCIS network are discussed in Sections 10.3, 10.4.1, and 8.5.5, respectively.

11.3.2 MASS ANNOUNCEMENT SYSTEM

The Mass Announcement System (MAS) provides the announcement capability that supports selected DIAL-IT service (see Section 2.5.1). This capability allows various sponsors to provide services in which a large number of calling customers can dial advertised numbers to listen to Public Announcement Service announcements, register their opinions via telephone calls (call counting), or talk to a celebrity—if they are one of the randomly selected callers who are connected (cut through).

The Mass Announcement System consists of a MAS frame connected to a 4ESS electronic switch. The MAS frame provides announcements to the 4ESS switch; the 4ESS switch then distributes them to the DIAL-IT service calls. Two MAS units are located on the MAS frame, each unit containing a disk on which the announcements are stored. The frame (see Figure 11-31) provides disk storage and control for eighty audio segments, each 30 seconds in duration. Announcements up to 300 seconds in length can be produced by linking a number of segments together.

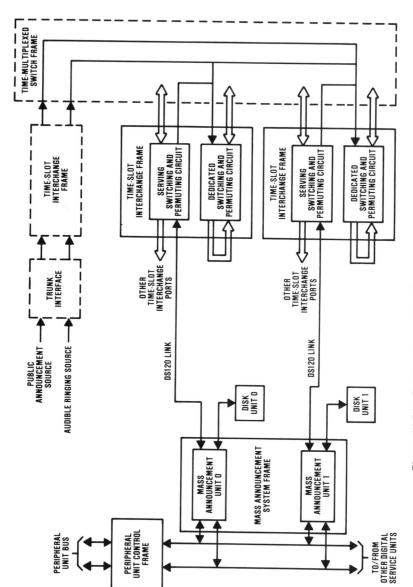

Figure 11-31. Configuration of the Mass Announcement System.

The two units contain identical recordings, thereby increasing reliability in case of failure of one of the units.

In the 4ESS electronic switch, terminations for announcements are provided by units known as dedicated *time-slot interchanges* (TSIs), which are discussed in Section 7.3.3. A dedicated TSI can connect any calling customer who reaches it with the announcement corresponding to the number that the caller has dialed. A dedicated TSI for MAS can terminate up to 896 calls simultaneously. Announcements can be associated with terminations in a completely flexible manner, permitting dynamic allocation of capacity within the 896 limit. Two dedicated TSI frames, one associated with each MAS unit, are required for each 4ESS switch equipped for MAS. This minimum equipment configuration provides announcement capability to 1792 callers. Additional dedicated units can be added to increase capacity.

Figure 11-32 shows a simplified event flow for a call using the MAS capability. A more detailed description is as follows:

- The caller dials either a 7- or 9-digit announcement number such as a *DIAL-IT* service number and is routed to a 4ESS switch equipped for MAS. The 4ESS switch recognizes the number as a request for a particular MAS announcement. It also recognizes whether call counting is associated with the number. Next, it determines whether cut-through service is in effect and, if it is, whether the call should be cut through or connected to the announcement.

- The 4ESS switch determines which dedicated TSI the call should be connected to in order to minimize the audible ring interval. If all terminations are busy on the first choice, the call will be connected to the second-choice dedicated TSI or, if no announcement terminations are available, to a busy tone. If the MAS equipment is out of service so no MAS announcement can be given, the call is connected to a no-circuit announcement from the office announcement machine.

- Finally, for successful calls, audible ring ends when the 4ESS switch connects the call to the announcement over the same dedicated TSI that provided the audible ring. Answer supervision then begins (unless specified otherwise). If the incoming trunk has centralized automatic message accounting (CAMA), CAMA billing is initiated (see Section 10.5.4).

- The customer can abandon the call at any time during the announcement. In the absence of early abandonment, the call is automatically removed from its MAS termination when the allocated playing time elapses.

The Mass Announcement System is an optional feature of the 4ESS switch. Offices with 4ESS switches equipped for MAS are strategically

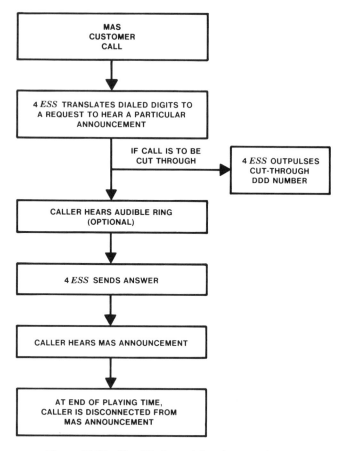

Figure 11-32. Simplified event flow for a MAS call.

located throughout the country (see Figure 11-33) so that sponsors can provide local, regional, or national service. Each of these offices is designated an *MAS node* and its associated calling region an *MAS island*.

11.4 MOBILE TELEPHONE SYSTEMS

The demand for telephone service to and from moving or temporarily stationary users continues to grow rapidly; much more than a convenience, this service can increase the efficiency of today's mobile society. Of the more than 100 million vehicles in the United States, about 10 million are equipped with 2-way radios (not counting citizens-band radios); however, in 1980, only about 100,000 could connect to the telephone network. Historically, the bulk of radio communication has been in the

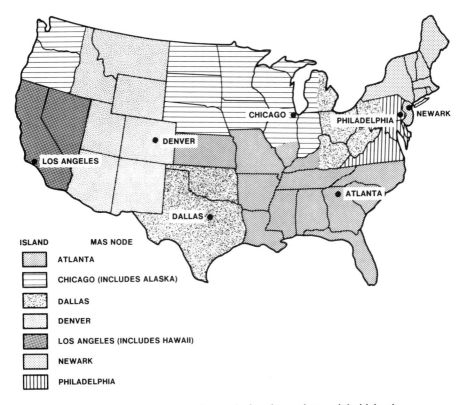

ISLAND   **MAS NODE**

ATLANTA

CHICAGO (INCLUDES ALASKA)

DALLAS

DENVER

LOS ANGELES (INCLUDES HAWAII)

NEWARK

PHILADELPHIA

Figure 11-33. National deployment of nodes and associated islands
at initiation of Mass Announcement System in 1980.

private, fleet-oriented area. Future growth is expected to shift toward the mobile telephone sector, which will be split about equally between radio common carriers and wire-line carriers (for example, the operating telephone companies). In addition to radio-equipped vehicles, service to hand-held portable units is becoming increasingly popular as the state of the art permits smaller, lighter designs.

The following sections review existing systems and examine newer systems now being deployed. The major emphasis will be on systems for land mobile (the largest market), although systems for airplanes, trains, and watercraft serve a small but important need.

The radio spectrum is a limited (but renewable) resource. Thus, the availability of radio channels is a fundamental concern for any mobile system. The FCC is responsible for making channels available to the public based on need. It sets standards (bandwidth, channel spacing, modulation methods, power, etc.) according to the state of the art.

11.4.1 LAND MOBILE TELEPHONE SYSTEMS

Channel Availability

Mobile telephone service began in the late 1940s. By the seventies, it included a total of thirty-three 2-way channels below 500 megahertz (MHz), as shown in Table 11-2. The 35-MHz band, which is not well suited to mobile service (because of propagation anomalies), is not heavily used. The other bands are fully utilized in the larger cities. In spite of this, the combination of few available channels per city and large demand has led to excessive blocking. The FCC's recent allocation of 666 channels at 850 MHz for use by cellular systems (described below) should change this situation. This allocation is split equally between wire-line and radio common carriers (each is allocated 333 channels). In many areas, the wire-line carrier will be the local operating company.

Use of conventional systems on the new channels would increase the traffic-handling capacity by a factor of about 10. The cellular approach, however, will increase the capacity by a factor of 100 or more. How this increase is achieved is discussed later in this section. The potential for very efficient use of so valuable and limited a resource as the frequency spectrum was a persuasive factor in the FCC's decision.

TABLE 11-2

MOBILE TELEPHONE ALLOCATIONS

Band* (MHz)	Number of Wire-Line Common-Carrier Channels	Channel Spacing (kHz)
35	10	40
150	11	30
450	12	25
850	333	30

*This is the nominal frequency that identifies the band. The frequencies allocated in the 850-MHz band, for example, are between 825 and 845 MHz and 870 and 890 MHz.

Transmission Considerations

Radio propagation over smooth earth can be described by an inverse power law; that is, the received signal varies as an inverse power of the distance. Unlike fixed radio systems (for example, broadcast television or the microwave systems described in Chapter 9), however, transmission to or from a moving user is subject to large, unpredictable, sometimes rapid fluctuations of both amplitude and phase caused by

- *Shadowing* — This impairment is caused by hills, buildings, dense forests, etc. It is reciprocal, affecting land-to-mobile and mobile-to-land transmission alike, and changes only slowly over tens of feet.

- *Multipath interference* — Because the transmitted signal may travel over multiple paths of differing loss and length, the received signal in mobile communications varies rapidly in both amplitude and phase as the multiple signals reinforce or cancel one another.[25]

- *Noise* — Other vehicles, electric power transmission, industrial processing, etc., create broadband noise that impairs the channel, especially at 150 MHz and below.

Because of these effects, radio channels can be used reliably to communicate at distances of only about 20 miles, and the same channel (frequency) cannot be reused for another talking path less than 75 miles away except by careful planning and design.

In a typical land-based radio system at 150 or 450 MHz, one channel comprises a single frequency-modulation (FM) transmitter with 50- to 250-watt output power, plus one or more receivers with 0.3- to 0.5-microvolt sensitivity. This equipment is coupled by receiver selection[26] and voice-processing circuitry into a control terminal that connects one or more of these channels to the telephone network (see Figure 11-34). The control terminal is housed in a local switching office. The radio equipment is housed near the mast and antenna, which are often on very tall buildings or a nearby hilltop.

Conventional System Operation

Originally, all mobile telephone systems operated manually, much as most private radio systems do today. A few of these early systems are still in use, but because they are obsolete, they will not be discussed here.

More recent systems (the MJ system at 150 MHz and the MK system at 450 MHz) provide automatic dial operation. Control equipment at the central office continually chooses an idle channel (if there is one) among the locally equipped complement of channels and marks it with an "idle" tone. All idle mobiles scan these channels and lock onto the one marked with the idle tone. All incoming and outgoing calls are then routed over this channel. Signaling in both directions uses slow-speed audio tone

[25] Section 6.3.6 discusses multipath fading, another term for interference.

[26] One of several receivers is selected for best reception from the mobile unit as it moves through the area.

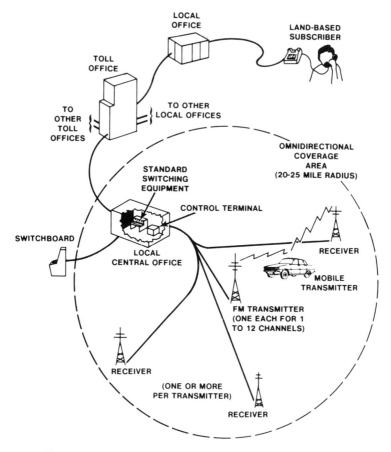

Figure 11-34. A manual, or MJ/MK, mobile telephone system.

pulses for user identification and for dialing. Compatibility with manual mobile units is maintained in many areas served by the automatic systems by providing mobile-service operators. Conversely, MJ and MK mobile units can operate in manual areas using manual procedures.

One desirable feature of a mobile telephone system is the ability to "roam"; that is, subscribers must be able to call and be called in cities other than their home areas. The numbering plan must be compatible with the North American numbering plan (see Section 4.3). Further, for land-originated calls, a routing plan must allow calls to be forwarded to the current location. In the MJ system, operators do this. Because of the availability of the MJ system to subscribers requiring the roam feature, the MK system need not be arranged for roaming.

Advanced Mobile Phone Service

Cellular Concept. Although the MJ and MK automatic systems offer some major improvements in call handling, the basic problems—few channels and the inefficient use of available channels—still limit the traffic capacity of these conventionally designed systems. Advanced Mobile Phone Service[27] overcomes these problems by using a novel cellular approach. It operates on frequencies in the 825- to 845-MHz and 870- to 890-MHz bands recently made available by the FCC. The large number of channels available in the new bands has made the cellular approach practical.

A cellular plan differs from a conventional one in that the planned reuse of channels makes interference, in addition to signal coverage, a primary concern of the designer. Quality calculations must take the statistical properties of interference into account, and the control plan must be robust enough to perform reliably in the face of interference. By placing base stations in a more or less regular grid (spacing them uniformly), the area to be served is partitioned into many roughly hexagonal cells, which are packed together to cover the region completely. Cell size is based on the traffic density expected in the area and can range from 1 to 10 miles in radius.

Up to fifty channels are assigned to each cell to achieve their regular reuse and to control interference between adjacent cells. This is illustrated in Figure 11-35, where cell A' can use the same channels as cell A. Because of the inverse power law of propagation, the spatial separation between cells A and A' can be made large enough to ensure statistically that a signal-to-interference ratio greater than or equal to 17 dB is maintained over 90 percent of the area. Maintenance of this ratio ensures that a majority of users will rate the service quality good or better.

Cellular systems also differ from conventional systems in two significant ways:

- High transmitted power and very tall antennas are not required.

- Wide FM deviation is permissible without causing significant levels of interference from adjacent channels.

The latter is responsible for the high voice quality and high signaling reliability of the Advanced Mobile Phone Service.

In any given area, both the size of the cells and the distance between cells using the same group of channels determine the efficiency with

---

[27] *Advanced Mobile Phone Service* (sometimes called *AMPS*) is a generic name referring to the cellular system concepts and the control algorithms used for mobile service in the United States in the 850-MHz band.

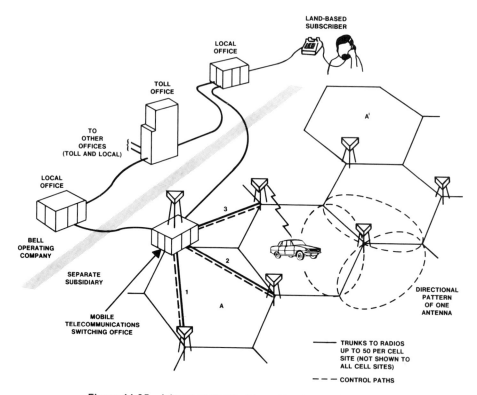

Figure 11-35. Advanced Mobile Phone Service system plan.

which frequencies can be reused. When a system is newly installed in an area (when large cells are serving only a few customers), frequency reuse is unnecessary. Later, as the service grows (a dense system will have many small cells and many customers), a given channel in a large city could be serving customers in twenty or more nonadjacent cells simultaneously. The cellular plan permits staged growth. To progress from the early to the more mature configuration over a period of years, new cell sites can be added halfway between existing cell sites in stages. Such a combination of newer, smaller cells and original, larger cells is shown in Figure 11-36.

One cellular system is the Western Electric *AUTOPLEX*-100. In this system, a mobile or portable unit in a given cell transmits to and receives from a cell site, or base station, on a channel assigned to that cell. In a mature system, these cell sites are located at alternate corners of each of the hexagonal cells as shown in Figure 11-36. Directional antennas at each cell site point toward the centers of the cells, and each site is connected by standard land transmission facilities to a 1*AESS* switching system and system controller equipped for Advanced Mobile Phone Service

522

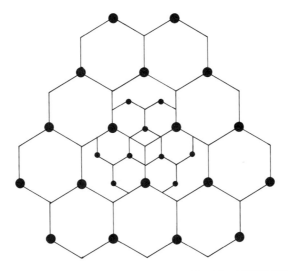

Figure 11-36. Staged growth of a Western Electric *AUTOPLEX* cellular radio system. (*AUTOPLEX* is a trademark of Western Electric.) Cell splitting, by adding new cell sites in areas of high demand, creates more traffic capacity.

operation (called a mobile telecommunications switching office, or MTSO). Start-up and small-city systems use a somewhat more conventional configuration with a single cell site at the center of each cell.

The efficient use of frequencies that results from the cellular approach permits Advanced Mobile Phone Service customers to enjoy a level of service almost unknown with present mobile telephone service. Grades of service of P(0.02)[28] are anticipated, compared to today's all-too-common P(0.5) or worse. At the same time, the number of customers in a large city can be increased from a maximum of about one thousand for a conventional system to several hundred thousand. Also, because of the stored-program control capability of MTSOs equipped with the 1A*ESS* system, Custom Calling Services and many other features can be offered, some unique to mobile service. Other, smaller switches provided by Western Electric or other vendors are also available to serve smaller cities and towns.

System Operation. Unlike the MJ and MK systems, Advanced Mobile Phone Service dedicates a special subset of the 333 allocated channels solely to signaling and control. Each mobile or portable unit is equipped with a frequency synthesizer (to generate any of the 333 channels) and a

---

[28] As described in Chapter 5, this means a 2-percent probability of blocking.

high-speed modem (10 kbps). When idle, a mobile unit chooses the "best" control channel to listen to (by measuring signal strength) and reads the high-speed messages coming over this channel. The messages include the identities of called mobiles, local general control information, channel assignments for active mobiles, and "filler" words to maintain synchronism. These data are made highly redundant to combat multi-path interference. A user is alerted to an incoming call when the mobile unit recognizes its identity code in the data message. From the user's standpoint, calls are initiated and received as they would be from any business or residence telephone.

As a mobile unit engaged in a call moves away from a cell site and its signal weakens, the MTSO will automatically instruct it to tune to a different frequency—one assigned to the newly entered cell. This is called *handoff*. The MTSO determines when handoff should occur by analyzing measurements of radio signal strength made by the present controlling cell site and by its neighbors. The returning instructions for handoff sent during a call must use the voice channel. The data regarding the new channel are sent rapidly (in about 50 milliseconds), and the entire retuning process takes only about 300 milliseconds.

In addition to channel assignment, other MTSO functions include maintaining a list of busy (that is, off-hook) mobile units and paging mobile units for which incoming calls are intended.

Regulatory Picture. The FCC intends cellular service to be regulated by competition, with two competing system providers in each large city: a wire-line carrier and a radio common carrier. To prevent any possible cross-subsidization or favoritism, the Bell operating companies must offer their cellular service through separate subsidiaries. These subsidiaries will be chiefly providers of service and, in fact, are currently barred from leasing or selling mobile or portable equipment. Such equipment will be sold by nonaffiliated enterprises or by American Bell Inc.

11.4.2 PAGING

Another radio service offering of the Bell System is a 1-way communication system that provides users with a personal number that identifies a signaling or alerting receiver, the *BELLBOY* radio paging set, that can conveniently be carried about. The Bell System originally developed this service during the 1960s. During the 1970s, radio common carriers also introduced this type of service under other trade names. Today the Bell System has a relatively small portion of this market, although it is a major market presence in many cities.

This service requires the mobile user to carry a small (3- to 4-ounce) pocket receiver. The fixed portion of the paging system comprises several transmitters and a control terminal. The person (usually a secretary) doing the paging uses *TOUCH-TONE* signaling to dial the number of the control terminal and then dials more digits to identify the specific user. The transmitters signal the user, who calls the secretary via some convenient land telephone to retrieve the message. In an urban area, several high-power transmitters are required to provide a signal strong enough to penetrate buildings and to activate the receiver's small (and hence, inefficient) antenna. Figure 11-37 is a schematic picture of the system.

All the new paging systems use high-speed digital signaling and can accommodate as many as 100,000 users on a single radio channel. The FCC recently allocated 120 channels at 930 MHz for paging-related services. This action is expected to spur rapid innovation and expansion of this type of service. Some Bell operating companies are already offering a service that uses a paging receiver with a visual display. A calling party equipped with *TOUCH-TONE* service can, via end-to-end signaling, furnish as many as ten digits to the paging controller. In this way, the telephone number of the calling party or some other message with a prearranged meaning can be displayed to the user. Other services either already available or under study include voice paging (the caller speaks to the subscriber) and nationwide paging (subscribers may be called in many cities across the country).

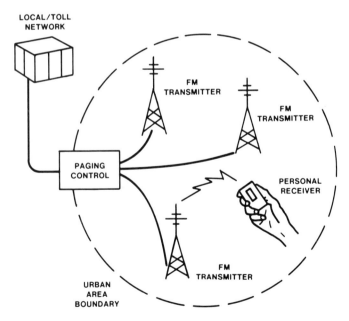

Figure 11-37. *BELLBOY* radio paging system plan.

11.5 VISUAL SYSTEMS

At the present time, the Bell System's principal involvement with television is through the cross-country distribution of television signals (see Section 6.1.3) for the broadcast industry, a private-line service (see Section 2.5.2). However, the Bell System has long been involved with video telephony, or interactive television, in which two or more parties see as well as talk to each other during the connection. The requirements for broadcast and interactive television differ and so do their systems. This section discusses interactive rather than broadcast television.

11.5.1 EARLY TELEVISION

Long-distance video telephony was first publicly demonstrated on April 7, 1927, when images were sent between Washington, D. C., and Bell Laboratories in New York City by wire and between an experimental station in Whippany, New Jersey, and New York City by radio. The images were crude by present standards: 50-line resolution, eighteen frames per second, at a bandwidth of 20 kHz. Two different receivers were designed for two different purposes. One produced a small image, approximately 2 by 2-½ inches, suitable for viewing by a single person, for use in private conversations. The other receiver produced a larger image, approximately 2 by 2-½ feet, intended for viewing by an audience of some size.

The success of this demonstration led, by 1930, to a 2-way experimental system in Manhattan, connecting AT&T headquarters at 195 Broadway with Bell Laboratories at 463 West Street. Further developments improved both equipment and image quality and required more bandwidth.

11.5.2 *PICTUREPHONE* VISUAL TELEPHONE SERVICE

Throughout its market trial from 1964 to 1975 (see Section 2.5.6), the *PICTUREPHONE* visual telephone service operated to the standards in Table 11-3. Also shown are the National Television Standards Committee (NTSC) standards for commercial television. The NTSC standards yield a resolution of about 140,000 picture elements, which is comparable to 16-millimeter movie film. Responding to the high cost of transmission bandwidth and to a reduced need for resolution in face-to-face video connections, the video telephone provided about one-fourth the resolution of the NTSC standard. The repetition rate (pictures displayed per second) was the same as the NTSC's. The aspect ratio (width-to-height ratio) was, however, quite different and better suited for use by individuals and for materials such as letters.

TABLE 11-3

NTSC BROADCAST TELEVISION STANDARDS
VERSUS BELL SYSTEM *PICTUREPHONE* STANDARDS

Characteristic	NTSC Standards	*PICTUREPHONE* Standards (Early 1970s)
Bandwidth (MHz)	4	1
Frames per second	30	30
Raster interlace	2:1	2:1
Lines per frame	525	267
Aspect ratio	4:3	11:10

Design of the equipment for the *PICTUREPHONE* visual telephone service was quite sophisticated. The compact unit contained both the camera tube and a 9-inch black-and-white picture tube. Among its features were an electronic zoom capability and automatic adjustment for light levels. Since the unit carried both picture and sound information, the camera required shock mounting to eliminate acoustic coupling from the microphone.

Ordinary telephone wires and newly developed broadband amplifiers were used for local transmission. Transmit and receive were on separate wire pairs. For longer distances, broadcast television channels could be used; a bandwidth of only 1 MHz was needed.

11.5.3 VIDEO TELECONFERENCING

In 1975, *PICTUREPHONE* meeting service was introduced on a trial basis. Rooms were set up to accommodate conferences of various sizes and were equipped with standard large monitors and several cameras. The NTSC standards were adopted.

Between 1975 and 1981, as part of the market trial, various room configurations were tested: both "public" rooms, eventually located in twelve cities, and private rooms on the premises of three customers. Information was gathered on such matters as customer acclimation to video technology, network size, business segmentation of potential users, adaptation to user needs, and exchange versus intercity offerings. A standard service is now offered under tariffs effective July 2, 1982.

Long-distance digital transmission at rates of 1.5 and 3 megabits per second (Mbps) is provided by the High-Speed Switched Digital Service (HSSDS). Connection from meeting rooms to HSSDS is provided by High-Capacity Terrestrial Digital Service (HCTDS, described in Section 2.5.4). The video coder/decoder (codec), a principal component of the meeting room equipment, reduces the transmission rate required for "conference grade" video to the range of 1.5 to 3 Mbps from the approximately 100 Mbps needed for broadcast quality video. It does this by digitally processing the meeting room video to remove redundancy, which is more prevalent in conference video than in some types of broadcast video.

At present, plans for PICTUREPHONE meeting service include public rooms in many cities, complemented by many more private rooms on customers' premises. Generally, a conference room contains a table for six people (see Figure 2-11). Each of three cameras is aimed at two of the participants. The system employs voice camera switching: The camera pointing at the speaker is the one in use at that moment. A fourth camera, the "overview," shows the whole conference room and is used, for example, when the local participants are listening to a speaker in the other conference room.

Three additional cameras are provided for graphics: one for images of sheets of paper or transparencies, one for slides, and one for a speaker standing at an easel or chalkboard. The last is a multipurpose camera, equipped with controllable pan, tilt, and zoom. All cameras and monitors have color capability. A preview monitor is provided to set up the graphics or slides or to aim the multipurpose camera properly before the images are transmitted to the other room. A video cassette recorder is also provided to record the meeting.

Room design for video teleconferencing represents a challenge in balancing light distribution and sound quality. Enough light must fall on the participants to provide a good camera image, but if too much falls on the monitors, the image contrast is degraded. The PICTUREPHONE meeting service audio system provides control of the line from one end or the other to avoid echoes. Voice transmission is effectively permitted in only one direction at any time by "squelching" the signal from the other direction by inserting loss in the signal path. In video teleconferencing, however, the amount of loss added has been reduced from that used in the 4A speakerphone to give a better feeling of interaction.

Many human factors experiments have been conducted at Bell Laboratories to evaluate various designs. Based on these evaluations and the information gained from the market trial, video teleconferencing should be both effective and pleasant.

11.5.4 RESEARCH ACTIVITIES

Research has produced two concepts of great interest. One is *motion compensation* (see Netravali and Robbins 1979), in which the picture processor algorithm attempts to predict the movement of a picture element from one frame to another. Computer simulations show that the concept is technically feasible and would reduce the bit rate required by a factor of 2 or more.

The other concept is *continuous presence* (see Larsen and Brown 1980). In one form, it replaces voice switching of the cameras with multiple cameras and multiple monitors, with two of the conference participants on each of the three camera-monitor pairs. Research has found that the processed signals for the three images can be statistically multiplexed,[29] so that the three images can be transmitted at less than three times the individual bit rates.

More work is needed on these concepts to determine if their benefits are worth their development and equipment costs and, if so, to reduce them to practice.

11.6 DATA COMMUNICATIONS SYSTEMS

The rapid growth of data communications in terms of both volume of data and diversity of application has resulted in the development of a variety of systems to meet customers' needs. Customers' data service requirements can be broadly classified as either **analog** or **digital**. *Analog* data transmission is important, but it represents a smaller volume than digital services. Included in the analog category are much of the current facsimile market and signals used in alarm and sensor-based systems.

Digital data can be further categorized by the transfer rates required as either low, medium, or high speed.

- Digital data transmitted at low speeds, generally at 300 bps or less, include telemetry, terminal-to-terminal message service, and terminal-to-computer services such as computer time sharing.

- Medium-speed transfer, generally from 300 bps through 9.6 kbps, can be accommodated on voiceband telephone facilities. Typical applications include terminal-to-computer transfers where the terminal may be a CRT or remote job-entry terminal. Credit checking or banking applications may operate in batch mode or real-time mode. Computer-to-computer transfer is also common in distributed data-processing systems. Other common applications are multiplexing of lower-speed data signals and digital facsimile.

- High-speed data are generally those transmitted at speeds above 9.6 kbps, for example, at 19.2 kbps, 56 kbps, and higher. Common applications at these speeds include computer-to-computer operation and multiplexing of lower-speed data.

[29] Section 6.5 discusses multiplexing.

Telephone-type facilities are still by far the most popular means of data transmission. These facilities include both the PSTN and leased private lines and private switched networks. Data sets (see Section 11.1.2) are required to condition the customer's data signal for transmission over these facilities. *DATAPHONE* II data sets also give private-line customers powerful diagnostic capabilities and control of their data communications systems.

A number of other transport services, some of which are described in Chapter 2, are available now or soon will be. The remainder of this chapter discusses the equipment and systems that support these services.

The **Digital Data System**, which is described in Section 11.6.1, provides *end-to-end digital connectivity* over dedicated facilities and supports *DATAPHONE* digital service. The preponderance of data communications, such as those between two digital computers or between computers and terminals, involves data that are inherently digital. With end-to-end digital connectivity, the data maintain their digital form throughout the transfer, and the efficiencies and cost savings discussed in Section 9.1.3 are obtained.

Other capabilities offering end-to-end digital connectivity are emerging. One of these, circuit-switched digital capability (CSDC), is described in Section 2.5.1. It will provide a 56-kbps end-to-end digital channel as an evolutionary capability of the PSTN. With the proliferation of digital facilities in the nationwide network, the integrated services digital network currently being planned will represent a further stage of evolution towards a network capable of meeting a wide range of telecommunications needs.

Section 11.6.2 describes **packet-switching systems**, and in particular the No. 1 PSS packet switch. Packet transport networks can serve those applications where bursts of data are to be transmitted with very short delays.

Finally, the *DATAPHONE* **Select-a-station service** (see Section 2.5.4) is an example of a capability designed to meet the needs of a fairly specialized segment of the data communications market. Section 11.6.3 provides a description of the system that provides this service.

11.6.1 DIGITAL DATA SYSTEM

The dedicated digital transport network used to provide and systematically support *DATAPHONE* digital service (described in Section 2.5.4) is called the *Digital Data System* (DDS).[30]

A typical point-to-point DDS channel is illustrated in Figure 11-38. Customers are connected to the DDS through a local office. Calls to

---

[30] Caution should be exercised in applying the abbreviation DDS. It must *not* be applied to the service because *DATAPHONE* digital service is a registered service mark.

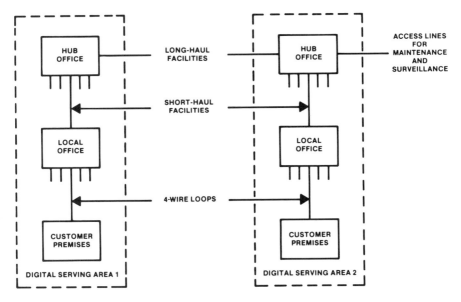

Figure 11-38. Typical point-to-point Digital Data System channel.

another metropolitan area, or digital serving area, are connected through hub offices. The location of DDS equipment and the stages of signal multiplexing are shown in Figure 11-39.

The *channel service unit* (CSU) is the first equipment unit on the network side of the customer interface. An additional signal-processing function can be provided by the customer or by common-carrier equipment to process control and information signals synchronously into a format compatible with the CSU interface. Using digital format, the CSU connects a customer's data communications equipment over two cable pairs to a local office. There, the line is terminated at an *office channel unit* (OCU) that regenerates the signal and prepares it for transmission through the multiplexing hierarchy[31] as outlined below. The CSU also provides the ability to test a DDS channel quickly and decisively up to the point of interface with the customer.

In the first stage of multiplexing, a number of customer data rates can be combined by a subrate multiplexer to form a basic 64-kbps (digital signal level 0 [DS0]) channel (see Table 11-4). A second stage of multiplexing combines up to twenty-four DS0 signals to form a 1.544-Mbps stream that corresponds to the DS1 signal of the time-division multiplex hierarchy. The DS1 signal is usually carried to the hub office over short-haul transmission facilities. Where the data traffic is very heavy, further

---

[31] Section 9.4.3 describes the digital time-division multiplex hierarchy.

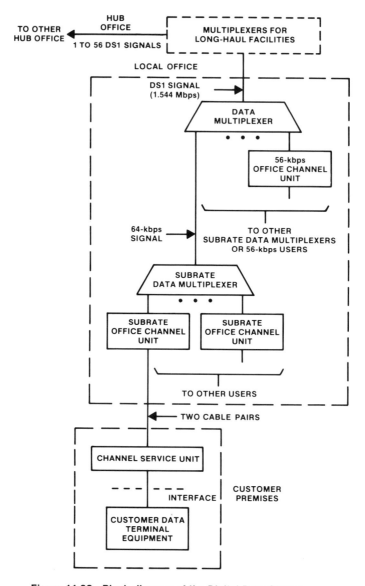

Figure 11-39. Block diagram of the Digital Data System.

multiplexing for short-haul facilities, which provide multiple DS1 signals (see Table 11-5), may be required.

Generally, the office chosen to be the hub will be one that serves a large number of data customers. This hub office also provides test access to individual data channels, cross-connecting facilities for efficient packing of customer data signals into various outgoing transmission facilities,

TABLE 11-4

SUBRATE MULTIPLEXER CAPABILITY

Bit Rate from CSU (kbps)	Maximum Number of Customer Signals
2.4	20
4.8	10
9.6	5
56.0	1

NOTE: Output is a 64-kbps DS0 signal, including byte-stuffing, framing, and control bits, where applicable.

and a highly stable timing source (derived from a master system clock) for the multiplexers at both this office and local offices. In turn, these local offices provide system clock information to individual station units on the customer's premises.

Most of the long-haul interhub transmission capacity at a DS1 rate has been derived from TD and TH microwave radio with *data under voice* (DUV) on the 1A Radio Digital System.[32] The rapid expansion of DDS in the 1980s, however, has exhausted most of the available facility capacity for DUV in many of the intercity routes. Table 11-5 summarizes a number of facility technologies available to provide long-haul and short-haul digital connectivity for DDS.

Arrangements are available to handle special cases. For example, a DS1 channel dedicated to DDS may not be economical in areas with a small number of customers. The percent fill on the channel may be too low. To meet this need, alternative arrangements of multiplexing equipment are available to mix digital data and voice customers. The most recent arrangement uses dataport units that plug into a D3 or D4 channel bank, in place of a voice channel unit, to derive a DS0 transmission channel. In a second instance, digital connectivity to the customer's geographic area may not exist. To defer the construction of new routes, arrangements are available that use analog facilities with the appropriate data sets to link these customer locations to the DDS network.

The interhub facilities are monitored full time on an in-service basis. For each DDS interhub link, one of the twenty-four DS0 channels is used to provide monitoring, to transmit results to a centralized data base in Chicago, and to remote alarms to appropriate centers for maintenance action. Therefore, in each intercity DS1 channel, only twenty-three DS0 channels are available to transmit customer data.

---

[32] Sections 9.3 and 9.4 describe these systems.

TABLE 11-5

DIGITAL DATA SYSTEM NETWORK
REPRESENTATIVE DIGITAL FACILITIES

Facility	Yield in DS1s
Long-Haul	
DUV on TD, TH radio*	1
DOV on L4 coaxial cable	2
DOV on L5 coaxial cable	4
DM12 on TD radio	12
DOM on L5	2
FT3C (etc.)†	56
Digital mastergroup on radio, coaxial cable, etc. (*future*)	4
DIM on AR6A	2
P140 on L4 coaxial cable†	84
DR6	56
TD45A	28
Short Haul	
T1	1
T1C, D†	2
T1OS	1
T2†	4
T4M†	168
6- or 11-GHz radio at 45 and 90 Mbps	28, 56
2-GHz radio	4

*With message load.

†Must be equipped with appropriate multiplexers.

DUV   Data under voice
DOV   Data over voice
DOM   Data on mastergroup
DIM   Data in the middle

Centralized testing, monitoring, and maintenance are the keys to achieving high-quality service for *DATAPHONE* digital service. Centralized test centers are the designated points of contact for customers. Test center personnel verify and sectionalize troubles while the customer is on the line. As seen in Section 13.3.3, which deals with network operations, this procedure is not generally followed in network maintenance, although it may be found in maintenance of customer loops. The procedure is an important factor in achieving guaranteed performance. A variety of operations centers and operations systems are involved in ensuring high-quality performance and rapid restoration of service.

A number of emerging services and capabilities such as circuit-switched digital capability and Basic Packet-Switching Service will use DDS network elements for connectivity for initial growth, expansion into new geographic areas, and/or data transport on an ongoing basis.

11.6.2 PACKET-SWITCHING SYSTEMS

In many data communications applications, data occur in bursts separated by idle periods, and the average data rate may be much lower than the peak rate. This type of "bursty" data can often be transmitted more economically by assembling the data into packets and interspersing packets from several channels on one physical communication path. A header is added to each packet to identify it. The contents of the header depend on the system used, but in general, the header must at least indicate the call of which it is a part and where it fits into the sequence of packets in a call.

A network can be formed to interconnect a number of users of packet switching. A packet switch sorts packets coming in on one circuit and switches them out to another circuit according to the header information in each packet. Potential savings from using a packet-switching network rather than direct connection of users are similar to those described in Section 3.1. The switch network trades the cost of the packet switch(es) against the reduced cost of transmission facilities and equipment for interfacing with the facilities. The packet-switching network becomes more economical as the number of user nodes and the distances between them increase.

Any type of traffic that has a sufficient peak-to-average information transfer rate is a candidate for packet switching. In the future, even voice messages may be assembled into packets. The immediate application of packet switching, however, is in data communications—primarily in the interconnections among computers. The first Bell System service designed for packet switching is the Basic Packet-Switching Service (BPSS) described in Section 2.5.4.

The No. 1 PSS Packet Switch

The packet-switching network supporting BPSS will use the No. 1 PSS packet switch. The architecture of the packet switch reflects the service objectives: high reliability and responsiveness at high capacity. One important characteristic of the architecture is that it is based or centered upon a Western Electric 3B20D computer. The 3B20D computer is designed to be out of service no more than 2 hours in 40 years. The reliability of this system is achieved by redundancy in hardware (duplex processor and disk, for example) and by diligence in software (the bulk of the operating system is devoted to maintaining the integrity of the system). All centralized functions associated with providing a packet-switching system are performed on this computer. For example, one function, routing, associates a physical path through the switch with lines to customers or trunks to other packet switches. Other functions include billing, traffic measurement and reporting, and system maintenance.

The capacity of the system to switch packets responsively is a function of the processing capacity available for packet switching. The architecture of the No. 1 PSS packet switch provides duplex processors as shown in Figure 11-40. Access lines to customers and trunks to other switches are connected to facility interface processors (FIPs), microprocessors especially designed for the system. The FIPs are connected to the duplex central processor (DCP). Technicians interact with the switch through CRT terminals connected to the DCP.

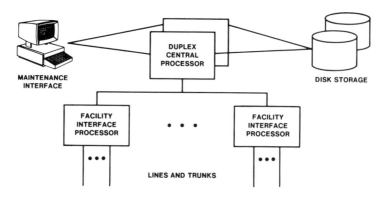

Figure 11-40. Architecture of No. 1 PSS packet switch.

An important characteristic of a packet-switching system is the protocol it implements. The No. 1 PSS packet switch implements the X.25—the packet-switching protocol[33] agreed to by the international standards organization of the Comité Consultatif International Télégraphique et Téléphonique (CCITT).

The X.25 protocol defines standards for three levels of communication between the data terminating equipment (the customer's terminal equipment) and the data communications equipment (the packet switch).

---

[33] Section 8.8 discusses data communications protocols and X.25.

- Level 1 (*physical* level) defines the electrical interface and is implemented in hardware.

- Level 2 (*link* level) processing for both trunks and access lines is done by firmware (code permanently placed in read-only memory) and special-purpose integrated circuits that are part of the FIP. The microprocessors that control the level 2 processing for each line or trunk operate asynchronously with respect to the level 3 processing on the FIP.

- Level 3 (*packet* level) processing for multiple lines and/or trunks is done by FIPs. The actual number supported depends on the traffic. The microprocessor chosen for the FIP does all the processing on data packets; packets associated with setting up calls are also processed by the DCP, where there is enough memory to store the routing tables.

The call setup rate of the No. 1 PSS packet switch is determined by the processing power of the 3B20D computer and is roughly 200 call setups per second. The capacity to handle data packets is primarily determined by the number of FIPs used and is about 1200 packets per second.

To achieve responsiveness for the large number of logical channels (simultaneous calls) that are supported on each line, the FIPs must have large amounts of memory to buffer packets. The 3B20D computer must also have large amounts of memory to store the routing information necessary to describe potential networks of users.

A characteristic that the system shares with other highly reliable systems is the amount of software devoted to maintaining the system. Of slightly more than one megabyte of object code for No. 1 PSS, only 200 kilobytes are transport (that is, X.25-related) functions. Administration and maintenance programs account for the other 80 percent.

An essential contributor to the high availability of the packet switch is the recovery strategy implemented by the software. When a failure is detected, the smallest portion of the system necessary for restoring service is reinitialized. The goal of the strategy is to minimize the effects of system recovery on customers. For the No. 1 PSS, levels of reinitialization result in:

1) clearing the call associated with the logical channel that was active when the error occurred

2) clearing all calls on the line that was active when the error occurred

3) clearing all calls on the switch

4) at the highest No. 1 PSS level, going through analogous levels of initialization provided by the operating system of the 3B20D computer.

Service Capabilities

To foster open interconnection of customer equipment, the No. 1 PSS packet switch is designed to adhere to international standards. It supports, for example, all essential services and facilities of the 1980 version of the CCITT X.25 protocol. It supports access line speeds of 9.6 and 56 kbps. Subscribers can interconnect their terminals, hosts, and office equipment effectively without large start-up costs or network management expenses. Key design goals include high availability, rapid data transfer (low end-to-end delay), high throughput (high average packet rate per channel), and high capacity (large number of packets per unit time over the switch as a whole). Operations centers and associated operations systems (see Chapter 15) provide both centralized and distributed support to respond to customer trouble reports most effectively.

The No. 1 PSS packet switch can support a common user network to provide a highly reliable, full-time computer communications network for users in need of basic data transport. A common user network allows intercorporate communications and facilitates resource sharing of value-added vendors who wish to market information services.

11.6.3 *DATAPHONE* **SELECT-A-STATION SERVICE IMPLEMENTATION**

DATAPHONE Select-a-station service is a voiceband private-line data service that is designed for applications—such as alarm service bureaus—in which a master station exchanges data with a number of remote stations, one at a time, usually in rapid sequence. The service allows 2-way transmission between the master and remote stations, but no direct transmission is possible between remote stations. Nor is broadcast communication possible between the master station and all remote stations. The security of this service makes it particularly well suited for alarm service bureaus. Connection control can only come from the master station, and all remote stations, other than the one connected at a particular time, are isolated from the connected path and from each other. This ensures that trouble in any leg cannot affect proper operation of the remainder of the circuit. This isolation of each point-to-point connection also ensures the privacy of communication between the master station and each remote station.

To implement *DATAPHONE* Select-a-station service, high-speed switches called *data station selectors* (DSSs) are located in the telephone company's central office building to connect the customer's master station with various remote stations (see Figure 11-41). Connection is established by the DSS stepping automatically in a fixed sequence or by the customer at the master station. The master station terminal is a minicomputer or a specially designed controller owned and operated by the alarm service

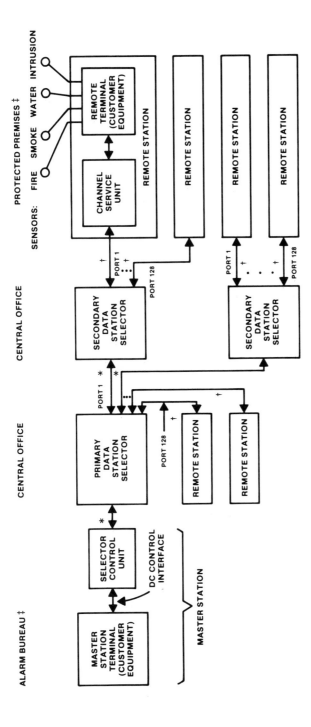

Figure 11-41. System configuration for *DATAPHONE* Select-a-station service.

* 4-WIRE VOICEBAND FACILITIES

† 2- OR 4-WIRE VOICEBAND FACILITIES

‡ CUSTOMER PREMISES

bureau. Control signaling consists of the transmission over the voiceband channel of control tones interspersed in time among the customer's data signals between the master station and the DSS. This inband signaling avoids the need for a separate control channel.

A *selector control unit* (SCU), located at the customer's master station, serves as the interface between the telephone company facilities and the customer-provided equipment. A multilead interface between the SCU and the terminal in the master station allows the customer various degrees of control over the duration and sequence of connection to the individual remote stations. The SCU will generate the necessary control signaling in response to the interface commands.

Once the connection to a particular remote station is made, the customer has full responsibility for the end-to-end transmission. The customer provides the terminal equipment that accomplishes the data exchange at both the master station and the remote sites. The master station can simply pick up an alarm signal, or it can transmit an activating signal to a remote terminal, causing the terminal to return a report. The remote terminal, supplied by the alarm bureau and located on the protected premises, has fire, smoke, or other sensors connected to it. Alarm signals from the terminal go through a channel-service-unit interface back over either 2- or 4-wire voiceband facilities to the master station.

In the customer-control version of the service, up to two DSSs, denoted as primary and a secondary, may be placed in tandem. Such tandeming may be desirable to allow geographically dispersed concentrations of stations to be served more economically. In this arrangement, the customer controls the sequence and interval for each primary port through the SCU and the primary DSS. The secondary DSS automatically steps through a fixed sequence to select associated remote stations. The number of secondaries that may be placed on a single primary is limited only by the number of output ports available on that primary; the maximum number is 128.

AUTHORS

M. D. Balkovic   F. G. Oram
H. J. Bouma   K. J. Pfeffer
R. Carlsen   P. T. Porter
D. C. Franke   S. P. Shramko
W. G. Heffron   R. W. Stubblefield

12

Common Systems

12.1 INTRODUCTION

In 1982, the Bell System network used approximately 20,000 telephone equipment buildings, including wire centers (customer-oriented), toll centers (long-distance oriented), transmission buildings, and radio relay stations. In each building, *common systems* provide power, interconnection, and environmental support for network elements associated with transmission and switching (see Figure 12-1). Since these systems are shared by switching equipment and transmission facilities, careful engineering is required. They must have the capacity to support growth as well as transitions from old to new systems with minimum service disruption. The major classes of common systems are **power systems, distributing frames**, and **equipment building systems**. This chapter discusses these major classes in detail and briefly discusses some **other common systems**—cable entrance facilities, cable pathways, and alarm systems.

12.2 POWER SYSTEMS

Power requirements for telephone company equipment buildings vary greatly depending mainly on the amount and type of equipment housed. The smallest buildings (for example, community dial offices or small repeater stations) may require less than 10 kilowatts of power, while a large central office building may require several thousand kilowatts.

A major function of power systems is to convert alternating current, supplied by an electric utility, to direct current, which is required for the proper operation of most devices and electrical circuits (relays, switches, electron tubes, transistors, integrated circuits, and even station sets) in telecommunications systems.

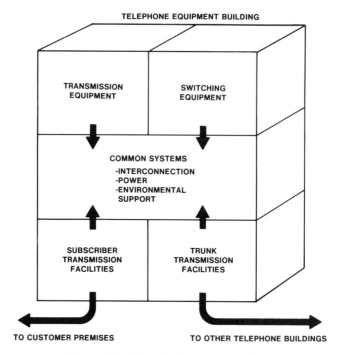

Figure 12-1. The role of common systems.

Many power systems also provide reserve energy storage to ensure service continuity if the normal energy supply is temporarily interrupted. Reserve energy applies generally to central office switching systems, transmission systems, and some computer-based operations systems. The reserve source must be available for instantaneous use since many calls in progress would be disconnected if power were interrupted even for only a few milliseconds. The widespread use of computer-controlled systems makes continuity of power even more important since information stored in volatile memory is lost during power interruptions and must be reloaded from backup memory before normal system operations can be resumed. Normally, telecommunications equipment installed on customers' premises (for example, private branch exchanges [PBXs], key telephone systems, and data sets) is not provided with standby power facilities, although such facilities are available for critical applications at extra cost. Section 12.2.7 describes the power systems available for customer-premises equipment.

Because the power facilities in a switching office or repeater station support equipment used in common by many customers, there is extensive redundancy so that the failure of a single power system component or unit does not disrupt system operation.

12.2.1 ENERGY SOURCES

The conventional energy source for most telecommunications equipment in the United States is alternating current purchased from an electric utility. It is modified and controlled as required for the specific telecommunications system.

Several unconventional power supply systems have been investigated, and some have been installed in Bell System field experiments or in remote systems. They include: continuously operated diesel-electric *alternators*[1] (installed at a remote microwave station in the Sierra Nevada Mountains), solar cells (used in a rural carrier field trial and as an auxiliary power source in a large toll center), propane-fueled thermoelectric generators (for remote repeaters and a digital radio system), and wind-driven generators to supplement the electric utility supply. Extensive research and development efforts by many organizations to develop new and improved alternate energy sources may eventually bring about their use in certain applications, such as in remote microwave repeater stations.

12.2.2 ENERGY STORAGE

Two types of energy storage systems are used extensively in the Bell System: **electrochemical cells**[2] and standby **alternator sets** powered by internal-combustion engines.

Electrochemical cells provide an instantaneous reserve source of dc power. One type, unsealed lead-acid cells, costs less than other electrochemical systems. In principle, they are similar to ordinary automobile batteries. However, the cells used for telecommunications service are designed for long life (typically 15 years or more) and engineered for long discharge times (hours) at moderate temperatures. Lead-acid cells are purchased in accordance with special design specifications and are available in a wide variety of sizes from 100 to 1680 ampere-hour nominal ratings.

In the early 1970s, Bell Laboratories developed a cylindrical, unsealed lead-acid cell that lasts longer and requires less maintenance. Detailed specifications for both materials and fabrication were required to gain these improvements. These cells are now being manufactured for the Bell System according to these specifications. Sealed nickel-cadmium and sealed lead-acid cells have been used in subscriber loop systems deployed in the outside plant environment to avoid the routine maintenance required with conventional lead-acid cells.

Lead-acid cells provide only short-term reserve power, so standby internal-combustion *engine-alternator sets* (like the ones shown in

---

[1] Machines that produce alternating current.

[2] Devices that supply electricity through chemical action.

Figure 12-2) are commonly installed in many telephone company buildings to provide a long-term reserve as described in Section 12.2.3. The
alternator, driven by the engine, produces alternating current to replace
electric utility power. For these systems, energy is stored in the form of a
liquid hydrocarbon fuel (for the engine). A typical communications
center may provide fuel storage for 2 days to 3 weeks of continuous
operation of essential equipment. This fuel supply can be replaced to
achieve unlimited reserve energy storage.

Engine-alternator sets use either *diesel engines* or *turbines*. Diesel
engines are generally used in sizes up to about 200 kilowatts, while gas
turbines are used in larger sizes—up to about 2000 kilowatts. Gas turbines are internal-combustion engines and are similar to jet aircraft turbines; the blades of the turbine are propelled by hot gases produced by
the combustion process. Gas turbines may burn different types of fuel; in
Bell System applications, they use diesel fuel. Gas turbine units are generally much lighter and smaller than comparable diesel engine sets and
can be installed in upper floors or on a building roof, whereas diesel
engine-alternator sets are generally installed in the basement or first
floor. In most buildings, a single turbine unit is installed. Multiple
installations may be used in larger buildings.

Figure 12-2. Standby engine-alternators. *Left*, 100-kilowatt diesel; *right*,
750-kilowatt turbines.

12.2.3 POWER SYSTEM OPERATION

The elements of a typical power system are shown in Figure 12-3. While
the dc power plant is common to all telecommunications equipment locations, the engine-alternator may be omitted where long-term reserve
power is not required. The dc-to-ac *inverter* may be omitted if there is no
essential equipment requiring ac power, and the dc-to-dc *converters* are

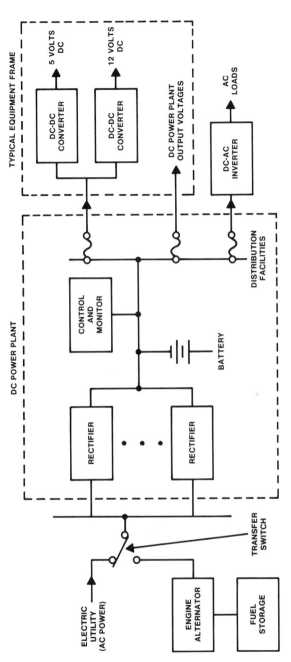

Figure 12-3. Typical power system.

provided where voltages not readily available from the dc power plant are required. Sections 12.2.5 and 12.2.6 discuss converters and inverters, respectively.

The dc power plant consists of four main elements: rectifiers; a group of lead-acid cells connected in series, commonly called a *battery*; control and monitoring equipment; and distribution facilities.

A typical dc power plant is shown in Figure 12-4. The power plant accepts alternating current when it is available and rectifies it to produce dc power that is then supplied to the telecommunications equipment. The battery provides a source of reserve dc power that is automatically supplied if the electric utility fails and continues until the lead-acid cells are discharged. In normal practice, the battery is selected to provide 3 to 8 hours of reserve time.

The engine-alternator set in a power system is normally idle and is started either manually or automatically after a disruption of the electric utility supply. Its output then replaces the utility supply, and the

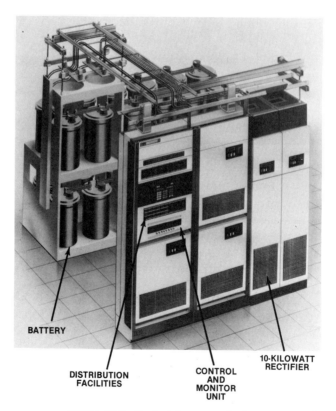

Figure 12-4. Typical dc power plant.

rectifiers again supply power to the telecommunications equipment and recharge the battery. The hours of battery reserve time and the choice of manual or automatic-start engine-alternator are determined by the operating company for each location, based on the history of commercial power outages and accessibility of the location to service personnel. An unattended microwave station on a remote mountaintop, for example, may have as much as 24 hours of battery reserve, as well as automatically operated engine-alternator sets.

In small community dial offices and readily accessible carrier repeater stations, a standby engine-alternator may not be furnished. In this case, the battery may be engineered to provide 24 hours or more of reserve. In the event of a long-term failure, a portable engine-alternator is brought to the site.

A number of dc power plants with different nominal output voltages for different applications have been developed. Table 12-1 identifies some of the common applications of these plants.

The description of a power system given above is adequate to illustrate its operation as a power conversion and energy storage system. However, the plants in Bell System installations include additional features to control the output voltage and maintain the battery in a proper state of charge.

TABLE 12-1

DC POWER PLANT OUTPUT VOLTAGES

Nominal Voltage (Volts)	Application
−24	Microwave radio, transmission multiplex
−48	Electromechanical and electronic switching systems, transmission systems colocated with switching systems
±130	Electron tubes, remote repeaters, telegraph systems
+140	4ESS switching equipment, L5 power feed stations, large toll centers

12.2.4 RECTIFIERS

Rectifiers convert alternating current to direct current. In normal operation, they supply dc power to the telecommunications equipment (the *load*) and maintain the battery in a fully charged state. In the event of an electric utility outage, some discharge of the battery will occur. When utility power is restored or if an engine-alternator set is activated, the rectifier output will recharge the battery and resume supplying power.

The equipment units are installed and connected in parallel to provide added capacity as required. Rectifiers with sizes up to about 80 kilowatts (1600 amperes at 48 volts) are in service in telecommunications power plants.

12.2.5 DC-TO-DC CONVERTERS

The introduction of new systems and devices has generated a need for voltages not readily available from a dc power plant. In particular, the large-scale use of semiconductor and integrated circuits, which require well-regulated supply voltages between 2 and 12 volts, has provided the impetus for the development of dc-to-dc converters to convert the output of a dc power plant to the required dc voltage.

Since the dc-to-dc converter interposes a major system element between the dc power plant and the load, the effect of a converter failure must be considered. A common approach is to supply a converter to power a single functional unit (for example, a channel bank, radio transmitter, or memory unit) and depend on system reliability arrangements, such as protection switching or redundancy, to maintain service if a converter fails. This approach may result in high costs, especially when many small functional units are required in a given system. In these cases, parallel operation of converters[3] may be more desirable. Many converter circuits include monitor, alarm, shutdown, and other peripheral functions to meet system needs.

Since the dc-to-dc converter must generally provide close control (regulation) of its output voltage, it should be physically close to the load to minimize *voltage drop*[4] especially in low-voltage applications. Small converters are often mounted on the same circuit pack as the load requiring the dc voltage. Medium-size units (20 to 500 watts) are designed as plug-in modules and are mounted in frames with the circuits they supply, as shown in Figure 12-3. Larger converters are generally mounted in separate bays in communication equipment rooms.

12.2.6 INVERTERS

Although most loads in telecommunications systems that require reserve power are operated by direct current, certain loads—such as motors in tape drives—require alternating current. Where such loads must operate

---

[3] Where the outputs of several converters (including a spare) are combined to supply all functional units. This is similar to the configuration of rectifiers in Figure 12-3.

[4] A reduction in voltage caused by current flow in the conductors between the power supply and the load.

during electric utility supply interruptions, solid-state dc-to-ac inverters provide reliable ac power. These units use semiconductor switching devices to convert the dc supply from a battery to alternating current.

The use of commercial computer systems for essential operations has increased the demand for continuous ac power since these systems generally are not designed for dc operation. Large uninterruptible power systems employing dedicated dc power plants and multiple inverters for redundancy are being used to fill these needs.

12.2.7 POWER FOR CUSTOMER-PREMISES EQUIPMENT

Most PBXs, key telephone systems, data stations, and other equipment on customers' premises do not have standby power facilities. These systems are energized from the electric utility supply through rectifiers to provide the dc voltages required. The rectifiers are often mounted in the cabinet that contains the communication equipment. If power fails, PBX systems automatically transfer selected stations to central office trunks to provide limited communication capability. The actual switching takes place when relays operated by PBX power release because of PBX power failure. In releasing, the relay contacts transfer the telephones to central office circuits. Until power is restored, service to the transferred telephones is much like the service provided to a single business or residence line.

In critical locations such as hospitals and emergency services, battery-inverter plants are furnished to provide reserve capability. In some locations, a standby engine-alternator set furnished by the customer may be used to power communication facilities and other important services if the utility supply fails.

12.3 DISTRIBUTING FRAMES

12.3.1 GENERAL DESCRIPTION

In a typical telephone equipment building (see Figure 12-5), the **distributing frame** is a common termination point for all equipment and facilities at that location. The facilities include subscriber cables to customer premises and trunk cables to other telephone equipment buildings; the equipment includes switching equipment (both the line and trunk sides of the switching network),[5] and a wide variety of transmission equipment including amplifiers, signal converters, and test access systems. By interconnecting specific equipment and facilities with distributing frame jumpers (see Section 12.3.2), basic and special telecommunications services can be provided. Distributing frames may also connect cables that are

---

[5] Section 7.3 discusses switching networks.

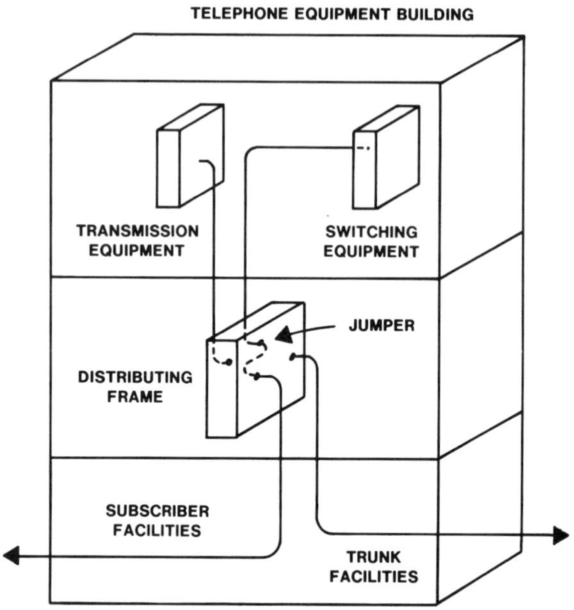

TELEPHONE EQUIPMENT BUILDING

TRANSMISSION EQUIPMENT

SWITCHING EQUIPMENT

JUMPER

DISTRIBUTING FRAME

SUBSCRIBER FACILITIES

TRUNK FACILITIES

Figure 12-5. Distributing frame role in the telecommunications network.

part of carrier systems to the appropriate transmission equipment for functions such as multiplexing[6] and demultiplexing.

Each wire center and toll center in the Bell System has its own distributing frame system. The size and complexity of these systems vary widely from small, single lineups to complex networks involving numerous large distributing frames interconnected by thousands of tie cable links. About 20,000 technicians are required to operate distributing frames in the Bell System.

12.3.2 DISTRIBUTING FRAME FUNCTIONS

Distributing frames provide three basic functions: cross-connection, electrical protection, and test access.

Cross-connection of two or more outside plant facility and office equipment terminations is required to provide service to a customer. Individual cross-connect wires called *jumpers*, which can remain in place for a few days to several years, join the termination points of cables representing particular facilities and equipment. To ensure efficient use

---

[6] Section 6.5 discusses multiplexing.

of facilities and equipment, Bell System policy has required total inter-connection flexibility on distributing frames (that is, any outside plant facility or piece of office equipment must be able to be connected to any other).

Distributing frames also provide electrical protection for equipment and personnel within a wire center. Fuse-like protector devices are mounted on distributing frames to guard against spurious voltages and currents. These electrical incursions could result from lightning strikes or power-line crosses anywhere in the outside plant.

In addition, distributing frames provide an access point where circuits can physically be opened and tested. Temporary disconnection of customers is accomplished by setting a *protector unit* (see Figure 12-6A) to an inactive position and does not require removal of the cross-connection.

12.3.3 DISTRIBUTING FRAME HARDWARE AND APPLICATION

Figure 12-6 shows conventional distributing frame hardware, which was the only kind used in the Bell System until 1964. As Figure 12-7 shows schematically, the conventional frame is a double-sided structure composed of vertical and horizontal parts. Typically, outside plant cables run to terminal apparatus mounted on the vertical side where electrical protection is provided. Cables from equipment located in the building terminate on the horizontal side. Jumpers between terminals on the vertical and horizontal sides run on shelves on the horizontal side and along vertical troughs to the vertical side termination.

The most widely used conventional frame in the Bell System is 11 to 14 feet high, and installations up to 400 feet long are in service. Two technicians are required to run a jumper, and ladders are needed to reach the upper shelves. Hence, these frames are somewhat unwieldy to operate. The low-profile conventional distributing frame was introduced in the early 1970s. It is about 8 feet high and is supported only at the floor. It conforms to the New Equipment-Building Systems (NEBS) standards (discussed in Section 12.4.5) and does not require ladders, although two technicians may still be needed to run jumpers.

Distributing frames may be interconnected for different applications to form distributing frame networks. The frames that make up these networks are given functional designations according to the type of plant facilities terminated on them. The functional frames in such networks are linked by *tie cables*.

A **combined main distributing frame** (CMDF) has both subscriber and trunk outside plant cable terminations (see Figure 12-8A). A **subscriber main distributing frame** (SMDF) terminates only customer-oriented cables, and a **trunk main distributing frame** (TMDF) terminates only interoffice cables and transmission facilities (see Figure 12-8B). Finally,

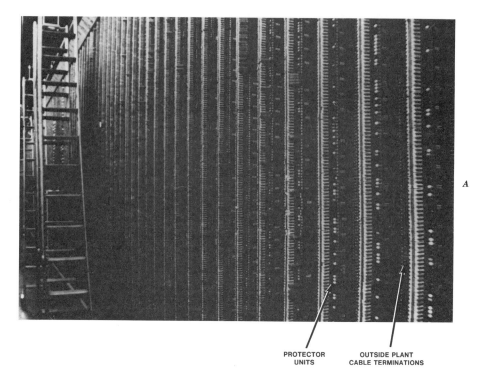

A

PROTECTOR
UNITS

OUTSIDE PLANT
CABLE TERMINATIONS

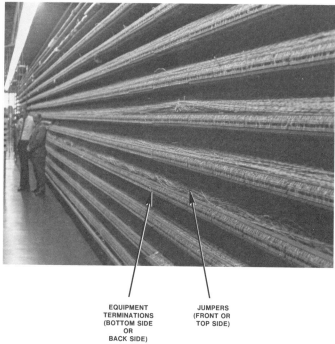

B

EQUIPMENT
TERMINATIONS
(BOTTOM SIDE
OR
BACK SIDE)

JUMPERS
(FRONT OR
TOP SIDE)

Figure 12-6. Conventional distributing frame. *A,* vertical side: termination of outside plant facilities on electrically protected connectors; *B,* horizontal side: termination of office equipment with associated jumpers.

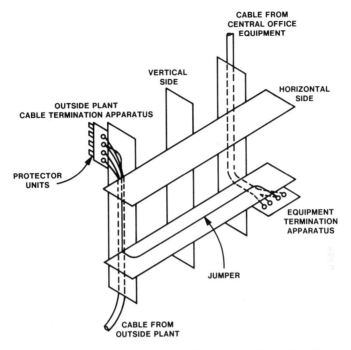

Figure 12-7. **Conventional frame hardware schematic diagram.** Terminations for equipment and outside plant cable run all along the frame, as shown in Figure 12-6.

an **intermediate distributing frame** (IDF) has no outside plant facilities terminated on it.

Either conventional or modular hardware may be used in distributing frame networks. Figure 12-9 shows one modular distributing frame, the Common Systems Main Interconnection Frame System (COSMIC). Modular frames were first introduced during the 1960s for use with 1ESS switching equipment and underwent a major redesign in the early 1970s.

Office equipment and outside plant facilities are cabled to alternate modules by means of the backplane of a modular frame. Each module contains numerous office equipment or outside plant facility terminations. Cross-connections are run on the front face of the frame. Vertical and horizontal troughs are provided to accommodate the cross-connections in an orderly manner. Modular frames are low, and one technician can run jumpers; thus, they offer potential operations efficiencies over conventional frames.

Modular frames in common use include: *ESS* switching equipment frames, the COSMIC frame (shown in Figure 12-9), and COSMIC II. The most widely used modular frame is the COSMIC frame, which comprises

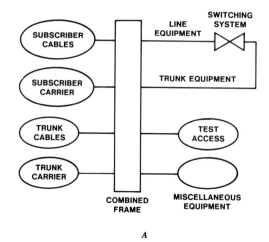

Figure 12-8. Typical distributing frame networks. *A,* combined main distributing frame; *B,* subscriber and trunk main distributing frames.

10,000-pair alternating modules of outside plant and office equipment. Because this COSMIC frame does not accommodate electrical protector units, a separate frame called the *protector frame* must also be used. Cables connect the protector frame and the COSMIC frame.

The COSMIC II distributing frame also uses alternating large modules of facilities and equipment. However, it differs from the earlier COSMIC in two important ways: (1) the separate protector frame is replaced by protected facility terminations on the rear of the outside plant modules,

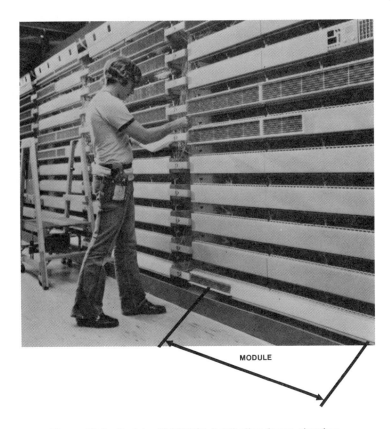

Figure 12-9. Modular (COSMIC) distributing frame showing modules for terminating outside plant facilities and office equipment.

and (2) it has a wider range of applications—it can be used as a subscriber, a trunk, or a combined main distributing frame, while COSMIC is always used as a subscriber frame.

12.3.4 DISTRIBUTING FRAME ADMINISTRATION AND ENGINEERING

Distributing frame administration includes recordkeeping and assignment of the facilities and equipment that provide telecommunications services. Distributing frame engineering provides adequate frame capacity and ensures systematic layout of facilities and equipment as the office grows.

Providing telecommunications services to customers involves a sequence of operations related to distributing frames. First, a customer request is transmitted to an assignment center where specific facilities and equipment are selected from lists of spares at the distributing frame. This assignment of circuit elements attempts to satisfy several requirements simultaneously, including circuit needs, equipment load balance (see Section 5.3.4), and short cross-connections (jumpers). After facilities and equipment are assigned, a service order is transmitted to the frame force. The frame force makes the necessary cross-connections, tests the circuits, and sends an order completion report to the assignment center.

It has become increasingly important during recent decades to attain short jumper assignments because distributing frames have been growing larger. Larger frames have more jumpers on their shelves and troughs, and the jumpers tend to congest midway along the frame's length. Figure 12-10 shows an overloaded main distributing frame in a central office. *Preferential assignment* (see Figure 12-11) is a method of attaining more short jumpers than would be obtained by randomly assigning from lists of spares. With preferential assignment, both the facilities and equipment sides (modules) of the distributing frame are administered in "zones" with assignment preferences from A to A, then A to B, then A to C, etc. This type of zoned administration can reduce the number of jumpers that cross midway along the frame, provided that the various types of facilities and equipment are distributed in all zones. This latter condition is accomplished by *distributing frame layout control.*

Figure 12-10. An overloaded main distributing frame in a central office.

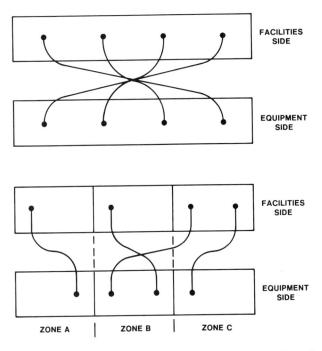

Figure 12-11. Conventional distributing frame assignment methods.
Top, random assignment; *bottom,* preferential assignment.

Layout control of terminations is an engineering technique that was introduced for conventional frames in the early 1970s with zoned layout schemes. The objective is to spread old and new terminations of all categories of equipment and facilities evenly over a distributing frame at any point in time. This can be accomplished with the aid of a *termination layout mask* (see Figure 12-12), which reserves particular regions of a distributing frame for specific termination categories.

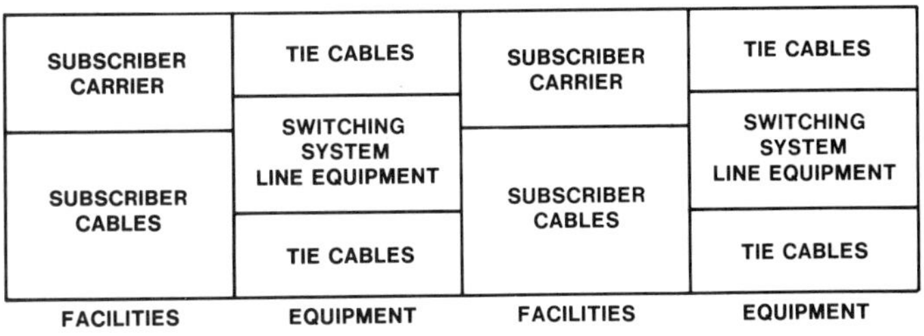

SUBSCRIBER CARRIER	TIE CABLES	SUBSCRIBER CARRIER	TIE CABLES
	SWITCHING SYSTEM LINE EQUIPMENT		SWITCHING SYSTEM LINE EQUIPMENT
SUBSCRIBER CABLES		SUBSCRIBER CABLES	
	TIE CABLES		TIE CABLES
FACILITIES	EQUIPMENT	FACILITIES	EQUIPMENT

Figure 12-12. Subscriber MDF termination layout mask.

Modular frames can be viewed as hardware manifestations of the zoned layout concept of conventional frames where the layout of facilities and equipment is reflected in the alternating physical hardware modules. Figure 12-13 shows a technician running a preferentially assigned short jumper on a layout-controlled modular frame.

Mechanized layout control was first introduced for COSMIC frames with a system called the *Program for Arrangement of Cables and Equipment* (PACE) and is now being extended to other applications with a new system called the *Mechanized Engineering and Layout for Distributing Frames* (MELD). Mechanized algorithms spread each type of facility and equipment uniformly when the frame is initially installed and as terminations are added over its service lifetime.

The manual administration process of maintaining facility and equipment inventories and assignments is complex. Insufficient coordination and followup can cause errors that result in wasted operations at the distributing frame. Poor records may mean that unused equipment and facilities are terminated at the distributing frame but not recognized as spares. In the manual mode, records are difficult to maintain and preferential assignment is almost impossible to implement; hence, efforts to mechanize record and assignment processes were begun in the early 1970s. The result has been implementation of substantial distributing frame administrative capabilities (including preferential assignment) in

Figure 12-13. Technician running a preferentially assigned short jumper on a COSMIC distributing frame.

computer-based systems such as the Computer System for Mainframe Operations (COSMOS) and the Trunks Integrated Records Keeping System (TIRKS).[7] With these operations systems, available equipment and facilities are also used more efficiently.

12.4 EQUIPMENT BUILDING SYSTEMS

12.4.1 TELEPHONE EQUIPMENT AREAS

Approximately 60 to 85 percent of the interior space in a telephone equipment building consists of equipment areas. These are large, usually windowless, partitionless rooms designed to contain the equipment, the appropriate cable support systems, and the air ducts for the environmental control the equipment needs to function (see Figure 12-14).

Equipment space and associated cabling are required for switching systems; transmission systems; and common systems such as power, main distributing frames, the cable entrance facility (see Section 12.5.1), and environmental support.

The switching and transmission equipment installed in telephone equipment buildings is usually mounted on steel framework lineups 1 to 2 feet deep and up to 6-½ feet wide (see Figure 12-15). The frames are installed side by side in lineups usually 30 to 50 feet long. Because the frames are large and heavy, they are bolted to the floor and/or to the

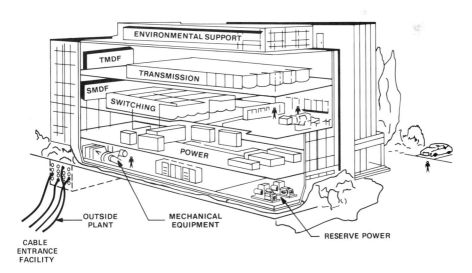

Figure 12-14. Equipment space in telephone equipment building.

---

[7] Section 14.2.1 describes TIRKS.

Figure 12-15. Equipment frames in telephone equipment buildings.
A, transmission equipment; B, crossbar switching equipment mounted on
11-½ foot high framework lineups; C, ESS switching equipment mounted on
7-foot high framework lineups.

ceiling with concrete-embedded anchors to eliminate vibration and to prevent toppling.

Typically, aisles 2 to 4 feet wide between lineups allow access to equipment for wiring and maintenance. Other steel structures above the equipment framework support the tons of cabling that interconnect the equipment and connect the equipment to the distributing frames (see Figure 12-16A). (Section 12.5.2 describes cable distribution systems.)

Main distributing frames (see Section 12.3) are also in the equipment area. In multistory offices, distributing frames may be located on several floors, requiring vertical access for cables (see Figure 12-16B). Many heavy bundles of interfloor cabling are used. In the equipment area, steel framework is positioned between floors to support the vertical cabling, and special reinforcement is used in the floors to maintain structural integrity at points of penetration.

The power systems (see Section 12.2) that provide the power conversion and uninterrupted energy source for the telephone equipment are also located in equipment areas but are usually located in power rooms separate from the transmission and switching equipment.

12.4.2 BUILDING ELECTRICAL SYSTEMS

In addition to the dc power systems that directly serve the telephone equipment, a telephone equipment building contains a number of other special electrical systems. These systems are designed for extreme reliability and high capacity; they use special control and protective circuits.

- The **ac power system**, fed by the electric utility, consists of service entrance, protection, distribution, and control apparatus. The system serves the building's electrical and mechanical systems, as well as the telephone power plants discussed in Section 12.2.

- Extensive **electrical grounding systems** are placed throughout each building to eliminate noise on lines, reduce high-speed data errors, and protect the telephone equipment from electrical short circuits and lightning strikes.

- **Shielding systems** are provided in offices located near sources of high-intensity electromagnetic or electrical fields such as television broadcasting stations, electrical power stations, or certain high-tension lines. Shielding prevents or reduces the penetration of electromagnetic fields into electronic equipment. Such penetration can cause malfunctions.

12.4.3 BUILDING MECHANICAL SYSTEMS

Equipment buildings have two general types of mechanical systems. The first type provides such environmental support as control of the temperature, regulation of the humidity of the surrounding air, and maintenance

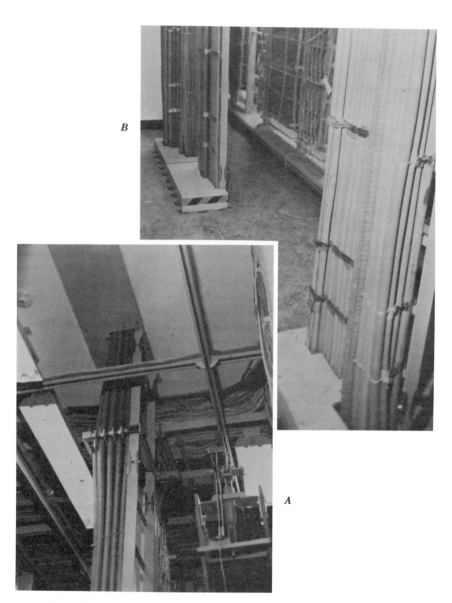

Figure 12-16. Cabling in telephone equipment buildings. *A*, routed above equipment frames; *B*, routed between floors.

of the appropriate amounts of outside air of high purity. The second type provides for vertical access in multifloor structures and for distribution or movement of air, water, and fuel. Mechanical equipment areas occupy up to 25 percent of the gross space, depending on the process cooling,[8] humidification, and air filtration needed by the telephone equipment and reserve power engines. Mechanical equipment systems include:

- **Cooling systems** designed to remove the heat released from telephone equipment. Recent developments that use solid-state devices have miniaturized telephone equipment components. The result is close-packed equipment that dissipates large amounts of heat and, therefore, requires exceptional amounts of cooling. For example, a 10,000-square-foot toll office will require up to 100 tons[9] of cooling capacity when it houses electronic transmission and switching equipment. Cooling an equivalent floor area for human comfort would rarely require more than 15 percent of this air-conditioning tonnage.

- Ventilating **fan systems** with enough capacity to maintain short-term ambient operating environments for the telephone equipment in the event of a cooling-system failure. While repairs are being made, these fan systems use outside air.

- High-capacity **air filtration systems** necessary to prevent dust and products of combustion from infiltrating the building and causing electrical contact failures.

- **Vertical access space** within the structure provided for water and drain lines, fuel lines and exhaust stacks for reserve power plants, and movement of large, bulky equipment assemblies between floors using freight elevators or hoisting shaftways.

The location of vertical runs is coordinated with the equipment plan for the building to ensure compatibility with the placement of future generations of equipment.

12.4.4 SPECIAL CONSTRUCTION

Another important characteristic of an equipment building or transmission station is the provision for expansion. If a horizontal addition to the building is anticipated, a rear or side wall must be designed so that it can be removed without interfering with the structural integrity of the roof

---

[8] A procedure for maintaining a cool environment for equipment rather than for people.

[9] In this usage, a *ton* is equal to 12,000 British thermal units (Btus) per hour.

and floors or with equipment assemblies that are operating to provide service. Extension of the air distribution ducts and refrigeration machinery must be accommodated. If vertical additions are anticipated, the footings, columns, and load-bearing walls must be adequate for the future configuration.

Special construction is also required where offices have roof-mounted microwave radio towers. These towers must be able to support several antennas and component assemblies weighing hundreds of tons. When a tower and antennas are on top of a building, the load is carried through the building to the foundation and requires a massive internal support system.

12.4.5 EQUIPMENT BUILDING SYSTEM STANDARDS

During its 100-year history, the Bell System has used three different sets of building design standards. The earliest equipment buildings were designed primarily for operator switchboards. In the mid-1920s, with the introduction of automatic switching, building standards were changed to accommodate 11-½ foot equipment framework lineups. Then, in the early 1970s, the standards were changed to provide 10 feet of clear height for equipment and cabling. These New Equipment-Building Systems (NEBS) standards were motivated by the trends in electronics that made it necessary to fit more compact designs with their higher heat and cabling requirements into existing space. A second objective was reducing costs in constructing new buildings. The NEBS documents include a set of coordinated specifications for equipment and buildings:

- **Equipment Design Standards** provide the spatial and environmental performance requirements (for example, air conditioning and lighting requirements) for all new equipment systems.

- **Building Engineering Standards** specify the planning and design of new buildings and additions and provide guidelines for the reuse of existing space to accommodate modern electronic equipment.

Equipment building standards are important not only to the performance of the facility but also to the cost of buildings, since the physical characteristics of telephone equipment have an effect on the design of the building intended to house that equipment. Frame height and cabling space, for example, control the "clear" ceiling height from the floor to the lowest overhead obstruction. Equipment weight determines the building's *live load*—the weight that foundations, columns, and floors must support in addition to their own weight. The amount and location of heat emanating from the equipment determines the size of the cooling plant and the location of the air ducts and *diffusers*.[10]

[10] Fixtures that attach to a duct and distribute air in aisles between equipment frames.

Problems can arise when equipment units differ greatly in physical characteristics. To avoid these problems, NEBS standards include the full range of spatial and environmental conditions. The requirements cover equipment frame areas, distributing frame areas, power equipment areas, operations systems areas, cable distribution systems, and cable entrance facilities. The environmental requirements are grouped according to functional effects and include fire resistance, grounding, radio-frequency interference, thermal effects, shock, vibration, earthquake, airborne contaminants, acoustical noise, and illumination. NEBS documents provide standards, design requirements, and planning guidelines for the building structure, for each equipment area in the building, and for all the building support systems. Use of the standards simplifies building design and equipment engineering, streamlines equipment and cable installation, and allows for flexibility in growth patterns. Equipment buildings vary widely in size and appearance, depending on the application, as the examples in Figure 12-17 show.

12.5 OTHER COMMON SYSTEMS

12.5.1 CABLE ENTRANCE FACILITY

A typical telephone equipment building must accommodate thousands of pairs of wires, coaxial cables, and optical fibers from outside plant transmission facilities. The cable entrance facility (CEF) provides an entrance area for all types of outside plant cables carrying subscriber lines and transmission facilities between equipment buildings. As illustrated in Figure 12-18, a typical cable entrance facility is a vault-like below-grade area. It is typically 12 to 15 feet high and 12 feet wide and runs the length of the building directly under the main distributing frame(s). It can be over 400 feet long. One or both of the end walls contain a conduit termination with a built-in gas-venting chamber that is used to prevent water and hazardous gas from entering the central office building.

12.5.2 CABLE DISTRIBUTION SYSTEMS

Switching and transmission systems in telephone buildings use multipair cables to connect submodules within a system or to interconnect that system with other (common) systems such as distributing frames and power. The overhead cable distribution systems and associated common hardware for the NEBS equipment take into account the requirements for cabling, cooling, assembling, lighting, and maintaining the equipment. Cable distribution systems are provided in modular arrangements to simplify engineering and installation. Although designed primarily for use in NEBS buildings, these arrangements can be modified to suit job conditions when NEBS equipment is installed in non-NEBS space.

Figure 12-17. NEBS equipment buildings. *A*, suburban wire center; *B*, metropolitan toll center; *C*, urban wire center.

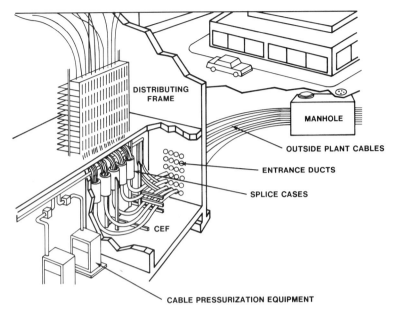

Figure 12-18. Cable entrance facility.

The overall coordination of the superstructure and common hardware required for NEBS equipment is achieved with the *Cable Pathways Plan.* This plan standardizes the maximum size and possible locations of cable racks and integrates the frame and aisle lighting system with the cable distribution system. The Cable Pathways Plan also incorporates the building columns, cable holes, cooling air diffusers, fire detectors, and access requirements in an overall allocation of space that minimizes possible conflicts throughout the life of the building.

12.5.3 ALARM SYSTEMS

Telephone buildings use a variety of alarm systems to indicate equipment failure and/or service interruption. Some typical examples are switching system alarms, transmission terminal alarms, utility power alarms, and dc power alarms. In addition, extensive detector, alarm, and control systems are employed throughout the building for protection against fires.

AUTHORS

E. J. Kovac
R. J. Skrabal
J. J. Stockert

PART FOUR

OPERATIONS

Part Four expands on the brief discussion of operating telephone company functions in Chapter 1. It examines how these companies manage their resources and interact with customers to provide services over the network.

Chapter 13 provides an overview of the major customer- and network-related operations. It emphasizes relationships, interfaces, and chronological sequences among a telephone company's major operations. Chapter 14 examines the role of operations systems in operating telephone companies. Rather than cataloging the many operations systems in use, the chapter illustrates the use of generic classes of systems and their impact on operations by describing the functional characteristics and benefits of a few selected systems. Chapter 15 describes the planning process that supports telephone companies in organizing their operations and the impact of this planning on the operating companies and on the development and evolution of operations systems. Part Three concludes with Chapter 16, which describes the activities and considerations involved in defining and maintaining desired levels of service and performance. This is an appropriate closure since meeting these levels is the central objective of telephone company operations.

13

Overview of
Telephone Company Operations

13.1 INTRODUCTION

This chapter outlines some of the many activities or functions required for operation of the telecommunications system. As indicated in the Foreword of this book, the material presented reflects the Bell System as it was at the end of 1982 and early in 1983. Then, and for many years prior, operation of the Bell System was a shared responsibility, with Bell operating companies responsible for operations in their territories and AT&T Long Lines responsible for interstate and international operations. Divestiture of the Bell operating companies from AT&T will result in changes in responsibility for certain operations functions and, in some cases, will change the way the functions are accomplished. However, the basic functions required will, by and large, remain the same, so that the material in this chapter will remain relevant from the standpoint of understanding what is involved in telephone company operations. Although the discussion focuses on Bell System operations, the activities described are also applicable, in a broad sense, to the many independent telephone companies.

Telephone company operations can be divided into three kinds of functions.

- **Provisioning** is the process of making the various telecommunications resources (such as switching systems, transmission facilities, and operators) available for telecommunications services. Provisioning includes forecasting the demand for service, determining the additions (or changes) to the network that will be needed, determining where and when they will be needed, and installing them.

- **Administration** covers a broad group of functions that sustain services once they have been provided. Administration generally consists of **network administration** and **service administration**. *Network administration* ensures that the network is used efficiently and that grade-of-service[1] objectives are met. *Service administration* includes such diverse functions as billing; collecting and counting coins from coin telephones; and, for customer switching systems, giving engineering and service evaluation assistance and keeping detailed engineering records.

- **Maintenance** operations ensure that network components work properly once they are installed. Maintenance includes the testing and repair activities that correct existing malfunctions (corrective maintenance) and those that prevent service-affecting malfunctions (preventive maintenance).

Bell System operations are complex and involve close to a million employees. To handle this enormous enterprise rationally and efficiently, operating company employees and the functions they perform have been grouped into *operations centers* (see Section 15.2.1). The people in an operations center report to a common manager and perform closely related functions for geographic areas that range from national to local in scope. In addition, operations centers may be customer specific or service specific. For example, a customer-specific operations center may serve residence customers; a service-specific operations center may be responsible for a service such as *PICTUREPHONE* meeting service.

The growing application of computer technology to operations is a significant factor in the productivity gains made by the Bell System. Computer-based *operations systems* (see Section 15.2.2) mechanize much of the routine, time-consuming, often tiresome tasks (such as recordkeeping) and enable people to do their jobs more accurately and efficiently. These systems may also accomplish very complex tasks that could not be done manually. (Chapter 14 discusses some of the major operations systems in detail.)

The application of people and machines to operations requires considerable ongoing effort to meet the rapid evolution of market opportunities and technology or, where possible, to anticipate this evolution. AT&T and Bell Laboratories are engaged in planning the evolution of operations to ensure that operations related to new products, services, and technology are efficient, that they benefit from continuing mechanization opportunities, and that they are responsive to market demands. Because the Bell System operation involves hundreds of operations centers and

[1] Section 5.2.2 discusses the *grade-of-service* concept.

systems, another goal is to see that the whole collection of people and machines work together. The *operations planning* process identifies the need for new or changed roles for existing operations centers and systems, as well as the need for entirely new centers or systems. Thus, it acts as a starting point for the development of methods and procedures that define how the people in the operations centers should perform their tasks (largely an AT&T function) and the design and development of operations systems (largely a Bell Laboratories function). Ultimately, the Bell operating companies carry out the plans by deploying the operations centers and systems in their service areas. (Chapter 15 discusses operations planning in detail.)

The rest of this chapter describes **customer-related operations** (Section 13.2) and **network-related operations**, the activities required to provide, administer, and maintain the network and its elements (Section 13.3). It is important to keep in mind that the network is a shared resource; thus, while Section 13.2 emphasizes customer contact operations, it frequently references the network operations needed to serve a customer.

Many other essential operations are not discussed here. For instance, one set of operations provides centralized support services to both customer-related operations and network operations. These operations exist in most large businesses and include such functions as providing administrative services (for example, payroll, comptroller, and legal services); operating motor vehicle pools; and managing real estate, inventory, and materials. Other operations are covered in other areas of the book; for example, Chapter 16 discusses the measurement of network performance and customer satisfaction, and Section 17.3 discusses the development of tariffs. Western Electric Company operations, while also essential to the Bell System, are beyond the scope of this book.

13.2 CUSTOMER-RELATED OPERATIONS

This section describes Bell System operations that are directly related to customer service. Because Bell System customers are so diverse in their needs and expectations, customers with similar needs are grouped into market segments, and operations are often adapted to these segments. Thus, a residence customer with simple service needs would be handled differently from a business customer with complex and sophisticated communications needs.

13.2.1 PROVISION OF SERVICE TO THE CUSTOMER

The *service-provisioning* process begins with the first contact between the customer and the telephone company representative to negotiate service. It ends with the satisfactory delivery of a product or service, timely billing for the service provided, and updating of all related records. Figure

13-1 shows the major functions of the provisioning process and the sequence in which they are performed.

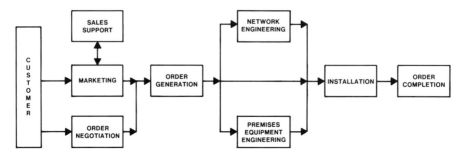

Figure 13-1. Service-provisioning operations—functional flow.

Order Negotiation and Marketing

The first phase of provisioning is order negotiation, that is, the negotiation of services through face-to-face contact or telephone contact initiated by either the customer, a Bell operating company (BOC), or AT&T Long Lines.

A *service representative* handles most customer-initiated requests either in a retail sales environment or over the telephone in a local service center. The service representative and the customer discuss the types of equipment,[2] services, and features available and agree on what is to be provided. To complete the interaction, the service representative must usually access the customer's current service and billing records if they exist. For a new customer, the service representative must obtain directory listing information (for example, name, address, and type of listing) and billing and credit information (for example, billing name and address and current employment status).

Customer-initiated requests for complex services, such as a private branch exchange (PBX) and data and private-line services, involve marketing functions. The marketing organization also initiates contact with selected accounts such as large business customers. It discusses problems

---

[2] Under provisions of FCC Computer Inquiry II (see Section 17.4.3), effective January 1, 1983, Bell operating companies can provide only that customer-premises equipment (CPE) in their inventory on that date. New CPE from AT&T must be furnished through a separate subsidiary. Customers may also obtain CPE from non-Bell System vendors. Neither the Bell operating companies nor Bell Laboratories can market or promote products of the subsidiary. CPE that is part of the embedded base of the Bell operating companies, that is, installed or in inventory on January 1, 1983, will be transferred to the subsidiary no later than the time of divestiture.

and concerns with the customer's decisionmakers and confirms the customer's interest in having the Bell System study and propose a communications system that would best resolve these problems. The customer makes a further commitment to provide the necessary information and resources to aid the marketing team during this process. Once the marketing representative secures this commitment, the sales support function helps the marketing team design a telecommunications system tailored to the customer's needs. Sales support organizations in the BOC help to acquire the necessary data (number of calls handled, peak hours, number of employees, etc.) to understand the customer's needs in more depth. For example, specific engineering centers dedicated to marketing may assist the sales support function by providing traffic engineering studies. These studies identify system usage patterns. Engineering centers may also provide presale design assistance to determine the customer's unique telecommunications requirements. Any subsequent negotiation that results from this activity (such as a change in the order) is also the responsibility of the marketing organization.

When the marketing organization, a service center, or a retail sales location completes a negotiation with a customer, a written *service order* implements the agreement. The service order, which describes the equipment or service to be provided and contains customer information, authorizes the various departments to do the work needed to complete the order. Customer information may include items such as the customer's name and address and billing name and address.

Order Generation

Service orders reflecting the customer's needs are formatted and entered into computer-based service-order processing systems that edit and validate the orders, distribute the order information to the appropriate work groups, and track the overall order process. Since processing of service orders involves the interaction of several departments within an operating company, tracking and coordinating the actions of all the work groups involved ensures that the service provisioning occurs by the promised date. The service-order processing system supports this tracking function by receiving status reports and reporting when the promised date is in jeopardy because other critical dates have not been met. Provisioning control begins at order generation (shown in Figure 13-1), and it continues until the order is completed.

The service-order processing system also maintains a pending-order file during provisioning to keep track of changes that occur as the order is being processed. For example, the customer may want to change the services ordered before the installation date, or a problem in the assigned facilities may require a new assignment. Incorporating these changes into the pending-order file ensures that the order information is accurate and current.

Network and Premises Equipment Engineering

This phase of service provisioning is unnecessary for certain orders, and when it is needed, the details may vary. A fundamental portion of network engineering—assignment—involves choosing a specific local loop and other items needed to fulfill the service order from inventories of available network equipment. For complex orders (for services such as tie lines, foreign exchange lines, and centrex), network engineering provides a detailed design for the service. (Section 13.3 describes these functions in more detail.) During this phase, assignment information is added to the service order.

Complex orders may require the engineering work group to design customer-premises equipment and/or circuit equipment, make cost estimates, order equipment and special software, and draw up installation specifications. These activities may begin with support of the initial marketing contact and may continue until the service is satisfactorily provided.

Installation

The hands-on work required to provide service may include central office work to connect a loop to the switching system, loop work to cross-connect cable pairs in the outside plant between the central office and the customer's premises, and premises work to install the terminal equipment. The premises work may range from installing a simple telephone set (which often takes less than an hour) to installing a complex customer switching system (which may take several weeks).

Since modular telephone sets have become available, the customer has become more involved in the installation process. Modular sets have plugs that allow easy installation; the customer can pick up the sets at a retail sales location, take them home, and plug them into standard jacks, thus reducing the need for premises visits by an installer.

Order Completion

When the service has been installed and tested and the customer verifies that it is in good working order, the service order is closed out, and the billing process is initiated. Any customer training required on the use of the service begins at this time.

Copies of the completed service order are forwarded to various organizations. Service representatives must have records in order to respond to future calls from the customer regarding service. The maintenance organization requires records of the type of service and central office connections to respond to repair calls. The Revenue Accounting Office uses service-order information for the billing process. White pages and Yellow Pages listings are prepared by a directory services organization from service-order information, and delivery of directories is arranged. Listing information must also be provided to the operator-services organization for the directory assistance function. The activities of these organizations are described in subsequent sections.

13.2.2 SERVICE ADMINISTRATION

The functions needed to sustain a service once an order is completed depend on the type of service. While all services require billing and recordkeeping functions, such functions are much more complicated for sophisticated business-customer services than for residence-customer services, and some administrative functions are unique to a service type (for example, collecting coins from public telephones).

The rest of Section 13.2.2 discusses billing and customer switching system administration. Sections 13.2.4 and 13.2.5 discuss service administration functions unique to Public Communications Services and Directory Services, respectively. Section 13.3 discusses network administration.

Billing and Resolution of Customer Billing Inquiries

The Revenue Accounting Office (RAO) is the operations center responsible for accumulating and processing billing information and for preparing bills. Bell System bills generally contain two types of charges: 1-time charges or a recurring flat monthly charge for equipment and services, and usage-sensitive charges based on the customer's use of the Bell System network. Information for the non-usage-sensitive portion of the bill comes from the service order, which specifies the services and equipment being provided to the customer. Information for the usage-sensitive portion comes from devices that record network usage, such as registers, and message accounting systems (see Section 10.5).

Billing functions include:

- reading and processing accounting records (primarily on magnetic tape) that identify calling and called telephone numbers, date, time of day, and duration of a direct distance dialing call.

- rating the call by calculating the distance between calling and called end offices and applying the appropriate rate based on distance, date, time of day, and call duration to determine the charge for the call. Operator-handled calls and Wide Area Telecommunications Services calls require specialized processing.

- posting the charged calls to the customer's account to accumulate charges during the billing period.

- incorporating the usage-sensitive and flat-rate charges into a customer's monthly bill.

Customers may pay their bills by mail, at public payment offices, or through agents such as banks. The RAO credits accounts on receipt of payment and keeps a record of accounts with unpaid balances. Service representatives receive lists of delinquent accounts and attempt to collect unpaid balances from customers.

Service representatives also generally handle customer billing inquiries. These inquiries may concern toll charges, charges for message-rate service, the balance due, or duplicate bill requests. The service representative investigates the inquiry and attempts to reach an agreement with the customer. An investigation may involve detailed analysis of the service being provided, current and past bills, and customer calling patterns. If an agreement modifies a bill, the RAO receives notification for billing record adjustment. (Figure 15-4 is a functional flow diagram of the Residence Customer Billing Inquiry Process.) For the larger business accounts, billing inquiries that are directly related to the types of services ordered are referred to the marketing organization for resolution.

Administration of Customer Switching Systems

Typically, service provided to most residence and small business customers remains unchanged for long periods of time and then changes only because of a move, which requires a new installation. In contrast, for large business customers with more complex communications requirements, demands change frequently. These demands are often met through the administration of the customer switching system (a PBX, for example).[3] Customer switching systems are designed and installed to meet current needs with some margin for growth, so the two major postinstallation activities associated with such systems are rearranging the system and capacity analysis.

A company reorganization, for example, could require the reassignment of telephone extension numbers. The older customer switching systems generally required a premises visit to make the cross-connections

---

[3] As discussed in footnote 2, limitations on the provision of customer-premises equipment were placed on Bell operating companies effective January 1, 1983.

needed to implement the change. With the newer, computer-based systems, clerks in centralized maintenance centers can perform rearrangements remotely by transmitting the appropriate coding information to the customer switching system over a dial-up data link.

Most rearrangements are handled by the service-order process, but a developing trend is to have customers become more involved in configuring their systems. For example, the customer may enter extension number and feature changes directly (as described in Section 11.2), reducing BOC involvement in short-term rearrangements and, therefore, reducing the time required to change a configuration. Equipment changes still require dispatching a craftsperson to the premises.

Periodic analysis of the load on a customer switching system determines whether the system is adequately handling it. The initiative for such a study may come from the marketing department or from the customer. When a capacity study is desired, a count of working customer stations is obtained and the load on the trunks to the central office is measured. Traffic information required for capacity analyses of computer-based customer switching systems is provided remotely by centralized operations systems at BOC locations. Such analyses may result in a recommendation from the marketing organization that the customer increase or decrease system capacity.

13.2.3 CUSTOMER-SERVICE MAINTENANCE OPERATIONS

Maintenance is divided into **preventive maintenance** and **corrective maintenance**. *Preventive maintenance* includes routine procedures to detect potential trouble conditions before they affect service. *Corrective maintenance* involves several separate, consecutive functions. These are described below and illustrated in Figure 13-2.

1) *Trouble detection* — recognizing that a trouble condition exists. Customers usually detect troubles in station equipment and some troubles in loops and report them to the telephone company repair service.

2) *Trouble notification* — alerting craft personnel to the existence and severity of a trouble condition so that corrective action may begin. Once the repair service attendant who is informed of the trouble condition has the associated information, craft personnel are notified.

3) *Trouble verification* — determining if a reported trouble condition still exists. An interval may occur between trouble notification and the start of the trouble-location function. Since experience has shown that many trouble indications are transient, correcting verified troubles receives first priority.

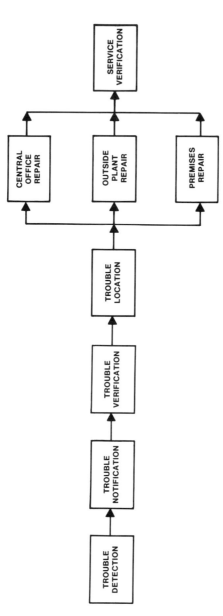

Figure 13-2. Customer-service maintenance operations — functional flow.

4) *Trouble location* — determining whether the trouble is in central office, loop, or premises equipment[4] so that the appropriate craft force can correct the trouble. This is often the most difficult and time-consuming step in the maintenance process.

5) *Trouble repair* — on-site repair of the defective unit is sometimes required. In other cases, the defective unit is replaced with a spare to correct the problem.

6) *Service verification* — after repair is complete, the craftsperson verifies clearance of the trouble condition.

The sequence of events for a typical customer-detected trouble is:

1) A customer reports a trouble to a BOC repair attendant.

2) The attendant asks the customer for the affected telephone number and a description of the trouble. The repair service attendant is supported by an operations system that provides information in real time, such as:

   a) customer name and address.

   b) service status (for example, whether service is disconnected, the telephone number is nonworking, or the number is affiliated with a telephone answering service).

   c) the date of the last trouble and the number of previous trouble reports if the current one is not the first.

   d) an appointment time that can be offered to the customer if a repair visit is necessary. The appointment time takes into account both the backlog of trouble reports awaiting dispatch and the size of the craft work force responsible for the corresponding repair coverage.

   e) results of any automatic tests (for example, shorted or open line or receiver off-hook).

   f) information on cable and other equipment failures that are affecting the particular customer's circuit.

Using this information, the attendant talks with the customer in an attempt to identify the cause of the trouble. The attendant then gives the customer a repair commitment time and generates a trouble report.

---

[4] In the case of special services, troubles may be traced to interoffice facilities.

3) The trouble report is reviewed, and if needed, additional tests are made to determine the location of the trouble so that the appropriate work force (central office, loop, or premises) can correct it. (Section 13.3.3 describes loop and central office repair.) Premises repair generally requires dispatching a craftsperson.

4) The typical repair sequence is completed when the craftsperson verifies that the trouble condition has been corrected, and the customer has been notified and is satisfied.

Current trends in maintenance operations include more customer involvement and remote testing and maintenance. Modular sets have made it possible for the customer to unplug a defective set and bring it to a customer service center for immediate replacement. As of May 1981, 51 percent of 11 million[5] home phones experiencing troubles in a 1-year period were being brought in for repair in association with a Defective Equipment Replacement Program (DERP). Troubles range from a burned-out bulb in a *PRINCESS* telephone to a damaged cord and a defective dial. Since 90 percent of the customers who participated in DERP indicated that they would do so again, DERP participation could increase from 51 percent to 80 percent without any additional incentive as more customers are referred to the program. Having customers bring defective phones to customer service centers reduces the need for craftspeople to visit customers' premises and is often more convenient and results in faster problem resolution.

Remote testing of computer-based customer switching systems (made possible by operations systems) helps in diagnosing faults and eliminating false dispatches. It also allows routine maintenance to be performed remotely so that troubles can be detected and corrected before they affect service.

13.2.4 OPERATIONS RELATED TO PUBLIC TELEPHONE SERVICE

Because the general public has access to the equipment, operations related to public telephone service differ from those associated with residence and business customers.

Provisioning

A specialized sales force handles sales of public telephones. As with other services, a service order authorizes the installation of a public telephone. A unique part of the installation may be the need to order and

---

[5] Out of a 100 million installed base.

install telephone booths. Booth installation is often subcontracted outside the Bell System.

Coin Collecting and Counting

Coins must be collected from public telephone stations before the boxes are too full to accept any more. However, if collection is scheduled too often, collection costs will increase. Therefore, computer-based operations systems aid collection by preparing lists of coin boxes that are candidates for collection, taking into account location and projected activity. The coin collection organization collects the coins, counts them, and enters the collection data (for example, time, amount, location) into the operations system. Discrepancies between actual and expected revenue are reported to a security group that investigates them and reports potential security problems.

Routine station inspections are also performed during collection, and out-of-service or hazardous conditions are reported immediately to the maintenance force. In addition, expanded collection duties require the collector to replace the directory if it is missing, outdated, or damaged. In some cases, directory replacement can be contracted outside the Bell System at lower cost.

Maintenance

A separate repair force is responsible for maintaining coin telephones. The public, Traffic Service Position System[6] operators, collection forces, or booth-cleaning personnel may report troubles. In addition, suspected trouble conditions are referred to the maintenance group when analysis of ongoing collection reports and coin deposit refund data indicates abnormalities. The maintenance group analyzes the trouble information received, dispatches the proper repair people when necessary, and submits booth repair orders to outside contractors as required.

The repair people dispatched to a public telephone site also perform other functions during their visits. They will replace directories if required and notify the collection force to make a special collection if they find a full coin box.

Public telephones must operate in a harsh environment. Heavy customer use and occasional misuse or abuse, exposure to the weather, and exposure to accidental damage (for example, being struck by an automobile) all contribute to the cost of maintenance. Vandalism is not a major

---

[6] Section 10.4.1 describes Traffic Service Position Systems.

problem, although some sites may suffer a high incidence. Total maintenance expense amounts to more than $250 million a year.

13.2.5 DIRECTORY SERVICE

In 1980, the Bell System published roughly 2130 different directories. Directory operations include white pages compilation, directory advertising sales, Yellow Pages production, directory printing, and directory delivery. The directory business is largely unregulated, and telephone company operations have evolved in different ways to take advantage of local opportunities. Operations systems, for example, may be supplied by vendors. Outside contractors may provide support to some portions of directory operations such as directory advertising sales.

Service-order information is used to add, delete, or change the listings in the white pages. Usually, the updated listing is stored and maintained in an operations system. Like service-order activity, this updating activity goes on daily throughout the year. Directories for a telephone company are published annually in a staggered manner, so that during any given week, directories for some areas are being prepared for the printer. Publishing the directory involves extracting and photocomposing listings from a data base.

Often, directory assistance operators use records from the white pages support system as a source of listings. Directory systems can produce paper or microfiche copies of listings for manual directory assistance operations or magnetic tape for computerized operations.

A major portion of directory work is the production of Yellow Pages directories. Annually, several months before each directory publication, sales operations for directory advertising will begin. A sales campaign starts with contracts containing the previous year's advertising together with any changes resulting from service-order activity. Sales people either visit or telephone directory advertisers and negotiate the details of advertisements with them. Display advertisements and art work are produced separately from the listings.

Yellow Pages production operations combine the directory advertising with listings to produce complete pages. An operations system maintains Yellow Pages listings in much the same way as white pages listings are maintained. Listings are photocomposed according to heading, and pages are laid out so that listings and advertising for each heading appear together.

Printing and distributing telephone directories is an enormous task. In 1980, the Bell System distributed about 275 million copies. Directory printing is usually done under contract by large commercial printers. Although Directory Service personnel maintain the delivery records, agents are hired to deliver the directories. The Directory Service organization provides the delivery agents with delivery route lists specifying

the number of directories to be delivered to each address (usually based
on the number of telephones).

13.3 NETWORK-RELATED OPERATIONS

Like customer-related operations, network operations consists of the same
three broad functional areas: **provisioning, administration,** and **maintenance**. The following sections discuss each of these areas and describe
many activities that make up network operations and the relationships
among them.

13.3.1 PROVISIONING

Network provisioning provides the types and quantities of network elements in the configurations needed to ensure economical, high-quality
service. Network-provisioning functions can be grouped into three
categories. Figure 13-3 shows how they are related in terms of the lead
time each requires.

- *Planning* — fundamental (long-term) planning for changes and growth
  in network structure and forecasting of the quantities and types of
  network elements required to provide service; current (short-term)
  planning to revise forecasts to reflect the actual evolution of the network structure.

- *Engineering* — specifying network elements and their configuration,
  accounting for unplanned, near-term needs.

- *Implementation* — installing telecommunications network components
  in response to the specifications of the long-term and near-term planning activities.

Planning. Starting with long-range forecasts of demands for existing and
new services, fundamental planning organizations develop plans for the
long-term evolution of the network. They evaluate the economics of
current alternatives and decisions and identify the sequence of projects
that should be implemented during the next 20 years (see Section 4.5).
The resulting view of the future network provides, in part, the basis for
shorter-range forecasting activities and provides information to be used
in the ongoing engineering and implementation of the network.

Current planning is more precise than fundamental planning. It usually provides estimates of network elements that will be required 4 to 6
years in the future based on historical traffic load data, growth projections, and an understanding of the present and projected structure of the
network. The results of current planning are provided to other planning
organizations. For example, they are the basis for engineering activities

that determine network changes needed to ensure timely and high-quality service.

Engineering. The objective of the engineering function is to specify network components and manage network investments in accordance with the plans and forecasts developed in the planning organizations. Engineering also takes into account near-term needs not fully anticipated by planning. Some specific objectives of the engineering function are:

- to determine the appropriate types and quantities of network elements

- to configure the network elements

- to design central office equipment and interoffice facilities and write specifications for the facilities and equipment

- to evaluate new products, technology, and services to be implemented in the network

- to develop plans for implementing new equipment and services.

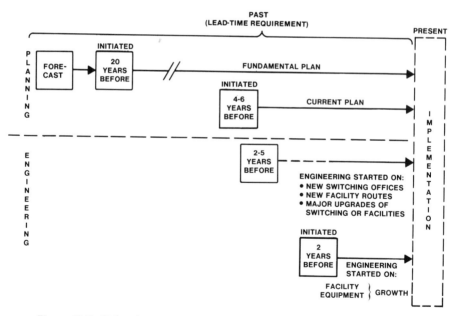

Figure 13-3. Network-provisioning activities and lead times required by each.

Implementation. The implementation aspect of provisioning can be divided into two parts: establishing network capacity and servicing the network[7] to maintain that capacity. Establishing network capacity means

---

[7] Network servicing is traditionally considered an administration activity. However, it is so closely coupled to the overall process of providing network capacity that it is discussed in the provisioning section.

installing additional equipment and facilities in response to the specifications of the planning and engineering organizations. Following installation, the new equipment and facilities are tested and made available for service as added capacity.

The objective of network-servicing implementation activities is to meet service demands by rearranging or connecting existing equipment and facilities. Service demands are originated by customers and by internal operating company activities (such as estimations of short-term trunk group size changes or routing alterations needed to resolve identified network capacity problems). For example, some types of network servicing may be done on a seasonal basis: It may be necessary to reconfigure equipment and facilities to provide more trunk capacity at a ski resort during the winter. In the summer, the configuration can be rearranged to provide trunk capacity elsewhere. Network-servicing activities can also be stimulated indirectly by rate and tariff changes requiring the installation or modification of billing equipment.

Provisioning the telecommunications network is a continual, complex set of interrelated activities. It is the performance of various disjointed tasks by separate organizations in several disciplines and the combination of the results of these tasks into a single plan.

Trunk, transmission facility, and switching equipment provisioning; operator-services provisioning; and common-systems provisioning are examined more closely in the following paragraphs.

Trunk Provisioning

The objective of trunk provisioning is to ensure that the proper numbers of trunks are provided where and when they are needed. Each stage of this provisioning process is coordinated with the planning, engineering, and resource implementation activities performed for facility provisioning (described later).

Long-range planning for trunk provisioning estimates how a portion of the traffic network—usually a metropolitan network or a numbering plan area (see Section 4.3)—will be configured 5 to 20 years in the future. (In this context, long-range planning does not include facility planning, which determines geographic routes and types of facilities to carry trunks.) Planning the topology of the network is based, first, on projections of traffic loads. This topology includes the numbers, types, and locations of the switching systems and the homing arrangements (see Section 4.2.1) whereby final trunk groups are established between lower and higher levels of the hierarchy. Then, the sizes of the other trunk groups in the network are estimated using the available projections of traffic load between points of the network. The resulting long-range plan embodies various traffic routing rules, such as whether local and toll traffic will flow through the same tandems in a metropolitan network and what sequences of alternate routes are to be used. Such planning does not involve commitments to spend money; its purpose is to ensure that the long-term consequences of current decisions are foreseen and that the evolution of the network proceeds smoothly and economically.

The current planning[8] interval, in contrast, extends from about 1 year to 4 or 6 years in the future. Six years is the approximate lead time required to establish a new switching location, including acquisition of land and construction of a building. The trunk forecast for each year specifies the circuits that will actually be needed for the next year's busy season to meet network objectives. The forecasting process prescribes the number of trunks required between switching systems. Two different approaches to trunk forecasting are in use: the *trunk-based method* and the *point-to-point method* (see Section 5.5).

Based on a careful forecast of traffic loads for each year, the current planning process determines the traffic routing rules for the network and increasingly detailed decisions about the functions of all the switching systems and lays out the trunk traffic network according to a detailed set of engineering rules. Given the trunk forecast and current facility and equipment plans, engineering groups generate work orders, and requisitions for the installation of the appropriate equipment and facilities are generated. The equipment and facilities are then installed and tested by installation groups.

When the required transmission facilities have been built, the appropriate circuits have been established, and the busy season has arrived, it is still unlikely that the network as engineered will exactly match the offered traffic. Thus, measurements of actual usage and counts of calls offered and overflowing are used to determine where to add or subtract circuits to meet the actual demand. A few trunks are generally removed from one group and added to another in various places, so that all final groups meet their objective of 1-percent blocking. The activity of determining what rearrangements to make is known as *trunk servicing*.

Transmission Facility and Switching Equipment Provisioning

The aim of transmission facility and switching equipment provisioning is to ensure that transmission facilities and switching equipment are available to meet the growing demand for communications services at the lowest possible long-term cost. Each stage of this provisioning process functions in coordination with the planning, engineering, and resource implementation activities performed for trunk provisioning just described.

Wire center studies are a good example of such provisioning activities. New wire centers must be planned well in advance to meet the growing

[8] Section 14.2.1 describes current planning in more detail.

demand for service. Typically, several alternative locations for the wire center may be proposed, and the serving area boundaries for each proposed location are determined for a particular time in the future. Then, year-by-year plans for establishing each alternative are developed. Each of the plans is actually a sequence of placements, removals, and rearrangements of facilities. The construction activity and capital expenditures of the plans are compared. Next, a detailed report of the most economical wire center plan (including customer assignments, loop costs, and trunk costs) is prepared. This plan must also be compatible with other factors, such as the availability of land for new buildings or floor space in existing buildings. Last, detailed growth patterns are developed by considering timing of cable additions, structural constraints, timing of transitions from previous methods of operation to new methods, sizing of central office equipment subject to economic constraint, and trunk network sizing. In this way, a comprehensive plan is produced for the efficient and economical coordinated growth of cable and switching equipment in the exchange area.

The engineering and implementation stages of transmission facility and switching equipment provisioning were covered in the section on trunk provisioning because they use information from trunk forecasts.

Operator-Services Provisioning

The operator-services facility administration and force management functions (see Section 13.3.2) provide the data that are used for operator-services planning, engineering, and implementation. Operator-services facilities engineering must provide adequate switching capacity and the proper number of operator positions to meet expected demand. As in all other operating company engineering projects, the goal is to specify equipment quantities to minimize capital and installation expense while meeting service objectives.

Operator-services switching systems are planned and engineered under assumptions much like those for local and toll switching systems. However, position engineering (the number of operator equipment positions) and operator staffing, or *forcing* (the number of operators), employ a blocked-calls-delayed queuing model, normally the Erlang C model (see Section 5.3.2). Operator-services traffic engineering is based on traffic measurements. These data are used to determine if service is adequate, to evaluate operator-services efficiency, and to provide engineering data used in planning for future needs. The primary traffic service measurements for operator systems are speed in answering the customer and efficiency in serving the request. (Efficiency is usually measured by the average time spent on a request.)

Operator-services implementation consists of those activities through which the operator force and its support systems are physically established. This latter activity entails installing operator-services equipment and facilities to provide sufficient traffic capacity to meet future demand.

Efforts to meet operating force requirements today and in the future include automation, consolidation, and control of demand. Computer-based systems such as the Automatic Intercept System (see Section 10.4.2) and the Traffic Service Position System (see Section 10.4.1 and Figure 10-16) have been developed to automate portions of the operator's job to decrease the work time per call.

Consolidation provides three benefits: It accommodates the large traffic volume needed to justify the initial cost of computer-based systems; it raises efficiency by substituting large groups of servers (operators) for dispersed smaller groups (therefore, service levels can be maintained while the productivity of the operators increases); and fewer operators are needed to provide night coverage.

The major approach used to control demand has been to charge higher rates for services requiring operator assistance. Another way is to improve network performance, thereby reducing traffic from customers having difficulty placing their calls.

Common-Systems Provisioning

Common-systems[9] provisioning ensures that adequate power, interconnection, and environmental support are available for switching equipment and transmission facilities over a long-term service lifetime (about 20 to 30 years).

Common-systems provisioning for a particular network equipment location includes plans for: power, distributing frames, the cable entrance facility, equipment space, and cable pathways. Figure 13-4 shows typical common-systems provisioning activities. (For discussion purposes, the activities are assigned step numbers that correspond to the circled numbers on the figure.)

Step 1 — Common-systems planners receive facility and equipment plan proposals from fundamental network planners. Plans for common systems are usually specified after the plans (or major alternatives) for facilities and equipment have been defined.

Step 2 — Common-systems planners consult with building planners and equipment and facility engineers about feasible alternatives. Common systems support a wide variety of transmission and switching equipment.

---

[9] Chapter 12 discusses common systems.

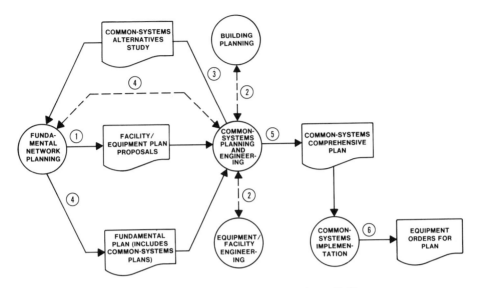

Figure 13-4. Common-systems provisioning activities.

At any time, much of this equipment is undergoing modernization and growth; hence, common-systems planning requires careful analysis of forecast data and plans from other areas.

Step 3 — Common-systems planners generate a common-systems alternatives study that includes economic evaluations. Typically, alternative plans for common systems involve varying amounts of rearrangement, addition, or replacement of existing capacity. Alternatives can be compared using present-worth evaluation of associated capital and expense items (see Section 18.3). Selection of an alternative is not based on common-systems economics alone, but on overall costs for various equipment and facilities alternatives.

Step 4 — Common-systems planners consult with fundamental network planners who generate a fundamental plan that includes common-systems plans. The common-systems plan summaries form a section of the fundamental plan for a telephone equipment building.

Step 5 — Common-systems planners develop a comprehensive plan that provides engineering details for the selected alternative. The comprehensive plan has a separate section for each of the common systems. It constitutes a detailed strategy by which common systems can be evolved from their existing condition to some planned "ultimate" configuration at the end of the planned service lifetime.

Step 6 — Implementation of common-systems plans is the responsibility of engineering organizations such as equipment engineering. Implementation engineers place orders with suppliers for common-systems equipment and arrange for its installation. When implementation cannot follow the comprehensive plan, implementation engineering may request a *variance*—an amendment to the plan made after approval by a common-systems planner. For example, a comprehensive plan may have been established for a central office, but added service requirements resulting from an unexpected increase in business activity may require equipment/facility upgrading that was not included in the original comprehensive plan.

13.3.2 NETWORK ADMINISTRATION

Network administration consists of ensuring efficient use of existing and planned components of the telecommunications network to provide pre-established levels of service to the customer. A variety of organizational structures may exist for performing the network administration job, but they have the same primary role—responsibility for the overall quality of service given by a specific portion of the telephone network.

The responsibilities for network administration are divided among seven major functional areas:

- data administration

- operator-services administration

- equipment utilization

- office status evaluation

- service problem analysis and corrective action

- transition management

- network management.

Data Administration

Data administration includes scheduling, recording, posting, and checking the validity of traffic and equipment data required to administer, evaluate, and engineer the switching system(s) and the associated trunking network properly. These data are generally obtained using automated data systems,[10] although manual register reading and recording on film may also be used. Data administration also involves reviewing the operation of installed measuring devices, establishing controls to

---

[10] These systems are part of the Total Network Data System (TNDS) described in Section 14.3.1.

ensure proper data collection, analyzing trouble-detection reports, and resolving any problem affecting data collection.

Operator-Services Administration

In the area of operator-services administration, two functions meet daily operating requirements. The first function, ensuring that sufficient operator facilities are available and working properly, is called *facility administration*. The second function, ensuring that sufficient operators are available, is called *force management*. Because of the high costs involved in providing operator services (in terms of equipment, facilities, and operator work force), both functions are closely monitored by Bell operating company management.

Both facility administration and force management start with the collection of traffic-usage and call-volume data from operator-services equipment and facilities (such as a Traffic Service Position System). Some of the traffic data and measurements are used by maintenance forces to maintain operator-services equipment and trunks. Operator services does not have a separate maintenance force but relies on regular central office maintenance forces.

Operator-services facility administration consists of analyzing volume and traffic-usage data in order to determine how well equipment is being used, to find potential maintenance problems, and to ensure proper load balance between operator offices. Facility administration also provides the data used for operator-services planning and engineering.

Operator force management is the continuous job of providing enough—but not too many—operators to meet established service levels. Operators work "tours" of from 5-½ to 8-½ hours. Given the volume (number of calls) and usage data[11] (the types of calls), a forecast of operator requirements is made by quarter-hours for the week 2 weeks ahead. These requirements are used to create a schedule of available tours and to allocate the tours to the administrative units that make up the serving team. The allocated schedule tells each manager how many operators to provide for each quarter-hour and which tours the operators are to work. Individual operators are then assigned to specific tours. Each day, a force manager monitors how well the planning anticipated the actual demand and makes adjustments as required. The adjustments take the form of projecting demand from intraday trends and calling out additional operators to meet unexpected demand or releasing an appropriate number of operators if demand is lower than anticipated. Fine tuning is practiced by adjusting lunches and reliefs and by rescheduling training or clerical functions to counter peak or slack periods.

---

[11] Usage data are similar to data used in facility administration.

Equipment Utilization

This aspect of network administration ensures that installed equipment is being utilized in the best way possible and that forecasts of equipment and trunk additions are consistent with projections of future capacity requirements. The administrator assesses how much each piece of equipment is being used (*loading*), evaluates the distribution of that use across components of the same type (*balancing*), and maintains up-to-date equipment, assignment, trunk group, and routing records. A variety of inputs, such as load-balance[12] data, forecasts, and engineered office capacities, are used to assign equipment and facilities to maintain objective levels of service as demand changes.

Office Status Evaluation

Another network administration responsibility is the daily analysis of the switching office status, including the integrated review of key data items on customer usage (CCS[13] per main station, call rates, holding times, etc.),[14] equipment loads, and traffic volumes. Overall office and component group capacities must be determined for daily review of load versus capacity. Growth sizing and scheduling are monitored to ensure that sufficient switching capacity will be available to meet forecast demand.

Service Problem Analysis and Corrective Action

Network administration encompasses responsibility for the identification (either through detection or reports of trouble), investigation, and resolution of service problems through daily surveillance of central office equipment and connecting trunk groups. A service problem is defined as any condition in which established service objectives are not met, with or without justification.

Some specific activities in this area include:

- diagnosing causes of service problems by evaluating service impairment indicators (such as excessive dial-tone delay) and coordinating with network service organizations to resolve problems

- coordinating with corrective action plans and providing frequent evaluation and updates of predicted service problems.

---

[12] Section 5.3.4 discusses load balancing.

[13] Hundred call seconds.

[14] Sections 4.4.1 and 5.2.1 discuss these concepts.

Transition Management

Transition management covers the analysis of plans for equipment additions, replacements, removals, and/or rearrangements. The administrator must evaluate the impact of this type of activity on service and ensure that procedures for transition will result in the desired equipment configurations with minimal equipment outages and service deterioration. Determining the maximum allowable quantity of equipment (by type and group) that could be removed from service for maintenance and/or transition activity during additions, rearrangements, or replacements is also a network administration responsibility. During transition, monitoring must be performed for the customer load, overall service, equipment outages, and the trunk and routing network to ensure that they adhere to the transition plan. Monitoring is also done to avoid, or at least limit, adverse service effects of the transition.

Network Management

As mentioned in Section 5.6, network management is the function that keeps the network operating near maximum efficiency (as defined by completed messages per unit of time) when unusual traffic patterns or equipment failures would otherwise cause network congestion and inefficiency. Network management involves the use of controls[15] to alter the normal network routing structure to respond to overloads or to anticipate them. The trend has been to automate these control actions as much as possible and to supplement the automatic controls with centralized computer-based systems. These systems provide status displays and reports that allow network management personnel to make decisions concerning when to implement manual controls and which controls are most appropriate. Figure 13-5 illustrates the type of information shown on a status display used by network management personnel.

A Network Management Center may be local, regional, or national. The local Network Management Center has responsibility for up to forty-eight switching offices and is supported by the Engineering and Administrative Data Acquisition System/Network Management (EADAS/NM).[16] There are eight regional Network Management Centers in the United States. Each is responsible for a specified large geographic area encompassing the control jurisdictions of several local centers and is also supported by EADAS/NM. The national Network Operations Center, located in Bedminster, New Jersey, is supported by the Network Operations Center System and is responsible for coordinating inter-regional network management controls.

---

[15] Section 5.6 discusses network management controls.

[16] EADAS/NM is part of TNDS, which is described in Section 14.3.1.

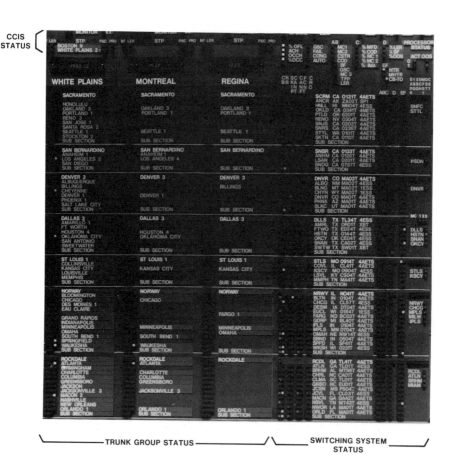

Figure 13-5. Typical information on a wallboard status display for network management. The portion of a large wall display shown uses colored indicators for various status information. At the top, status of signal transfer points and other elements of the common-channel interoffice signaling network is shown. Trunk groups are in columns by region (only three are shown) and grouped vertically according to where (which region) the group terminates. Associated indicators show whether trunk groups are in a normal or overload condition and what controls are in effect. At the right are toll switching systems, with colored indicators for measures of congestion and the status of controls implemented on traffic to or from each system.

Although day-to-day network management is handled best by automatic controls, manual network management controls are also needed to handle unusual situations requiring flexibility and human judgment. A fundamental network management function performed by people is preplanning. Preplanning develops a plan of controls for handling anticipated overloads such as those that occur, for example, on Mother's Day (intertoll congestion) or as a result of a localized natural disaster (focused overload).[17]

---

[17] Section 5.6 discusses overloads.

13.3.3 MAINTENANCE OF THE NETWORK

As discussed in Section 13.1, maintenance involves both preventing potential troubles and correcting existing ones. Preventive maintenance, as noted in Section 13.2.3, includes routine procedures, such as testing, to detect and eliminate potential trouble conditions before they affect service. It also includes such activities as lubrication, cleaning, and adjusting of equipment.

On a functional level, corrective maintenance procedures for customer and network operations are very similar and involve the same sequence of events. However, unlike customer service troubles, most network troubles (for example, switching system or transmission facility troubles) are detected automatically.

Every telecommunications system is planned and designed to achieve an economic balance between system reliability, preventive maintenance, and corrective maintenance. A system can be designed to have very high reliability and, thus, require very little maintenance. However, a less reliable system—that is, a system that is less expensive but requires more maintenance support—might be more economical. Also, maintenance planning organizations at Bell Laboratories interact with various development areas during the design stages for new equipment and maintenance systems. This interaction ensures that new network components are designed with ease of maintenance in mind and are introduced with a viable maintenance plan.

Network Service Centers

Systems for switching system maintenance, trunk maintenance, and carrier system maintenance are each designed to support one portion of the telecommunications network. To ensure that the overall network provides good service, Network Service Centers have been established. Their major inputs are reports of troubles encountered by operators, either directly or through customer reports, and reports from other Bell System employees. The reports are analyzed for patterns pointing to specific problems in the network (such as a large number of reports resulting from attempts to reach a certain location). These problems can be referred to the local repair forces. To aid in the analysis of trouble reports, computer-based support systems[18] have been developed to sort, format, forward, and examine trouble reports from the entire country for standard errors. These systems also manipulate data for analysis and return suspected trouble patterns to Network Service Centers in each operating company.

---

[18] Such as the Network Operations Trouble Information System (NOTIS), the Network Service Center System (NSCS), and the Automated Trouble Reporting System (ATRS).

Figure 13-6 shows the relationship between Network Service Centers and the loop, switching system, trunk, carrier system, and special-services circuit maintenance functions. The following sections discuss each of these functions in more detail.

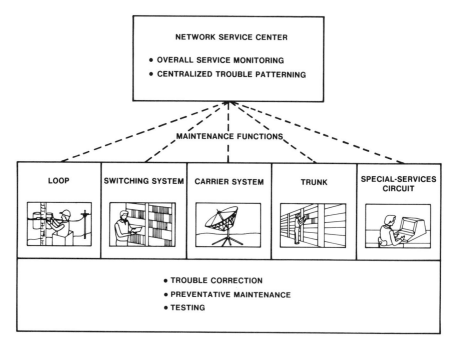

Figure 13-6. Network maintenance functions.

Loop Maintenance

The customer is the major source of trouble reports[19] in loop equipment. Because a customer location is permanently associated with a particular loop, the customer will always be affected by a loop trouble and will be aware of the trouble until it is corrected. Therefore, prompt correction of loop problems is extremely important.

In loop maintenance development, the emphasis has been on improving the efficiency of craft forces through mechanized aids and better testing facilities. Surveillance systems have been developed for automatic line-insulation testing and for cable-pressure monitoring to detect incipient trouble conditions before customer service is affected. However, once a customer loop problem has been identified and located by the various

---

[19] Section 13.2.3 describes the operations involved in processing a customer trouble report.

systems in use, it is still necessary, in many cases, for a repair person with a portable test set to go to the main distributing frame or out on a cable route to pinpoint the specific trouble location.

Switching System Maintenance

In most central office buildings, craft personnel perform a variety of tasks including maintenance, rearrangements, and cross-connects (see Section 12.3) on frames. In very large buildings with several switching systems, however, the assignment of craft personnel dedicated only to maintenance tasks is justified. In such cases, these same people may have maintenance responsibilities for transmission and signaling equipment located in the building as well. In contrast, for small switching offices that are not staffed for one or more shifts per day, system alarms indicating trouble conditions or failures are transmitted to a centralized location to initiate maintenance action.

The maintenance procedures associated with electromechanical and electronic switching systems are substantially different. Because they experience minor trouble conditions more frequently, electromechanical switching systems require more extensive periodic maintenance. Electronic switching systems are relatively trouble free, but when they do fail, the trouble symptoms are usually more difficult to diagnose, and some failures can affect a large number of customers.

Because of their basic reliability and because their maintenance is automated, electronic switching systems do not afford the craft force much practice on difficult trouble analysis and location tasks. The number of highly skilled craftspeople capable of diagnosing certain electronic switching system troubles is limited, making centralized maintenance and self-diagnostics a necessity. For electromechanical systems, the reasons for centralization have not been so compelling.

With centralized maintenance for switching systems, the majority of the maintenance functions are performed remotely from a suitably equipped Switching Control Center (SCC).[20] Regularly scheduled on-site maintenance for individual central offices can often be reduced, and necessary on-site work can be performed on a dispatch basis from a central pool.

Trunk Maintenance

Before direct distance dialing became available, a customer making a long-distance call had to speak to an operator first. The operator would then set up the connection and speak to operators at the other end. If a

---

[20] Section 14.3.2 describes the Switching Control Center System, which supports SCCs for electronic offices.

particular trunk was not operating properly, the operator could identify it and report it. Today, most connections are established without the aid of an operator, and the particular trunks used by a customer are not identifiable once the customer hangs up even if the customer reports the trouble immediately. Consequently, an important source of trouble reports on trunks has been lost. As a result, trunk maintenance today depends primarily on routine testing of trunks, generally done automatically by one of several computer-based systems.[21]

Carrier System Maintenance

Carrier system maintenance is handled quite differently from trunk maintenance. Combining many circuits on a single carrier facility necessitates the use of sophisticated alarm and surveillance techniques to ensure that service to a large number of customers is not impaired if a failure occurs. In addition, periodic testing and maintenance are usually performed to ensure that the carrier systems are operating properly. However, with newer, more reliable solid-state systems, the trend is toward reducing the amount and frequency of periodic maintenance.

In analog carrier systems, alarms indicating trouble conditions are activated when control and operating signals[22] fall outside a predetermined range. The alarm signals are displayed in the central maintenance location responsible for that system. In addition, certain analog systems are monitored at intermediate repeater points. In the event of a failure in a repeater station, an alarm[23] that identifies the station is transmitted back to the central maintenance location. The central operator can then request, via electrical signals (telemetry), an examination of the station for a limited amount of detailed information about the failure.

Maintenance operations for digital carrier systems are generally centralized within a geographic region. A computer-based system[24] has been developed to support the trouble identification and location functions. In addition, a new generation of maintenance features is being implemented to provide automatic equipment surveillance alarms and automatic back-up lines (a spare maintenance line between repeaters where the failure occurred) within a carrier system. Thus, for digital systems, continuous

---

[21] The Centralized Automatic Reporting on Trunks (CAROT) system is an example.

[22] These are also called *pilot tones* and are located between standard channel frequency allocations.

[23] The Telecommunications Alarm Surveillance and Control (TASC) System.

[24] The T-Carrier Administration System (TCAS).

monitoring of system performance has generally replaced periodic maintenance.

A major difference in the monitoring of analog versus digital carrier systems is that a direct measurement of system performance is available for the digital systems. The performance of a digital carrier system can be monitored by computer-based operations systems to measure the bit error rate[25] while the carrier system is in service. A report can be made when the quality becomes degraded. With analog carrier systems, performance is determined indirectly by measuring a number of impairments such as random noise and impulse noise.

Special-Services Circuit Maintenance

Special services require circuit layout, transmission, or signaling features in combinations that are not found in the PSTN. (Chapter 3 gives some examples.) These configurations present a particular maintenance problem in that diverse maintenance abilities are required to handle the particular characteristics of each configuration.

One important concept in special-services maintenance is the use of consolidated equipment arrangements. For example, the voice-frequency facility terminal concept, presented in Section 9.2.2, consolidates transmission, signaling, and circuit maintenance access functions in one equipment assembly, eliminating the need for intermediate distributing frames and complex office wiring. The design and structure of such arrangements incorporate maintenance objectives from the start. The use of functional plug-in units makes the initial engineering and subsequent maintenance of the circuits much simpler.

Another major concept is the use of computer-based systems[26] through which personnel can access and perform sophisticated tests on most types of special-services circuits. In particular, with these operations systems, a special-services circuit can be tested by one person, eliminating the time and expense associated with providing personnel at both ends of the circuit to be tested. One-person testing is a Bell System goal attainable through the use of the current technology. Remote digital access for testing of special-services circuits is provided by the Digital Access and Cross-Connect System (DACS), discussed in Section 9.4.3.[27]

---

[25] A measurement of the number of bits that are perceived to be incorrect at the receiving end.

[26] An example is the Switched Access Remote Test System (SARTS).

[27] Martz and Osofsky 1982 discusses the impact of DACS on operations.

Conclusion

The activities discussed in the network operations section of this chapter are crucial to keeping the network operating at acceptable standards so that customers are provided with consistently high-quality service. However, while the provisioning and administrative activities are completely transparent to customers, customers may interact with maintenance personnel to some degree (for example, when there is trouble on a line). Maintenance operations of the future will detect service degradation for repair before a service-affecting trouble becomes apparent to the customer. This improved detection will result from increased automation of maintenance activities.

AUTHORS

B. R. Barrall   *E. J. Kovac*
J. A. DeMaio   *R. E. Mallon*
L. L. Desmond   *H. I. Rothrock*
B. S. Eldred   *P. F. Wainwright*
J. A. Fitzgerald

14

Computer-Based Systems
for Operations

14.1 INTRODUCTION

In the mid-1960s, computer technology had developed to the point where it could be applied to operations in many ways other than in traditional accounting functions. Over the years, the term *operations systems* has come into use to cover these applications.

Operations systems have a variety of objectives and corresponding benefits. Sometimes, the primary objective is to provide analysis support for Bell operating company (BOC) personnel, and the benefit is a more thorough, accurate, and timely analysis capability. Most systems contain a data base, and in some cases, mechanization of the data base is the primary objective, relieving telephone company personnel of the need to maintain certain records manually. Both of these benefits provide expense savings. In many cases, the main benefit of the system is in reducing capital investment for additional equipment by providing more accurate records of what capital equipment is already available to meet a current need.[1] A related benefit provided by some systems is assistance in planning the use of new capital investment to meet the various equipment needs in a timely, cost-effective manner. In addition to mechanizing manual tasks, operations systems facilitate the introduction of network capabilities and services, some of which might not be feasible if planning, administration, and maintenance were manual functions.

Operations systems have yielded a considerable reduction in clerical and recordkeeping personnel. This is partly due to the mechanization of tasks and partly due to personnel economies resulting from the centralization of work groups that some operations systems allow. These systems

---

[1] This type of benefit is referred to as *equipment and facility recovery.*

relieve employees of tedious and repetitive clerical tasks, freeing them for functions that require judgment and personal interaction, such as:

- establishing an acceptable level for the risk that an inventory of spares will be depleted

- forecasting growth in number of subscribers in a particular area and in their calling characteristics

- shipping, receiving, and inserting plug-in units.

Table 14-1 shows the growth expected in the use of mechanized systems between 1981 and 1985. The BOCs use over 400 such systems. The development of these systems involves Bell Laboratories, AT&T, Western Electric, the BOCs,[2] and non—Bell System organizations, such as Tymshare, McDonnell Douglas Automation, and National CSS.

Most operations systems are designed to run on computer systems owned or rented by the operating telephone companies. Most of the smaller systems are designed for Digital Equipment Corporation or Hewlett-Packard computers; most of the large ones are designed for IBM^3-compatible or $UNIVAC^4$ computers. A small number of operations systems run on central, time-sharing systems. Access to the time-sharing systems is usually obtained through dial-up, switched facilities.

TABLE 14-1

MECHANIZATION OF BOC OPERATIONS

	1981	1985 (Est.)	Growth (%)
Employees (thousands)	841	886	5
Terminals (thousands)	114	180	60
Minicomputers	4200	5800	40
Large mainframe computers	315	500	60
Terminals per employee	0.14	0.20	50

NOTE: The 1985 estimate does not account for the effects of divestiture on operations.

[2] Section 14.5 discusses the roles of these organizations.

[3] Registered trademark of IBM Corporation.

[4] Registered trademark of Sperry Corporation.

Sections 14.2 through 14.4 describe a few representative systems to give a general idea of their functions, operations, and benefits. The systems described fall into three categories:

- recordkeeping and order processing

- equipment surveillance, maintenance, administration, and control

- planning and engineering.

Section 14.5 discusses development and support of operations systems.

14.2 RECORDKEEPING AND ORDER PROCESSING

Some of the largest scale operations systems are used for recordkeeping and order processing, massive tasks for operating telephone companies. The following sections describe three such systems. The **Trunks Integrated Records Keeping System** and the **Plug-In Inventory Control System** support network operations related to growth and change in the network by providing an accurate record of circuits and components in use and available for use. The **Premises Information System** supports customer-related operations by providing rapid access to information concerning equipment located on a customer's premises.

14.2.1 TRUNKS INTEGRATED RECORDS KEEPING SYSTEM (TIRKS)

The Role of TIRKS in Circuit Provisioning

In recent years, the scope and complexity of circuit provisioning has increased substantially. Large growth and huge Bell System investment in equipment and facilities have required a typical operating company to administer millions of pieces of equipment, facilities, and circuits and process thousands of orders. The complexity and interdependence of records needed by the network has increased because of technological advances, such as digital switching systems, and intelligent network elements, such as the Digital Access and Cross-Connect System (see Section 9.4.3). As a result, manual recordkeeping systems have become inadequate. TIRKS was developed to mechanize the circuit-provisioning process. It is deployed in eighteen BOCs to mechanize two aspects of circuit provisioning: daily circuit provisioning and current planning (see Figure 14-1).[5]

---

[5] Section 13.3.1 discusses provisioning for network operations.

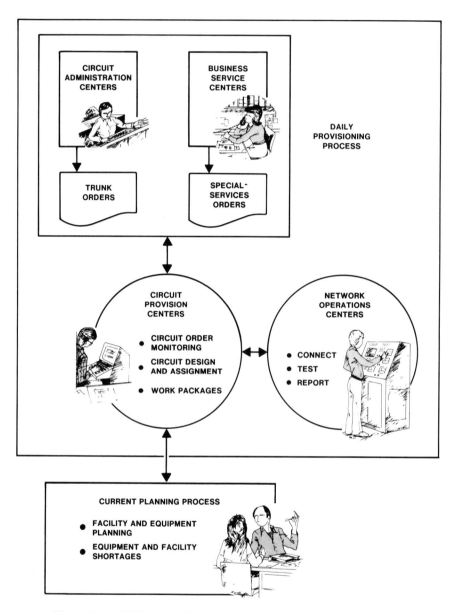

Figure 14-1. TIRKS circuit-provisioning process. (Section 14.3.1 discusses the functions of Circuit Administration Centers and Network Operations Centers.)

Daily provisioning consists of processing orders to satisfy customer needs for special-services circuits and processing orders initiated for message trunks and carrier systems for the public switched telephone network (PSTN). The daily circuit-provisioning process begins at various *operations centers*[6] and flows to *Circuit Provision Centers* (CPCs), which track the orders, design the circuits, and assign the components using TIRKS. The CPC prepares *work packages* and distributes them to the technicians working in the field who are responsible for implementing the work packages and for reporting completion of their work.

Current planning determines the equipment and facility requirements for future new circuits. The current-planning process apportions forecasts for circuits among the circuit designs planned for new circuits. For example, given a variety of designs for a particular type[7] of circuit, the number of circuits of that type forecast will be divided among the possible designs.

The same designs are used in both daily provisioning and current planning, and all the operations are accomplished using TIRKS.

Components

TIRKS comprises five major interacting component systems: the **circuit order control (COC) system**, the **equipment (E1) system**, the **facility (F1) system**, the **circuit (C1) system**, and the **Facility and Equipment Planning System** (FEPS). Figure 14-2 shows the relationship between COC and C1, E1, and F1.

COC controls telephone company message trunk, special-services, and carrier system orders by tracking critical dates along the life cycle of an order as it flows from the source to the Circuit Provision Center and on to the field forces. It produces daily, scheduled, unscheduled, and on-demand reports to provide management with the current status of all circuit orders. It also provides data to other TIRKS component systems to update the assignment status of equipment, facilities, and circuits as orders are processed.

C1 is the heart of TIRKS. Using basic facility information (such as location and type), it automatically determines the types of equipment required for a given circuit (circuit design), assigns the equipment and facilities needed, determines levels at the various *transmission level points* (see Section 8.6.1) on the circuit, specifies the test requirements, and establishes circuit records for the circuits. When the automated design process is complete, C1 reformats the circuit records into work packages

---

[6] Operating company work centers. See Sections 13.1 and 15.2.1.

[7] Data circuits, foreign exchange lines, and message trunk circuits are examples of different types of circuits.

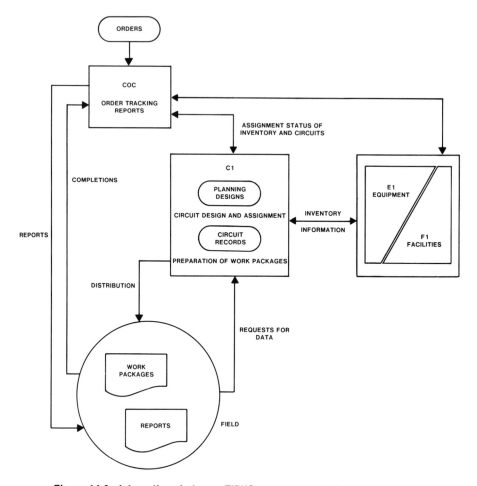

Figure 14-2. Interactions between TIRKS components in daily provisioning.

needed for the installation of the circuit and distributes them to the field forces at various work locations. Once the circuit record is established, C1 maintains the record for future additions or changes.

E1 and F1 are the two major inventory component systems in TIRKS. E1 contains telephone company equipment inventory and assignment records and pending equipment orders. The records indicate the number of various equipment types that are *spares* (that is, available for assignment) and circuit identification for equipment already assigned. F1 contains telephone company cable and carrier (that is, transmission facility) inventory and assignment records. C1 accesses both the E1 equipment inventory and the F1 cable and carrier inventory.

FEPS supports the current planning process (see Figure 14-3). The planning process determines the transmission facilities and equipment that will be required for new services, which can then be compared to inventory to determine possible shortages. FEPS helps planners use the information in the E1, F1, and C1 data bases along with forecasts of future growth to allocate existing inventories efficiently, to determine future facility and equipment requirements, and to update planning designs.

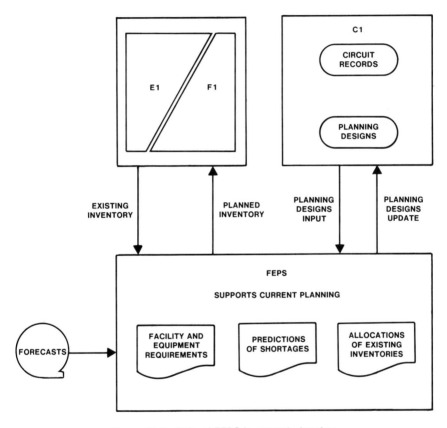

Figure 14-3. Role of FEPS in current planning.

Interfaces

TIRKS is the master recordkeeping system for the network. Consequently, it must interface with many other operations systems, processes, and various new intelligent network elements that rely on TIRKS for records. A number of these interfaces are provided by the extensive

reports that are available from TIRKS; others are on-line. These interfaces are being developed in accordance with an overall operations plan
(see Sections 14.2.2 and 15.4).

Implementation

TIRKS is an on-line system that uses *IBM* 370-compatible hardware and
direct-access storage devices. Configuration planning is required to determine the computer system equipment needed to support TIRKS. The precise hardware configuration for a given operating company depends on
the company's inventory universe and the volume of activity affecting
that inventory.

Cathode-ray tube (CRT) video display terminals and teletypewriters
(TTYs) are used to enter and retrieve data. On-line information inquiries
and data-base changes use Teletype Corporation *DATASPEED* Model 40
display monitors (or the equivalent) for input and output. Reports are
automatically distributed to teletypewriters and medium- or high-speed
printers.

Conversion from a large, manual data base to TIRKS is a large undertaking and may take several years in an operating company. TIRKS
conversion systems have been designed to perform a number of interim
functions. For example, they prepare the data base for implementing the
fully automated system and provide early benefits in capital savings as a
result of finding unused equipment and facilities and increasing their
utilization.[8]

Benefits

TIRKS benefits BOCs in three areas:

- *Improved service to customers.* Through its single, integrated data base,
  TIRKS serves the total circuit-provisioning process—from initial planning of circuit components through revisions to the eventual retire-
  ment of components—which results in more accurate, efficient, and
  timely circuit provisioning. Through its order control procedures,
  TIRKS can track critical dates and alert management when completions of orders may be in jeopardy.

- *Capital and expense savings.* Capital savings result from improved planning and more efficient use of equipment and facilities. Expense savings result from the mechanization of Circuit Provision Center operations. TIRKS also eliminates the need to spend time verifying data

---

[8] The ratio of working equipment to working-plus-spare equipment.

from inadequate records. The work force is used more efficiently because TIRKS issues more accurate work orders.

- *Better management control.* TIRKS provides an integrated records base with consistent terminology and format and can generate detailed and summary reports on demand and on a scheduled basis for different operating company levels.

14.2.2 PLUG-IN INVENTORY CONTROL SYSTEM (PICS)

The Bell System spends over $1 billion per month to add and replace equipment. A large portion of this capital investment is for switching and transmission systems. The physical design of these systems has changed with time, as discussed in Chapters 9, 10, and 11. Early systems, such as the No. 5 Crossbar and the 701 private branch exchange (PBX), were configured in large racks and "hardwired" in place. During the mid-1960s, electronic switching systems and transmission systems, such as T-carrier systems, that use large quantities of plug-in equipment were introduced. Little attention was paid to inventory control because of the rate at which growth was occurring. As a result, most central offices accumulated significant numbers of spare plug-in units, the existence of which was unknown to equipment engineers responsible for ordering maintenance and growth spares. This "lost" spare equipment was generally scattered throughout central offices, often unused even by resident technicians. New equipment was often requested and ordered from suppliers when the very plug-in equipment needed was nearby. It became evident that the effective management of inventories of this size required the use of a mechanized operations system, and the development of PICS began.

PICS assists operating telephone company personnel with inventory management and materials management. *Inventory managers* establish corporate policies for the types of equipment and for equipment utilization, assist engineering organizations in introducing new types of equipment while phasing out older types, and set utilization goals that balance service objectives (see Section 5.2.2 and Chapter 16) and carrying charges on spare equipment. *Materials managers* work to achieve utilization goals. They acquire spare equipment for growth and maintenance purposes (*growth spares* are used to provide additional service such as the expansion of a switching system; *maintenance spares* serve as replacements for equipment that has failed while in use). They also administer a hierarchy of locations used for storing spare equipment, ranging from large warehouses to small, unattended transmission system locations.

The following paragraphs describe a system that administers all types of central office equipment—PICS with Detailed Continuing Property Records (PICS/DCPR).

PICS/DCPR

About 50 million plug-in units are in use in the Bell System. From the viewpoint of PICS, these plug-in units represent about 100,000 different types of equipment.

During the design of PICS, it also became clear that a detailed investment record of equipment by categories would be needed for regulatory purposes. It was quickly recognized that an equipment inventory was a prerequisite for a detailed investment or property record. The DCPR portion of PICS/DCPR serves as a detailed investment data base supporting telephone company accounting records for all types of central office plug-in and "hardwired" equipment.

Implementation. PICS/DCPR provides software, data bases, administrative procedures, and workflows to accomplish its goals of increasing utilization, decreasing manual effort, and providing a detailed supporting record for a telephone company's investment. First, PICS/DCPR establishes two new functional entities in each company: the *Plug-In Administrator* (PIA) and the *central stock*. The PIA becomes the materials manager, responsible for acquiring equipment, distributing it as needed to field locations, repairing it, and accounting for it. To assist the PIA, the central stock is created. The central stock is a warehouse where spare equipment is consolidated and managed. To implement PICS/DCPR, a telephone company must create a new management-level job (the PIA), provide a clerical staff, set up a central stock, and provide a computer environment for the system software. The PIA then recalls all excess plug-in units. Some, but not all, units are returned to the central stock. The PIA then conducts complete physical inventories of the central stock and all field locations, returning excess spare equipment from the field to the central stock. At this point, the plug-in inventory is established, and the PIA uses on-line terminal programs to order, receipt, bill, move, repair, retire, and track equipment. A mechanized link between the PIA and the central stock provides warehouse personnel with instructions for shipping or receiving stock from the field.

As part of the inventory process, cost data based on average yearly prices are supplied. These data make up the DCPR for plug-in equipment. The data are supplied by equipment engineers for both plug-in and hardwired equipment on a *going-forward basis*[9] as it is purchased. A portion of the system provides an annual reconciliation between the DCPR and the company books to ensure that the DCPR remains accurate.

---

[9] Because of difficulties in identifying and pricing older hardwired equipment, no physical inventory occurs. Equipment details are captured on a going-forward basis from the start up of PICS/DCPR.

Subsystems. PICS/DCPR software comprises five subsystems: the **plug-in inventory subsystem**, the **inventory management subsystem**, the **plug-in DCPR subsystem**, the **hardwired DCPR subsystem**, and the **reference file subsystem**.

The *plug-in inventory subsystem* maintains order, repair, and inventory records for all types of plug-in equipment. It spans seven processing areas:

- acquisition (order control, logging, cancellation, equipment receipts)

- movement (between locations; connections and disconnections, for example, from maintenance spare to growth spare)

- maintenance activity (logging defectives, tracking repair orders, junking plug-in units)

- circuit order control (to interface with circuit-provisioning systems like TIRKS)

- inventory adjustments (to compensate for shortages and surpluses)

- "back-order" control (to track items on back order)

- miscellaneous (scan/search of inventory data base; printing of shipping information at central stock, etc.).

The *inventory management subsystem* provides the PIA with mechanized processes to assist in these tasks:

- recommending levels for maintenance spares based on the rate at which equipment is returned because of failure

- recommending when orders should be made based on historical demand, known scheduled growth, or both

- monitoring supplier performance so that proper order lead time may be taken into account

- reporting on equipment overstock or understock by location and equipment type.

Specific models and algorithms exist for these activities for both central stock and field locations. At the PIA's request, the subsystem will shift stock from overstocked locations to those that are understocked or back-ordered.

The *plug-in DCPR subsystem* provides the processes required to maintain investment records for plug-in units, including:

- daily processing (loading cost data from purchases, updating investment as a result of movement, adjusting for shortages and surpluses)

- billing adjustments and corrections

- monthly processing (to reconcile plug-in equipment billing between the DCPR and accounting, to notify accounting of changes in investment because of plug-in equipment movement and retirements)

- annual processing (to reconcile the DCPR with the company books)

- miscellaneous (on-line display/search capability, auditing records for tracking the source of investment change, etc.).

The *hardwired DCPR subsystem* maintains detailed accounting records for hardwired central office equipment. Processing is analogous to that of the plug-in DCPR subsystem, except that it is equipment engineers, instead of the PIA organization, who supply the input. Because of difficulties involved in identifying and pricing older hardwired equipment, no physical inventory occurs. Equipment details—records of the various parts of larger systems, such as an electronic switching system— are captured on a going-forward basis from the startup of PICS/DCPR. Investment for prior years is maintained on an undetailed basis; that is, for equipment in existence before the startup of PICS/DCPR, total equipment investment is collected together by year for recordkeeping purposes. Another distinction between the plug-in and hardwired DCPR concerns accounting rules. Hardwired equipment may not be moved from location to location without a complicated accounting process. Plug-in units, by nature of their mobility, were exempted from this process by the Federal Communications Commission (FCC) by special agreement.

The *reference file subsystem* provides and maintains reference data used by all the other subsystems, including:

- equipment reference data, such as the J-number and list number from the J-specification for the equipment, common-language equipment identification codes, and availability designations[10]

- location reference data, such as address, telephone number, and *common-language location identification code* designation—a 10-character code name that identifies any physical location within the Bell System

- DCPR reference data, such as account code and equipment category number (for both plug-in and hardwired equipment)

---

[10] J-numbers and the list numbers associated with them identify equipment. The nature of systems or equipment units is described in J-specifications, which also explain applications of equipment units or systems. Common-language equipment identification codes identify equipment by family (for example, T-carrier), subfamily (for example, channel units), and type (for example, a specific channel unit).

- on-line security data that prevents movement of equipment between unauthorized locations

- audit reference data, such as the date and time a data base was updated and the terminal that originated the update. (These data track changes made to all reference data bases.)

PICS/DCPR subsystems are developed and maintained by Bell Laboratories. Both on-line and batch programs are contained in PICS/DCPR. Software packages containing executable program code and ancillary data needed to run the system are periodically released to the telephone companies. PICS/DCPR runs on *IBM*-compatible computing equipment operating with the *IBM* Information Management System data-base manager with teleprocessing options.

Interfaces. PICS/DCPR is an interdepartmental system. Interfaces exchange data with organizations within a company in the areas of property and cost accounting, division of revenues, depreciation accounting, and service cost accounting. Interfaces have also been provided to TIRKS and other circuit-provisioning systems, to Bell Laboratories for the on-going update of reference data, and to other BOCs that jointly own equipment. Initially, the interface between PICS/DCPR and TIRKS was designed for manual operation, where plug-in requirements are forwarded to the PIA concurrent with the issuance of a TIRKS work order. Ultimately, the process will be automated, and the systems will interact with each other so that the specific plug-in equipment item will be recorded on the work order before issuance to the field.

Benefits. All of the Bell System has now implemented PICS/DCPR. Significant economic benefits arise as a result of reduced capital investment needs as spare equipment is located and effectively reused. Obsolete equipment is junked and eliminated from a company's investment base. These actions increase the utilization of plug-in equipment. Many companies have improved their utilization from approximately 70 percent to 90 percent or more with PICS/DCPR. By mechanizing many actions that were previously done manually, staff has been either reduced or made more productive. Secondary benefits include faster provision of equipment, a basis for asset verification (the DCPR), and improved record accuracy.

Various economic modeling techniques allow estimation of PICS/DCPR's economic benefits. The total benefit over the period 1969 through 1990 is estimated at $1.1 billion.[11] (This figure includes Bell Laboratories development costs.)

---

[11] Cumulative discounted funds flow (see Section 18.3.3) based on a January 1980 study.

14.2.3 PREMISES INFORMATION SYSTEM (PREMIS)

Each BOC has thousands of *service representatives*. One of their functions is to respond to requests for telephone service from residence customers. To ensure customer satisfaction, each transaction must be completed as quickly and as accurately as possible. Speed and accuracy are also important to an operating company since a faster, more accurate transaction means lower costs for both customers and the company.

Developed by Bell Laboratories in cooperation with AT&T, PREMIS provides the service representative with fast, convenient access to the information needed to respond to service requests. The need for a PREMIS-like system grew in the mid-1970s when efforts were being directed toward reducing operating costs by decreasing residence installation visits. The installation of modular jacks in residences and the institution of PhoneCenter Stores, where customers could pick up sets, were both part of this effort. As a result, the need for maintaining records of jacking arrangements[12] arose and led to the development of PREMIS.

PREMIS was also developed in response to the need for address standardization—a need that has grown as Bell operating companies mechanize operations. Computer-based systems are used by the different departments of a Bell operating company in processing the customer *service order* prepared by the service representative. Mechanization requires that the customer address used in records be standardized, because the service order (with its address information) feeds other downstream systems. Precision in the way that living units (such as houses or apartments) are represented—along with address standardization—promotes the automatic reuse and reassignment of facilities. This benefit affords savings to a Bell operating company by reducing the time and labor needed for assignment and installation of outside plant facilities.

PREMIS has three mechanized data bases for use by service representatives in a telephone company:

- address data

- a credit file

- a list of available telephone numbers.

These data bases ensure greater accuracy than the paper records, microfiche, and other information sources previously used.

PREMIS has one other major feature that serves the *Loop Assignment Center* (LAC) rather than the service representative. This feature, called *PREMIS/LAC*, is an extension of the address data base and provides for the storage of outside plant facility data at each address entry. PREMIS/LAC is described later in this section.

---

[12] The number of telephone jacks and their locations in a living unit.

Customer Service Tasks and PREMIS Applications

PREMIS helps service representatives handle the following transactions:

* new service—a customer wishes to initiate service

* relocation of service—a customer wishes to disconnect service at the existing address and initiate service, perhaps with changes, at a new address. Service may exist or may have previously existed at the new address for another customer.

These transactions involve several tasks: determining the customer's correct address, negotiating service features, negotiating a service date, checking the customer's credit status, and selecting a telephone number. Some of these tasks provide information needed for the preparation of the service order that will be used by other departments in the telephone company to provide service. PREMIS support for these service representative tasks is described below.

Determining the Customer's Correct Address — The major feature of PREMIS is its address-related and address-keyable information. When an order for new or relocated service is being taken, the customer gives the service representative the new address, which the service representative then keys into PREMIS using a video display terminal, as shown in Figure 14-4. Address-related information is stored in the *address data base* (see Figure 14-5).

When an input request does not contain an accurate or complete address, PREMIS displays information that can be used to query the customer. For example, in a transaction with a customer who requests service for a residence at 125 Main Street, PREMIS would prompt for more information if it had data for both 125 East Main Street and 125 West Main Street. This prompting capability ranges from a list of street names that partially match the input to a list of apartment numbers for a specific address. In each situation, PREMIS provides selected information designed to direct the service representative's attention to the specific problem with the input.[13]

When the address matches information contained in the address data base, PREMIS responds with the full address (community, state, and zip code) and information about the geographic area that is needed on the service order. This includes wire center, exchange area, tax area, directory group, and the service features available for that area. This same display will also include the existing or previous customer's name and telephone number, the modular jacking arrangement at the address,

---

[13] As a further example, for a request for service at a new residence, PREMIS would compare the address provided by the customer to its data base of valid addresses (streets and numbers).

Figure 14-4. Service representatives with video display terminals for PREMIS.

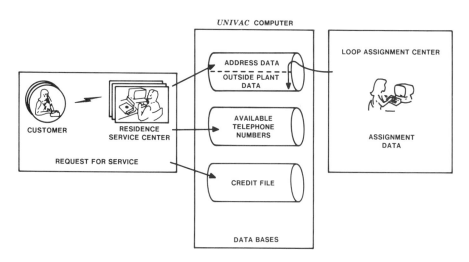

Figure 14-5. PREMIS data bases.

and an indication of whether a connected outside plant loop from the address back to the central office was left in place.

If the previous service at the address was discontinued, the reason for disconnect and the disconnect date are also displayed. This helps identify customers who are trying to re-establish service at the same address when previous service was disconnected for nonpayment.

Before PREMIS was developed, service representatives had to use multiple paper sources to obtain the data PREMIS now provides in the address data base. Without PREMIS, the service representative typically would obtain address information from a book called the *Street Address Guide* (SAG); modular jack information from a microfiche file; other information from paper records (previous disconnects, for example); and some information, such as available service features, from the service representative's handbook.

Through its address-keyable data base, PREMIS offers the following advantages:

• increased address accuracy on orders for new service because a mechanized SAG can be kept more accurate and up to date. This accuracy reduces installation delays and correction costs that ripple through every other mechanized system fed by the service order.

• reduction in the number of installer visits. The modular jacking information available from PREMIS is more complete than the file of disconnect service orders that it usually replaces. With PREMIS, the service representative has jacking information available for a new customer at a given address whether or not the existing customer at that address has placed an order to disconnect the service. The capability of locating unnumbered addresses also helps reduce installer visits. Formerly, installer visits were required to locate the *terminal box*.[14] With PREMIS, if a new customer can supply the previous customer's name or telephone number, a service representative may be able to identify an unnumbered living unit and the records of its associated facilities.

Negotiating Service Features — When a service representative uses PREMIS to check an address, PREMIS also indicates the service features that can be sold at that address, providing useful information for discussing these with a customer.

Negotiating a Service Date — PREMIS also indicates whether the outside plant loop back to the central office has been left in place.[15] If so, an

---

[14] Outside plant where pairs of cables are accessible by installers.

[15] When outside plant has been left in place, more detailed information is stored in the address data base, although the service representative does not use this information. This information is used by the Loop Assignment Center.

installer will not be needed to connect service, and an earlier installation date can be offered to the customer. Thus, additional revenue is generated for each additional day of service, and customer satisfaction is increased.

Checking a Customer's Credit Status — Although a service representative can use the address-keyable data base to determine if service has been disconnected at a residence for nonpayment, PREMIS also maintains a name-keyable file of customers with outstanding debts to the telephone company, whose service has been disconnected. The service representative enters the new customer's name into PREMIS for comparison with the credit file. If there is a customer in the credit file with the same name, information on this customer is displayed, including the address where the disconnect was made and the customer's social security number (for identification). Thus, at the time that the order is taken, the service representative can determine if the new customer has an outstanding debt and can require payment of the past bill and a deposit before installing service. Before PREMIS was developed, many telephone companies maintained a paper file of disconnect service orders. All new service orders were checked against this file, but often not until service had already been established. Although the elimination of the paper file provides clerical savings, prevention of potential loss of revenue is the major benefit.

Selecting a Telephone Number — PREMIS contains a file of available telephone numbers, from which the service representative requests a telephone number for a specific address. Formerly, each day, clerks in a central location prepared paper lists of available telephone numbers and distributed them to service representatives.

Not only does PREMIS reduce the errors that result when paper records are used for recording and assigning telephone numbers, but it also reduces the total number of telephone numbers that need to be available in an area. With PREMIS, many residence service centers or service representatives can access the same centralized list. Before PREMIS, different lists of available telephone numbers were prepared for each location to avoid assigning the same telephone number to different customers.

PREMIS reads the available telephone numbers from a magnetic tape supplied by the Computer System for Mainframe Operations (COSMOS)—a computer system developed by Bell Laboratories that stores the full inventory of telephone numbers. For locations where COSMOS is not used, the numbers can be input by a clerk using a maintenance transaction.

PREMIS/LAC Data

Another major feature of PREMIS is the mechanized *dedicated plant assign-ment card* (DPAC) capability used by the Loop Assignment Center. Each operating company LAC maintains records of addresses where the outside plant loop facilities (for example, cable pairs and terminal boxes) are dedicated (left permanently in place). These records are organized and accessed by address; thus, the DPAC data can be added to the PREMIS address data base, and the paper DPAC file can be eliminated. Outside plant data may be maintained in PREMIS for a line that is currently working at an address or for a nonworking line where the outside plant loop has been left in place for reuse by the next customer. This mechanized capability is more accurate than the paper records it replaces, and it reduces both errors and the outside plant assigner's time. In promoting the reuse of facilities, this feature also reduces the number of installation visits.

System Characteristics

PREMIS is an on-line interactive system whose prime users are service representatives interacting with customers. During peak hours on peak days, thousands of service representatives may be interacting with the system at the same time. Therefore, PREMIS has fast response-time requirements and a particular need for simplicity and clarity of input and output.

The PREMIS system has a centralized data base and uses a large main-frame computer—the *UNIVAC* 1100 series. Depending on the size of the Bell operating company, either one or two computers will serve the entire company. The terminal used by the service representative is a *DATASPEED* teletypewriter Model 40/4 or 4540. The service representative who uses PREMIS is often accessing other systems as well, so PREMIS is designed to share the terminal with these other systems via a communication network. To free the PREMIS computer for heavy on-line usage, update of the data base is done overnight (in batch) with selected data from each day's service orders. The data are extracted from the telephone company's mechanized service order distribution system and supplied to PREMIS on a tape.

PREMIS was piloted in South Central Bell in January 1979, and by mid-1983, it was being used by nine BOCs.

14.3 EQUIPMENT MAINTENANCE, ADMINISTRATION, AND CONTROL

The previous section described systems concerned with recordkeeping. This section describes systems that gather and process data from operational network components (such as switching systems and transmission facilities) to improve the administration, maintenance, or control of the network or its components. The systems in this category provide information to support a number of operating company functions, including:

- long-range and current planning for growth and rearrangement of the network

- ordering and installing equipment and facilities

- adding, deleting, and rearranging circuits to meet changing traffic demands

- managing the flow of traffic in the network to control overload conditions

- monitoring the performance of transmission systems, switching systems, and trunks and facilitating maintenance by improving the reporting and control of equipment problems.

The **Total Network Data System**, described in Section 14.3.1, is actually a large and complex set of coordinated systems. It supports a broad range of operating company activities that depend on accurate traffic data. The **Switching Control Center System**, discussed in Section 14.3.2, supports surveillance, maintenance, and control activities for network switching systems and transmission terminal equipment. This system illustrates the advantages of combining a computer-based system with the concept of centralized maintenance.

14.3.1 TOTAL NETWORK DATA SYSTEM (TNDS)

The Bell System communications network contains over 11,000 switching systems that are connected by over 7.5 million trunks arranged in 400,000 separate trunk groups. Both the general growth of telecommunications and the replacement of older switching and transmission systems by newer, more versatile ones necessitate a construction program for the Bell System that exceeds $3.5 billion each year just for the trunks of the PSTN. Because such large capital investment is involved, adequate planning for the future growth of the trunking and switching networks requires significant attention to traffic data collection, administration, and engineering. Measurements of current traffic levels are the basis on which current network performance is assessed and future growth is planned. These measurements (see Chapter 5) are made in terms of both trunk usage and switching system operation. For the average operating company, these evaluation and planning functions can require the collection, processing, and distribution of over 50 million pieces of traffic data per week.

TNDS is a family of operations systems that work together to mechanize data gathering and reporting. TNDS consists of both manual procedures and computer systems that provide operating company managers with comprehensive, timely, and accurate network information that helps them to analyze network operation in several ways. TNDS supports operations centers responsible for administration of the trunking network, network data collection, daily surveillance of the load on the switching network, the utilization of equipment by the switching network, and the design of local and central office switching equipment to meet future service demands.

TNDS Modules

TNDS comprises several major component systems or *modules* that operate at various locations. Modules that collect and format traffic data typically use dedicated minicomputers and are located at an operating company's computation center, referred to as a *minicomputer maintenance center*. Special data lines link these dedicated minicomputers to associated measuring equipment or directly to the switching systems located at central offices. Other TNDS modules generate engineering and administrative reports on switching systems and on the trunking network of message trunks that interconnects them. Most of these modules run on general-purpose computers located in the operating companies, although some run on AT&T computers centrally located for use by all the Bell operating companies. Depending on their needs, operating companies may use all of the TNDS modules or various combinations of them.

TNDS Functions

TNDS modules perform four basic processes: data acquisition, central office equipment reporting, trunk network reporting, and system performance measurement. Figure 14-6 shows the overall flow of information in TNDS between the systems that perform these processes, and Table 14-2 summarizes the component systems.

Data Acquisition. Switching offices—both electronic and electromechanical—provide traffic data in terms of peg count, overflow, and usage measurements.[16] In electromechanical offices, a specialized data collection device, called a *traffic usage recorder*, scans trunks and other switching components periodically (every 100 seconds) and counts how many are busy. In electronic offices, the data are collected by the switching system's central processor. As shown in Figure 14-6, these traffic data are transmitted by most switching systems to the first of the TNDS systems—the *Engineering and Administrative Data Acquisition System* (EADAS). EADAS is the major data-collecting system and runs on a dedicated minicomputer at an operating company's *Network Data Collection*

---

[16] Section 5.7.1 defines these traffic measurement parameters.

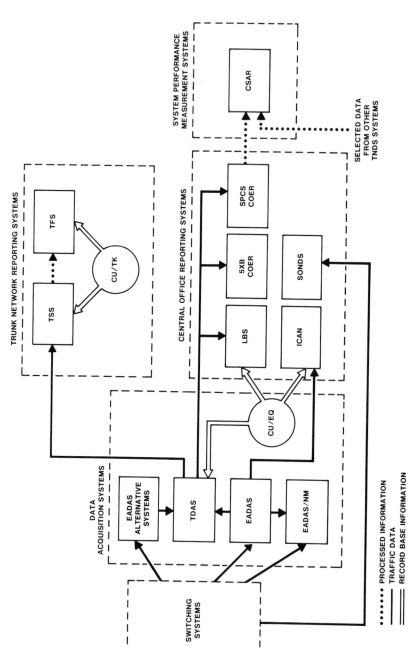

Figure 14-6. Data flow among TNDS systems.

Center. Each EADAS serves up to fifty switching offices. Some companies use other data-collecting systems that are developed locally or supplied by vendors.[17] Interfaces are provided so that measurement data gathered on these systems can be processed downstream. Two large toll switching systems—the 4*ESS* system and the No. 4A Crossbar[18]—collect their own data and do not interact with EADAS or alternative data acquisition systems. They provide their data directly to the TNDS component systems downstream from EADAS.

When EADAS receives traffic data from switching offices, it assembles and summarizes the data for processing by the other downstream TNDS systems, and it processes some of the data in near real time to provide hourly and half-hourly reports for network administrators.

Network administrators use EADAS reports to determine the quality of service (for example, in terms of dial-tone delay) and to identify switching problems, such as failure to complete calls. EADAS also makes additional real-time information available to these administrators by providing traffic data history that covers up to 48 hours. The data history provides flexibility (via a module called *NORGEN*—Network Operations Report Generator) so that administrators can tailor their requests for information to determine critical quantities such as dial-tone delay, holding time for certain units of switching equipment, and overflow on important trunk groups.

EADAS forwards traffic data to three other TNDS systems by data links or magnetic tape (see Figure 14-6). One of these systems—the *Traffic Data Administration System* (TDAS)—formats the traffic data for use by most of the other downstream systems. The other two systems—the *Individual Circuit Analysis* (ICAN) program and the *Engineering and Administrative Data Acquisition System/Network Management* (EADAS/NM)—use data directly from EADAS. ICAN is one of the central office reporting systems discussed in **Central Office Reporting Process** below.

EADAS/NM also receives data from those switching systems that do not interface with EADAS. It monitors switching systems and trunk groups designated by network managers and reports existing or anticipated congestion on a display board at operating company *Network Management Centers.* Network management personnel use EADAS/NM to analyze problems in near real time to determine their location and causes. For problems that require national coordination, EADAS/NM provides information to the AT&T Long Lines *Network Operations Center* at Bedminster, New Jersey. This center is supported by a *Network Operations Center System* (NOCS), which performs functions similar to EADAS/NM on a

[17] For simplicity, only the term EADAS will be used in describing these acquisition functions. But the possible use of alternative systems should be understood.

[18] No. 4A Crossbar Systems currently in service use a peripheral bus computer (PBC) to collect data. This is an adjunct to the electronic translator system. (See Section 10.2.5.)

national scale. Like EADAS, EADAS/NM uses dedicated minicomputers to provide users with interactive real-time response and control.

TDAS, the second module in the data acquisition process, accepts data from EADAS, local vendor systems, and large toll switching systems on a weekly basis as magnetic tape. TDAS functions primarily as a warehouse and distribution facility for the traffic data and runs as a batch system at an operating company's computation center.

TDAS treats its data-acquisition job as a basic order/inventory system, in which the traffic data collected represent the inventory, and traffic measurement requests for specific data represent the orders. Traffic measurement requests, prepared manually and sent to TDAS by operating company personnel, are stored in a master data base called *common update/equipment* (CU/EQ) that shares information with some of the other central office reporting systems. This shared information for each central office is necessary to ensure the correct association between recorded traffic data and the switching or trunking elements being measured. CU/EQ runs as a batch system in the same computer as TDAS. CU/EQ is updated regularly with batch transactions to keep it current with changes in the physical arrangements of the central office switching machines. This ensures that recorded measurements are treated consistently in each of the reporting systems that use CU/EQ records.

The traffic data (the inventory) processed through TDAS are matched against the traffic measurement requests (the orders) stored in CU/EQ. When the data necessary to fill an order have been received, a weekly data summary (printed or on magnetic tape) is sent to the personnel requesting it for use in preparation of an engineering or administrative report. The availability of summarized traffic data to downstream systems completes the TNDS data acquisition function. The remaining TNDS systems, shown in Figure 14-6 and described below, help managers analyze the data that have been gathered.

Central Office Reporting Process. Five TNDS engineering and administrative systems provide operating company personnel with reports about central office switching equipment. These component systems run either as batch processes on operating company computers or interactively at a centralized AT&T (mainframe) computer center (see Table 14-2). The systems are:

- the Load Balance System (LBS)

- the No. 5 Crossbar Central Office Equipment Reports (5XB COER)

- the Stored-Program Control Systems Central Office Equipment Reports (SPCS COER)

- the Individual Circuit Analysis (ICAN) program

- the Small Office Network Data System (SONDS).

The first three systems receive their traffic data directly from the Traffic Data Administration System (TDAS). ICAN receives data from EADAS independently of TDAS and uses the CU/EQ data base for some of its reference information. SONDS collects its own data from small step-by-step offices independently of both EADAS and TDAS (as shown in Figure 14-6).

LBS is a batch-executed system that helps assure the network administrator that traffic loads in each switching system are uniformly distributed. LBS analyzes the traffic data to establish traffic loads on each *line group* of a switching system.[19] The resulting reports are used by the *Network Administration Center* to determine the lightly loaded line groups to which new subscriber lines can be assigned. LBS also calculates load balance indices for each system and aggregates the results for the entire operating company.

The 5XB COER and the SPCS COER systems provide information on common-control switching equipment operation for different types of switching systems. 5XB COER is a batch-executed system that runs on an operating company's (mainframe) computer, while SPCS COER is an interactive system that runs on a centralized AT&T mainframe computer. Both analyze traffic data to determine how heavily various switching system components are used and to measure certain service parameters such as dial-tone delay. Network administrators use these service and analysis reports to monitor day-to-day switching system performance, diagnose potential switching malfunctions, and help predict future service needs. Traffic engineers rely on these equipment utilization reports to assess switching office capacity and to forecast equipment requirements. To be most useful, service and traffic load measurements must be made during the busiest periods of the day and year. Both 5XB COER and SPCS COER produce busy hour and busy season[20] reports to meet that demand.

The ICAN program detects electromechanical switching system equipment faults by identifying abnormal load patterns on individual circuits within a circuit group. These faults, for example, include defective circuits that prevent customer calls from being completed. ICAN produces a series of reports used by the Network Administration Center to analyze

---

[19] A *line group* is a collection of subscriber lines that share a concentration module of the switching system. Section 5.3.4 discusses load balancing, and Section 7.3 discusses the concept of concentration.

[20] Section 5.2.3 discusses these measurement periods.

the individual circuits and to verify that such circuits are being correctly associated with their respective groups.[21]

SONDS differs from the other office equipment reporting systems in that it performs a full range of data manipulation functions. It provides a number of TNDS features economically for the smaller electromechanical step-by-step offices. SONDS collects traffic data directly from the offices being measured, processes the data, and automatically distributes weekly, monthly, exception, and on-demand reports to managers at the Network Administration Centers via dial-up terminals. SONDS runs on an interactive basis at a centralized AT&T mainframe computer.

Trunk Network Reporting Process. Shown in Figure 14-6 are three TNDS systems that support trunk servicing and forecasting at the *Circuit Administration Center*. They are all batch programs run on an operating company's computer and include:

- Common update/trunking (CU/TK)

- Trunk Servicing System (TSS)

- Trunk Forecasting System (TFS).

CU/TK is the data-base system that contains the trunking network information (such as alternate routes) and other information (such as trunk group identification and trunks in service) required by TSS and TFS. Personnel in the Circuit Administration Center update the CU/TK data base regularly to keep it current with changes in the physical arrangements of the trunks and switching machines in the central offices.

This updating process includes maintaining office growth information and a "common-language" (standard) circuit identification (see Section 14.2.2) of all circuits for individual switching machines. Such maintenance is necessary to ensure that traffic data provided by the Traffic Data Administration System (TDAS) will be correctly associated with the proper trunking and switching configuration when it is processed by TSS and TFS.

TSS helps trunk administrators develop short-term plans and determine the number of circuits required in a trunk group. It processes traffic data from TDAS and computes the offered load for each trunk group, that is, the amount of traffic that would have been carried had the number of circuits been large enough to handle the load without trunk blocking. (Chapter 5 discusses these traffic-related parameters.) Using offered load on a per-trunk-group basis, TSS calculates the number of trunks theoretically required to handle that traffic load at a designated grade of service.

---

[21] Section 3.2 discusses trunks, special-services circuits, and trunk groups.

TSS produces weekly reports showing which trunk groups have too many trunks (under-utilization) and which trunk groups have too few trunks and are performing below the grade-of-service objective. Personnel in the Circuit Administration Center will use this information to issue trunk orders that either add or disconnect trunks.

The traffic load data computed by TSS are also used to support the trunk forecasting function performed by TFS. These data, along with information on the network configuration and forecasting parameters stored in the CU/TK data base, are used in long-term construction planning for new trunks. TFS forecasts message trunk requirements for the next 5 years as the fundamental input to the planning process that leads to the provisioning of additional facilities.

System Performance Measurement Process. Traffic data flow through TNDS provides operating company decisionmakers with timely and accurate information so that they can deal with short- and long-term issues. Because of the complex steps and the number of systems involved in providing the various engineering and administrative reports, a separate reporting system assesses TNDS performance and identifies potential problem areas. The *Centralized System for Analysis and Reporting* (CSAR) is designed to monitor and measure how well data are being processed through TNDS. CSAR collects and analyzes data from other TNDS systems. (CSAR does not now analyze data from EADAS/NM, SONDS, or TFS.) CSAR provides operating company personnel at Network Data Collection Centers, Network Administration Centers, and Circuit Administration Centers with quantitative measures of the accuracy, timeliness, and completeness of the TNDS data flow and the consistency of the TNDS record bases. CSAR also furnishes enough information to locate and identify a data collection problem, for example, if EADAS receives no data for several days on some equipment component in a particular 5XB COER office.

In addition to assisting each operating company in monitoring the overall operation of TNDS, CSAR summarizes the results for that company as input to the TNDS Performance Measurement Plan (TPMP). TPMP results are published monthly by AT&T.

CSAR runs as a centralized on-line interactive system at an AT&T computer center. At the conclusion of each run of a TNDS system at the BOC, data required by CSAR are placed into a special file. At the appropriate time, these files are merged and transferred to the AT&T computer center. CSAR performs the proper associations and analyzes each system's results. Operating company managers access the resulting information using dial-up terminals from their own work locations. Reports can be arranged in a number of formats that provide details on overall TNDS performance or individual system effectiveness. These reports also help to identify and resolve specific problems. Reports can be arranged to reflect performance on a level as broad as a total company or as specific as an individual trunk group or switching system. Such reports enable managers to analyze TNDS operation on an end-to-end basis, a process that would otherwise involve significant effort to summarize and correlate information across the various system boundaries, if done manually.

Near-Term Evolution of TNDS

Since its inception in the late 1960s, TNDS has evolved into a large, mature system meeting operating company needs for traffic data collection and processing. In its early stages, TNDS represented the first set of tightly integrated operations systems and eliminated the need for the extensive manual recording, tabulating, and summarizing that had characterized the traffic measurement, collection, and analysis process thus far.

New algorithms and better techniques are being developed to improve the operational and reporting characteristics of TNDS and to keep pace with advances in the state of technology and with the needs of its users. Some of these improvements will mechanize the user interface to include on-line access to the data base and processed results generated by the various batch component systems. Because TNDS is an integrated system, the systems-planning effort associated with the continual introduction of new systems, features, and generic programs served as a forerunner of the main Bell System operations planning activities. TNDS is now a key element of the Total Network Operations Plan (see Chapter 15), and its future evolution is being carefully coordinated through this plan.

14.3.2 SWITCHING CONTROL CENTER SYSTEM (SCCS)

The *Switching Control Center* concept evolved within the telephone companies starting in the late 1960s. The primary motivation was to centralize the administration, maintenance, and control of the 1ESS switching system that was being rapidly deployed within the Bell System. By grouping personnel performing the same and related functions, centralization results in more efficient use of people as well as other advantages such as cross-training.

The 1ESS system, a *Stored-Program Control System* (SPCS), is far more sophisticated than its electromechanical predecessors and requires different training for maintenance personnel. Maintenance of 1ESS switching equipment is supported by a *master control center* (MCC), a frame of equipment in a 1ESS system with lamps to indicate the current state of the office equipment and keys for operating controls (see Figure 14-7). Maintenance data are available at a maintenance teletype and are

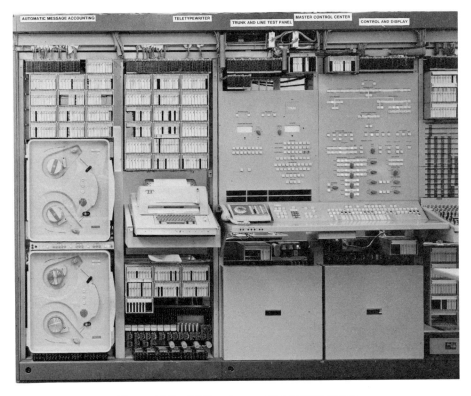

Figure 14-7. 1*ESS* system master control center.

used by a technician to request diagnostics and to remove and restore equipment units to service. Because the 1*ESS* system has a highly reliable duplex nature, an office can operate unattended, either entirely or for substantial parts of each day.

Unattended operation is both feasible and practical primarily because the 1*ESS* system maintenance data can be sent to a remotely located maintenance teletype. In addition, the functions of the MCC for the 1*ESS* system can be performed remotely using specially designed interface equipment that works with standard telemetry equipment. These capabilities allow the remote administration, monitoring, and control of 1*ESS* switching equipment from a centralized maintenance center—a Switching Control Center (SCC).

With teletype channels connected to a centralized location, it is economically attractive to add a minicomputer system at the SCC to interface with the 1*ESS* system since the cost of the computer system can be shared among several switching systems. This minicomputer system, the *Computer Subsystem* (CSS), and the equipment units that remote the MCC capability make up the *Switching Control Center System* (SCCS). The SCCS has been broadened to handle other SPCSs besides the 1*ESS* system.

The CSS can support a number of SCCs. In a typical Bell operating company application, the computer is located in the minicomputer maintenance center. A number of SPCSs within the same geographic area can be supported from the center, so that the centralized work force can collaborate on maintaining any that need immediate attention. The use of a centrally located common pool of highly skilled technicians offers clear economic and technical advantages. Further, when an on-site visit is needed, for example, to replace a faulty circuit pack, a technician can be dispatched from the center.

System Operation

Figure 14-8 is a schematic representation of two Switching Control Centers (A and B) served by one Computer Subsystem and connected to several Stored-Program Control Systems.

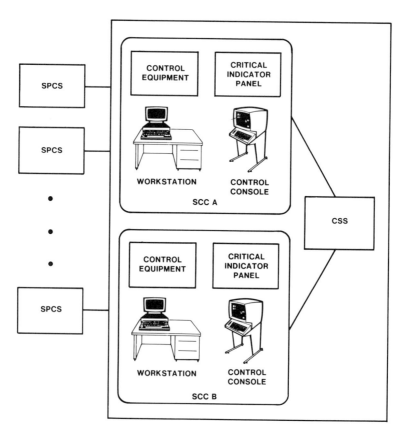

Figure 14-8. Two Switching Control Centers served by a Computer Subsystem.

Maintenance and administrative data are sent to the SCCS from each switching system connected to it. The control equipment provides the interface for all forms of data received from the switching systems. The MCC data on equipment status from all connected SPCSs are displayed on indicator lamps on a wall-mounted critical indicator panel within view of SCC personnel. In addition, a minicomputer-based control console also receives the MCC data. Using the console, a technician can operate MCC keys remotely (that is, at the SCCS).

During any day, an SPCS may generate the equivalent of a hundred or more pages of teletype messages. These data are of various types, for example, equipment status data, administrative data, diagnostic results data, data on abnormal conditions, and audit data. The SCCS computer receives them, logs them on disk, performs a number of operations on them, and takes various actions depending on the information contained in the messages. All the received data can be viewed on a workstation CRT either immediately or later, since a long-term history of the logged data is kept. In addition to viewing and analyzing the data, a technician using a workstation terminal can directly communicate with any connected SPCS and can remotely execute any command that is available locally at the SPCS. For instance, a technician at the SCC can request an SPCS to run a program to diagnose a piece of equipment, remove or restore a trunk, etc.

The number of switching offices that may be connected to an SCCS depends on the size of the offices and the amount of data transmitted. Typically, about fifteen offices are connected, although it is possible to handle thirty or more offices. Generally, this number would include a mix of stored-program control switching system types located within a geographic area surrounding the SCC. This proximity permits maintenance personnel to be dispatched to an office as needed. Generally, the goal is to locate the SCC within a half-hour dispatch time from any office it serves.

The number of offices and corresponding volume of maintenance activity also determine the number of workstation terminals and control consoles required for effective operation. Typically, an SCCS may include fifteen to twenty workstations and two or three consoles. The SCCS automatically gathers two basic types of real-time data from each monitored office—status data reflecting the state of the lamps on the MCC and teletype message data (which also appear on the maintenance teletype in the switching office). These two types of data are transmitted over separate channels to the SCCS.

If a major alarm is triggered at a switching office, an alarm will sound within seconds at the SCC. Further, the major alarm will cause an update of the status of the office on the critical indicator panel and display a specific description of the alarm condition on a CRT alarm monitor at a workstation. Whenever an alarm occurs, the computer displays that condition only on the CRT monitor that is assigned to a particular set of switching offices. This assignment of responsibility allows technicians to work in their areas of specialization.

In response to the alarm, a technician might use a CRT workstation terminal to access CSS software tools to analyze the situation and choose an appropriate action, which might consist of sending a specific teletype command from the workstation terminal (through the CSS) directly to the SPCS involved. In a situation where an SPCS system restart is required or the teletype channel is unavailable, a control console in the SCCS may be used to operate MCC keys remotely. Thus, technicians can monitor and control an office and have the same capabilities from the SCC that they would have at the MCC in the switching office.

Added Software Capabilities

In addition to remoting the capabilities that are available locally, the SCCS computer provides many additional software tools that simplify and improve the maintenance and operation of the SPCSs. These software tools fall into four broad classes: **enhanced alarming, interaction with message history, mechanization of craft functions,** and **support for SPCS administration**.

Enhanced Alarming. The SCCS does much more than simply reproduce the alarms generated locally at an SPCS. It may also facilitate the use of the incoming data, where necessary, by generating a failure description in a form that is easily interpreted. In addition, it provides many real-time analysis techniques. For example, the SCCS may generate alarms for conditions that are not detectable by the SPCS itself (that is, when there are more than a certain number of messages in a given interval).

Interaction with Message History. After an alarm occurs, additional information may be needed to determine the appropriate action to be taken. To ease information gathering, the SCCS computer provides many tools to aid in sifting through the large amounts of historical information related to the SPCS. For instance, one may use the CSS to "browse" through the history data; filter out information relating to the failure at hand; sort that information by failure type, equipment unit type, unit number, etc.; and analyze the results by looking for subtle patterns that may pinpoint the cause of the problem.

Mechanization of Craft Functions. In some cases, the handling of certain conditions has become so routine that it can be reasonably mechanized. In these cases, the SCCS computer may intercept alarms and status messages as they are received and perform functions that were formerly done manually to analyze and respond to problems. For instance, the system is capable of responding automatically when a trunk problem is detected. The SCCS computer can input messages to the SPCS to remove the trunk from service and test it via the teletype channel. Depending on the results of the test, the SCCS may either return the trunk to service or print a *trouble ticket*, with which a technician may be dispatched to the office to fix the trunk. The process of removing a trunk, restoring it, testing it, and writing the trouble ticket—previously a manual function—is totally automated by the SCCS.

Support for SPCS Administration. Finally, the SCCS provides tools for mechanizing and streamlining the basic administration of the SPCSs. For example, the CSS can send SPCS data to operations centers and other operations systems, as required, without user intervention. Furthermore, time-consuming recordkeeping functions like the tracking of SPCS program changes have been simplified with the SCCS.

In summary, personnel at the SCC, aided by direct access to powerful software tools, have complete maintenance and administrative control of the remote switching offices under their jurisdiction. Dispatching a technician to an office is required only when a problem is identified that requires local maintenance action, such as replacing a circuit pack.

Present Status

The Switching Control Center System was first introduced into the Bell System in 1974. It worked initially with the 1*ESS* system and provided increased benefits by reducing the size of the maintenance force required; by improving responsiveness through mechanized functions; and by providing sophisticated, more efficient support tools. The SCCS received almost immediate acceptance by the BOCs. In subsequent reissues of the Computer Subsystem software, features were added to centralize the maintenance and control of the other SPCS types as quickly as possible.

At the end of 1980, over 200 SCCSs were in operation within the Bell System supporting almost 300 SCCs. Using the latest available software package, the SCCS can maintain not only the entire *ESS* system family, but also the Traffic Service Position System (TSPS), the Automatic Intercept System (AIS), and several other auxiliary processor systems.[22] In addition, SCCS now has the capability to support network transmission terminal equipment.

The number of available SCCS software features has grown enormously since the first SCCS was introduced. Today, sophisticated programs are available for many diverse kinds of operations. For example, within seconds, an SCCS program can analyze and pinpoint a component failure in the complex switching network of an *ESS* system. Without

---

[22] Section 10.4 discusses TSPS and AIS.

these programs, the problem might go undetected for a much longer time because of the sheer volume of messages that would have to be analyzed.

Future Plans

With the economic benefits resulting from the SCC concept and the rapid deployment of SCCSs within the BOCs, the future will most certainly involve expansion of the role of the SCCS. In keeping with the philosophy of centralized maintenance for all switching systems for efficiency and economy, SCCSs are expected to accommodate new SPCS types as they are introduced into the Bell System. In addition, development of additional software analysis programs will certainly continue.

Finally, with the growth of operations systems, there is an increasing need to share data among them. For instance, data collected by an SCCS from a particular switching office may also be needed by another operations system or another operations center. Today, the SCCS supports this data distribution function to some extent. For example, it notifies the T-Carrier Administration System (TCAS), an operations system responsible for T-carrier alarms, of certain types of carrier problems. SCCS also interfaces with the Centralized Automatic Trouble Locating and Analysis System (CATLAS), an operations system that automates trouble location procedures that identify faulty circuit packs in an SPCS when trouble is detected and diagnosed.

Interaction with other operations functions will grow. For example, a secure, automatic dial-up capability is being added to the SCCS to provide direct communication to BOC security bureaus and call annoyance bureaus. In this way, data about a suspect call detected at a switching office can be routed automatically to a security center via an SCCS.

14.4 PLANNING AND ENGINEERING

As described in Chapter 13, planning and engineering are extremely important telephone company functions. They involve several thousand people in the BOCs whose actions affect the entire construction budget.

Some of the areas for which planning and engineering functions are required are the overall network; buildings, including power and common equipment; transmission facilities; and switching equipment. The application of operations systems has provided significant benefits in all these areas.

This section describes examples of operations systems that support planning and engineering functions for local switching equipment and for transmission facilities, respectively—the **Central Office Equipment**

Engineering System (COEES) and the **Metropolitan Area Transmission Facility Analysis Program** (MATFAP)—and two similar programs.

14.4.1 CENTRAL OFFICE EQUIPMENT ENGINEERING SYSTEM (COEES)

COEES is the standard system for planning and engineering local switching equipment. It is a time-sharing system that runs on *DEC PDP*-10 computers, centrally located and accessed via the computing system vendor's communication network. Figure 14-9 is an information flow diagram of the system.

COEES contains component systems for step-by-step switching systems, crossbar switching systems, the 1/1A*ESS* and the 2/2B*ESS* switching systems.[23] Each component system has a different capability. The one that is the most complete—the component for the 1/1A*ESS* system—is described here.

COEES uses a 4-year planning horizon. It is assumed that each year has a busy season. For each of the four busy seasons, the following data,

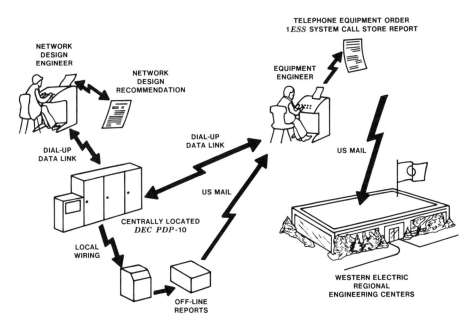

Figure 14-9. COEES information flow diagram.

[23] The necessary functions for the 5*ESS* system are performed by a similar system called the *Digital Ordering and Planning System.*

obtained from company forecasts, is stored in the COEES data base for
each local switching office:

- number of lines of all types

- number of trunks of all types

- average call rate per line and trunk

- average usage per line and trunk

- all special features, signaling types, etc., required.

COEES determines the quantity of each type of equipment in the
office needed to satisfy the forecasted load at objective service levels;
determines an estimated price for engineering, procuring, and installing
the equipment addition needed to reach the required level; and then
compares the *present-worth* costs (see Section 18.3) of each of the eight
different ways of satisfying the office needs (that is, four 1-year jobs, a 1-
year job followed by a 3-year job, a single 4-year job, etc.). It displays all
eight alternatives for the network design engineer to review, including
the present-worth penalty or benefit of each alternative.

An important capability of the system is *sensitivity analysis*, where network design engineers vary different parameters to determine the sensitivity of the costs to that parameter. The following kinds of changes, for
instance, might be considered:

- The call rate (calls per line per busy hour) increases 10 percent.

- The proportion of lines with Speed Calling increases 20 percent.

- The number of centrex lines increases 15 percent.

The network design engineer may decide to change certain equipment
quantities (such as the number of incoming registers or the number of
intraoffice trunks) depending on the results of the sensitivity analysis.
The final result of the network design engineer's activity is a network
design recommendation.

After the network design recommendation is complete, a telephone
equipment order must be prepared. This task requires additional detail
because some simplifying assumptions that can be made for planning
purposes cannot be made for ordering purposes. The equipment
engineer provides the required detail, and COEES prints out an order,
ready to be mailed to the equipment supplier.

Another important output from COEES is the 1ESS system *call store*[24]
report. This report, prepared at the request of the network design

[24] The *call store* is used to store information needed by the 1ESS system to process calls in
progress and certain other information that is likely to change with time. (The 1ESS
system is described in Section 10.3.3.)

engineer and based on inputs provided by that engineer, provides information to the equipment supplier (Western Electric) on the amount of memory to allocate to the various functions of the call store. For instance, there are many registers (areas of memory) that provide such functions as storage of dialed digits or billing information. The appropriate sizes of these registers are a function of the traffic characteristics of the office.

All BOCs and several non–Bell System companies use COEES. Studies have shown that it can save about forty person-hours per telephone equipment order for a typical switching office growth job, compared to the amount of time that would be spent if the same functions were performed manually. There is also strong evidence that it results in more accurately sized offices.

14.4.2 FACILITY NETWORK PLANNING PROGRAMS

The Metropolitan Area Transmission Facility Analysis Program (MATFAP) is a computer program that aids in facility planning. Using present worth of future expenses and other measures, MATFAP analyzes the alternatives available to an operating company for its future transmission equipment and facilities. By combining trunk and special-services circuit forecasts with switching plans, network configuration, cost data, and engineering rules, MATFAP identifies what transmission plant will be needed at various locations and when it will be needed. It also determines the economic consequences of particular facility and/or equipment selections and routing choices and provides the least-cost assignment of circuits to each facility as a guide to the circuit-provisioning process. As its name implies, MATFAP is oriented towards metropolitan networks and towards equipment and facilities found there, such as voice-frequency plug-in units and cables, digital terminals and multiplex equipment, the Digital Access and Cross-Connect System, and digital carrier systems such as T1, T1C, and FT3.[25]

MATFAP provides two benefits. First, it helps automate the transmission-planning process which, in terms of data handling alone, could otherwise impose enormous engineering staff requirements. Atlanta, Georgia, for example, is a moderate-size metropolitan area. Over 4000 circuit groups must be routed on more than 100 links between sixty buildings. Then, facilities and equipment must be selected, the cable and office equipment additions must be sized, and their costs must be calculated over multiple time periods. The hardware possibilities to be considered include several different types and gauges of wire pair cables, a growing number of digital facilities, and a wide variety of voice-frequency equipment and digital terminals.

---

[25] Section 9.4 describes these systems.

Second, MATFAP deals with the entire network and, thus, takes into account economies that cannot be identified by analyses restricted to end-to-end circuit studies, link-by-link investigations, or office-by-office evaluations. One example of this networking advantage is afforded by the program's *minimum-cost routing* algorithm, which assigns forecasted circuits to paths other than the shortest ones if long-term savings in facility expenditures can be obtained. Another example is MATFAP's *circuit segregation* algorithm, which determines the cost advantage of digital carrier trunks (see Section 8.6.3) that can serve message trunks terminating on the 1/1AESS system. Because any special-services circuits also provided from such an office cannot use digital carrier trunks, they require different digital terminals and separate (that is, additional) digital facilities. The overall economics in an office depend on the numbers of each type of circuit, their rate of growth, and the timing of digital carrier trunk installations in other offices to which the circuits are connected. MATFAP accounts for all these factors over the length of the study and allows the planner to find the *best time*[26] to install digital carrier trunks in each office and the best circuit growth policy. MATFAP also balances circuit loads on high-capacity digital lines with the additional multiplex equipment that may be required for high *line fill*.

Because of the large amount of input to MATFAP, data are usually forwarded on tape to the computing center where the program is run. Editing can be done remotely, however, using Western Electric's Remote Data Entry System (RDES).

The *Outstate Facility Network Planning System* (OFNPS) and the *Intercity Facility Relief Planning System* (IFRPS) are similar in function to MATFAP but are tailored to meet the needs of rural and toll networks, respectively. For example, OFNPS contains a decision aid that identifies strategies for introduction of digital facilities in a predominantly analog (N-carrier) network; IFRPS deals with radio and coaxial cable but not with voice-frequency facilities.

14.5 DEVELOPMENT AND SUPPORT

A major software development effort was begun in 1967 with the formation of the *Business Information Systems Programs* (BISP) area, first as part of AT&T and later transferred to Bell Laboratories. This area was given responsibility for the design of a set of computer systems (then called *Business Information Systems*) to support all nonfinancial aspects of the telephone business. Many of these have become operations systems.

From the beginning, Western Electric played a significant role in a set of operations systems collectively called the *EPLANS*[27] Computer Program

---

[26] The installation time that results in the least cost over the length of the study; that is, the installation time that will be least expensive for meeting the need.

[27] Service mark of Western Electric Co.

Service. In recent years, Western Electric has become more involved in operations system maintenance and user support and began, in 1981, to assume funding responsibility for some BIS products.

As the number of operations systems grew, it became clear that the various operations systems should not be considered separately; they could, and should, have interactions and interfaces. Consequently, about 1975, in coordination with AT&T and the BOCs, Bell Laboratories began to develop a series of plans for network and customer operations. The goal of the plans is to establish a framework for the consistent development and application of operations systems for the maximum benefit of the operating telephone companies. (Chapter 15 discusses the need for planning, the planning process, and the impact of the plans in more detail.)

The coordinated design and application approach permits the functions of several operations systems to be interconnected. For example, when the Total Network Data System (TNDS) detects a need for an additional trunk in a trunk group, it might automatically send a message to the Trunks Integrated Records Keeping System (TIRKS), which would assign necessary equipment and facilities and send cross-connect instructions to frame technicians. It could also send an order to the Plug-In Inventory Control System (PICS), which could print a shipping order for plug-ins; to the Remote Memory Administration System (RMAS), which changes translations in the switching systems; and to the Centralized Automatic Reporting on Trunks (CAROT) system, which arranges for testing the new trunk. The only actions requiring human intervention would be shipping and installing the plug-in equipment and making all frame connections.

Similar direct interactions between other operations systems would allow computers to do all straight-forward, clerical tasks. People would concentrate on those tasks that require judgment, such as creating forecasts, approving expenditures, and isolating failures; and those that require manual dexterity, such as inserting plug-in equipment, installing new equipment, and making cross-connections.

AUTHORS

L. R. Bowyer
J. S. Fleischman
L. P. Hawkins
R. E. Machol
S. K. Stearns
H. R. Westerman
W. J. Zide

15

Operations Planning

15.1 INTRODUCTION

Operations in the Bell System (see Chapter 13) are influenced by many factors, as shown in Figure 15-1. These factors are continuously changing, and the operations they influence must respond to the changes. One way operations are evolving is that, as a result of the introduction of computer-based operations systems, they are becoming more sophisticated. These systems, some of which are described in Chapter 14, can make operations more effective and improve customer service. But the introduction of any operations system affects the related operations. Specifically, an operations system takes over functions that were previously done manually and, in addition, often performs new or different functions that were previously not feasible. Consequently, changes must occur in the functions performed by people and the interactions among people responsible for different functions.

Operations planning is the key to operating effectively in this changing environment. Operations planning ensures that changes in the roles and responsibilities of people are linked to changes in the nature of the telephone business. It further ensures that operations-related functions are assigned to people and systems in ways that realize potentials for greater efficiency and better customer service. As the number of operations systems deployed in the Bell System increases, the role of operations planning becomes more important.

This chapter provides a basic understanding of operations planning and its effect on operations now and in the future. The rest of this section describes early work directed at developing an efficient operations plan for a specific geographic area of one operating company and progress in developing operations plans for all Bell System operations. Subsequent sections describe the contents of operations plans, how the plans

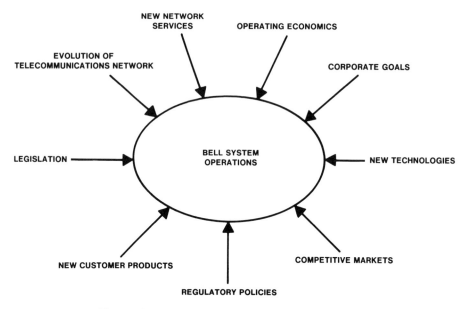

Figure 15-1. Factors affecting Bell System operations.

aid operations in the operating companies, the effect of operations planning on the development of operations systems, and the planning for an *operations systems network* to provide efficient transfer of information between multiple operations systems and between operations systems and a user's computer terminals.

This chapter describes operations planning as it was conducted in the Bell System before divestiture. While operations planning is expected to continue in each of the separated entities, its scope and approach are necessarily changed to comply with the provisions of the 1982 Modification of Final Judgment.

15.1.1 INITIAL OPERATIONS-PLANNING EFFORTS

Planning for specific operations (for example, efficient collection schedules for coin telephones) has gone on for years in the Bell System. The need for planning on a broader scale arose when Bell Laboratories began to apply computer technology to telephone company operations. Around 1970, it was recognized that computer technology would profoundly affect operations even as it made those operations more effective. And, indeed, as the Bell System developed and implemented more and more operations systems, opportunities for new ways of operating emerged. To understand and take advantage of the opportunities, AT&T, Bell Laboratories, and Bell of Pennsylvania undertook a joint study in

1972. The study addressed the question of how to apply new operations systems to the network most effectively in a specific operating area—metropolitan Philadelphia. This was the first time all the functions necessary for network operations were considered together in an overall picture.

The formal result of this study was a plan for integrating and applying operations systems to the network operations expected in Philadelphia in 1980. Reorganization of the work force into operations centers formed a major part of the plan. (An *operations center* consists of a group of people, reporting to a common manager, who perform a set of related functions for a specific geographic area, group of customers, or service.) All the jobs in each kind of operations center were essential for effective functioning of the network. The plan described how the people in the different jobs would work together within and between operations centers to make the best possible use of the capabilities afforded by operations systems.

A significant finding of the study was that integration of the new operations systems and centers and their coordinated deployment and use were essential if the full benefits of computer technology were to be realized for Bell System customers. AT&T informed the Bell operating companies (BOCs) of the study results so that they could perform similar studies for selected parts of their operating areas.

15.1.2 PLANS FOR BELL SYSTEM OPERATIONS

The Philadelphia study had addressed the application of existing or soon-to-be-available operations systems in an actual operations environment. AT&T and Bell Laboratories used the experience gained during the study and the results of the BOC studies to construct a new kind of plan for the Bell System. The goal was to formulate a fundamental, long-range network operations plan generally applicable to the entire Bell System—a plan that would define the evolution of operations systems and centers to meet changing nationwide network operations needs. The new effort was a major step forward in that it aimed at a synthesis of new operational patterns and new operations systems capabilities.

The result, in 1978, of three years of detailed interdisciplinary analysis and planning by AT&T and Bell Laboratories was Issue 1 of the *Total Network Operations Plan* (TNOP). Since then, new issues have been released periodically to reflect advances in telecommunications technology and operations systems technology, new network services, and new operations concepts.

In the late 1970s, to complement the planning for network operations, broad technical planning for customer-services operations was begun. AT&T and Bell Laboratories produced four additional operations plans that were interlocked with the plan for network operations presented in

TNOP. These four plans presented an integrated picture of operations systems and operations centers for customer services.

The plans for network and customer-services operations have become the "road maps" for Bell System operations. AT&T and Bell Laboratories use them to guide the planning and development of operations systems, centers, and work methods and as the basis for detailed studies of future operations system architecture and work force job design.

The BOCs assist AT&T and Bell Laboratories in preparing Bell System operations plans by participating in field studies, periodic reviews, resource-sharing programs, and Bell System—wide conferences. The operating companies then use the Bell System plans as guides in formulating localized operations plans for the configuration of centers and systems in their specific geographic areas.

Thus, operations planning has become a coordinated, disciplined procedure involving close cooperation between AT&T, Bell Laboratories, and the BOCs.

15.2 ELEMENTS OF BELL SYSTEM OPERATIONS PLANS

Bell System operations plans include functional descriptions of operations centers and operations systems and show, by means of operations processes, how they interact to perform various operations. *Operations processes* are detailed accounts of the sequential roles of centers and systems in producing a specific result. The Trunk Addition Process in TNOP, for instance, describes how systems and centers interact in adding a trunk to the telecommunications network.

Operations plans are generally presented as both short-term and long-term views. The short-term view describes operations 1 to 2 years into the future and includes only existing operations systems. It is useful for guiding the development of detailed procedures for personnel in operations centers and for providing a transition strategy toward operations in the long-term view. The long-term view, representing a "target" plan, usually describes operations 4 to 7 years into the future. It includes the directions in which operations centers are planned to evolve to meet future needs, the changes planned for established operations systems through new software program modules, and the proposed functions of new operations systems currently being defined.

The approach used to formulate Bell System operations plans, which is shown in Figure 15-2, consists of the following five major steps:

1) Identify all essential processes within the scope of the plan and the result of each process. For TNOP, this includes all processes necessary to provision, administer, and maintain the telecommunications network. This first step identifies *what* processes are necessary; the remaining steps in the planning approach are concerned with determining *how* processes achieve desired results.

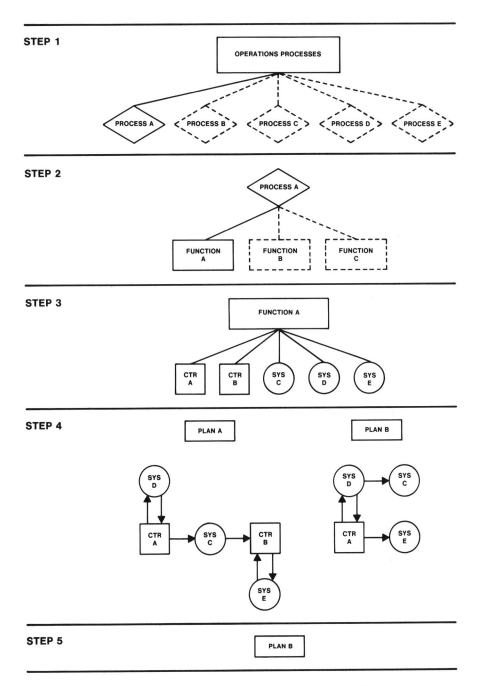

Figure 15-2. Formulation of Bell System operations plans.

2) Break each process down into the major functions that must be performed in order to achieve the desired result. For the Trunk Addition Process mentioned above, major functions include: generating the order for a trunk to be added, designing the trunk, installing the trunk, testing the trunk and putting it into service, and updating all relevant network data bases.

3) For each process, identify operations centers that are candidates to perform the activities involved in each major function and identify operations systems that will provide the mechanization needed to support the centers. Appropriate systems may already exist, or it may be possible to design software enhancements to increase the capabilities of existing systems. Alternatively, the opportunity for mechanization might lead to the development of entirely new operations systems.

4) Generate alternative plans for systems and centers to perform the activities involved in each major function for each process.

5) Evaluate alternatives based on such factors as operational viability, commonality between processes, and economics and select a plan.

Bell System operations plans are usually reviewed annually to take into account new technical concepts and current corporate policies.

15.2.1 OPERATIONS CENTERS

As stated earlier, an *operations center* is a group of people, reporting to a common manager, who perform a set of related functions for a specific geographic area, group of customers, or service. Centralizing related functions in operations centers takes full advantage of the power of operations systems and has the added benefit of presenting opportunities to improve the quality of the work people do. An example is the centralization of trunk testing in the Switching Control Center, which is made possible by the deployment of the Centralized Automatic Reporting on Trunks (CAROT) system. CAROT is an operations system that tests trunks at electromechanical and electronic switching systems and sends its findings to a remote computer terminal. A technician at that terminal is relieved of the tedious, time-consuming task of manually testing trunks and can concentrate on the challenge of problem analysis and solution. Centralizing these terminals at a Switching Control Center (see Figure 15-3), to which data from several switching offices are routed, gives a technician frequent opportunities to handle varied and interesting problems. The technician also has access to several computerized tools to help in problem analysis and correction.

Bell System operations plans describe more than eighty types of operations centers. The description of each type includes the functions

Figure 15-3. A typical Switching Control Center. One of New York Telephone's Switching Control Centers is the hub of local switching office surveillance, testing, and maintenance for over a dozen central offices. Switching equipment technicians are assisted in their work by computer-based operations systems.

performed in the center, the operations systems used, the data bases needed (such as trunk circuit layout records in the Circuit Provision Center), and the other operations centers with which the center exchanges information.

15.2.2 OPERATIONS SYSTEMS

As defined in Chapter 14, *operations systems* are computer-based tools that Bell System employees use in performing many operations activities. The support functions provided by well over 100 operations systems may be reflected in one Bell System operations plan such as TNOP. Some of the systems are fully developed and available to the BOCs within the short-term view presented in the plans. Others may be in initial development but are expected to be available during the long-term view. The long-term view also describes enhancements to currently available operations systems that enable them to support future network services, telecommunications technology, or operating methods.

The long-term view may include conceptual operations systems. A *conceptual operations system* is a set of closely related functions that are candidates for mechanization. The capabilities assigned to a conceptual system may be implemented in a variety of ways: A new stand-alone operations system may be developed; the functions may be incorporated as enhancements to existing systems, or some of the functions may remain manual if detailed economic analyses in the future do not justify the development of mechanization. The conceptual systems provide the basis for further study by suggesting how functions might be mechanized and integrated into overall operations processes.

15.2.3 OPERATIONS PROCESSES

Telephone company operations are described in the form of *operations processes*. As mentioned earlier, identifying operations processes is the first step in formulating a Bell System operations plan. As plan formulation continues, each process is further described until it is in the form of a detailed account of the sequential roles of centers and systems in producing a specific result.

Bell System operations plans describe about eighty operations processes. Each process is depicted in a diagram accompanied by a narrative explanation of the work flow, time sequence, and logical structure of the process. In the diagram, the process is typically broken down into twenty to forty activities assigned to systems and centers and the interfaces between the systems and centers.

Figure 15-4 illustrates a relatively simple process, the Residence Customer Billing Inquiry Process for 1985.[1] The figure shows the sequence of centers and systems involved in responding to questions from residential customers about their telephone bills. Other operations processes describe the maintenance of an electronic switching office, the installation of complex equipment for a business customer, and the maintenance of a specific network service. By associating required functions with operations systems and centers, processes such as these establish functional requirements for operations systems (see Section 15.4) and provide the BOCs with guidelines for their own operations (see Section 15.3).

15.2.4 MODEL AREAS AND MODEL COMPANIES

Because of their scope, Bell System operations plans cannot address the special circumstances of each BOC. Fortunately, however, many telephone company operating areas throughout the Bell System have similar characteristics (customer population density and number of telephone lines, for example) and can be divided into groups on that basis. By using a few model areas and companies to represent these groups, the Bell System operations plans provide some practical examples for the BOCs to use in formulating their own operations plans.

For example, one model area is a large, densely populated metropolitan region with about 2 million telephone lines (for example, Chicago, Cleveland, Detroit). Another is a large, primarily rural state having no metropolitan area with more than 200,000 telephone lines and including large sections served by independent telephone companies (for example,

---

[1] This process and others will change as the result of judicial and legislative actions.

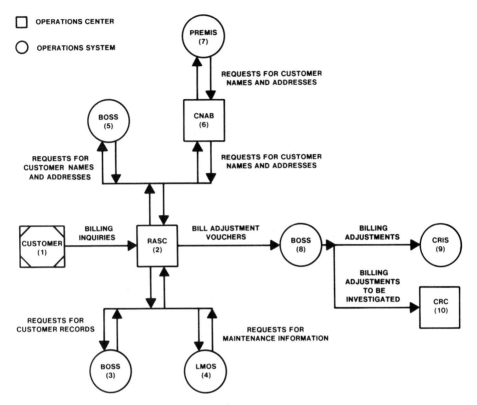

OPERATIONS CENTER

OPERATIONS SYSTEM

PREMIS (7)

REQUESTS FOR CUSTOMER
NAMES AND ADDRESSES

BOSS (5)

CNAB (6)

REQUESTS FOR
CUSTOMER NAMES
AND ADDRESSES

REQUESTS FOR CUSTOMER
NAMES AND ADDRESSES

CUSTOMER (1)

BILLING
INQUIRIES

RASC (2)

BILL ADJUSTMENT
VOUCHERS

BOSS (8)

BILLING
ADJUSTMENTS

CRIS (9)

BILLING
ADJUSTMENTS
TO BE
INVESTIGATED

CRC (10)

REQUESTS FOR
CUSTOMER RECORDS

REQUESTS FOR
MAINTENANCE INFORMATION

BOSS (3)

LMOS (4)

Figure 15-4. Residence Customer Billing Inquiry Process—1985. The billing inquiry process begins when a customer (1) calls the Residence Account Service Center (RASC) (2) with a question about a bill. The customer remains on the line while the telephone company employee in the RASC accesses customer records from the Billing and Order Support System (BOSS) (3), maintenance outages from the Loop Maintenance Operations System (LMOS) (4), and customer name and address from BOSS (5) or from the Customer Name and Address Bureau (CNAB) (6). CNAB is contacted for name and address information for outlying geographic areas not covered by the BOSS data base and accesses the Premises Information System (PREMIS) (7) for this information.

If bill adjustment is required, the employee in the RASC accesses BOSS (8) to generate an adjustment voucher. BOSS sends the voucher to the Customer Records Information System (CRIS) (9), which updates the customer-billing data base. If bill adjustments must be investigated, BOSS routes them to the Customer Record Center (CRC) (10).

When the inquiry process is complete, the telephone company employee answers the customer's question or indicates the need for further investigation based on process results.

Wyoming, Montana, South Dakota). A third model area describes states or portions of large states with moderately sized urban centers (Georgia or Illinois excluding Chicago are examples). New York City, with over 4 million telephone lines, is treated separately. In addition, the basic model areas can be grouped to form model companies.

The model areas and companies serve as prototypes for ongoing operations planning in each BOC by suggesting how operations centers and systems can be deployed in different environments. Bell Laboratories operations planners use the model areas and companies as the basis for economic and operational studies analyzing alternative configurations of operations systems. The models are also the basis for estimating data traffic for the planning of the future operations systems network (see Section 15.5) that will enhance communication capabilities among the centers and systems.

15.3 IMPACT OF OPERATIONS PLANS ON BELL OPERATING COMPANIES

15.3.1 IMPROVING OPERATIONS

The capabilities and features of switching and transmission systems are continuously evolving to provide new customer services and increased operational efficiency. Operations plans guide Bell Laboratories in coordinating the operations system enhancements necessary to support this evolution.

With the centralization of operations system enhancement at Bell Laboratories, operations systems are kept compatible with rapidly changing telecommunications equipment technologies on a Bell System—wide basis. In turn, the BOCs can keep up with technological change in a timely and efficient manner because they do not have to spend time and money individually arranging for enhancements. Most enhancements are software program changes provided by Bell Laboratories. For instance, when a switching system whose trunks are tested by the Centralized Automatic Reporting on Trunks (CAROT) system is modified, Bell Laboratories only needs to develop a new software program for CAROT. The software is thoroughly tested by Bell Laboratories and then made available to the BOCs. The program is then installed in existing trunk-testing computers across the country, providing quick and economic deployment of new capabilities.

By keeping the family of operations systems in step with network innovations, Bell Laboratories also enables the BOCs to maintain stability and consistency in their operations—a technician uses CAROT to conduct a trunk test in the same way regardless of the vintage of equipment being tested. Thus, it is possible to provide a continuous, consistent user-computer interface in a changing environment.

15.3.2 OPERATING NEW NETWORK SERVICES

A major application of Bell System operations plans is in planning for the implementation and operation of new network services. The set of operations processes in the plans is the standard with which operations for new services must be compatible, so that operation of a new service will be consistent throughout the country.

Automated Calling Card Service (see Section 2.5.1), a nationwide service introduced in the early 1980s, is an example. Customers who subscribe to the service can dial calling card calls, collect calls, and calls billed to a third number (that is, calls billed to a number other than the calling or the called number) without operator assistance. The service is offered by each BOC, yet a customer need not be within the geographic area of the customer's "home" BOC to use the service. Each customer is assigned a calling card number that is stored in one of several data bases across the country. The data bases are maintained by AT&T Long Lines as part of the stored-program control/common-channel interoffice signaling[2] network. Regardless of the area from which a customer places a call, there is an accessible data base that can be checked to validate the customer's calling card number.

The nationwide scope of this service poses special maintenance problems. What happens if a salesperson who lives in San Francisco is on a business trip in Cleveland and has difficulty completing a call using Automated Calling Card Service? Whom does the person call for help: the operator, the local BOC business office in Cleveland, the local telephone repair service bureau found in the front of the telephone directory, or the person's "home" telephone company in California? How does the BOC employee receiving the call respond? Operations planners must consider these and many other questions in advance. The planners must provide directions for BOC employees, plans for the operation of supporting systems and centers, and guidelines for customer education.

Before any new service is formally offered, a maintenance plan is devised and tested. Its objective is to establish a uniform set of maintenance processes nationwide so that, when the service is implemented, it operates smoothly and consistently throughout the country. For example, a customer using the Automated Calling Card Service follows the same simple procedure for reporting trouble in all areas of the country. In addition, the BOC employee in Cleveland who receives the trouble call follows the same procedures for resolving that trouble as BOC employees in San Francisco, even though the source of the problem may be in a data base owned by the Pacific Telephone and Telegraph Company.

Furthermore, these maintenance functions are incorporated into existing operations centers and systems within the framework set by the Bell System operations plans. For Automated Calling Card Service, maintenance procedures for the already established Automatic Intercept System (see Section 10.4.2) were used as the basis for the new service.

Planning before new services are introduced helps ensure that operations for new services will fit into the well-established, stable operations picture and will be integrated with operations for other services with

---

[2] Sections 8.4 and 8.5 discuss common-channel interoffice signaling.

similar maintenance requirements. Thus, the effect of new services on current operations is reduced, and services can be made available to customers more quickly than if each operating company had to plan individually for the operation of a new service.

15.3.3 OPERATIONS PLANNING IN THE BOCS

In addition to the operations-planning activity at AT&T and Bell Laboratories, each BOC is involved in its own operations planning. In contrast to the planning at AT&T and Bell Laboratories, which is primarily concerned with the future structure and configuration of operations systems and centers, planning in the BOCs focuses on adapting the concepts presented in the Bell System operations plans to their own situations.

The major emphases of BOC operations planning are:

- determining the present status of operations

- evaluating the economics and the operational feasibility of alternative ways of implementing operations systems and centers

- budgeting capital and expense commitments

- developing transition plans to achieve the objectives of the Bell System operations plans

- overseeing short-term planning and implementation of operations systems, centers, and processes

- providing input to the formulation of the Bell System plans based on BOC experience.

In the BOCs, the operations-planning process for individual projects, such as the deployment of a new center or system, generally consists of four phases (see Figure 15-5):

1) strategic, or long-term, planning

2) tactical, or short-term, planning

3) implementation

4) ongoing support.

Strategic, or Long-Term, Planning

This phase answers the question: What should be implemented? During long-term planning, the BOCs adapt corporate goals and objectives, such as those presented in Bell System operations plans and the Bell System Corporate Plan,[3] to their local operations environment.

---

[3] The Bell System Corporate Plan sets objectives for the BOCs in the areas of service, investment, expense, work force, the deployment of operations systems and centers, and the use of administrative programs.

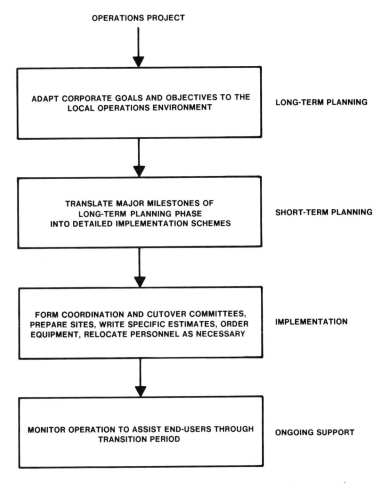

OPERATIONS PROJECT

ADAPT CORPORATE GOALS AND OBJECTIVES TO THE LOCAL OPERATIONS ENVIRONMENT — LONG-TERM PLANNING

TRANSLATE MAJOR MILESTONES OF LONG-TERM PLANNING PHASE INTO DETAILED IMPLEMENTATION SCHEMES — SHORT-TERM PLANNING

FORM COORDINATION AND CUTOVER COMMITTEES, PREPARE SITES, WRITE SPECIFIC ESTIMATES, ORDER EQUIPMENT, RELOCATE PERSONNEL AS NECESSARY — IMPLEMENTATION

MONITOR OPERATION TO ASSIST END-USERS THROUGH TRANSITION PERIOD — ONGOING SUPPORT

Figure 15-5. Operations planning in the Bell operating companies.

Before conducting long-term planning studies, BOC operations planners must prepare an inventory of the present status of operations in their company. This is especially critical because most operations projects will affect many facets of existing operations. The present status includes the extent of deployment of operations systems and operations centers, the company's current organizational structure for operations, the current distribution of functions among operations centers, and basic data on the telecommunications network within the company. This present-status inventory provides information for all future operations planning.

Using the present-status inventory as a basis, various organizations within a BOC initiate proposals for new projects or modifications to existing projects. These proposals may involve, for example, implementing a new or enhanced operations center or system recommended in a Bell System operations plan; instituting a new way of maintaining new network equipment or service offerings; or adding data-switching capability to the operations systems network (see Section 15.5).

Operations planners screen proposals and conduct evaluation studies 2 to 6 years before proposed deployment of a project. The studies generate alternatives, and operations planners select a course of action based on economic, operational, and technical considerations. Often, the planners must coordinate their activities with corporate plans for the telecommunications network to ensure that proper operations systems and centers are deployed to support growth or replacement of telecommunications equipment. In fact, operations planners may initiate plans for the deployment of some telecommunications equipment that may have significant operational benefits, such as the Digital Access and Cross-Connect System (see Section 9.4.3).

The course of action selected includes a transition strategy with major milestones specified for the tactical-planning and implementation phases of the project. Once this course of action is approved by BOC management and the financial requirements are met by BOC expense and construction budgets, the project proceeds to the tactical-planning phase.

Tactical, or Short-Term, Planning

This phase addresses the question: How should the project be implemented? It typically occurs up to 2 years before project deployment. During tactical planning, the objectives and major milestones of the long-term planning phase are translated into detailed implementation schemes. These schemes may include number and location of proposed operations systems and centers, personnel and operational transition plans, and year-by-year deployment schedules.

A detailed economic study may be conducted to supplement the evaluation from the long-term planning phase. This study includes final size and cost relative to physical layout, hardware, software, and telecommunications equipment required. Necessary methods of operation, training, and performance measurement plans are also provided. During this phase, BOC planners may actively consult with project managers from AT&T and Bell Laboratories on technical matters.

Implementation

This phase proceeds according to the specifications resulting from tactical planning. Typically, it includes forming coordination and cutover committees,[4] preparing sites, writing specific estimates, ordering equipment,

---

[4] Cutover committees are responsible for planning and executing actions required to effect the actual change (*cutover*) to use of a new system or system configuration.

and relocating personnel as necessary. The implementation team may include representatives from engineering, information systems, and end-user[5] organizations. The equipment supplier or purchasing agent may become actively involved with the details of equipment installation.

Ongoing Support

Once the project is implemented and acceptance testing is complete, the end-users assume responsibility for the ongoing operation. For a new operations system or center, BOC operations planners—sometimes with the assistance of Bell Laboratories and AT&T planners—may monitor the operation to assist the end-users through the transition period.

15.4 IMPACT OF OPERATIONS PLANS ON THE DEVELOPMENT OF OPERATIONS SYSTEMS

Bell System operations plans guide Bell Laboratories in evolving the whole family of operations systems in concert. Proposals for new operations systems or enhancements to existing systems are viewed in terms of their contribution to the overall plan. The objective is to maximize the contribution of the whole set of standard operations systems by ordering the assignment of functions among related systems.

The Engineering and Administrative Data Acquisition System (EADAS)[6] is an example of how this philosophy is applied. EADAS was originally designed to supply traffic data on an hourly, daily, and monthly basis for long-term traffic engineering for the network. Bell Laboratories modified the system to supply traffic data at 30-second and 5-minute intervals for short-term network management purposes, and an interface was built to the Network Management operations system (called EADAS/NM). If EADAS had been viewed narrowly from the perspective of its original purpose, its potential adaptability for short-term use might not have been recognized. But, from the perspective of overall network operations planning, modification of this traffic data collection system was the most effective way to provide certain network management capabilities.

Bell Laboratories has adopted a structured approach to planning the integration of operations systems. To ensure that each operations system

---

[5] The end-users are the personnel constituting the operations centers supported by the operations system being implemented.

[6] EADAS is part of the Total Network Data System (TNDS) described in Section 14.3.1.

is developed and modified to fit the overall Bell System plan, planning is required at several levels. These levels include:

- defining design requirements for individual operations systems and allocating functions between operations systems and telecommunications systems

- grouping operations systems into specific operations centers, including defining the roles of proposed or existing systems in order to provide efficient support for the operation of a center

- specifying the interactions between groupings of related centers and the operations systems with which they interact

- integrating the entire family of operations centers and systems into the Bell System operations plans.

15.5 OPERATIONS SYSTEMS NETWORK PLANNING

The number of operations system computers (including minicomputers and large mainframe computers) in use throughout the Bell System in 1981 was estimated at about 4500. These computers are accessed by about 114,000 computer terminals dispersed among the operations centers. These numbers are expected to increase to about 6300 computers and 180,000 terminals by 1985 as implementation of operations plans matures.

To provide more efficient transfer of information among operations systems and between operations systems and terminals in operations centers, Bell Laboratories is planning the operations systems network illustrated in Figure 15-6. When fully deployed in the mid-1980s, this network will provide a flexible, cost-effective national network of operations systems, computer terminals, and data communications capabilities. It will carry the over 2.5 trillion characters of information required each month to operate the nationwide telecommunications network. In fact, it can be viewed as a "second network," parallel to the telecommunications network whose operation it supports.

The operations systems network will provide many new benefits to nationwide network operations. For example, network operations tasks often require Bell System technicians to access several computer systems. To repair a circuit, a technician may first need to access a record of the circuit from TIRKS, the Trunks Integrated Records Keeping System (see Section 14.2.1), and then perform a test using the CAROT system mentioned earlier in this chapter. With the operations systems network, the technician will be able to access these systems and others, as needed, from a single computer terminal. Bell Laboratories is in the process of deploying this capability using a data communications system called the *Bell Administrative Network Communications System*, designed as part of the operations systems network plan.

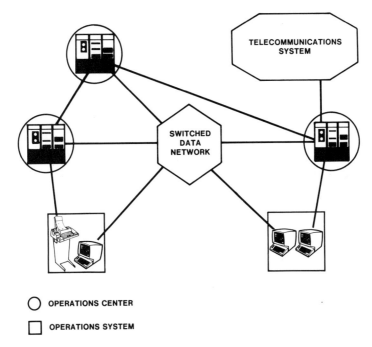

OPERATIONS CENTER

OPERATIONS SYSTEM

Figure 15-6. Operations systems network. The operations systems network includes the collection of operations systems, communications terminals at operations centers, and the switched and direct communications links interconnecting them and connecting them to a variety of telecommunications systems, such as an electronic switching system.

In the future, the flexibility afforded by the operations systems network in rapidly switching data between operations systems computers and terminals will permit the Bell System to take advantage of advances in distributed-processing technology. It will be possible for many different operations systems to share data bases to a much greater extent than is possible today. The result will be even more economical computer support, using more timely and accurate data.

Planning is also underway to incorporate "intelligent terminals" (computer terminals with self-contained processing capabilities) into the operations systems network. With the integration of this technology, it will be possible to shift much of the processing done today in centralized computer systems to a technician's terminal. This will make more efficient use of computer resources and the computer communications network and will provide even more rapid response to a technician's inquiries. This new technology will also allow individual technicians to customize computer displays into a form that they find personally convenient, both increasing the technicians' effectiveness and making the human-machine interface more personal.

15.6 SUMMARY

Operations in the Bell System are becoming more sophisticated through the increasing use of advanced computer technology. Operations must respond to a rapidly changing environment as the result of the evolution of the telecommunications network as it embraces new technologies, the development and marketing of new network services and customer products, and competitive and regulatory forces. All these factors change the way operations are performed.

Operations planning is a relatively new and expanding cooperative effort among Bell Laboratories, AT&T, and the Bell operating companies to guide the evolution of operations in the Bell System. Centralized planning for the entire Bell System ensures a consistent, high standard of telecommunications services throughout the nation and the rapid introduction of new services and technologies into the network. The operations systems network will allow an even more effective use of operations systems in the mid-1980s and beyond.

AUTHOR

B. A. Newman

16

Evaluation of Service and Performance

16.1 THE SERVICE EVALUATION CONCEPT

When purchasing a new car, a customer considers not only price but the quality of the car as a whole as indicated by mileage, warranties, safety reports, and repair costs. Similarly, communications customers are concerned not only with their monthly bills, but increasingly, with the availability and suitability of their telephone service. To basic voice residence customers, quality may simply mean that they are nearly always able to place calls as desired, that they are able to hear and be heard well on a given connection, and that they receive appropriate aid from the telephone company when they have a problem. To the sophisticated business customer, quality may mean the ability to transmit error-free data or to transmit clear images over an electronic blackboard or other visual telephone service. Acceptable quality, then, depends on the degree to which the delivered service satisfies customer expectations. Service evaluation, thus, becomes a matter of

1) characterizing what customers expect of a service and what can be provided, based on current network performance

2) setting service and performance objectives based on these expectations

3) assessing conformance to these objectives

4) taking action to modify (usually improve) service or to re-examine service and performance characteristics and objectives, if objectives are not being met.

The service evaluation process is shown in Figure 16-1 and described
briefly below.

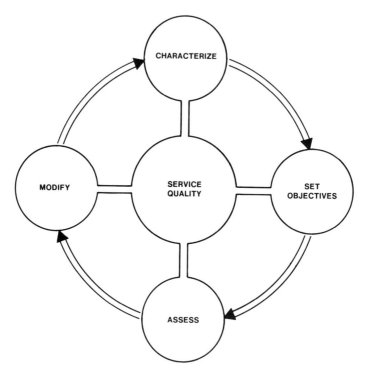

Figure 16-1. The service evaluation process.

In characterizing what is expected of a service, the first step is to identify which service characteristics most affect the opinions and actions of
the people using the service and the sensitivity of customers to different
values of each characteristic. For example, the customer's satisfaction
with a given connection for basic voice service is largely determined by
the amounts of loss, noise, and echo[1] present on the connection. The
next step is to relate the relevant service characteristics to physical characteristics of the network, that is, the network performance parameters that
affect the service characteristics. Some relationships between service
characteristics and network performance parameters are direct, but others
are not. After the appropriate performance parameters have been determined, the levels of these parameters in the network are measured.

[1] Section 6.6.1 discusses these transmission impairments.

Network performance characterization studies have typically been conducted jointly by Bell Laboratories with AT&T and the operating companies. Section 16.2 discusses part of the characterization process—the measurement of performance parameters—in more detail.

The network characterization process feeds the process of setting objectives by providing information on service and performance characteristics that are required to meet customer expectations. Marketing studies may also provide information on customer expectations and needs, particularly in the case of a new service. It is not difficult to see, however, that as quality increases, costs also usually increase. This is acceptable only as long as the increased cost is offset by an increased net value to the customer of the product for which the customer is willing to pay an increased price. An important element in setting service objectives and performance objectives, then, is striking a balance between the best possible service and the minimum possible cost. Section 16.3 describes in detail the process of setting performance objectives.

Section 16.4 discusses various means by which the Bell System has routinely assessed conformance to its service and performance objectives. Since people and equipment interact to deliver service to the customer, service and performance measurements have ranged from customer opinion surveys to employee evaluations to direct measurement of network and component operation. In addition to helping assess service and performance, the measurements themselves indicate strategies that might be used to correct certain problems, thus providing for feedback in the service evaluation process.

Service evaluation is, thus, a dynamic, iterative process and must be flexible enough to respond to new service definitions, the introduction of new technology, and changing customer expectations.

16.2 NETWORK CHARACTERIZATION

As indicated in Section 16.1, network characterization involves several steps that describe, analyze, and measure relevant customer-perceivable service characteristics and related network performance parameters. This section discusses the rationale for network characterization and focuses on the process of measuring network performance parameters.

16.2.1 THE NEED FOR NETWORK CHARACTERIZATION STUDIES

Traditionally, Bell Laboratories has conducted field measurement studies of performance on existing network services. The results have been used to set performance objectives, to determine whether performance is adequate for existing and proposed new services, and to indicate where changes in network design or objectives could lead to improved performance or lower costs. While these traditional reasons remain, two new forces make the need for characterization more critical:

- *Rapid growth of new services and technologies* — Rapid advances in switching, transmission, and computing technology have spurred development of a host of new voice, data, and video communications services. During field trials and early introductions of new network services and technologies, field characterization studies provide the data needed to set realistic performance objectives and supply developers and systems engineers with essential feedback. Examples of new services are circuit-switched digital capability (see Section 2.5.1) and teleconferencing (see Sections 2.5.6 and 11.5.3). New technologies include low bit-rate voice and packet-switching networks (see Section 5.8).

- *New network structure* — The 1982 Modification of Final Judgment[2] effectively mandates the restructuring of the Bell System network into a set of separate components, or "piece-parts," connected together at well-defined interfaces to provide end-to-end service. Examples of components are the networks provided by interexchange carriers and the access and intraexchange networks provided by Bell operating companies. The components will be individually specified, administered, and sold. They will each require separate performance characterization so that those who file tariffs and maintain the components and those who assemble services from the components can enter into meaningful, performance-based contract negotiations.

  For example, an interexchange carrier wishing to provide telephone service to end-users may pay the local operating company for access to the local network in order to provide a path from the customers' premises to the long-distance access point. In return for access charges, the interexchange carrier would require appropriate access performance levels from the local telephone companies to ensure high-quality end-to-end service. The telephone company would have responsibility for maintaining the agreed-upon performance levels.

16.2.2 THE RESULTS OF MODERN CHARACTERIZATION STUDIES

As shown in Figure 16-2, network *performance planners* study performance issues related to new services and technologies and network restructuring, initiate characterization studies as appropriate, and identify key parameters (for example, call-setup delay and transmission-related

---

[2] See Sections 1.3.1 and 17.4.4.

parameters) to be characterized. *Characterization planners* apply advanced measurement technology, statistical sampling methods, and data analysis techniques to the specified parameters in field characterization studies. As indicated in Figure 16-2, the primary outputs from the studies are

- *Performance-related documents* — Statistical performance descriptions obtained in traditional characterization studies have appeared in articles in technical journals (such as *The Bell System Technical Journal*), public reference documents, and internal Bell System Practices (BSPs). To meet the needs described in Section 16.2.1, current characterization

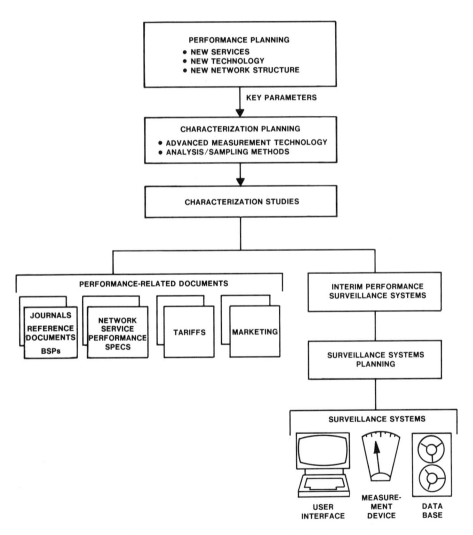

Figure 16-2. Overview of network characterization process.

studies also supply information for internal network service performance specifications, tariff preparation, and marketing campaigns and
literature. These new applications for performance information are
especially important in the newly competitive arena for telecommunications services.

- *Interim performance surveillance systems* — After its design and initial
  use in characterization studies, a new performance measurement system may serve as an interim performance surveillance system for a
  service. In the first year or two of the new service offering, the
  interim system provides the performance information needed to
  ensure that published performance specifications are being met.
  Meanwhile, experience with the interim system is used in the
  development of the surveillance system that will ultimately replace
  the interim system.

16.2.3 MECHANIZING PERFORMANCE CHARACTERIZATION STUDIES

Traditionally, a separate measurement system has been designed for each
characterization study, used once, and then dismantled. Technological
advances and the rapid growth in the number of new services has made
it both possible and necessary to develop generic, reusable measurement
capabilities. Figure 16-3 shows one such system, the Automatic System
for Performance Evaluation of the Network (ASPEN), in its first
application—a comprehensive characterization of end office—to—end
office transmission performance in the Bell System network.

The key elements of the ASPEN system are remote test modules
(RTMs) and a central computer. The RTMs can be moved to different
locations from which they act as "robot customers." In the End-Office
Connection Study illustrated in the figure, RTMs were placed at selected
end offices throughout the country. Under the control of the central
computer, the test nodes placed calls to one another at all hours and
made over twenty-five measurements, mostly transmission-related, during
each connection. (Amplitude and phase response versus frequency,
noise, hits, and call-cutoff rate[3] are examples of the parameters measured.)

After each call, the test nodes sent the measurement data to the central computer to be screened, stored, and analyzed. For this study,
central-computer control programs were run using the *UNIX*[4] operating

---

[3] The probability that a call in progress will be interrupted or terminated other than by
actions of the calling or called party. Section 6.6 discusses amplitude and phase response
versus frequency, noise, and hits.

[4] Trademark of Bell Laboratories.

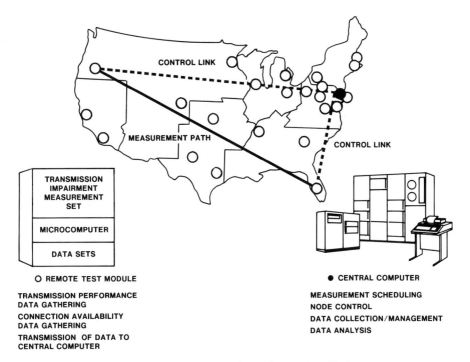

Figure 16-3. ASPEN End-Office Connection Study.

system, and measurements were stored using POLARIS, a data-base management system originated at New York Telephone Company. The principal data analysis tool used was the "S" Data Analysis System developed at Bell Laboratories.

Variations of the original ASPEN design are being applied in new-service performance characterization studies for which the robot customer structure is appropriate.

16.3 SETTING PERFORMANCE OBJECTIVES

16.3.1 OVERVIEW

Performance objectives must ensure satisfactory levels of service to customers with disparate needs, while being cost-effective for the service provider. Moreover, customers' needs and perceptions with respect to service change with time. These factors dictate that setting objectives for the performance of elements of the telephone network be an iterative process. The major steps in this process are:

1) dividing the network into components of manageable size and describing and quantifying the operation or performance of each component using mathematical models based on measurements made during the characterization process

2) determining user opinion and acceptance levels by subjective testing (or other laboratory testing if the "users" are equipment such as data sets)

3) determining grade-of-service distributions

4) formulating performance objectives.

Iterations in this objective-setting process are generally stimulated by the availability of new technology, which leads to proposals for new services or cost-saving changes to the network. A proposal is first evaluated by sensitivity studies of models of the existing network. These studies develop estimates of operation or performance effects of the proposal, the impact on service, and the cost of achieving those effects. Given acceptable results, a trial consisting of a small-scale application of the proposal may be conducted. Here, grade-of-service ratings are determined and performance objectives are formulated. If the results of the trial application are satisfactory, the proposal may be implemented throughout the system. Once the changes have been incorporated in the telephone network, performance or operational measures are again characterized, mathematical models are refined, new grade-of-service ratings are determined, and objectives are reformulated.

16.3.2 CREATING PERFORMANCE MODELS

Performance models facilitate analysis of proposals for new services or new network technology because they are more tractable and less expensive than field trials for evaluating alternatives. Generally, mathematical models allow better understanding and more precise definitions of processes and the characteristics of facilities than do word descriptions. The models must be sufficiently detailed to account for all factors that significantly affect or contribute to the performance of the telephone network. When general models do not exist (for example, for new services under consideration as public offerings), they may be created as described in the following paragraphs.

The first step is to specify the nature and features of the proposed service or technology, the facilities that will be involved, the performance parameters that must be controlled in order to render satisfactory service, and the characteristics of the geographical area where it would be implemented. The next step is to collect all information pertinent to the cause-and-effect relationships among these factors. This step may include searching sources (such as Bell Laboratories new-system design information), referring to operating telephone company measurements of facilities, and if necessary, planning and conducting a new characterization study. Mathematical expressions are then generated for each of the parameters under consideration. These expressions are defined in statistical terms based on standard techniques of data analysis, for instance, correlation analyses relating performance parameters and facility descriptors. The set of mathematical expressions derived constitutes the mathematical model.

Preparing a performance model for transmission parameters affecting voice telephone service (loss, noise, and echo) in the public switched telephone network (PSTN) provides an example of this process. A call using the PSTN involves the telephone sets on the customers' premises, the communication paths (the loops) to local class 5 offices, and a number of interconnecting links (interoffice trunks) and switching systems.[5] The actual route chosen for a particular call through the network could be one of several possible combinations of links. Surveys indicate that transmission parameters are determined primarily by trunk length; therefore, a performance model for transmission parameters requires a way of expressing the expected route length for a call and combining that with the relationship between trunk length and each parameter.

Computer simulation is used to generate routing data weighted by the probability of occurrence of a given route. These routing data provide information as to the number of trunks and the length in airline miles of each trunk in a connection between two class 5 offices. Transmission characteristics are then derived by combining this information with trunk statistics based on system-wide measurements of transmission parameters.

16.3.3 ASSEMBLING CUSTOMER OPINION MODELS

Customer opinions and perceptions of the quality of a telecommunications service influence the demand for the service. Given the increasing role of competition in telecommunications, customer perceptions may also influence the customer's choice among alternative services. Therefore, it is important to appraise customer satisfaction with offered services.

The models of customer opinions quantify subjective opinion of performance conditions by discrete ratings (for example, excellent, good, fair, poor, unsatisfactory). These ratings, which are obtained primarily from

---

[5] The toll network is currently designed in a hierarchical structure as described in Section 4.2.1.

subjective tests, are first tabulated and then converted to mathematical expressions of the form:

$$P(R|X).$$

In quantifying customer satisfaction with the transmission quality of a telephone call, for example, $P(R|X)$ is the conditional probability that a customer will rate a call in opinion category R, given a value X of the stimulus. This function can be estimated from measurement of the responses of groups of subjects to various controlled levels of the given stimulus (noise, loss, echo, etc.).

As Cavanaugh, Hatch, and Sullivan (1976) found, opinions depend on various factors such as the subject group, the time, and the test environment (that is, whether the test was conducted during conversations in a laboratory or during normal telephone calls). Furthermore, the effects of various stimuli may be interdependent. In testing the effects of noise and loss, for example, both the noise and loss stimuli must be varied so that different combinations of noise and loss can be observed.[6] However, by taking into account the type and size of the sample and controlling for the influences of other pertinent factors and stimuli, customer opinions that reflect the distribution of telephone users can be tabulated.

The techniques described above have been used to prepare customer opinion models for transmission parameters (loss, noise, echo, etc.) and call-setup parameters (dial-tone delay, postdialing delay, etc.)[7] for voice communications. Similar techniques are now being applied to data transmission and graphics services.

The diagrams in Figure 16-4 show how variations in loss, noise, and echo in transmission paths affect customer opinion about the quality of transmission. In the upper diagram, the contours represent the percentage of laboratory subjects who rate simulated telephone service as "good or better" for varying combinations of values for overall loss and noise. Loss is measured acoustically in decibels (dB) from the input of the talker's transmitter to the output of the listener's receiver. Noise is measured in dBrnC, that is, dB above a reference noise (rn) of 10^{-12} watts, weighted to emphasize noise at frequencies where noise is relatively more annoying (C-message weighting).[8] For example, with a loss of 15 dB and noise of 30 dBrnC (*dashed lines in Figure 16-4*), 75 percent of telephone customers rate the transmission as good or better on a rating scale whose categories are excellent, good, fair, poor, and unsatisfactory.

---

[6] The text in Section 6.6.1 covering message circuit noise gives more information on the effects of noise and loss combinations.

[7] Section 8.3 discusses dial-tone delay and postdialing delay.

[8] Section 6.6.1 discusses C-message weighting.

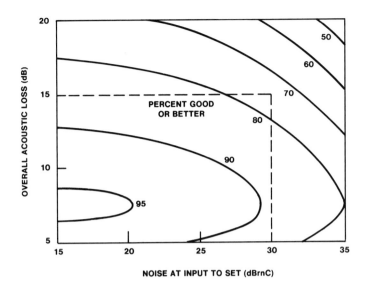

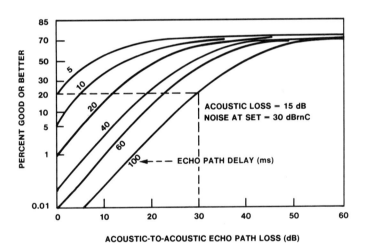

Figure 16-4. Effect of transmission parameters on customer opinion.

If loss and noise for the primary signal path are held constant at the values shown by the dashed lines, opinion is then a function of loss and delay in the echo path (shown in the lower diagram in Figure 16-4). For example, for a 30-dB echo path loss and a 100-millisecond delay in the round-trip echo path, only 20 percent of telephone users find the quality of transmission good or better.

16.3.4 DETERMINING GRADE-OF-SERVICE RATINGS

Mathematical models of performance and models of customer opinion are combined to obtain grade-of-service ratings for specific aspects of telephone service.

Grade of service is defined (for a single parameter, X) by the integral

$$\int_{-\infty}^{+\infty}(R\,|X)f(X)dx,$$

where $P(R\,|X)$ is the conditional probability that a customer will rate a call in category R, given a value X of the stimulus, and $f(X)$ is the probability density function of obtaining that stimulus. (Multiple impairments are handled similarly with multiple integrals and joint density functions.)

The process used to determine grade of service is illustrated in Figure 16-5. Controlled subjective tests provide the needed opinion curves, $P(R\,|X)$, as discussed in Section 16.3.3, and characterization studies, together with a mathematical model such as the one described in Section 16.3.2, provide the required performance distributions, $f(X)$.

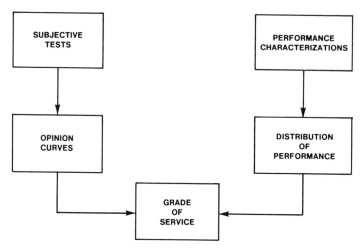

Figure 16-5. Grade-of-service determination.

Grade-of-service ratings are essential criteria used in formulating objectives for various performance parameters. Figure 16-6 shows that as loss in a telephone connection increases, the grade of service for loss and noise (in terms of the number of people who rate transmission as good or better) decreases (*dotted curve*). On the other hand, this same increase in loss increases the echo path loss, thus improving the grade of service for

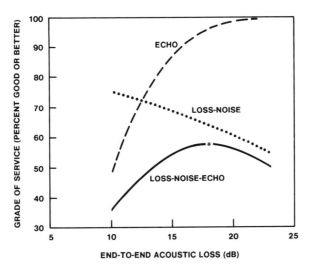

Figure 16-6. Example of loss-noise-echo grade of service
(for connection length of 1270 miles).

echo (*dashed curve*). The result is an optimum value of loss for which the highest values of the combined loss-noise-echo grade of service (*solid curve*) are achieved. The value of optimum loss is a function of connection length because the delay is longer on long connections, and thus, more loss is required to control echo. For the connection length illustrated (1270 airline miles), an end-to-end acoustic loss of approximately 18 dB (*the asterisk in the figure*) results in the optimum loss-noise-echo grade of service.

16.3.5 FORMULATING OBJECTIVES

An overall objective of the Bell System has been to provide high-quality telephone service at low cost. As technology provides new methods and facilities, the quality and ease of use of telephone services should improve, and costs should be reduced. However, as services improve, customers' expectations increase.

An important consideration in formulating objectives, then, has been balancing desired service levels and incremental costs. This requires identifying performance regions of diminishing returns (that is, the points where the increase in customer satisfaction is not commensurate with the increase in dollar investment needed to improve performance) or where savings can be realized without significant reductions in customer satisfaction. Thus, formulating objectives takes into account: (1) the grade of service being provided, including areas where performance is lower or higher than objectives; (2) the costs of providing services,

675

including costs of correcting poor performance or savings possible without degrading performance; (3) trends in customer opinions; and (4) economic trends.

In addition, the end-to-end objectives have been divided and allocated to each component of the physical plant forming the connections (stations sets, loops, switching offices, and trunks) so as to ensure that the overall objectives are met and to minimize differences between local and long-distance calls. These objectives have been applied in preparing the instructions used by operating company personnel in designing new installations and making additions to existing ones. The performance limits developed here are also useful for maintaining the systems in the field.

Examples of parameters for which performance objectives are established include: signaling (dial-tone delay and postdialing delay), availability and reliability (blocking and cutoff), and transmission (loss, noise, and echo for voice quality; bit error rate for data transmission; and graphics quality).

16.4 MEASUREMENT AND CONTROL OF SERVICE AND PERFORMANCE

16.4.1 MEASUREMENTS IN THE BELL SYSTEM

The Bell System has traditionally monitored its own effectiveness through a system of comprehensive measurements. These measurements were designed to assess compliance with preset objectives and to detect unsatisfactory cost, service, and performance levels. Some of the measurements have been applied across all Bell Operating Companies (BOCs) in an attempt to achieve common Bell System goals.

As is the case with most corporations, the Bell System has monitored its financial effectiveness by means of a number of financial measurements. Some of these measurements apply at the corporate, or macro, level—profitability and rates of return are examples. Others, particularly those relating to cost, also apply to the micro level, that is, to organizations in the lower levels of the corporate hierarchy. The Bell System, however, has also used macro and micro measurements extensively to reflect service and performance levels (see Militzer 1980), a rare practice among large corporations.

Service measurements reflect aspects of operations perceivable by the customer. *Performance measurements* reflect whether the intended operation of a Bell System operational unit (for example, an operations center or an operations process) is meeting objectives. Service and performance measurements are combined in the overall process of measuring, assessing, and reporting both the quality of service provided to customers and the efficiency of the system and its associated operations.

In the Bell System, service and performance micro measurements specific to an operational unit have often been combined into one or more measurement plans (sometimes called *indices* or *index plans*). Since 1980, approximately forty major service and performance measurement plans have been in use in the Bell System network segment alone; a similar number serve the other segments.[9] (Section 16.4.5 describes typical Bell System measurement plans.) In addition, suitable aggregations of the micro measurements at the operational unit level are summarized into results applicable to higher levels of Bell System management.

Those planning measurements of service and performance face many challenges. Different classes of customers have different needs and expectations. Business customers usually place a high value on fast reaction time and reliable workmanship in both installation and repair. Residence customers tend to be more tolerant regarding reaction time and more concerned with courteous, helpful service and neat workmanship. Customers located in different geographic regions in the United States can have different levels of expectations. Finally, customer attitudes and expectations toward telecommunications tend to change with time. A generation more acclimated to advanced technology is likely to expect more from the telecommunications network.

In addition, measurements should foster rather than restrict efficiency and productivity. Performance criteria should not be too rigidly determined and, moreover, should be in step with the changing needs and objectives of the Bell System. In the late 1970s, for example, as a part of a greater emphasis on customers' perceptions of service, many measurement plans included customers' subjective reactions. Planners must also cope with the accelerating pace of new technology, which imposes new requirements on the performance of equipment and operations systems. New technology may also permit the measurement of performance parameters that could not be measured previously or the more accurate or convenient measurement of existing parameters. In addition, competition creates the need to re-examine existing performance objectives continually.

In summary, the changing nature of customer expectations and the evolving nature of telecommunications dictate that service and performance measurement planning be dynamic. To meet these challenges, service specifications and performance measurements constantly undergo re-evaluation and modification.

---

[9] Much of the collection and compilation of measurement results is automatically carried out by operations systems.

16.4.2 THE MEASUREMENT OF SERVICE AND PERFORMANCE

Because the Bell System has traditionally stressed the quality of service provided to customers, it has always carefully monitored reports from customers concerning unsatisfactory service. These reports reflect *customer actions* and represent a type of service measurement. When a customer reports a trouble to a repair center, for example, a trouble report is created. Repair efforts aimed at correcting the trouble are then initiated and tracked, and the final disposition recorded. The trouble report and the ensuing effort to rectify the trouble provide the data for a number of micro measurements, both on the nature of customer actions and on the performance of the Bell System's maintenance effort.

Significant attention is also paid to *customer attitudes,* that is, expectations of the customer (see McDade 1979). An example of a plan that measures customer attitudes is the Telephone Service Attitude Measurement Program described in Section 16.4.5.

Service can also be evaluated by measuring the Bell System operations that provide service. From this point of view, service operations can be divided as follows:

- *Installation* — providing what the customer wants. Measurements here include whether the work is done properly and by the date agreed upon.

- *Availability* — ensuring that the service or equipment for which the customer has paid will be ready for use. Dial-tone delay and network blockage are examples of measurements in this category.

- *Suitability* — ensuring that the contracted service meets objectives. Transmission performance measurements are included in this category.

- *Billing Integrity* — ensuring that the customer is not charged an erroneous amount. Measurements in this area include automatic message accounting (AMA)[10] errors.

16.4.3 THE CONTROL OF SERVICE AND PERFORMANCE

Measurements provide visibility into service and performance, but visibility is not enough. Follow-up activities designed to correct substandard service and performance are necessary. Both the set of directed activities

---

[10] Section 10.5 discusses AMA.

(control actions) and the set of methods (control procedures) that direct these activities are encompassed by the term *control.*

Ideally, control is based on several considerations. First, most measurements are observed effects. They are a quantification of symptoms rather than causes. The first stage of control, then, is to identify symptoms and diagnose the cause of a problem. Once the cause is known, which control parameter must be adjusted and to what extent to overcome the problem in a cost-effective way can be determined. Control, therefore, requires information beyond that needed for evaluation: It requires efficient analytical methods for diagnosis, and it further requires procedures for best rectifying the problem.

To facilitate control, many measurement plans in the Bell System have had a multitiered structure. The measurement components making up the official measurement plans are at the top tier. They are used to assess service and performance improvement or degradation. The next lower tier consists of measurements supporting the official ones. These are typically more detailed measurements that are not components of the measurement plans but can furnish diagnostic aid. These measurements are sometimes called *performance indicators,* and they frequently accompany the official measurements. In many recent cases, the supporting measurements have been further supported by computer algorithms and documented procedures to analyze additional data.

16.4.4 THE ADMINISTRATION OF MEASUREMENT RESULTS

In the Bell System, official measurements results have been collected monthly from all the BOCs. As mentioned previously, many of the results extend down to the level of individual work centers. At the same time, most results have been aggregated and apply to higher level management units up to the company level. Measurement results have facilitated comparison among BOCs, and they have been published and widely distributed throughout the Bell System. This section examines the measurement methodology and the mechanized system used by the Bell System for handling measurement results.

The method typically used to present measurement results divides service and performance levels into four bands. The "H" (high) band is applied to cases where service and performance levels are so high that they may not be cost-effective. The "O" (objective) band is applied to levels of service and performance that meet objectives. The "L" (low) band is applied to levels lower than objectives that may require future action, and the "U" (unsatisfactory) band is applied to levels that require immediate corrective action.

**Measurement Collection, Compilation, and Publication —
The Centralized Results System**

The significant task of managing the large amount of measurement data in the Bell System was made easier by the introduction of an operations system known as the *Centralized Results System* (CRS). CRS is a management information system that automates the collection, analysis, and publication of many measurement results. In many cases, the analysis includes banding.

CRS supports the BOCs by gathering data at the lowest organizational level. It can receive the monthly raw measurement results from BOCs either as terminal input or by a direct computer link. CRS checks the received data for validity and summarizes and compiles them into a form suitable for dissemination as official measurement plans. Based on the organizational structure of each individual BOC, CRS also accumulates and aggregates official measurement plan results up through the company level. Each company can retrieve its own compiled results interactively from CRS.

16.4.5 TYPICAL MEASUREMENT PLANS

This section describes two typical measurement plans. The **Telephone Service Attitude Measurement (TELSAM) Program** measures customer attitudes about service. The **1/1A***ESS* **Network Switching Performance Measurement Plan (NSPMP)**, measures network component performance.

TELSAM

TELSAM is currently the primary method used to measure customer attitudes about Bell System service. It is based on short (typically, 3 minutes), direct interviews with customers by telephone. Trained personnel under contract to outside research firms carry out the interviews using specially prepared questionnaires. Areas covered by TELSAM questionnaires include Service Centers, Installation Services, Repair Services, and Operator Services.

The Residence Installation questionnaire is typical. The customer to be interviewed is selected from a random sample of those who have had a telephone or line installed recently. After ensuring that the person receiving the call was the one who had the service experience, the interviewer asks whether:

- the customer got the desired type of equipment and telephone

- the customer got the desired type of line service (that is, private line, measured service, or party line)

- the work was done neatly and in the way that the customer wanted

- the installer was courteous

- the appointed installation date was set properly by the business office

- the appointment was kept

- it was necessary to have someone come out again for the work

- the installation is working satisfactorily

- overall, the customer is satisfied or dissatisfied.

The questionnaire is designed so that if in answering a specific question, a customer expresses dissatisfaction, the interviewer is guided to ask a series of further questions to gain a deeper understanding of the cause of the displeasure.

Because of the cost of interviewing customers and the resultant limit on the number of samples taken, most TELSAM results are compiled at the area[11] level. A number of operating telephone companies, however, have elected to sample at lower levels to gain a better picture of district-to-district variability.

1/1AESS NSPMP

This measurement plan is designed to monitor the performance of the 1/1AESS switching equipment. Since a switching system sets up calls, the plan is also viewed as one that monitors the quality of customer service provided by the 1/1AESS switching equipment.

The plan has a 2-tiered structure. Each tier includes micro measurements judged to be important in evaluating switching service and performance. Ten *measured components* are at the top tier. These are the key measurements that collectively reflect the service and performance of switching equipment. The next tier consists of eighteen *performance indicators*. These more detailed measurements are designed to support the measured components in pinpointing specific potential trouble spots when corrective action is necessary. (The document describing the 1/1AESS NSPMP instructs craftspersons to monitor a number of even more detailed items not specifically reported by the plan.) Finally, for control purposes, detailed diagnostic procedures or computer algorithms are available for the more complex measured categories such as Transmitter Timeouts (described below).

---

[11] Typically, a significant part of a state.

Table 16-1 shows the items in the measured-component portion of the standard report for the plan. The components are grouped into four categories:

- *Machine Access* — measurements designed to reflect difficulties experienced by the customer in obtaining service from the switching equipment. For example, Receiver Overflow measures the number of times that all of the digit receivers (the equipment in the 1/1A*ESS* switch used to collect the digits dialed by the customer) are busy. Restore Verify Failure indicates the number of times that a customer's off-hook signal is not properly detected by the switching equipment.

- *Machine Switching* — measurements of customers' call attempts (or incoming call attempts from another switch) that failed during call processing. For example, Transmitter Timeouts registers failures between the measured switch and a far-end switch when the former attempts to outpulse digits to the latter.

TABLE 16-1

NSPMP MEASURED COMPONENTS

Category	Weight
MACHINE ACCESS	
Dial-Tone Speed	15
Receiver Overflow	5
Restore Verify Failure	5
MACHINE SWITCHING	
Transmitter Timeouts	10
Office Overflow	15
FCG* & Supervisory Failures	15
Receiver Timeouts	10
Equipment Irregularities	5
BILLING	
Lost Billing	10
CUSTOMER REPORTS	
Code 5 and 8 Equipment	10

*False cross or ground; a test for a false ground on T or R leads or a cross between the leads.

- *Billing* — AMA entries that cannot be billed because of AMA tape problems.

- *Customer Reports* — counts of those customer-reported troubles that are attributable to the switching equipment. Those trouble reports that result in a "trouble found" (code 5) are counted separately from those that result in "no trouble found" (code 8).

To report NSPMP results, the normalized,[12] measured value for each component is first banded using the banding scheme described above. Then, a weighted average of the measured components is obtained by multiplying the measured value by the weights shown in Table 16-1. The products are added to arrive at a band for the overall switching office performance. In the results summarized for higher company levels, the number and percent of offices in each band are given for each component as well as for the overall office performance.

16.4.6 FUTURE TRENDS

Effective measurement becomes even more important in a postdivestiture era of increasing competition and technological change. Each operating company must individually choose the most cost-effective measurements to determine its own internal efficiency. Furthermore, since no one company will be responsible for complete end-to-end telephone service, the focus of the specific items to be measured will be shifted toward quality assessment and control of the part of the network administered by the company. The postdivestiture measurement activities will focus on the following areas:

- *Increased measurement mechanization* — The measurement process itself must be cost-effective. New technology needs to be explored and a highly mechanized approach is increasingly necessary. Automated systems capable of remote performance measurements, such as the ASPEN system described in Section 16.2.3, are likely to become extremely useful. To provide complete service and performance monitoring, this class of system may need to be supported by operations systems capable of measuring end-to-end performance, such as the No. 2 Service Evaluation System (see Hester 1982),

- *Added attention toward company and network interfaces* — With each company responsible for a portion of the network, there will be many complex interfaces occurring at the company boundaries. They will

---

[12] Each component is normalized by dividing the number of occurrences by a factor related to the appropriate equipment usage or activities.

encompass operations such as installation and repair in addition to interfaces between network components. Tools and methods will have to be found to delineate the interfaces clearly and to measure the service and performance related to such interfaces.

- *More detailed performance specifications* — With communications service provided by a number of companies in tandem, the performance to be expected from each company must now be clearly stipulated. This means a more careful identification of those parameters that are crucial in determining whether service or performance falls within acceptable bounds. Setting performance objectives for these parameters and determining how best to measure the parameters will take on added significance.

AUTHORS

H. Aldermeshian
D. P. Duncan
J. C. Hsu
E. Jeffers

D. G. Leeper
P. Lopiparo
K. I. Park
K. C. Szelag

PART FIVE

ENVIRONMENT
AND EVOLUTION

The first four parts of this book describe the Bell System, its resources, and the way those resources are managed through operations to provide service to customers. This part examines the evolution of products and services in the Bell System and the environment in which it has occurred.

Chapter 17 discusses the external factors that have had a significant effect on the Bell System operations. It assesses the constraints and challenges of regulation, tariffs, and competition, and examines the nature of relationships between the Bell System and independent telephone companies and other common carriers. Chapter 18 addresses internal processes and considerations that have affected the evolution of new and modified products and services and discusses how the corporate units of the Bell System have interacted as part of this process.

17

The Environment

17.1 INTRODUCTION

On March 7, 1876, Alexander Graham Bell was granted a patent for the invention of the telephone. Two years later, nearly 11,000 telephones were in service in the United States, and a 5-telephone central office had been installed in Washington, D.C. Now, more than 100 years later, over 500 million telephones provide voice and data communications worldwide. The environment in which this progress has occurred is constantly changing; actions taken by all three branches of government have been instrumental in defining changes in response to such factors as new technology, marketplace demands, and changing social forces and customer perceptions.

This chapter focuses on the regulatory and competitive aspects of the evolving telecommunications environment and the development of the Bell System[1] as it responded to opportunities and restraints of this environment. To put the following discussion in perspective, Table 17-1 provides a brief, selective list of significant events related to regulation and competition.

17.2 REGULATION

In general, regulation has been employed as a substitute for competition in markets where competition either does not exist or where its existence was judged not to be in the best interests of the public. While this chapter discusses regulation and competition separately, they are closely linked, and the extent to which competition may exist depends on the degree of regulation imposed.

---

[1] On January 1, 1984, the Bell System in its present form will no longer exist as a result of the 1982 Modification of Final Judgment.

Early government efforts to regulate business were not specifically directed towards the telecommunications industry, but many were precursors of telecommunications regulations that followed. The following sections discuss some of the pertinent events that have affected the regulation of the telecommunications industry.

17.2.1 THE INTERSTATE COMMERCE ACT

The Interstate Commerce Act of 1887 established the *Interstate Commerce Commission* (ICC) to regulate interstate carriers, of which railroads were the first. The ICC was empowered to inquire into the management of interstate carriers, to examine records and documents, to summon witnesses, and to use the federal courts for proceedings resulting from investigations. Practices such as *pooling*,[2] special rates, rebates, and discrimination against persons, places, or commodities were outlawed by the commission. The ICC also made it unlawful for a carrier to charge more for short hauls than for long hauls if certain conditions were similar and when the longer haul included the shorter haul.

In 1910, the Mann-Elkins Act was added to the Interstate Commerce Act enlarging ICC responsibilities to include regulation of the telecommunications industry. With this act, Congress recognized the *natural monopoly* characteristics of the telephone business, that is, that the public would not benefit from competition among telephone companies serving the same area.

17.2.2 THE SHERMAN AND CLAYTON ANTITRUST ACTS

A few years after the legislation that produced the ICC was enacted, major legislation of a different type was passed—the Sherman and Clayton Antitrust Acts.

In 1879, more than forty individual companies joined together to form the Standard Oil Trust.[3] About the same time, other large companies combined to form monopolies in other commodities. Under pressure from several states in which the trusts did business, the courts moved against them using legislation that existed under common law, and by the 1890s,

---

[2] An informal agreement by which a group of individuals pledged among themselves to maintain prices or divide markets. Self-interest usually demanded that those pledges be kept, but since the agreements were voluntary and therefore unenforceable under common law, they were often broken.

[3] A *trust* was an agreement between individual companies to place their collective stock in the hands of a group of elected trustees. The trustees were empowered to vote the stocks of all member companies, to evaluate the properties that made up the combination of companies, and to issue *trust certificates* on the basis of which properties were divided.

the trusts began to dissolve. Far from disappearing, however, they gave way to *holding companies*[4] and consolidation.

The *Sherman Antitrust Act*, enacted in 1890, was designed to curb the activities of these trusts and holding companies. It forbade restraint and monopolization of trade and placed the responsibility of enforcement with the federal government. The Sherman Act, however, was not actively enforced until Theodore Roosevelt's administration (1901-1909). Then, the energies of the Department of Justice were directed at vigorously enforcing it.

In October 1914, under Woodrow Wilson's administration, the *Clayton Antitrust Act* was passed to supplement the Sherman Act. The Clayton Act banned price discrimination, anticompetitive mergers, *interlocking directorates*,[5] and exclusive-dealing arrangements. Most important, it permitted private individuals to file suits seeking damages as a result of violations of its provisions.

17.2.3 THE KINGSBURY COMMITMENT AND THE GRAHAM-WILLIS ACT

Thus far, none of the attempts to regulate business had been devised specifically to control the Bell System. In fact, as mentioned earlier, Mann-Elkins actually recognized the natural monopolistic character of the telephone business. But, in 1910, AT&T acquired control of the Western Union Telegraph Company by a stock purchase. Three years later, when Wilson assumed the presidency of the United States and started to fulfill a campaign promise to destroy private monopolies and trusts, the Department of Justice considered an antitrust action against the Bell System. This action was averted when AT&T Vice-President Nathan C. Kingsbury committed the Bell System to a course of conduct that would satisfy government complaints. The *Kingsbury Commitment* specified that the Bell System would relinquish its Western Union stock (which it did in 1914), stop further acquisitions of competing independent telephone companies (except when in the public interest, and then only with ICC approval), and interconnect the Bell operating companies and independent telephone companies with its long-distance lines.

Acting under the authority of a joint resolution of Congress during World War I, President Wilson declared that as of July 31, 1918, the nation's telephone and telegraph systems would be put under the control

---

[4] A *holding company* either bought all the properties of its several subsidiaries or acquired control of each company through purchase of a majority of the stock. In the latter case, consolidation was effected through the election of common directors who sat on the boards of all the participating companies.

[5] Management of a number of separate corporations by the same or nearly the same group of directors.

of the federal government, specifically, the Post Office Department. On August 1, 1918, Postmaster General A. S. Burleson assumed supervision, control, and operation of those properties and continued in that role until control was returned to private ownership on August 1, 1919.

The *Graham-Willis Act* of 1921 affirmed that the Bell System had become a national resource and allowed its existence as a natural monopoly because it provided a unique technology. During the 1920s, the Bell System emerged in its modern form when the numerous associated operating telephone companies merged into larger regional groups.

17.2.4 THE FEDERAL COMMUNICATIONS COMMISSION

With the Federal Communications Act of 1934, Congress established *universal telephone service*[6] as a national goal. This act created the Federal Communications Commission (FCC) and initiated the modern regulatory environment for the telecommunications industry.

The FCC was given the primary responsibility for regulating the rates and conditions of interstate, international, and marine communications. Immediately after its creation, the FCC ordered an intense investigation of the Bell System and its operations. As a result of this investigation, the Department of Justice began to develop a major antitrust suit against the Bell System. With the imminence of World War II, however, the Department of Justice postponed any action because it considered the Bell System vital to national defense.

17.2.5 STATE REGULATORY COMMISSIONS

From 1935 through 1968, the philosophy of regulation matured. State regulatory bodies generally referred to as *public utilities commissions* (PUCs), but with different names in different states, regulated the rates and conditions of intrastate communications for the common carriers in their jurisdictions, just as the FCC regulated rates and conditions of interstate service. The basic task assigned to the PUCs was to establish rate systems to promote the public good and the goal of universal service while providing sufficient revenues so suppliers could meet the costs of doing business in the state they served. Methods were established to distribute revenues among participating companies for calls that use the facilities of more than one carrier. The goal of universal service was a factor in formulating these distribution methods. (Section 17.3 provides a more detailed discussion of rates and revenues.)

---

[6] The goal of universal service was to make basic telephone service available at an affordable price, anywhere in the nation.

17.2.6 THE "ABOVE-890" RULING

Technological developments that emerged from World War II—many based on the work of Bell Laboratories—were applied to areas such as customer terminal equipment, data services, and microwave communications after the war.

Before 1949, the FCC had assigned the microwave spectrum to the telecommunications common carriers and had not allowed private systems to connect with the public network. After the war, microwave technology was in demand by businesses and industries located in remote areas without telecommunications services, so the FCC began leasing private microwave systems on a case-by-case basis to serve areas where there were no common carriers.

The major common carriers, including the Bell System, objected that these leased microwave systems would congest the airwaves and result in poor service. Many private companies, however, saw the potential savings involved in using microwave systems for telephone service and were eager to exploit a promising market. As the demand for this market grew, the FCC relaxed its restrictions on the private companies.

By the end of 1951, 13,000 route miles of private microwave systems were already in place or being started. In 1959, over common-carrier objections, the FCC decided that there were ample frequencies higher than 890 megahertz (MHz) to accommodate those who wished to construct their own microwave systems.

In that ruling, referred to as the "Above-890" Ruling, the FCC decided in favor of licensing private intercity microwave systems for both voice and data transmission. Then, in the Specialized Common Carrier Decision of 1969, the FCC decreed that the new microwave companies be allowed to compete with existing (and regulated) telephone companies in the sale of private network transmission services.

17.2.7 FCC INTERCONNECTION RULINGS

In 1947, the FCC permitted use of customer-provided recording devices but ordered that direct electrical connection of such devices to the telephone network must be through protecting arrangements provided and maintained by the telephone company.

A variety of interconnection devices, many of which were foreign made, became available during the late 1950s. They were designed to attach to existing telephone sets or to be used as terminal equipment themselves. The major common carriers maintained that it would be impossible to ensure efficient telephone service if devices supplied by firms with no legal responsibility for the quality of service were attached to the network by customers; interconnection of such devices could increase the network's operating costs and disrupt its efficiency. This could be particularly damaging in times of emergency.

The FCC supported this position and refused to allow the use of interconnection devices, but it was overruled by the court of appeals in the case of the *Hush-A-Phone* device, a small cup-like nonelectrical handset attachment that enhanced privacy when talking. Then, in 1968, the FCC ruled in favor of Carter Electronics, a Texas firm that made a mobile radio device that could be acoustically coupled to the common carrier voice telephone network. This device, called the *Carterfone*, was primarily being sold to oil exploration and drilling companies for use by field engineers in remote areas.

This ruling by the FCC was a landmark: It set in motion the forces of deregulation and led to intense competition because, unlike the Hush-A-Phone Decision, the Carterfone ruling permitted the direct electrical attachment of devices to the telephone company's equipment provided that the operation of the network was not adversely affected (see Section 17.4.2). The FCC recognized the concern for potential adverse effects on the network and on the quality of service as a result of the attachment of customer-provided equipment and contemplated the continued use of network control signaling apparatus provided by a common carrier. Consequently, the FCC approved *tariffs* (see Section 17.3) requiring the use of a protective coupler between customer-provided equipment and the network. In November 1975, the FCC issued a report and an order instituting a registration program. Under that program, carrier-provided protective coupling devices are no longer necessary if the customer-provided equipment is registered with the FCC or uses a registered protective coupling device. (Section 8.7 discusses interfaces for interconnection.)

17.2.8 THE 1956 CONSENT DECREE

In 1949, the Department of Justice filed against the Bell System the antitrust suit that it had postponed because of World War II. In particular, the suit attempted to force AT&T to divest itself of Western Electric. The suit was settled by the *1956 Consent Decree*, which allowed AT&T to retain ownership of Western Electric as long as Western Electric manufactured only products used by the Bell operating companies.[7] The decree also specified that Bell operating companies confine their activities, with some exceptions, to providing telecommunications services under regulation.

---

[7] Exceptions were made for government-sponsored work or cases where no competitive products existed. Also, the Bell System was allowed to continue to provide the artificial larynx (see Section 2.2.5) because it was felt that no competitive commercial supplier would undertake the task.

As a result, the Bell System was effectively barred from providing commercial data-processing services. The Bell System was also required to decline royalties on its then-existing patents and to license all future patents to any applicant on a nondiscriminatory basis at reasonable royalty rates.[8]

The 1956 Consent Decree became a large part of the telecommunications environment in which the Bell System had to operate. Although the decree had a major impact on the scope of its corporate activities, the Bell System was able to maintain its corporate identity and *vertical integration*.[9] The decree remained in force until implementation of the 1982 Modification of Final Judgment (see Section 17.4.4) brought an end to both the period during which regulation was the primary controller of the telecommunications business and the Bell System itself.

In the years following the 1956 Consent Decree, advances in signaling and data-processing capabilities brought telecommunications and data processing closer together. In 1971, because of the potential impact of the 1956 Consent Decree on the ability of the Bell System to provide useful new services, the FCC began the first of two computer inquiries to explore the link between telecommunications services and data-processing services. Because of their significant impact on competition in the modern telecommunications environment, the results of these computer inquiries and subsequent legislation are discussed in Section 17.4.

17.3 TARIFFS AND RATE SETTING

17.3.1 THE ELEMENTS OF A TARIFF

Tariffs are one manifestation of state and federal regulation of the telephone industry. A tariff describes a service, the rate that may be charged for the service, and the regulations under which that service can be provided. It is a set of terms between the carrier and the customer that must be submitted to a regulatory body before any new or changed services can be provided. A tariff, for example, may state the nature of the service, the class of customer to which given rates apply, the availability of the service in defined areas, the method of measuring the service (for example, initial and overtime minutes on long-distance calls), and the method of computing customer charges. Figure 17-1 shows part of a simple tariff.

---

[8] This had, in fact, been Bell System policy even before the decree.

[9] A *vertically integrated* business is concerned with all the processes involved in a product or service.

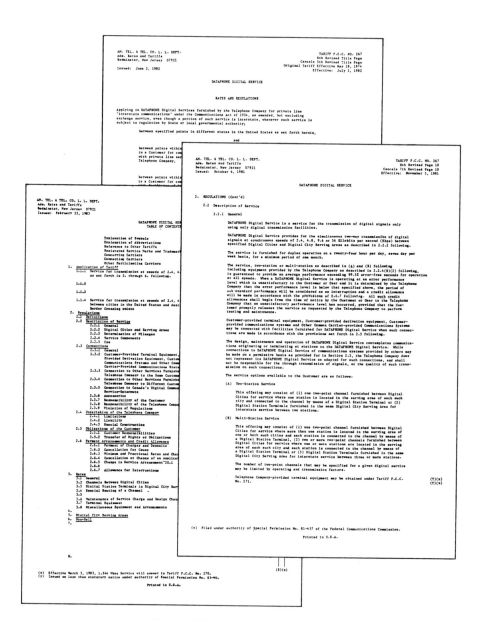

Figure 17-1. Part of a typical tariff.

Local tariffs are specific to each of the fifty sovereign states. Because of that and the diversity of regulatory history among the states, individual state regulatory agencies take different approaches.[10] Uniform nationwide tariffing of intrastate services has not been contemplated under these circumstances.

17.3.2 THE DEVELOPMENT OF A TARIFF

A common carrier submits the basic elements of a proposed tariff—the rates and regulations—and supporting information to the appropriate agency, usually called a *commission*. (The tariff must be clear and concise; by precedent, any ambiguity is interpreted in favor of the customer.) The commission, customer, and others (such as possible competitors) have a period of time in which to study the terms (this period can last from one day to several years). During this period, notice of the proposed tariff is given to all subscribers who may be affected. Unless the commission chooses to suspend it, the tariff goes into effect as filed at the end of the study period. A carrier who wishes to implement a tariff earlier may file an interim request with the commission to do so.

A carrier files a *concurrence* if it decides to adopt the rates and regulations of an existing tariff filed by another carrier, rather than preparing a new tariff. Once a carrier submits a concurrence request, the carrier is bound by all subsequent amendments made to those rates and regulations. Generally, independent telephone companies and local Bell operating companies concur in applicable FCC tariffs.

17.3.3 RATE SETTING

The overall rate structure for all tariffed services balances the common carrier's need to provide an adequate return on investment with the need to ensure the lowest overall cost to customers. Several principles based on social policy have been used in rate setting. They include: **flat rates versus usage-sensitive rates, value-of-service pricing**, and **nationwide rate averaging**. These principles and two other factors—**separations** and **settlements**—are discussed below.[11]

---

[10] At various times, and in Texas until recently, some states did not have a state regulatory agency. There were many agencies of municipal or county governments leading to even greater diversity.

[11] Rate-setting principles and practices are undergoing significant changes in response to competition and actions of the courts, regulatory agencies, and Congress.

Flat Versus Usage-Sensitive Rates

Historically, flat rates for local service have encouraged the use of telephones and helped to achieve the goal of universal service. With flat rates, a group of customers are charged the same rate regardless of the number of local calls they make. Until recent technological advances in billing, flat rates were also the most economically feasible method of rate setting. Usage-sensitive rates, the alternative, were first introduced for toll calls where the higher revenues could justify and offset a higher measurement and billing cost. These rates relate directly to a customer's use of equipment and service—those who use less, pay less. In the 1980s, with the goal of universal service met and with pressure caused by increasing competition to relate charges to cost of service more closely, the trend is toward usage-sensitive rates.

Value-of-Service Pricing

Value-of-service pricing is another rate-setting principle. The value of service is directly related to the customer density in the local calling area, frequency of use, and importance to the customer. With this method, each group or class of customers is charged relative to the value of the service to that group. For example, since city customers can reach more local telephones than rural customers, and they are likely to place and receive more calls; they pay more for local service. Also, since the calling rate per line is higher, and the telephone service is more important for businesses than for individuals; business rates are higher than residential rates.

A significant exception to the value-of-service concept is the provision of low-cost telephone service for the elderly, the hearing-impaired, and the economically disadvantaged. This exception emphasizes the fact that under regulation, the setting of telephone rates reflects social policies as determined by the regulators.

Value-of-service pricing has been another factor used by the regulatory commissions to achieve the goal of universal service: providing basic telephone service at a low cost. As a result of this philosophy, services such as *TOUCH-TONE* calling have been judged to be nonessential. To keep the rates for basic service low, commissions set the rates for these nonessential services above their cost.

Nationwide Rate Averaging

Nationwide rate averaging has been used to establish rates for interstate telephone service. Under rate averaging, similar services and like calling distances carry the same charges even though the costs of interstate service to the common carrier may vary widely; for example, routes in low-density areas have much higher costs than routes in high-density areas.

Nationwide averaging also makes a broader range of telecommunications services available to a variety of industrial, commercial, and residential users at affordable rates. The advent of competition for interstate services, however, has required a departure from total nationwide averaging. In 1974, for example, to meet competition in private-line services, AT&T attempted to restructure rates for voice-grade private lines. The "hi-lo" rate structure proposed at that time specified lower rates for high-density routes and higher rates for low-density routes; however, this deaveraged structure was not approved by the FCC.

Separations

Another factor considered in rate setting is a legal requirement called *separations*. The separations process specifies that the costs of providing telephone service be appropriately distributed between *interstate services*, which are under the jurisdiction of the FCC, and *intrastate services*, which are under the jurisdiction of the state commissions. The separations process was designed to compensate the carriers for the costs of providing the services, including common costs (such as the cost of customer-line and local switching equipment that is used for interstate and intrastate long-distance service as well as for local service). These costs were allocated on the basis of an appropriate measure, such as minute-miles of use. Separations procedures reflect average costs and, where based on time calculations, treat all minutes, peak and off-peak, equally.

Settlements

While separations procedures define how costs are distributed between interstate and intrastate traffic, *settlements* define how revenues of a single call are split among different companies (both Bell and independent) involved in that connection. A settlement is an accounting procedure based on the total investment in telephone equipment; a company's total investment determines the base for allowed earnings (called the *rate base*). This emphasizes the importance of ownership of terminal equipment on customer premises, particularly for smaller, independent telephone companies for whom such equipment may be a large percentage of their total equipment investment.

17.4 COMPETITION

Competition existed in the telecommunications business for the first several decades after the telephone became available. Later, there were several decades when the Bell System and independent telephone companies, operating as a cooperative partnership, provided essentially all services with little or no competition. In recent years, however, technology and social pressures have combined to promote a new competitive era. This section discusses some of the important issues and events in the evolution of competition.

17.4.1 EARLY YEARS

After Alexander Graham Bell was granted patents for the telephone and before the first Bell Telephone Company was formed in 1877, Bell reportedly offered his telephone invention to Western Union for $100,000. Western Union declined but soon began their own venture in telephony using a transmitter designed by Thomas Edison and a receiver developed by Elisha Gray. In December 1877, Western Union founded a subsidiary, the American Speaking Telephone Company, to market their telephone. This subsidiary and another Western Union subsidiary, the Gold and Stock Telegraph Company, established in 1878, were the first of the Bell Telephone Company's many competitors in the telephone business.

The telecommunications industry's first major lawsuit involving competition arose out of public demand for voice communication rather than coded message communication. In 1878, the Bell Telephone Company sued the American Speaking Telephone Company for Edison and Gray's infringement of their patents (see Table 17-1). The suit was settled on November 10, 1879: Western Union agreed to stay out of the telephone business, and Bell agreed to stay out of the public message telegraph field in territories already occupied by Western Union.

During the early days of the telephone industry, equipment was made for Bell-licensed companies by a variety of manufacturers. But in February 1882, American Bell (a name adopted by the Bell Telephone Company in 1880) and Western Electric (formed in 1872) entered into an agreement that made Western Electric the sole supplier of Bell telephones and equipment. Meanwhile, several Bell-licensed telephone companies were beginning to consolidate during the early 1880s, forming the early building blocks of the Bell System.

In February 1885, the American Telephone and Telegraph (AT&T) Company was incorporated to establish telephone communications for cities on the American continent and elsewhere around the world by wire, cable, and "other appropriate means." Theodore Vail, AT&T's first president, later expressed the concept of universal service by stating the following goals: "one policy, one system, universal service."

On March 7, 1893, the first Bell patent expired, making it possible for anyone to sell telephone equipment and services without infringing on the legal monopoly afforded Bell's basic patents.[12] As a result of this

---

[12] Previous to this expiration, the validity of the Bell patent had been upheld by a 5-4 vote of the Supreme Court in the famous *Telephone Cases, 126 U.S.* The Bell patent was involved in approximately 600 patent infringement suits, in all of which the Bell interests were successful.

event and an ever-increasing demand for voice communication, independent telephone companies proliferated. They first appeared in rural areas not served by Bell companies, but by the mid-1890s, they had moved into the cities, where they were in competition with each other as well as with Bell companies. The following decades saw the continuing growth of independent telephone companies and technological advances that made new types of equipment and new services possible. As a result of various government actions discussed earlier, the telephone companies (both Bell and independents) were recognized as a natural monopoly and operated essentially free of competition until the interconnect issue arose in the 1950s.

17.4.2 INTERCONNECTION

Neither the Hush-a-Phone nor the Carterfone equipment introduced electrical signals to the network. The FCC's Carterfone ruling (1968), however, permitted direct electrical attachment to telephone company equipment, and this decision led to new *interconnection tariffs*. Those tariffs were more liberal than required by the Carterfone ruling. They allowed customers to provide their own telephone instruments— including key telephone systems (KTSs) and private branch exchanges (PBXs)[13] through carrier-provided network control signaling equipment—and permitted, as well, unrestricted interconnection of all devices (acoustically or otherwise indirectly connected) to the network.

Before the Carterfone ruling, terminal equipment manufacturers other than Western Electric had found their major market among the independent telephone companies and Bell operating telephone companies. But the ruling and ensuing tariffs allowing interconnection of any device created a large, new class of customer—the business community—for terminal equipment manufacturers. Competition to manufacture and distribute interconnection devices grew, and many companies, all unregulated, became involved. The Bell System, however, was still restrained by regulation.

Foreign manufacturers were also quick to move into the American customer terminal equipment market. They did so by means of interconnect "suppliers," or distributors, which were retail companies formed to serve as outlets for Japanese and European terminal equipment. These distributors were able to purchase the foreign equipment at favorable prices because of lower foreign labor costs and sell or lease the equipment at prices that compared favorably with the cost of leasing the equipment from regulated common carriers. Interconnection distributors made

---

[13] Section 11.2 describes PBXs and KTSs.

rapid gains in areas with large business populations, and by the second
half of the 1970s, they had captured a fair portion of the PBX and KTS
markets.

The interconnect companies opposed the requirement of carrier-
provided protective connecting arrangements, and for years, the FCC
grappled with the question of how best to boost competition in the terminal equipment market without permitting indiscriminate interconnection
that could cause harm to the network. Late in 1975, the FCC finally
issued an order in *Docket 19528* establishing a registration program under
which manufacturers could design equipment to comply with specific
technical requirements, register it, and sell it for direct connection to the
network.

17.4.3 MERGING DATA PROCESSING WITH
TELECOMMUNICATIONS – FCC INQUIRIES

The same technology—electronics—that led to the development of
diverse customer terminal equipment also led to innovations in the data-
processing industry. Because of the 1956 Consent Decree, however, the
Bell System was effectively prohibited from participating in the data-
processing industry. The FCC, determined to explore the link between
telecommunications services and data-processing services, began the first
of two computer inquiries in 1968.

During *Computer Inquiry I* (completed in 1971, with the report issued
in 1972), the FCC took a definitional approach to communications and
distinguished between *telecommunications services* and *data-processing services*. Those services defined as telecommunications would be provided
by the regulated telecommunications industry, and those defined as data-
processing would be provided by the unregulated data-processing industry. Services judged to be hybrids would be left for ad hoc determination. As a result, the inquiry identified only telecommunications and
certain hybrids of telecommunications-related data processing (including
some sophisticated high-speed data transmission services) as legitimate
areas for Bell System participation.

By 1976, however, both customer demands and technology had
changed further, and the FCC began a second inquiry. *Computer Inquiry II*
lasted from 1976 to 1980 and was intended to formulate a policy that
would allow regulated carriers (like the Bell System) to benefit from the
new technology in electronics while preventing them from directly competing with the unregulated data-processing industry. In contrast to the
definitional approach taken in Computer Inquiry I, a structural approach
was taken in Computer Inquiry II. In formulating the policy for this
inquiry, the FCC established the structural differences between *basic* services, which would be provided under tariff, and *enhanced* services, which
would be provided under regulation but not tariffed.

Computer Inquiry II stated that the Bell System must offer any *enhanced services* and all terminal equipment through a "fully separated" subsidiary, whose activities, while still subject to regulation (thus complying with the 1956 Consent Decree), would not be subject to tariffing. The provision of *basic services* by the common carriers, however, would remain subject to tariffing. In June 1982, in response to this decision, AT&T organized American Bell Inc. to provide enhanced services; and in January 1983, American Bell Inc. began offering new terminal equipment to telecommunications customers.

17.4.4 THE 1982 MODIFICATION OF FINAL JUDGMENT

In parallel with the computer inquiries, the Department of Justice was pursuing an antitrust suit against AT&T, Western Electric, and Bell Laboratories. The suit had been initiated in 1974 and sought complete dismemberment of the Bell System. It went to trial in early 1981, but the trial was dramatically stopped on January 8, 1982, by an agreement to dismiss the suit and submit to the court a modification of the 1956 Consent Decree. This modification removed from AT&T the limitations that it could only enter markets subject to regulation and, thus, allowed AT&T to take advantage of the new technology, much of which it had created.

The federal government, however, exacted a high price for this change: the breakup of the Bell System. The rationale behind the 1956 decree had been that telecommunications was to be provided under regulation, and since the Bell System would enjoy a monopoly status in telecommunications, it should be limited to regulated activities. Changes in technology and FCC decisions allowing unregulated companies to compete for the telecommunications market had made that approach obsolete. The Modification of Final Judgment considers only the provision of local exchange service as a true natural monopoly; everything else is considered subject to competitive market forces—eventually if not immediately. Accordingly, local exchange companies that provide *local service* are to be divested from AT&T, and AT&T will retain the competitive *interexchange* service and the competitive manufacture and provision of telephone equipment.

While subscribers have gained the choice of competitive equipment suppliers and competitive interexchange carriers, the common carrier approach to telecommunications as an overall system entity with end-to-end responsibility has been lost. Also lost is the integration within Bell Laboratories of the interests of the users and operators of services.

AUTHOR

J.W. Falk

18

Evolution of
Products and Services

18.1 OVERVIEW OF PRODUCT AND SERVICE EVOLUTION

Products and services offered by the Bell System have evolved over the years in response to new markets and changing technology. This chapter discusses some of the major activities and considerations involved in developing and introducing new products and services.

The evolution of a product or service (from conception through final phasing out of manufacture or deployment) may be divided into four major stages: **concept formation, development, introduction into the market,** and **mature product management.** This section briefly describes the activities in each stage and the roles of the participants as a background for the rest of the chapter. Since this book describes the Bell System prior to divestiture (see Section 17.4.4), the discussion in this chapter reflects the integrated nature of these activities as they have existed for many years. After divestiture, the introduction of products and services will be carried out in a different way, will involve different participants, and will depend on the type of product and the customer. A description of these future processes is beyond the scope of this book.

There are a number of differences in detail between the evolution of a new product and that of a new service, but the gross features of the process are the same. The description of the evolutionary process in this section uses a hypothetical new service as an example.

18.1.1 CONCEPT FORMATION

Concept formation for a new service can begin in different ways. For example, through an analysis of the technological evolution of the Bell System network, a Bell Laboratories engineer may realize the capability for a new service. Or, through an analysis of customer needs, an AT&T or Bell operating company marketing manager may realize that an opportunity for a new service exists. In any case, the next step is to determine the potential commercial success of the service. This analysis is generally carried out by the marketing managers in consultation with network design engineers from AT&T and the Bell operating companies and with design and systems engineers from Bell Laboratories. Figure 18-1 illustrates the process schematically, and the next few paragraphs describe it.

The first requirement in analyzing the service is to define its functions and technological requirements. This requires consideration of both the needs of the market and the present and future technological capabilities of the Bell System network. Understanding market needs is crucial since a product or service, no matter how well designed technically, will be a commercial failure unless customers buy it. Information about the market potential for the service comes from a variety of sources, for example, the analysis of sales of similar Bell System and competitive services, discussions with members of Bell operating company and AT&T sales forces, consultations with industry and marketing experts, and direct interviews with customers. From the analysis of such data, the marketing manager attempts to "segment" the market (that is, to identify those customers most likely to buy the service). The marketing manager then plans appropriate marketing strategies for each segment and also attempts to estimate the expected revenues from the service.

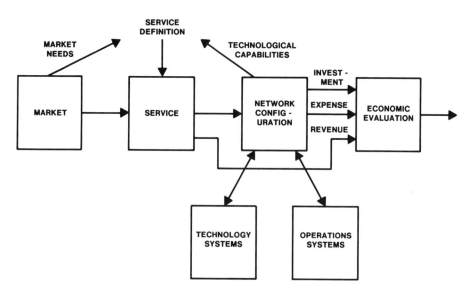

Figure 18-1. The service planning process.

Existing and projected capabilities of the Bell System network must also be evaluated to determine whether the network is likely to contain the components and operations systems necessary to support the service and what expense and investment costs will be incurred in providing them. This evaluation typically relies on analysis and support from Bell Laboratories design and systems engineers and Western Electric engineers and product-line managers for the network components.

The last part of the service analysis includes an economic evaluation of the potential profitability of the service. The evaluation combines estimates of revenues from the market analysis and estimates of expense and investment from the analysis of the network requirements. In the concept formation stage, these estimates may be relatively crude. If the economic evaluation indicates that the service is likely to be profitable, if the service fits into the overall business strategies of the Bell System, and if there are no regulatory or legal constraints to prevent its deployment, the formal development stage begins.

18.1.2 DEVELOPMENT

This stage also involves the steps indicated in Figure 18-1, but now each step involves much more detail. Systems engineers complete a detailed specification of the service describing its features, the network components and operations systems required to support it, and how to facilitate its use by customers.[1] The required modifications and additions to the hardware and software of both network components and operating systems must be designed, built, and tested. This is the primary responsibility of development engineers and usually involves a substantial investment in both human and financial resources. As the service becomes more clearly defined, marketing managers attempt to determine the customer demand for the service more accurately. This generally involves direct interviews with customers. During this period, there tends to be frequent interaction between systems engineers and marketing managers. As market research begins to relate projected demand to the precise features and price of the service, modifications in the design may be necessary. Systems engineers and development engineers will define the desired modifications, which may require further development and testing. For example, particularly costly features for which there is insufficient demand may be redesigned or eliminated to lower the price of the service. With a more precise picture of the projected demand for the service, further economic analysis can be done. If the service still looks profitable, the project moves into the third major stage where plans are made for introducing the service.

---

[1] The last aspect is the responsibility of human factors engineering (see Section 18.6).

18.1.3 INTRODUCTION INTO THE MARKET

On the technical side, the service is tested for quality and reliability, final refinements are made in the design, and a detailed plan for deployment is prepared. On the marketing side, final plans for introducing the service are drawn up. These plans generally involve offering the service first in a few test areas (one or two operating companies) and, if the service is successful in those areas, gradually introducing it across the country. Final market surveys are conducted by the marketing managers to determine a suggested price for the service, based on such factors as an analysis of how the price affects demand in general (which considers existing competitive services) and how the price affects demand for other Bell System services.

Next, AT&T and Bell operating company marketing managers determine where the service should be offered and propose a tariff. The tariff must be approved by the Federal Communications Commission (FCC) or the appropriate state public utilities commissions before the service can be offered (see Section 17.3). The AT&T and Bell operating company sales force develop plans for the sale of the service to customers, and advertising managers develop promotion campaigns.

After the service is introduced, the initial sales are monitored to determine whether the marketing strategy is successful and how it might be improved. This may suggest modifications in the service, which would involve coordination with the design engineers. If the service introduction is successful, the process moves into the last stage.

18.1.4 MATURE PRODUCT MANAGEMENT

During mature product management, marketing tends to be the dominant function. Marketing managers at AT&T and the Bell operating companies track the sales of the service and its competitive position in the market and determine whether changes in the marketing strategy are necessary. For example, change in demand for the service due to the introduction of a competitive service might be countered by a price change, an advertising campaign, or a change in the service features. Technical adjustments may occasionally be required if use of the service by the public indicates flaws in the design. The service will eventually be withdrawn from the market when it becomes obsolete or when it becomes only marginally profitable and the resources devoted to it would be more profitably allocated to other services.

As mentioned earlier, the description of the evolution of a new service given above also applies, for the most part, to the evolution of a new product, although details will be different. For instance, the participation of the marketing and planning organizations of Western Electric is a crucial component of new product evolution. Also, central for hardware products are issues concerned with product manufacture, such as the design and construction of production equipment and systems and the development of quality standards and controls. The rest of this chapter provides details about specific areas involved in the evolution process.

18.2 MARKETING

Numerous studies (see Urban and Hauser 1980) indicate that, while technical problems and changes in the business environment account for some commercial failures of products, a large proportion of failures can be attributed to mistakes in *marketing*, that is, those activities that influence the flow of goods and services from producers to consumers. Since the success of the Bell System, like that of any firm, depends on the commercial success of its products and services, a proper approach to marketing is critical.

A modern approach to marketing is based on the idea that a key concern of a business is to determine the needs and wants of its markets and then to provide products and services that deliver the desired satisfactions, consistent with the goals and objectives of the business. Therefore, marketing should enter into all phases of product development from conception to final delivery to the customer. This approach has been adopted by the Bell System, and it tends to increase the chance for the commercial success of Bell System products and services.

18.2.1 MARKETING CONCEPTS

The role of marketing in the evolution of Bell System products and services is to develop strategies for dealing with marketing opportunities and to translate those strategies into marketing plans. Three important concepts used in developing marketing plans are **market segmentation, the marketing mix,** and the **product life cycle.**

Market segmentation is a way of analyzing and treating the markets for a product. It takes into account the fact that markets are rarely homogeneous; that is, different components or segments of a market may have different needs for or interests in a product. It is, therefore, frequently useful to identify those segments for which the products have the greatest potential appeal and devote most of the marketing effort toward them.

A number of variables are considered in classifying market segments. The most important distinction is customer type: Is the customer a consumer, a business, or a government agency? Variables that are sometimes useful for segmenting the consumer market are geographical location, product or service usage, demographics (age, sex, income, occupation, family size, etc.), and psychographics (attitudes, beliefs, personality, life style, etc.). Geographical location and usage are also useful for business customer segmentation, as are such variables as company size, company type, and purchasing behavior (for example, what organizations within the business make the buying decisions).

The *marketing mix* refers to those variables that can be adjusted by a marketing manager to attract a target segment. Four types of variables are usually identified.

- **Product** — the actual or perceived attributes of the product itself. This includes not only the physical attributes of the product but also such aspects as the reputation of the vendor, accompanying service contracts, and warranties.

- **Price** — the price and payment terms of the product. This involves not only the total amount paid for the product but also such concerns as whether the product is leased or sold, credit terms, and sale or volume discounts. The prices for regulated services must be approved by the appropriate regulatory bodies such as the FCC and state public utilities commissions.

- **Place** — the channels of distribution for the product. This involves deciding who will sell the product to the end-users. The Bell System sells products and services both through its own sales force and through non–Bell System vendors.

- **Promotion** — the advertising and face-to-face selling of the products. The goal of promotion is to communicate the potential benefits of the product to the customer.

The market for long-distance calling provides an example of the application of segmentation and marketing mix strategies. A gross segmentation of this market might distinguish between business and residential customers and, for each of these two categories, might distinguish between light and heavy users of the service. A possible strategy for the light-user segments is to promote the benefits of long-distance calling through advertising in order to stimulate more usage. Such a strategy is known as *market penetration*. For the heavy-user segments, a *product development* strategy (that is, the development of a new service to fill unmet needs) might be appropriate. The offering of Wide Area Telecommunications Services and private-line services are examples of successful applications of the product development strategy to the business heavy-user segment.

The particular market segmentation and marketing mix strategies adopted for a product depend on the phase of the *product's life cycle*. Figure 18-2 shows the four stages of a typical product life cycle: introduction, growth, maturity, and decline. The maximum sales and length of a product life cycle curve depend strongly on the nature of the market and the effectiveness of the marketing strategy. In relatively noncompetitive

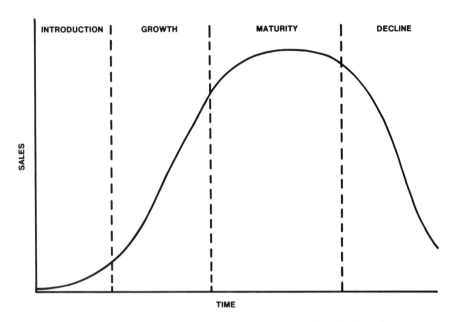

Figure 18-2. Typical product life cycle. During the introduction phase, sales increase slowly, since it usually takes some time for the product to be accepted. If the product is successful, it moves into a rapid growth phase. Maturity occurs when sales begin to level off; this generally occurs because similar products have been introduced, either by competitors or the firm itself, and they take sales away from the original product. Finally, a decline phase occurs as the product becomes obsolete and is replaced by newer products that better meet market needs.

markets, a product life cycle can be decades long. Until recently, Bell System products have tended to have long life cycles; however, a highly competitive and innovative market frequently reduces the length of life cycles to only a few years.

18.2.2 OTHER MARKETING CONSIDERATIONS

Timing (that is, when the product is introduced) also has a significant effect on its success. If an unfilled need exists in a market, the first product available that fills this need has strong sales potential. A product that is introduced too late must compete with existing products, and the firm must develop a marketing mix or segmentation strategy that gives it a competitive edge in order for the product to succeed. A product can also be introduced too early, before customers feel a need for it.

The customer's perception of a product relative to competitive products is critical for sales. The analysis of this perception and the adjustment of marketing mix variables to modify this perception and, thereby, improve sales is known as *product positioning*. One way to think of product positions is in terms of a perceptual map. There are often only a few attributes of competing products that cause them to be differentially perceived, and there are a number of sophisticated techniques (see Green and Tull 1978) that can be used to discover these attributes and place the products on a perceptual map.

It is also important to know where on such a map the customers' ideal products would lie. If these ideal products tend to form clusters for different customers, there is a natural way to segment the market. Optimal product positioning might be near a cluster of ideal points, particularly if the relevant customers happen to be heavy users.

The relationships among the total sets of products offered by the same company are also important. (Related products are grouped together in *product lines*, and all the product lines together form the firm's *product mix*.) For example, it is important to know whether new products will be perceived so similarly to other products that they will draw sales from the other products (*cannibalization*). It is also important to realize that the funds used for the research, development, and initial marketing of a new product come from the profits derived from sales of the successful products already on the market. The management of these interrelations is a critical part of marketing planning and is known as *product portfolio management*.

18.2.3 BELL SYSTEM MARKETING ORGANIZATIONS

During the last few years, the organizational structure of Bell System marketing has been involved in a rapid evolution. This evolution is, in part, a response to the increasingly competitive telecommunications environment. In addition, Computer Inquiry II and the Modification of Final Judgment are having a major impact on the Bell System marketing structure. Therefore, a detailed account of current Bell System marketing organizations would soon be outdated. Nevertheless, some general features are likely to persist into the near future, and they are described below.

Marketing organizations currently exist in all the major Bell System divisions except Bell Laboratories. They are generally subdivided both by line of business and basic marketing function. Typically, a *line of business* is a grouping of similar market segments or products. In the past, for example, separate marketing organizations were associated with residence customers and business customers. The most common functional units are:

- **Market management** organizations that plan and control the marketing effort within the various market segments. They attempt to assess the needs of these segments, to look for unmet needs that might be addressed by new products or services, to work with research and development organizations to develop new product or service ideas, and to monitor the competition.

- **Product (service) development** organizations that work with development engineers to ensure that new products and services will be commercially successful and to plan product introduction strategies.

- **Product (service) management** organizations that manage existing products and services. They develop and monitor marketing strategies and coordinate pricing, distribution, and promotion.

- The **sales** force that typically presents whole product or service lines to Bell System customers.

- **Advertising** organizations that work with advertising agencies to develop campaigns to promote Bell System products and services.

- **Market research** organizations that support the activities of the other organizations by suggesting possible market research to solve specific marketing problems, designing and analyzing market surveys, arranging for the services of outside market research companies, and developing marketing data bases.

Despite the current lack of explicit marketing organizations at Bell Laboratories, marketing has a major impact on research and development. Bell Laboratories interacts with marketing organizations in other Bell System companies during product and service development. In addition, a number of Bell Laboratories organizations are engaged in performing marketing studies and developing marketing analysis techniques. The close cooperation among companies within the Bell System has been a major contributor to commercial success.

18.3 ECONOMIC EVALUATION

18.3.1 OVERVIEW

In order to maintain high-quality telecommunications at the lowest cost, the Bell System continually develops, manufactures, and deploys new *products*. For a particular product, this process is called a *project*. There are many ways to characterize projects. For example, a project can be characterized by ultimate *product*: a (hardware) item, such as a new type of repeater or a *PRINCESS* telephone; a service, such as Call Forwarding; a method, such as a network planning procedure; or an operations system, such as Centralized Automatic Reporting on Trunks (CAROT), a computerized system for testing trunks.

Projects can also be characterized by the ultimate *user*: a customer using a service such as Call Forwarding or a product such as a *PRINCESS* telephone or a Bell operating company using a new type of repeater, a new method for network planning, or an operations system such as CAROT.

From yet another viewpoint, a project can be characterized in terms of the nature of its *financial impact* on the Bell System. It may be **revenue-producing, expense-saving,** or **capital-saving**. Projects that fall within any of these specific categories usually follow a general pattern of economic behavior:

- *Revenue-producing* projects normally require an investment in order to offer a new service or to increase the desirability of an existing service. Moreover, this type of project usually increases annual expenses, for example, to maintain equipment or to pay the personnel needed to provide a new service.

- *Expense-saving* projects often require an investment to decrease operating costs, usually by reducing the need for personnel. For example, Bell Laboratories has developed a number of computer-based operations systems (some of which are described in Chapter 14), requiring an investment in a computer and software, to assist the Bell operating companies in handling a variety of operations functions more efficiently. Expense-saving projects normally have little or no impact on revenues.

- *Capital-saving* projects normally result in a new product that performs the same function as a more expensive existing product. Expenditures to purchase the new product will be less than those required to purchase the existing product, thereby yielding capital savings. Quite often the new product will also have reduced maintenance expenses (primarily savings in maintenance personnel) because it is based on a newer, more reliable technology. New software tools (for example, an inventory control system such as the Plug-In Inventory Control System [PICS][2]) or new methods can also reduce the need for capital expenditures. In most cases, this type of project has little or no impact on revenues.

Figure 18-3 is a matrix representing ways to classify a project. Of course, not all combinations are currently possible projects in the Bell System; for instance, operations systems are not provided for subscribers at the present time.

---

[2] See Section 14.2.2.

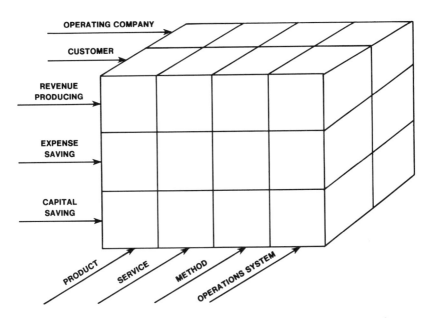

Figure 18-3. Project classification by type, user, and economic impact.

Several projects often compete for limited resources. Although many other factors may influence decisions, a critical question is: Which projects will have the most favorable economic impact? Since it is costly and difficult to reverse a business decision once it has been implemented, the process of bringing out a new product or service includes both initial and ongoing evaluation of the economic viability of a project. Figure 18-4 represents this continuous project evaluation, beginning with an initial concept driven by technology and market need and continuing into the mature, product management phase after the project has been implemented. As a project evolves, the attendant economic studies reflect an improvement in both the quality and quantity of the input data.

Project evaluation plays another, indirect role that should not be overlooked. By analyzing the economic impact of a project, designers and planners become aware of the economic consequences of each of their decisions. Thus, optimizing the economic viability of a project by frequent evaluation plays a crucial role in shaping its ultimate properties.

Economic evaluations are made at various points throughout the life of a project both jointly and separately by Bell Laboratories, Western Electric, AT&T, and companies providing products and services to customers (for example, Bell operating companies). Because of the changing regulatory environment and resulting competition, project evaluation has become more important in recent years. One consequence is a trend

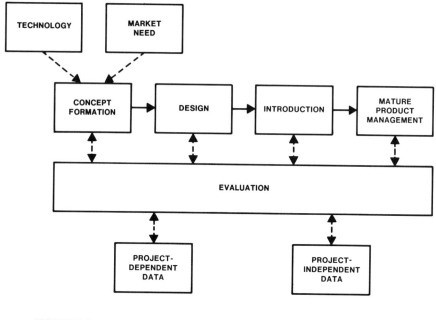

TECHNOLOGY

MARKET NEED

CONCEPT FORMATION

DESIGN

INTRODUCTION

MATURE PRODUCT MANAGEMENT

EVALUATION

PROJECT-DEPENDENT DATA

PROJECT-INDEPENDENT DATA

——— PRODUCT FLOW

— — — INFORMATION FLOW

Figure 18-4. Ongoing economic evaluation over the life of a project.

toward uniformity and consistency in evaluation methodology throughout the Bell System. Such a methodology ensures that project evaluations can be compared, which yields these benefits:

• Comparing different projects at the same point in time allows a ranking in the order of economic desirability. This provides a basis for assigning a priority to development or investment funds.

• Comparing updated views of a given project to earlier views (or to actual data) can often indicate the need for revised strategy in project development, manufacturing, or marketing. Significant differences can also indicate weaknesses in forecasting procedures and can help to improve future economic studies.

The rest of this chapter discusses in detail the three major sequential steps in the economic evaluation of a project: **preparing input data, calculating economic measures,** and **presenting and interpreting results**.

716

18.3.2 PREPARING INPUT DATA

Preparing input data is normally the most difficult part of project evaluation because the data must be tailored to each specific project, and the process requires forecasting benefits and costs, often far into the future.

A "perfect" analysis would require information describing the revenues, expenses, investments, and retirements associated with the project in question and every project it affects, from the date of the study (or the date of the first project expenditure) until the last item of equipment in the study is retired. The average service lifetime of equipment in the Bell System ranges from 5 years for computers and motor vehicles to 54 years for underground conduit. The weighted average lifetime of all equipment in the Bell System is 18 years. Moreover, individual equipment items often remain in service far past the average lifetime. In most cases, the uncertainties associated with any long-range forecast are aggravated further by the accelerating pace of technology and the effects of competition.

Project evaluation requires two types of input data:

- **Project-dependent data** include future tariffs, service demands, equipment costs (manufacturer's price and installation costs), personnel requirements, etc. These data obviously vary from project to project. They are supplied by the project specialist responsible for a particular analysis, who in turn, obtains them from the responsible organizations at Bell Laboratories, AT&T, Western Electric, and the Bell operating companies. It is important for the analyst to follow standard guidelines to ensure that assumptions included in the preparation of these data are consistent from project to project.

- **Project-independent data** include data that do not depend on forecasts associated with a particular project. They include financial factors (for example, tax rates or discount rates), depreciation lifetimes (discussed in Section 18.3.3), inflation rates, salaries by wage category, and other data describing the Bell System or describing the characteristics of general categories of equipment or personnel. These *standard factors* tend to have the same value for all projects. For this reason, a standard-factor file can be developed and maintained for use by all analysts. Maintaining a standard-factor file helps to achieve comparability of project evaluation results. Of course, the standard factors should be modified whenever they do not accurately describe a project.

A common analysis technique is to perform an *incremental study*. The word *incremental* refers to the change in forecasted quantities (such as revenues, expenses, and investments) resulting from the development, manufacture, and implementation of the new project. This change is *the value of a quantity if the future includes the project under study minus the value of the quantity without the project.* The view of the future without the proposed project is sometimes referred to as the *baseline* view. In an incremental study, quantities that are unaffected by the new project need not enter the analysis since they disappear in the comparison. For example, overhead costs not directly assignable to the project are not included in incremental calculations. The economic evaluation of a project must take into account the impact of the project under study on other existing and future projects. For example, if sales of an existing product are reduced because of the introduction of a new product, the incremental sales of the existing product are negative.

18.3.3 CALCULATING ECONOMIC MEASURES

Just as preparing input data requires a uniformity of assumptions used to forecast data, calculating economic measures requires a uniformity of assumptions used to model the Bell System. It is not possible to replicate the accounting structure, the tax structure, and other aspects of the Bell System. Therefore, a simplified model must be used to approximate the economic impact of a project. Consistent use of a model enhances comparability of different projects at the same point in time or of the same project at different points in time.

A large degree of consistency in this area is achieved by the use of software systems that have been developed in different parts of the Bell System. Some of the major systems currently in use are:

- The **Economic Impact Study System** (EISS), used for project economic evaluation and special studies at Bell Laboratories. In addition, EISS is currently the AT&T Long Lines standard tool for economic evaluation.

- The **Integrated Planning and Analysis** (IPLAN) system, used mainly for project economic evaluation at Western Electric.

- The **Capital Utilization Criteria** (CUCRIT) program, used mainly for project economic evaluation and capital budgeting and planning at AT&T and in the operating companies.

Besides these general economic analysis programs, other software tools have been developed in many areas of the Bell System to meet specialized needs. The input requirements and output formats of EISS, IPLAN, CUCRIT, and other tools reflect the specific needs of the user community each supports (for example, project development, manufacture, or software tool implementation).

A project can be viewed in terms of three different types of economic measures: **cash flow, regulatory,** and **accounting.** The measures chosen depend on the purpose of the description. The models used to calculate measures representing these viewpoints can be constructed so that the results are consistent and complementary.

Cash Flow Measures

Cash flow measures are calculated using data that describe the actual movement of funds into and out of a company. To decide whether a product should be developed, its impact on the cash flow of the manufacturer as well as the operating companies (if they are the purchasers) must be determined.

There are four components of cash flow used to describe a project. A particular project can require input data for some or all of them.

- **Revenues** are the payment (or promise of payment) received by a company from a customer in exchange for goods or services.

- **Expenses** are costs incurred by the day-to-day operation of the equipment and facilities required to provide a product or service. Purchased materials with an expected useful life of less than a year are considered supplies, and their costs are treated as expenses. However, the costs (such as depreciation) of owning other equipment and facilities are capital costs and are not included in cash flow measures.

- **Current taxes** are liabilities to taxing authorities incurred in a given year. The most important type is income taxes payable to federal, state, and local authorities.

- **Capital investment,** mentioned earlier, is the initial 1-time expenditure to acquire and install equipment and facilities. It includes the labor required to plan and install new equipment as well as the price of the equipment.

Measures based on these components are used to describe the economic impact of a project on the cash flow of a company. Calculating cash flow and other measures that are linearly dependent on the input data with incremental input data gives the change in these measures due to the implementation of a project. That is, incremental input produces incremental measures.

The following paragraphs describe four cash flow measures: **pretax funds, posttax cash flow, cumulative discounted cash flow,** and **internal rate of return.**

Pretax Funds. Pretax funds is an annual measure that describes the difference between cash inflows and cash outflows due to a project before any tax effects are included. It represents the inherent cash flow character of a project based on the revenues and costs it generates: *Pretax funds equals revenues minus expenses minus capital investment.*

Posttax Cash Flow. Like pretax funds, posttax cash flow represents differences between cash inflows and cash outflows, except it takes taxes into account: *Posttax cash flow equals pretax funds minus current taxes.* If the incremental posttax cash flow due to a project is positive in a given year, the project is providing funds to pay interest to debt holders and dividends to shareowners. The posttax cash flow of a company in excess of dividends and interest represents cash that is available to be used immediately as needed. In this sense, posttax cash flow is an annual measure of the impact of a project on the short-range economic viability of a company.

Current taxes, as mentioned above, are the tax liabilities payable in a given year. *Current taxes are calculated by subtracting any realized tax credits[3] from the product of the tax rate and taxable income (revenues minus allowed deductions).*

Allowed deductions include expenses, tax depreciation, interest on debt, and other taxes paid, such as gross receipts, property, state, and local. (A more detailed discussion of taxes would include salvage, cost of removal, taxable gain, interest charged during construction, and other effects.)

Tax depreciation is a deduction associated with a capital investment during each year after it is placed in service until the investment is fully depreciated. Normally, this happens when its age is equal to its tax life (the length of time depreciation can be deducted) but, because of salvage value, full depreciation can occur sooner. Tax depreciation is an allocation of a portion of the initial capital investment to each year of its tax life. The sum of yearly tax depreciation over the tax life of a project is equal to the original investment less salvage value. In many instances, tax depreciation for current taxes is calculated with an accelerated schedule. This means that the allocation of the cost of capital equipment over the life of the equipment is greater in the early years of the investment than in the later years; thus, the tax deduction is larger, and taxes are less in the early years. The total depreciation with an accelerated schedule is the same as with a straight-line schedule (equal depreciation each year), but the timing is different. The overall effect is to defer taxes until the later years of the investment, which yields an overall benefit to the taxpayer (the company) because of the *time value of money* (discussed below).

---

[3] An important tax credit for the Bell System is the investment tax credit that results from the purchase of capital equipment. The amount of the allowed credit is determined by the useful life of the investment.

Interest on debt, another allowed deduction, arises from one of the external sources of funds for the Bell System—the sale of debt. The other external source is the sale of equity, or ownership, in the Bell System, which is represented by shares of stock. Funds from debt are obtained by borrowing money or selling bonds. The debt ratio (ratio of debt to total assets) for the Bell System is about 0.45, that is, on the average, 45 percent of every investment is financed by selling debt. The Bell System, like any other company, has contractual agreements to make regular interest payments to debt holders for the use of their funds. When a company cannot meet scheduled interest payments, it is bankrupt. On the other hand, a company's dividend payments to shareowners are a matter of corporate policy.

Cumulative Discounted Cash Flow. Some projects require capital investments in the early years of their implementation in order to generate revenues or expense savings in later years. In these cases, there is a period when the project decreases the cash flow of the company followed by a period when the project increases the cash flow of the company. The year-by-year value of the annual posttax cash flow is important because it describes the timing as well as the magnitude of the economic impact of a project. Another significant measure of economic impact is the overall cumulative effect of a project, including the effect of the time value of money, that is, the earning power of money over time.

For example, if \$1.00 is invested in a savings account at 10 percent simple annual interest, the principal plus interest will equal \$1.10 one year later. In other words, \$1.00 *now* is equivalent to \$1.10 one year later.

Amounts of money associated with different points in time must, therefore, be adjusted before they can be combined. One way to make this adjustment is to discount, or *present worth,* amounts occurring at different times in the future to an equivalent present value. After this adjustment, the resulting items of money are comparable and can be added meaningfully. In applying the discounting concept, future worth (FW) is translated to the equivalent present worth (PW) by:

$$PW = \frac{FW}{(1+d)^n}$$

where n is the number of periods separating the points in time associated with PW and FW, and d is the discount rate per period. In general, the discount rate is determined by a company's cost of capital.

When there is a time sequence of annual cash flow values, the concept of discounting is expanded to a cumulative discounted value. Each annual value is discounted to the same point in time, and these discounted values are then summed to form the cumulative present worth or the cumulative discounted value. Thus, for a sequence of values, $[x_k]$, for periods, k, from 0 to n, the cumulative present worth (in time period 0) at discount rate d is given by:

$$PW_d^n[x_k] = \sum_{k=0}^{n} \frac{x_k}{(1+d)^k} \cdot$$

Applying this calculation to annual cash flows yields the *cumulative discounted cash flow*, or CDCF. This measure represents the cumulative impact of a project on the treasury of a company, taking into account the time value of money over an interval of time. The information in a CDCF is different from but complementary to the information in a plot of posttax cash flows. The CDCF is often represented by a curve as a function of time. Each point on the curve represents the cumulative present worth of cash flows occurring up to that time. Figure 18-5 is an example of a typical CDCF curve.

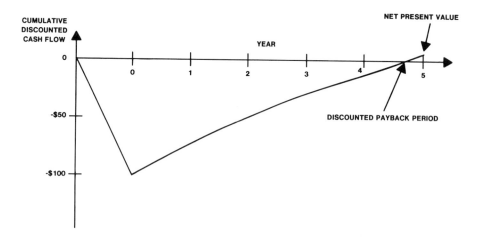

YEAR	DISCOUNT FACTOR @12%	POSTTAX CASH FLOW	ANNUAL DISCOUNTED CASH FLOW	CUMULATIVE DISCOUNTED CASH FLOW
0	1	-$100.00	-$100.00	-$100.00
1	.8929	29.00	25.89	-74.11
2	.7972	29.00	23.12	-50.99
3	.7118	29.00	20.64	-30.35
4	.6355	28.50	18.11	-12.14
5	.5675	28.50	16.17	(NPV) 3.93

Figure 18-5. Simple example of a CDCF curve. (Assuming an initial expenditure of $100 in year 0, followed by the posttax flows shown in the third column.)

Two points on the CDCF curve have special significance:

- **Net present value (NPV).** NPV is the value of the CDCF at the end of the project life, that is, the cumulative impact of a project on the cash flows of a company (including the time value of money) over the full life of the project. A positive NPV indicates that the overall impact of a project is beneficial. (Of course, other economic aspects, such as the amount of uncertainty in forecasted data or timing of the expected cash flows, must also be considered to evaluate the worth of a project.)

- **Discounted payback period (DPP).** All projects require an initial investment in design and development, and many projects require an initial capital investment to achieve future benefits. Initially, the CDCF is negative. However, as the benefits associated with the project begin to accrue, the CDCF becomes less negative and, for an overall profitable project, eventually becomes positive. The DPP is the time required for the CDCF to achieve a positive value. After the DPP, an ongoing project will have paid back the initial investment and will be making money available for day-to-day operations. In general, the longer the discounted payback period, the greater the risk associated with the project. The reason for this is that the CDCF is calculated using forecasted data, which normally become less certain as a forecast is extended farther into the future.

Internal Rate of Return (IROR). The internal rate of return is the discount rate that, together with the estimated annual posttax cash flows of a project, would yield an NPV equal to zero. Projects requiring an initial investment to achieve future benefits are profitable if the IROR is greater than the cost of money used to finance the project. In other words, even if a company had to pay a rate equal to the IROR for the use of investor funds to finance a project, it would still "break even" (NPV=0). If a company is, in fact, paying less than the IROR for the use of capital, the project will be profitable (NPV>0).

Regulatory Measures

Cash flow measures are calculated using real cash flows: revenues, expenses, current taxes, and investments. These items represent funds actually received or paid out in a given year. However, another class of measures depends on rules and definitions prescribed by regulatory and accounting practices. The rest of this section discusses **revenue requirements** and **contribution**, both regulatory measures, and **net income**, an accounting measure.

Revenue Requirements. Revenue requirements is a measure of the revenues needed by a company to sustain its operations. In a regulatory environment, the rates a company is allowed to charge for its products and services are set by regulatory agencies so that the company's revenues are equal to its revenue requirements.

The impact of a project on the revenue requirements of a regulated company is important because of the fundamental role of this measure in determining the allowed revenues of the company. Projects associated with existing products or services should reduce revenue requirements relative to a baseline without the project. This will reduce the overall need to increase rates resulting from the pressures of inflation and other sources. For a new customer product or service, the new investments and expenses normally required would increase the revenue requirements but would be offset by the new revenues.

Revenue requirements are determined as follows: *Revenue requirements equal expenses plus required earnings plus book depreciation plus operating taxes based on revenues equal to revenue requirements (revenue requirements taxes).*

The expenses used to calculate the impact of a project on revenue requirements are the same (the operating costs) as those used to calculate the impact of a project on cash flows. But the other components of revenue requirements are fundamentally different from the components of cash flow measures:

- **Required earnings.** Revenue requirements include a charge, or return, required by investors for the use of their funds (equity or debt). The overall required earnings used to determine revenue requirements are calculated by *multiplying a rate of return times a rate base.* The rate base is the *current value* (defined as the *first cost minus accumulated depreciation*) of existing capital investments. The allowed rate of return is set by regulatory agencies, which also decide exactly what can be included in the rate base. It is principally through this mechanism that regulatory agencies influence revenue requirements and, thus, control the overall rates charged to subscribers for products and services.

- **Book depreciation.** Book depreciation is sometimes classified as an operating expense in the calculation of revenue requirements or net income (see below). However, it is important to realize that book depreciation is not a cash flow but a regulatory or accounting artifact that distributes or allocates the impact of an investment (a real cash flow) over the life of the investment. Regulatory policy requires that book depreciation be calculated on a straight-line basis with an equal amount allocated to each year of the life of the project.

  One consequence of this procedure is to "smooth" the impact of an investment on revenue requirements and eliminate abrupt year-to-year variations due to investments with useful lives of several years.

- **Revenue requirements taxes.** Like the taxes used to calculate cash flows (current taxes), operating taxes for revenue requirements depend on revenues, tax deductions, and tax credits. However, these components differ from those used to calculate current taxes in important ways:

— Taxes on revenues in excess of revenue requirements are not included.

— The depreciation used as a deduction for revenue requirements taxes is not the same as the tax depreciation used to calculate current taxes. The calculation of book depreciation as a deduction in revenue requirements taxes is on a straight-line basis. Depreciation as a deduction for current taxes often uses an accelerated schedule. Moreover, the assumed life of an investment can be different for these two calculations.[4] The depreciation rate for revenue requirements taxes is determined by the average service life of the equipment in its capital account. For current taxes, the tax rate is specified by tax regulation.

— Finally, there is a difference in the method of treating investment tax credits in current taxes and in revenue requirement taxes. For current taxes, the entire tax credit associated with an investment is taken in the year of the investment. For revenue requirement taxes, the same tax credit is allocated equally to each year of the life of the investment.

Contribution. Contribution is a regulatory measure defined as the *cumulative discounted value of revenues minus revenue requirements over the life of a project*. This measure is analogous to the net present value (NPV) from the cash flow point of view. When the NPV and the contribution associated with a project are calculated in a consistent manner and with a discount rate equal to the composite cost of money, they are proportional by a constant that depends on the tax rates:

$$NPV = (1\text{-}T)\ (1\text{-}g)\ (\text{Contribution}),$$

where T is the income tax rate, and g is the gross receipts tax rate (a percentage of the gross revenues).

Both measures describe a project over its entire lifetime and should be expected to give consistent answers. (That is, a collection of projects should always be ranked in the same order whether the ranking criterion uses a cash flow or a regulatory measure.)

The relationship between the two cumulative discounted measures, NPV and contribution, cannot be extended to their annual counterparts.

---

[4] The original investment value may also differ for the two calculations because certain costs may be included in one calculation and not the other.

On a year-by-year basis, there is no correspondence between annual post-tax cash flows and revenue requirements.

Accounting Measures — Net Income

The income statement of a company presents the course of business over a period of time from an accountant's point of view. Quarterly income statements for AT&T and its consolidated subsidiaries can be found, for example, in the AT&T *Share Owners Newsletter;* yearly income statements are normally an essential part of any company's annual report. Values in the income statement describe the company's *profit* in terms of revenues and costs as defined by accounting rules and regulations. Analysts and investors use the income statement, together with a statement of a company's assets and liabilities at a point in time and a description of cash flows during the period of the income statement, to assess the financial and economic well-being of a company. Therefore, the year-by-year impact of a project on the income statement is an important aspect of the economic evaluation of a project.

Net income is the final figure (the "bottom line") in an income statement: *Net income equals revenues minus expenses minus book depreciation minus operating taxes minus interest.* Net income shares many features with the annual revenue minus revenue requirements for a project. For example, both depend on revenues, expenses, operating taxes, and book depreciation. In both cases, taxes and book depreciation are bookkeeping artifacts defined by regulatory or accounting practices. However, there are also important differences in the calculation of net income and contribution:

- The taxes used to calculate net income include not only taxes on revenues equal to revenue requirements, but also operating taxes incurred by the total revenues. Thus, net income is fully a posttax measure.

- Net income accounts only for the interest paid to the debt holder rather than the return required by the *composite investor* (a weighted average of debt and equity holders). It is the earnings available to the equity holder after debt obligations have been satisfied. In fact, the annual earnings per common share (commonly used as a measure of corporate performance by financial analysts) is calculated by *subtracting the preferred dividend requirements from the net income and dividing by the number of common shares outstanding.* Part of net income is paid directly to the owners of a firm (the shareowners) in the form of dividend payments. The remainder, by definition, is reinvested (retained earnings). Since retained earnings increase the total assets of a company, the value of each share of outstanding stock is also increased. (This *book value* of a share of stock should not be confused with the *market value* of a share, which depends on many other factors.) Thus, the shareowners receive net income either directly as dividend payments or indirectly as retained earnings that increase the value of their stock.

18.3.4 PRESENTING AND INTERPRETING RESULTS

The final stage of project evaluation is presenting and interpreting the results of the calculation of economic measures. Standardization and uniformity of format and content are essential characteristics of this stage because they enhance the efficiency of incorporating the results of an economic analysis into management decisions. Preparing a standard report including tables, graphs, and supporting text ensures that three criteria are met:

- The report will be complete.

- Reports prepared for different projects will be directly comparable.

- Reports prepared at different times will be directly comparable.

As with the preparation of input data and the calculation of economic measures, the trend in the Bell System has been toward standardization in reporting. For example, at Bell Laboratories, a Bell Laboratories Executive Summary, describing the economic impact of a proposed project on the Bell System, is currently required for all funds authorized or reauthorized by Western Electric for specific development projects.

The Bell Laboratories Executive Summary is a concise document that meets the three criteria stated above. It consists of five principal parts:

1) **Project description.** The project description presents a summary of the objectives of the project and a brief functional description of the hardware and software products expected to result from its development and implementation. The project is characterized in terms of its principal economic nature (capital saving, expense saving, or revenue producing), and any existing or potential future projects that would be significantly affected by implementation of the project are identified.

2) **Project implementation assumptions.** Project implementation assumptions identify and describe the major assumptions included in the economic evaluation of the project. For most projects, the types of information that need to be discussed include demand, prices and tariffs, time periods (lifetimes and study periods), and personnel levels. Besides these project-dependent data, any changes to standard factors such as inflation rates, wages, or financial factors must also be explained.

3) **Display of economic measures.** Figure 18-6 shows the display format implemented in the Bell Laboratories Executive Summary for a typical project. These are specified measures displayed in a required format. Charts A and B describe the impact of a project on the Bell operating companies in terms of cash flow and regulatory measures. Chart C describes the project from the point of

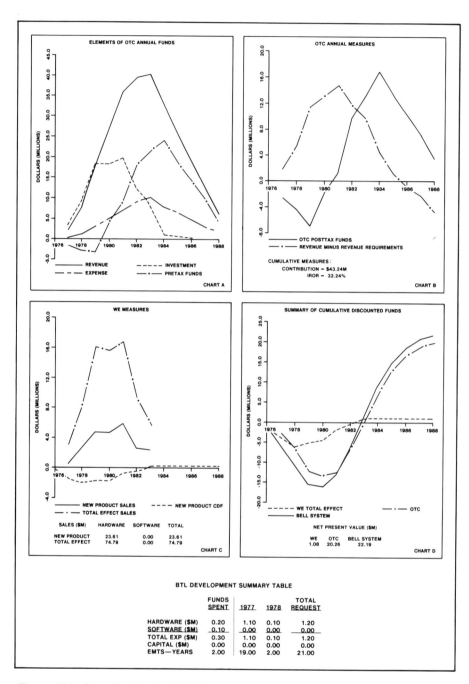

Figure 18-6. Example of a standard display in the Bell Laboratories Executive Summary.

view of Western Electric, and the table describes the impact of the
project on Bell Laboratories. Finally, chart D plots the impact of
the project on the year-by-year value of the cumulative discounted
cash flow for the operating companies, Western Electric, and the
consolidated Bell System. The net present values for these three
curves are explicitly displayed on this chart. The information in
Figure 18-6 was chosen to satisfy the needs of the Bell Laboratories
Executive Summary; for other standard reports, different measures
or different formats might be more appropriate.

4) **Interpretation of economic measures.** The analyst should interpret and explain the significance of the results of the calculation of
economic measures. The salient features of the analysis should be
addressed and overall conclusions summarized.

5) **Sensitivity to alternative assumptions.** An essential part of any
economic evaluation is an analysis of the sensitivity of the results
to the input data. This includes an analysis of the risks associated
with project implementation. Both pessimistic and optimistic versions of the input data should be used to complement the results
found in the nominal study based on the most likely values of the
input data.

18.4 APPLICATION OF NEW TECHNOLOGY

The success of the Bell System has always depended on products and services fed by a stream of technological advances. In some cases, the
advances have reduced the first cost or the operating expenses. Sometimes technology has led to the introduction of an entirely new product
or service or to the addition of new features at negligible additional costs.
This dependence on technology requires a close cooperation between
those who plan and develop the products and services and the scientists
and engineers who develop the underlying technologies. In the Bell System, the overall process has been described by Ian M. Ross, current
president of Bell Laboratories, as "organized creation and application of
technology."

The following sections discuss the coupling of technology and market
needs and describe two examples of the effect of new technology on
products and services.

18.4.1 MATCHING TECHNOLOGY TO BELL SYSTEM NEEDS

Creating a new product or service is virtually synonymous with the
application of new or improved technology. But, while technology
shapes new markets, it must itself be shaped by the marketplace.

The process of developing new technology is somewhat analogous to using two teams, one on each side of a mountain, to drill a tunnel through the mountain. Both teams are ready to burrow ahead, but without communication and coordination, their efforts to construct a useful product (in this case, a tunnel) would prove futile. Specifically, the optimal point from which to dig may appear quite different if judged with the limited vision available on only one side of the mountain. So it is with new telecommunications technology. The technology team must be aware of product or service needs where the fruits of its work might be used. Likewise, the team responsible for the development of a new product or service must be aware of the potential of new technologies that it could successfully apply. The most useful advances and the most successful products arise from the combined and closely coordinated efforts of technologists and system designers.

Matching new technology to market needs in the Bell System has been an important role of Bell Laboratories systems engineering and development groups. Interactions with AT&T and operating companies provide information on market needs. Interactions with technology areas in Bell Laboratories and knowledge of new technology developed elsewhere provide the means for meeting those needs. The latter interactions also may result in ideas for new, marketable products and services. Because many organizations in the Bell System are involved in the overall process, committees and forums have been established to help bridge organizational and geographic boundaries.

18.4.2 EXAMPLES OF APPLYING NEW TECHNOLOGIES

No other recent technology can match the impact that integrated circuits have had on Bell System products and services. By packing more and more functions onto a single piece of silicon, microelectronics[5] technology has led directly to equipment that is smaller, less costly, cheaper to operate, and more reliable.

While successive generations of integrated circuits, such as memories, can provide the same electrical function at progressively lower costs, these cost reductions can be overshadowed by the savings that they trigger in other parts of the product. Packing more onto silicon reduces the number of circuit boards (and the associated costs of assembly and testing), connectors, backplane wiring, and frames. In addition, operating expenses become lower as a result of higher system reliability and lower requirements for operating power and cooling. Thus, even if there were

---

[5] *Scientific American* 1977 is an excellent introduction to the technology and applications of microelectronics.

no reduction in the cost per function of the integrated circuits themselves, there would still be an incentive to use devices of smaller size and increased capability to achieve lower equipment costs.

More frequently, modification of an existing product represents an opportunity to use "cheap" silicon to add new features. Often this adds intelligence to the network and provides more capabilities for the customers. Some products and services become economically feasible only as the cost per electronic function decreases sufficiently.

Thus, integrated circuits can reduce product costs and operating expenses, can provide new features at negligible additional cost, or can make a new product or service feasible. The following examples illustrate most of these characteristics.

Cost Reduction Example — Memory for Electronic Switching Systems

A dramatic example of the results brought about by advances in microelectronics is the evolution of electronic switching system memory portrayed in Figure 18-7.

By the mid-1970s, semiconductor memories, which had been deemed too expensive at the 1-kilobit (1K) level, finally displaced magnetic memories. First came the 4K random-access memory (RAM), introduced into the main store for the 1AESS switching equipment in 1977. As a result of advances in semiconductor technology, it was supplanted by the 16K RAM in 1978. In 1981, the 64K RAM took over; the 256K memory will be introduced in 1984. While the device costs (measured in cents per bit) were dropping significantly, the cost of memory (measured at the system level) was dropping even more rapidly. By eliminating frames, boards, and backplane wiring, the system costs for memory dropped by a factor of more than 30 from 1977 to 1981. As memory costs decline, more memory tends to be used. This is partly due to the shift to higher-level languages for the stored-program control—in effect, trading relatively expensive software development for cheaper hardware.

But cost savings are not the whole story. Figure 18-7 shows an important reduction in power dissipation accompanied by an increase in speed. Both characteristics reflect the improvements that result from shrinking the dimension of the circuits on the devices. In addition, it is well known that the reliability of connections is higher on silicon than in any other part of the system. Thus, moving a connection from a printed wiring board to a silicon chip improves the system reliability.

It might be tempting to conclude that in semiconductor memory design, the technologist could simply press ahead to higher levels of integration without consulting system designers. After all, the benefits of previous designs are obvious. But there are important tradeoffs to be considered in designing these memories. One of them is the balance

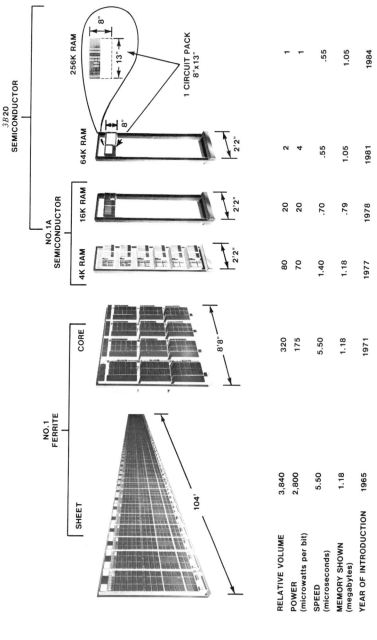

	SHEET	CORE	4K RAM	16K RAM	64K RAM
RELATIVE VOLUME	3,840	320	80	20	2
POWER (microwatts per bit)	2,800	175	70	20	4
SPEED (microseconds)	5.50	5.50	1.40	.70	.55
MEMORY SHOWN (megabytes)	1.18	1.18	1.18	.79	1.05
YEAR OF INTRODUCTION	1965	1971	1977	1978	1981

256K RAM: 1 / 1 / .55 / 1.05 / 1984

Figure 18-7. Electronic switching system memory evolution.

between speed and power. If the device designer knows the speed requirements of the system, the memory can be tailored appropriately. A device operating at a speed that is extremely high would lead to a wasteful dissipation of power. Should the memory be organized to output a single bit or a full byte? Again, the answer lies with the system designer. It is the ongoing dialogue at Bell Laboratories between the system designer and the technologist that has been a key element in bringing the appropriate innovative changes to the Bell System.

New System Example — Digital Access and Cross-Connect System

Assessing precisely the impact of new technology in shaping new systems is more difficult than evaluating cost reductions. It is seldom possible to identify clear-cut cases where a new development leads immediately to a new product or service. More frequently, someone realizes that both technology and market potential exist. The pocket calculator and video games are examples from the consumer market. Each appeared when the evolution of microelectronics had reached the point where most of the functions could be performed by a few devices, bringing the cost of the product into a marketable range. Analogously, the Digital Access and Cross-Connect System (DACS)[6] was developed when large-scale integration became the state of the art, and it became feasible to place a very important function, the *time-slot interchange* (TSI), on a single chip of silicon. This provided connecting and test access functions that were much improved and less costly at a time when there was a growing need to handle the large volume of circuit rearrangement.

In DACS, cross-connection from incoming to outgoing circuits is accomplished by a combination of time-division and space-division switching. The former is performed by the TSI chip (see Figure 18-8), which takes a stream of twenty-four 12-bit words and rearranges them into 256 time slots of a data bus. (Section 7.3.3 discusses the operation of a TSI.)

An examination of Figure 18-8 suggests why the device only became feasible in the very early 1980s: The chip accommodates almost 4000 bits of "static" metal-oxide semiconductor memory. In addition, even more area is taken up by the control logic. This level of complexity represented the state of the art at that time. In fact, when first manufactured, the TSI chip was Western Electric's largest integrated circuit.

The TSI chip is not the only complex integrated circuit in DACS. In combination with other large-scale integrated circuits, it makes possible a compact and flexible hardware design that is expected to create an attractive market for the system.

---

[6] See Section 9.4.3.

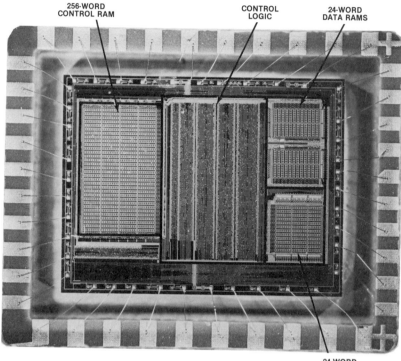

256-WORD
CONTROL RAM

CONTROL
LOGIC

24-WORD
DATA RAMS

24-WORD
ALTERNATE
MESSAGE STORE

Figure 18-8. The Bell System's largest very large-scale integrated chip in 1980
(429 by 328 mils)—a time-slot interchange (TSI).

18.5 INTEGRATION OF NEW AND OLD SYSTEMS

18.5.1 THE NATURE OF THE PROBLEM

The tradition of designing telephone equipment for long life has
influenced depreciation rates and the tariffs approved by public utility
commissions. However, tariffs have recently given consideration to
obsolescence as well as equipment life, because equipment offering new
technology, economic savings, and new features usually phases out old
equipment. Since the rate of replacement is usually much slower than
the evolution of new designs, though, several generations of equipment
will usually be in operation simultaneously. Local switching systems are
a good example of this phenomenon. As seen in Figure 18-9, from the
1930s to the 1980s, telephone lines have been served by various switching

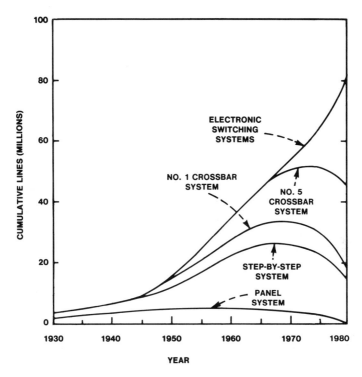

**Figure 18-9. Automatic switching systems
serving Bell System lines, 1930-1980.**

systems. The birth, growth, and decline of switching technologies is evident, along with the fact that a mixture of switching technologies will
usually exist at any point in time. Eventually, old technologies are
entirely replaced by newer systems.[7]

The oldest type of Western Electric switching system currently in service in the Bell System is the step-by-step, which was first used in the
Bell System in 1919. The continued serviceability of this older equipment
along with limitations on capital for its replacement are the primary reasons for its continued presence in the system. However, step-by-step systems are expected to be phased out completely by about 1990. New electronic switching systems (both analog and digital) are currently being
used for the expansion of service as well as for replacement. Since
different electronic switching designs are being developed with time,
various electronic switching systems coexist in service just as various
electromechanical switching systems still do.

---

[7] The last of the panel systems was retired in 1982.

Each new design of telecommunications equipment must consider several factors: (1) new switching and transmission systems must be designed so that interfaces with the variety of systems in service will function properly; (2) new designs may be restricted by building characteristics such as floor loading, frame height, temperature variations, or electrical induction; and (3) customers will react to changes in operations. Standardization of interfaces, environments, and operations allows for a smooth introduction of new equipment but, at the same time, may restrict the economies and service features that new technologies might realize.

Most new services must be made available in areas served by old and new equipment and at the same price to the customer. The biggest challenge of a new service feature is to retrofit it into the old equipment without losing money. As a specific example of integration of a new system, the following section discusses some of the interface problems that had to be resolved when TOUCH-TONE calling was introduced.

18.5.2 SOME PROBLEMS IN INTERFACING TOUCH-TONE SERVICE WITH OLDER EQUIPMENT

The rotary dial represents a highly successful coupling of human capability and telephone technology. For reliability and economy, the rotary dial has been unsurpassed since the earliest days of automatic switching, and it is more reliable than a pushbutton dial. The convenience of pushbuttons, however, was widely recognized. The advent of the transistor offered an economically feasible way to implement pushbutton dialing. TOUCH-TONE dialing was introduced in 1963. It can be tariffed as a premium service because it is a convenience to the customer. Since it is faster than rotary dialing, it ties up registers for shorter intervals in common-control offices. It also provides two additional characters (# and *), which are useful in providing new capabilities.

However, as a premium service, TOUCH-TONE dialing had to be made available on individual lines in areas that were predominantly rotary dial. It had to be possible to get service from any type of central office to avoid number changes for customers. Common-control offices that receive customer-dialed digits in register circuits had to be designed to receive signals from TOUCH-TONE telephones as well as rotary-dial signals. Conversion from rotary-dial service to TOUCH-TONE service required the installation of a TOUCH-TONE telephone and a central office change to access an appropriate register. To avoid problems that would occur if these two events were not synchronized, registers that accept signals from either rotary or TOUCH-TONE dialing were designed. As an added benefit, the registers accommodate party lines with a mix of TOUCH-TONE telephones and rotary-dial station sets or customers who have both types of telephones on their premises.

Because step-by-step switching systems use direct control of the switches from the station, conversion to TOUCH-TONE dialing is difficult and relatively expensive. All *TOUCH-TONE* telephone (or mixed) lines must be grouped together and new switching stages introduced so that these lines may access appropriate registers. The registers must then generate dial pulses to drive the succeeding switches.

Another problem in introducing *TOUCH-TONE* service is related to dial tone. When dial tone is sent to the customer, some of its energy is reflected back to the central office by the station set, and in early installations, this reduced the sensitivity of the receiver at the central office to the first digit dialed. It was necessary to use a new dial-tone signal composed of frequencies that did not interfere with *TOUCH-TONE* telephone operation in all central offices accommodating *TOUCH-TONE* service. The new signal also had to be similar enough to the old one to be accepted as a dial tone by customers.

18.6 HUMAN FACTORS IN THE BELL SYSTEM

The Bell System, historically, has recognized the importance of factoring people's capabilities and behavior into the design of telephone networks, customer equipment, and services. As early as the 1920s, human hearing and speech characteristics were studied in order to set standards for transmission quality. Subsequently, Bell Laboratories engineers and behavioral scientists also became concerned with considerations such as the shape, size, and weight of telephone sets; the design of telephone dials; and, later, the development of direct distance dialing codes.

As the value of *human factors* became more apparent, activities expanded to include the consideration of Bell System employees—technicians, business office employees, telephone operators, and managers—as well as customers. Today, human factors specialists are helping to improve the productivity and effectiveness of Bell System employees by contributing to the design of computer-based operations systems, automated operator systems, stored-program control systems and services, and customer-service procedures.

Bell Laboratories does more human factors work than any other organization in the world outside of the United States government.[8] The following sections provide a brief description of the human factors discipline at Bell Laboratories and some illustrative applications.

18.6.1 THE HUMAN FACTORS DISCIPLINE

The goal of human factors at Bell Laboratories is to ensure that Bell System products and services are designed for the people who use them. For customers, this means ensuring that a product or service satisfies customer needs and preferences, operates easily, and is fully supported by good

---

[8] *Bell Labs News*, March 1, 1982, p. 4.

instructions and service delivery systems. For employees, it means allocating system tasks among people and system components and ensuring that employees can perform their duties efficiently and effectively in rewarding and meaningful jobs. To deal with these issues, human factors specialists must have two types of knowledge. They must know about people—their perceptual skills, their learning abilities, their physical characteristics, how they make decisions—and they must know methods for analyzing human performance in systems. Because of the knowledge required, many human factors specialists have been trained in experimental psychology, which emphasizes methods for measuring and analyzing human performance.

Some of the information used in human factors work is available in handbooks and other references, particularly facts that are useful in physical design (see Figure 18-10). But, because of the absence of a large body

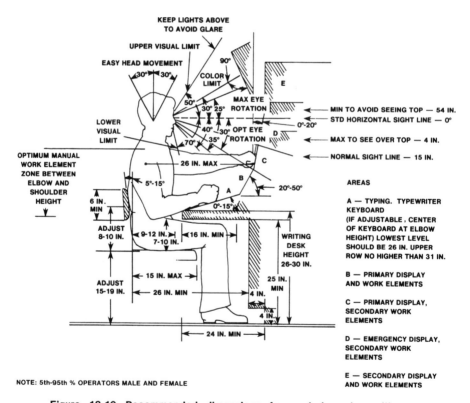

Figure 18-10. Recommended dimensions for seated work position.
(Example of information available from human factors reference sources.) *Redrawn from Van Cott and Kinkade 1972, which is based on data from Dreyfuss 1959, Kennedy and Bates 1965, and Woodson and Conover 1964.*

of useful theory, much design information must be obtained on a case-by-case basis. Data collection techniques used in human factors analysis include laboratory experiments, field observation studies, evaluations of alternative designs, and user interviews.

18.6.2 DESIGNING FOR CUSTOMER SERVICES

Until recently, customer interactions with the telephone system have been relatively simple, involving only placing and receiving voice calls. But advancements in technology have provided the opportunity to create products and services of great complexity. And, as services become more complex, the interactions between the services and their users become more complicated. Human factors specialists have the responsibility of minimizing the problems this can create.

For example, Advance Calling is a potential new service that has been installed for field trials. This service enables a customer to record a voice message, address it to any telephone that can be dialed directly, and specify the time at which the message is to be delivered—all from a standard telephone set. A list of the human factors questions that arise in designing such a service would include:

- In what sequence should the necessary information (voice message, time, destination telephone number) be entered, or is it unimportant?

- Should the time and destination telephone number be repeated back to the customer for verification? (This complicates the interaction.)

- Should the recorded message be replayed to the customer, and if found unacceptable, should the system permit recording of a new message without requiring the customer to hang up and start over?

- Should messages be accepted for delivery during late night hours? Should attempts to deliver messages scheduled earlier be continued into late night hours?

- What sorts of instructions should be provided for message delivery time entry? Should a 12- or 24-hour clock be used? If a 12-hour clock is used, how should morning and afternoon be specified? If a message is to cross time zones, which time should the customer specify?

- Should written instructions be available to the customer? What should they emphasize?

- How many times should the message be played to the recipient?

- How good must the voice quality of the recorded message be?

- How can the instructions be designed to satisfy the needs of both inexperienced and regular users?

The human factors specialists who actually addressed these questions used a variety of methods. First, they used logic and experience to identify the most important questions and some possible solutions. A simulation was then constructed that enabled potential customers to attempt to place advance calls. Errors, other difficulties, and user reactions were measured and led to changes and additional testing. Laboratory tests were performed to assess the effect of various digital encoding rates[9] on users' abilities to recognize callers and understand their messages. Surveys were conducted of customer knowledge of 24-hour "military" time and the differences between time zones. Studies previously conducted for a related service provided information about recipients' problems and the number of times a message should be replayed to the recipient.

Because of the close interaction between the procedures the customer must follow and the hardware and software, design decisions resulting from these studies were made in close cooperation with both the systems engineering and development areas. And, because laboratory analysis does not perfectly reflect real life, the initial introduction of Advance Calling included a built-in feature for tracking customer performance.

Many other recently developed customer systems have undergone intensive human factors study, and each had its own unique problems. In Advanced Mobile Phone Service (see Section 11.4.1), questions of the impact of mobile telephone usage on driver performance required in-car studies (see Figure 18-11). Design of the Automated Coin Toll System (ACTS)[10] required studies to determine the length of time needed to insert coins and the sorts of situations in which the customer should be connected with a live operator. Automated Calling Card Service (see Section 11.3.1), PICTUREPHONE meeting service (see Section 11.5.3), the HORIZON communications system, and other recently developed products and services have received similar scrutiny to ensure their "friendliness" to the user.

18.6.3 DESIGNING FOR EMPLOYEES

Human factors has been especially relevant to the design of large operating company data-base systems, such as the Trunks Integrated Records Keeping System (TIRKS),[11] which replace tedious and labor-intensive manual operations. Such systems depend critically upon the accuracy and timeliness of employee-entered data. They also depend on a reasonable allocation of tasks between people and computers. For example, routine data handling and error checking are best done by a computer, while

[9] The caller's message is encoded and stored in digital form.

[10] See Section 10.4.1.

[11] See Section 14.2.1.

Figure 18-11. Human factors simulation. Laboratory and field simulations play a significant role in the design of user interfaces. Here, human factors engineers use a television camera to study the ease of placing and receiving calls while driving. Tests showed that placing a call with Advanced Mobile Phone Service telephones is as easy as tuning a car radio.

people are better at handling the more complicated and less predictable parts of the job (and find them more interesting).

The human factors role in designing such a system includes:

- analyzing the organizational environment in which the system must operate
- analyzing the tasks to be performed
- dividing tasks between people and machines
- defining required employee skill levels
- specifying training and on-the-job documentation.

The success of human factors work in the design of operations systems has led to involvement with more traditional employee groups as well. Below are just a few examples:

- designing and evaluating new hardware for the outside plant, including cable cross-connection terminals and interfaces, splicing equipment, and tools

- designing and modifying the Traffic Service Position System (TSPS)[12] operator console and working on computerized directory assistance systems

- developing improved computerized tools for outside plant planning and engineering (see Section 14.4) and systems for detecting and locating faults in local loops.

Future challenges in employee human factors rival those on the customer side. Telecommunications is becoming more computerized and less manual, and Bell System employees must make the transition. At the same time, there is a growing emphasis on the quality of work life, which is substantially determined by the tasks one performs. The challenge of human factors work on employee issues will be to improve employee productivity and the quality of work life as telecommunications technology becomes more mechanized and complex.

18.7 QUALITY ASSURANCE

18.7.1 THE ROLE OF QUALITY ASSURANCE

Primary responsibility for the quality of products used throughout the Bell System resides with the organizations directly involved in their design, development, manufacture, or supply. Bell Laboratories development organizations are responsible for the quality of design, including specifying requirements and evaluating performance. Western Electric, in carrying out its supply functions, is responsible for the quality of manufacture and repair, including engineering and installation. In addition, Western Electric and other suppliers are directly responsible for the operation of effective, on-line quality control programs.

The distinction between quality control and quality assurance is important. *Quality control* is the set of procedures used by the supplier to provide sufficient process control over the machines, personnel, and material necessary to meet acceptable quality criteria consistently and economically. *Quality assurance*, on the other hand, provides continuing independent verification of satisfactory results from the standpoint of the customer-user. The quality assurance system is structured to determine the effectiveness of quality controls used by the designer and producer

---

[12] See Section 10.4.1.

and is effected through feedback directed to the responsible management of Bell Laboratories, Western Electric, and outside suppliers.[13]

This discussion focuses on the quality assurance system. The basic system consists of two major elements (as shown in Figure 18-12). The first is a highly structured *quality audit*—a system of inspections carried out within Western Electric and at the interfaces between Western Electric and the Bell operating companies. The second element, *quality assurance monitoring*, involves a variety of activities that confirm the effectiveness of prior controls or identify problems requiring corrective action. Section 18.7.2 discusses rating of quality through formal, ongoing **quality assurance audits**, and Section 18.7.3 discusses **quality assurance monitoring**. (More detailed information and numerous examples of the wide variety of recent quality assurance activities can be found in a series of articles in the *Bell Laboratories Record* October through December 1978. Peters and Karraker 1975 contains a description of the Western Electric role in assuring quality.)

Periodic reports highlighting quality problems identified by the quality assurance activities are distributed within Bell Laboratories and, more importantly, to the upper management of Western Electric. Prompt response by Bell Laboratories and Western Electric management resolves the majority of quality problems. Occasionally, corrective actions require additional technical effort involving the Bell Laboratories Quality Assurance Center.

Many products are purchased outside the Bell System for use by the Bell operating companies or for assembly into products manufactured by Western Electric. Usually these products are procured under specifications provided by Bell Laboratories or Western Electric, and quality inspections are performed at the source. In addition, a Quality Surveillance System (QSS), under the administration and guidance of AT&T, has been developed for non—Western Electric (*general-trade*) communications products purchased directly by the Bell operating companies. The basic responsibilities of the QSS include: (1) technical evaluation and determination of risk involved in product use (for example, safety, service, economy), (2) evaluation of the supplier, (3) evaluation of product quality and reliability, (4) an inspection program (including criteria for

---

[13] Many of the basics of both quality control and quality assurance originated within the Bell System. W. A. Shewhart of Bell Laboratories first published the concept of a control chart, widely used in quality control throughout the world (see Shewhart 1926). Shewhart, Dodge, and others at Bell Laboratories continued to develop the statistical foundations of quality assurance auditing and refinements in the implementation of quality assurance through the 1950s. Recent contributions have been towards more economical and statistically powerful tools. **Additional Reading** for Chapter 18 at the end of the book lists documents that contain more information on the development of quality control and quality assurance.

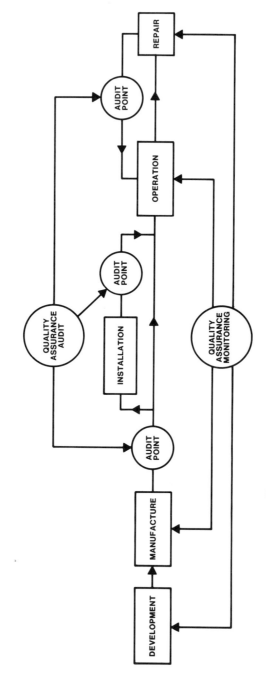

Figure 18-12. Major elements of the quality assurance system.

setting standards, scope of inspection, and acceptance sampling procedures), and (5) field performance monitoring. Although the basic elements of the general-trade QSS are similar to the major functions of the quality assurance program for Western Electric products, significant differences in implementation details may exist because of the proprietary nature of general-trade designs and manufacturing processes. (AT&T 1983 contains additional information on the Quality Surveillance System.)

18.7.2 QUALITY ASSURANCE AUDITING

Quality assurance audits are conducted on products manufactured, installed, and repaired by Western Electric. For example, Western Electric quality assurance personnel take samples of units from the ends of production lines and subject them to extensive tests and inspections. Sampling and inspection are repeated, and the results accumulated over a rating period (currently about six weeks). The Bell Laboratories Quality Assurance Center compares the cumulative statistics to quality standards specified for each product. These quality standards are intended to minimize total cost to the Bell System, that is, to ensure an economic level of quality that is consistent with the costs of manufacture and the subsequent costs of use.

Data on products are gathered from the numerous Western Electric locations at *audit points* (shown in Figure 18-12) and are transmitted to a computerized data base maintained by the Quality Assurance Center. If the data confirm that quality is at a satisfactory level, the results are recorded and filed for future reference. If the data indicate with a reasonably high degree of certainty that the product is not meeting expectations, the results are reported to management and included in an exception report. The audit data are also used to conduct special studies and to analyze quality problems in detail when the need arises. The following paragraphs describe the auditing process.

Audit Structures

In general, quality assurance auditing for a product addresses these basic questions:

1) Was it made correctly?

2) Will it operate properly?

3) Will it continue to operate for a reasonable period of time?

Since it is neither feasible nor desirable to formally rate the quality of each unique product in the many product lines manufactured by Western Electric, products are grouped into rating classes according to major system, similarity of function, Western Electric location, etc. For example, exchange area cable is grouped into three major categories—pulp, PIC[14] air-core, and waterproof—and rated accordingly at each of five manufacturing locations. Within each of these categories, the audit results are often subdivided according to the quality characteristics examined. For example, each exchange area cable category is rated by (1) electrical transmission parameters and (2) visual and mechanical construction characteristics. In total, there are about 2000 subdivisions called *scoring classes*. The audit for each of these classes is a highly structured system of checks and inspections consisting of these basic procedures:

- **Checking** — defining the scope of an audit: tests to be performed and attributes to be checked on each unit of the sample

- **Appraising** — classifying defects found during the checking, and assessing their seriousness: procedures to be used for declaring a product to be nonconforming and subject to corrective action

- **Establishing standards** — determining the expected level of quality

- **Sampling** — selecting units to be inspected

- **Rating** — comparing audit results with established standards and assessing the statistical significance of departures

- **Reporting** — portraying rating results in regularly issued reports and highlighting quality problems for corrective action.

The following paragraphs discuss these procedures in more detail.

Checking. The tests, checks, and inspections performed on each unit of a sample are intended to verify design specifications, engineering requirements (specified or implied in applicable product drawings), generally accepted good workmanship and manufacturing practices, and basic design or application considerations (such as system performance and function, reliability, appearance, life, interchangeability, maintainability, and cost of use or repair). Specific product requirements and testing procedures are derived from a variety of other sources and include visual, mechanical, and electrical tests.

Appraising. Bell System products are currently rated in terms of three alternative quantities: defects per unit, percentage defective, or demerits per unit. A *defect* is defined as a failure to meet a requirement, for example, a measurement exceeding a specified limit. In a defects-per-unit audit, the average number of defects per unit is determined over the rating period. In a percentage-defective audit, a unit of the sample is considered defective if it contains one or more defects, and the percentage of

---

[14] Plastic-insulated cable.

defective units in the total sample is the measure of quality. In a
demerits-per-unit audit, defect seriousness is quantized, as shown in Figure 18-13, according to the classification guidelines summarized in
Table 18-1. The numerical sum of the demerits observed during the
rating period normalized by the number of units sampled is the measure
of quality.

When many different defect items are included in the scope of an
audit, there is some risk that the existence of major defects will be
masked in an overall rating. To mitigate this risk, the Bell Laboratories
Quality Assurance Center frequently specifies a dual rating structure,
wherein the major defect types are rated separately, in addition to the
overall composite rating of the product. Such dual rating is appropriate
when, for instance, Bell operating company service needs specify relatively few incidences of major (class A and B) defects (discussed in more
detail in **Establishing Standards** below).

In another type of dual rating structure (for example, in the auditing
of electronic switching systems and circuit packs), test results and workmanship are rated separately. The quality of workmanship during
manufacture of a product is generally considered to be an indicator of
long-term reliability. Serious defects of workmanship (such as poor
solder connections) can affect operability in service and may not be
readily detected by electrical tests at the time of inspection. This was

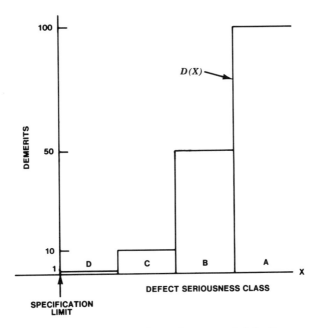

Figure 18-13. Assessing seriousness of defects.

TABLE 18-1

DEFECT SERIOUSNESS CLASSIFICATIONS

Type	Class	
	A **100 Demerits**	**B** **50 Demerits**
Major	1. Will surely cause an operating failure of the unit in service. 2. Will surely cause intermittent operating trouble. 3. Will render unit totally unfit for service. 4. Is apt to cause personal injury or property damage under normal conditions of use.	1. Will probably cause an operating failure of the unit in service. 2. Will surely cause trouble less serious than an operating failure, such as substandard performance. 3. Will surely involve increased maintenance or decreased life. 4. Will cause a major increase in installation effort. 5. Has extreme defects of appearance or finish.
	C **10 Demerits**	**D** **1 Demerit**
Minor	1. May possibly cause an operating failure of the unit in service. 2. Is likely to cause trouble of a nature less serious than an operating failure, such as substandard performance. 3. Is likely to involve increased maintenance or decreased life. 4. Has significant defects of appearance or finish.	1. Will not affect operation, maintenance, or life of the unit in service. 2. Has minor defects of appearance, finish, or workmanship.

particularly true for the older, wired electromechanical equipment. However, with modern, miniaturized electronic equipment, test-type auditing has gained increased importance and become efficient to implement, particularly with systems combining hardware and software.

Another type of audit, involving a check of early-life failures and longer term reliability performance, has also gained prominence. In this type of audit, devices (such as integrated circuits), circuit packs, or systems (such as complete private branch exchanges) are tested for a specified period of time, and the cumulative test failures, indicative of in-service trouble rates, are used as the measure of the level of quality.

Establishing Standards. Usually it is neither feasible nor economical to expect zero defects during continuing production (except for safety-related items). A quality standard is, therefore, an estimate of the economically optimum value of the expected level of the measure of quality. Responsibility for specifying these standards resides with the Bell Laboratories Quality Assurance Center. Setting standards should be an independent and unbiased process, balancing the possibly conflicting pressures of production, the parochial views of design and development, the difficulties encountered during installation, the strong desire to hold down the costs of field operation, and of course, the expectations of the customers or end-users of the products and services. The process of setting a standard also should recognize the scope and structure of the audit to which it applies.

For example, to set standards for a demerits-type audit (where the seriousness of the defects is quantized), in which there are X_A, X_B, X_C, X_D of type A, B, C, and D defects, respectively, the total number of demerits (TD) will be:

$$TD = 100X_A + 50X_B + 10X_C + X_D,$$

where the Xs are Poisson[15] random variables. If λ_A, λ_B, λ_C, and λ_D are used to denote the means of the respective distributions when the production process is at standard, then, setting standards in demerits-type audits usually consists of specifying λ_A, λ_B, λ_C, and λ_D. This is accomplished by taking into account the impact of the different levels of defect seriousness as specified in Table 18-1 on the Bell operating companies. In addition, manufacturing process capabilities must be considered. These studies are frequently carried out empirically, using historical audit data on similar or closely related products. Such an approach is called the *base period study* method, and it presumes that over a period of time the conflicting forces of operating company desires for improved quality and

---

[15] Chapter 5 discusses Poisson distribution.

the producer's attempts to hold down costs have reached, or at least closely approached, an equilibrium value of the level of quality.

This equilibrium model for the optimum cost standard is depicted in Figure 18-14. The standard is defined to be the quality level yielding minimum total cost, that is, the minimum of the sum: *field cost plus manufacturing cost*.

A viable static model is obtained by recognizing that field cost increases as quality degrades. Over the range of interest, the field cost is assumed to increase directly as a power function of the quality level. Manufacturing cost is assumed to increase as quality improves, and this relationship can reasonably be modeled as an inverse power function of the quality level.

Assigning actual field cost values can be very difficult in practice. These costs should take into account both the cost of repair or replacement and the more intangible cost of adverse customer reaction or product reputation.

For major defects, as defined in Table 18-1, it is highly likely that repair activities will be required. The cost of these activities can vary widely, depending on the type of product, location, etc. The extent of adverse customer reaction to, for example, failure of a packaged electronic product will depend on the importance of the system in which it is used (how many customers are affected when the system experiences a service

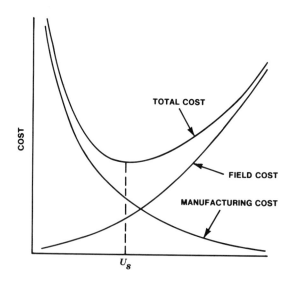

Figure 18-14. Equilibrium model for the optimum cost standard quality, U_s.

problem) and the criticality of the function of the unit (how seriously a customer's service is affected).

Since precision in these cost estimates usually is not feasible, the parameter values are quantized. For example, the constant of proportionality for the field cost, C_f, may take on three relative levels: high, normal, and low, depending on the relative weighting of repair costs and the estimated degree of adverse customer reaction as determined by system importance and criticality of the product's function.

Sampling. Thorough inspection of a single telephone set may cost, at least, several dollars; other more complex equipment may cost hundreds of dollars to inspect. Therefore, it is not feasible to audit each unit when large numbers of units may be manufactured or installed under similar, usually well-controlled conditions. Hence, the optimum quality assurance strategy is to perform thorough tests on a few units and use the results to indicate the quality of the total production. The audits, in general, are designed to monitor the long-term quality of a continuous stream of products, rather than the acceptability of any given lot.

Since a sample is used to estimate the overall level of quality for a scoring class, it should be representative of the ongoing production. This is achieved by randomly selecting samples of units from randomly selected lots suitably distributed over the full rating period. The optimum sample size is determined by an economic model that forms a basis for the *Universal Sampling Plan* (USP), described in detail in Hoadley 1981b.

The fundamental concept underlying the USP is that more effective audits enhance the information feedback loops and, therefore, result in better quality due to management actions. Improved quality reduces field maintenance cost, but this must be balanced against increased audit cost if larger samples are selected. The economic model underlying USP portrays this tradeoff. An optimum sampling plan will minimize the total cost: $K_b(audit\ cost)\ plus\ field\ cost$, where the constant K_b is determined by a constraint on the overall auditing budget.[16] An audit cost can be modeled[17] as a fixed or overhead cost plus a cost that depends on the sample size, n, and the cost per sample, C_i.

The field maintenance cost affected by the audit can be represented as: *field cost equals defects per unit produced times* NC_f, where N is the number of units produced during the quality rating period and C_f is the field cost per defect sent to the field.

---

[16] The complete problem is to minimize the sum of audit costs and field costs over all scoring classes, subject to a budget constraint on the total auditing activity.

[17] The economic models used for sampling and setting standards are separable; audit costs are assumed to be independent of standards and manufacturing costs (mentioned in **Establishing Standards**) independent of sampling.

The probability of detecting poor quality will depend on the statistical procedures used for estimating the quality level and on the threshold selected for deciding that the quality is unsatisfactory. To a good approximation,[18] this detection probability will be proportional to the extent of the poor quality (measured by Θ_p, the multiple of the defects per unit produced at standard quality), the sample size (n), and the expected defect level per unit as inspected in the audit (U_s), or

$$Pr[\text{detecting poor quality} \mid \text{true quality is poor}] = K_d n \Theta_p U_s,$$

where K_d is the constant of proportionality.

Starting with these considerations, a model for total cost can be derived, which is then minimized with respect to n to yield the optimum sample size formula:[19]

$$e = nU_s = \sqrt{K_b K_d^{-1} \, PrNU_s^*}$$

expressed in terms of the *sample expectancy, e.* The sample expectancy e, or nU_s, is the total number of defects expected in the sample if production is operating at standard. The risk factor or cost ratio, $r = C_f/C_a$, where $C_a = C_i/U_s$ can be interpreted as the (incremental) cost to sample and inspect enough units to get one expected defect if production were operating at standard. P is the probability that the true quality is poor, and the poor quality is detected.

Current values of the constants relating to budget restrictions and detection threshold, K_b and K_d, have been determined empirically to be 5 and 1/10, respectively. Figure 18-15 shows the universal sampling curve using these values.

Rating and Reporting. The rating of quality consists of estimating the true level of quality, based on results of the quality assurance audit samples, and then determining whether the quality level is satisfactory, based on a statistical comparison of the estimate with the standard or expected level of quality for each scoring class. The Quality Measurement Plan (QMP)[20] provides a uniform and consistent set of procedures for the rating and reporting of quality.

[18] More complicated models of audit detection power do not significantly affect the *optimum* sample size result. With the approximations used, the sample size result is within a few percent of the more accurate value, over the range of interest.

[19] The square root dependence on production (N) has been used in quality assurance sampling since at least 1930.

[20] See Hoadley 1981a.

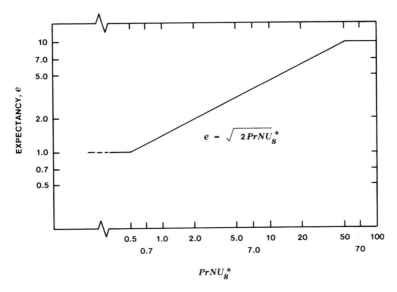

EXPECTANCY, e

$$e = \sqrt{2\,PrNU_s^*}$$

$PrNU_s^*$

Figure 18-15. Universal sampling curve.

It is convenient to describe the (relative) observed level of quality in a particular rating period, t, by the *sample index*, I_t, defined as the ratio of the observed quality level in period t to the sample expectancy (e, as in **Sampling** above). An index of unity means that the observed quality in that period is at standard, and an index greater than one means that the observed level is worse than standard. If T denotes the current rating period, then I_T could be interpreted as a simple estimate[21] of the true quality Θ_T with a sampling variance σ^2.

Each production process is generating a random time series of process indices $\{\Theta_t\}$. If the random process has a fixed (but unknown) process average index Θ and process variance γ^2, at least over a reasonably short interval of time, then one possible estimate of the process average index, taking into account the information contained in the audit results of the previous rating periods, is a weighted long-run average

$$\overline{\Theta} = \sum_t w_t I_t,$$

where the weights are inversely proportional to the individual period variances.

---

[21] This is, in fact, the statistic used in the older t-rate method of rating quality. (See Dodge 1928, pp. 350-368 and Dodge and Torrey 1956, pp. 5-12.)

For rating purposes, information on the long-run average and the current sample index are combined to form an estimate of current quality:[22]

$$\hat{\Theta}_T = W\overline{\Theta} + (1-W)I_T,$$

where

$$W = \frac{\sigma^2}{\sigma^2 + \gamma^2} = \frac{sampling\ variance}{(sampling\ variance) + (process\ variance)}.$$

When the sampling variance is large or when the process variance is small (the process is relatively stable), W will be close to unity, and most of the weight will be on the long-run average, as should be expected. Conversely, when the process variance is large, W will be small, and most of the weight will be on the current sample index.

In practice, W must be estimated from the data. In addition, an interval estimate of the current population index is needed for rating purposes to compare the best measure $(\hat{\Theta}_T)$ against the quality standard.

A Gamma distribution is used to approximate the current produced quality (see Hoadley 1981a) as shown in Figure 18-16. This distribution

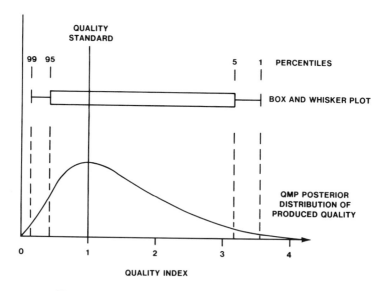

Figure 18-16. Estimated distribution of produced quality
and corresponding box and whisker plot.

[22] This result is derived from a Bayesian model as in Hoadley 1981a.

can be represented by a "box and whisker" plot, where the box extends from the 5th to the 95th percentiles, and the whiskers extend to the 1st and 99th percentiles. This plot forms the basis for quality assurance reporting, as shown in Figure 18-17. When the value of the 99th percentile exceeds unity, the product is declared below normal, since a posteriori, there is at least a 99-percent chance that Θ_T is larger than one, that is, that the level of quality of the production in the current rating period exceeds (is worse than) the standard. Similarly, the rating class is declared to be on alert when the 95th (but not the 99th) percentile exceeds unity.

Nonconformance

The quality assurance function is distinct from quality control and usually cannot serve as a screen of defective products on a lot-by-lot basis. However, when strong statistical evidence indicates that the quality of a product has fallen substantially below standard, the product is identified as nonconforming. In this situation, the Bell Laboratories Quality Assurance Center determines whether the affected product can be shipped to the Bell operating companies or must be reworked or discarded. The role of the quality control process is adequate protection for the customer, and the quality assurance procedures are designed to monitor this control process.

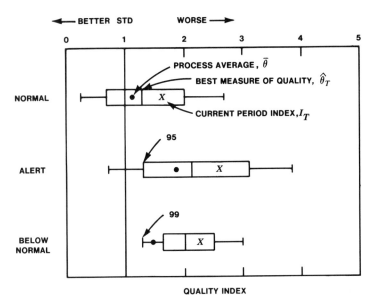

Figure 18-17. Quality reporting criteria.

Evidence for a nonconformance usually originates from the audit; nonconformance procedures are thus subsidiary to and a by-product of the audit procedures. Such procedures are useful primarily where there is a gross quality problem that the producer is unaware of, has misevaluated, or has chosen to ignore. It provides only marginal protection for the customer on a lot-by-lot basis, since all lots may not be sampled in the quality assurance audit, the audit subsample chosen from a selected lot may be small, and the trigger threshold used in declaring the nonconformance is usually severe in order to control the false-alarm probability.

Trigger thresholds, called *allowance numbers*, are specified for the various product lines, frequently for major defect types and minor defect types separately, and sometimes individually for various categories of defects.

When a nonconforming condition is identified, an attempt is usually made to determine the extent of the problem, so that large quantities of a defective product do not appear in the field.

It should be emphasized that the primary objective of the nonconformance procedure is to stimulate corrective action in the material inspection process, the manufacturing or assembly process, and the quality control process. A nonconformance is in effect until the cause of the defects is identified and corrected, with independent verification by the appropriate quality assurance organization.

18.7.3 QUALITY ASSURANCE MONITORING

Quality assurance auditing alone is not always sufficient to ensure adequate quality of the final product; hence, the audit is supplemented by various quality-monitoring activities. As shown in Figure 18-12, monitoring occurs throughout all stages of the development, manufacture, operating, and repair processes to verify that adequate controls have been maintained. In addition, numerous tracking studies are conducted in the Bell operating companies to determine the performance of the products quantitatively under actual field conditions. The quality assurance activities that continue throughout the life of a product are described below.

Surveillance

Continuing surveillance usually occurs when available information indicates that chronic quality problems exist in some phase of the production processes or when a formal audit of the end product is either not appropriate or is, by itself, deemed not to provide an adequate measure of quality. Examples might be components or piece parts manufactured at one location and subject to only a partial inspection after assembly into a finished product at another location.

Surveillance ordinarily includes:

- reviewing engineering and manufacturing information with regard to completeness and clarity

- verifying that all engineering requirements are checked or otherwise accounted for in a satisfactory manner, including process requirements and generally accepted standards of good workmanship

- verifying that test sets are calibrated, maintained, and operated in accordance with specifications

- reviewing the quality control program and its effectiveness

- reviewing the adequacy of manufacturing facilities (machines, tools, ovens, test sets, gauges, etc.) and their maintenance

- reviewing packaging and shipping operations.

Surveys

Quality assurance surveys comprise 1-time, objective, in-depth reviews of factors affecting initial quality, performance, and service life of products or services provided to the Bell operating companies. They are most often applied to areas where severe quality problems are known to exist or may be anticipated, including situations where extensive field problems have been reported or recurrent poor quality levels are evident (for example, from the audit). Quality assurance surveys may also be conducted on newly introduced products or when manufacture of a product is transferred to a different location. The intent is to determine if all necessary steps have been taken to obtain and maintain satisfactory quality and to identify specific conditions that may affect the adequacy of quality. Surveys are usually the joint effort of Western Electric and Bell Laboratories personnel directly involved with the affected product lines.

Field Representatives

To expedite the flow of information between Bell Laboratories and the Bell operating companies, the Bell Laboratories Quality Assurance Center maintains a staff of eighteen field representatives and fifteen assistants at the company locations shown in Figure 18-18. They constantly monitor the operating side of the business and participate in field evaluations and reliability studies. Their activities take them into all parts of the telephone plant, bring them into close contact with personnel at all levels, and expose them to service problems, operating company needs, and equipment performance.

One of their major functions is alerting Bell Laboratories and Western Electric to quality problems in the field. For situations that require

Figure 18-18. Locations of Bell Laboratories field representatives.

extended investigations and possibly coordination of several organizations, Bell operating companies submit *engineering complaints* (discussed below) that are reviewed and tracked by field representatives. In addition, field representatives are available for technical consultation concerning the engineering, operation, and maintenance of Bell System communications equipment. In their interface role, they represent the Bell Laboratories viewpoint to the Bell operating companies, and equally as important, they represent the Bell operating company viewpoint to Bell Laboratories.

Engineering Complaints

This mechanism is a formal feedback path between the user and the designer or manufacturer. It serves as an additional source of information concerning field problems and supports the standard supply contract between the Bell operating companies and Western Electric (see Section 1.2.3).

Currently, about eight thousand complaints are submitted annually; about 70 percent are handled by Western Electric and the rest by Bell Laboratories. Complaints sent to Bell Laboratories are reviewed and distributed to the responsible development organizations by the Quality Assurance Center. Bell Laboratories designers report their findings and corrective action in a final report, which is reviewed by the Quality Assurance Center and the appropriate field representative before distribution to the operating company. As part of the complaint operation, the

Quality Assurance Center issues, through the field representatives, summary reports generated from the computerized engineering complaint data base. These documents alert the Bell operating companies to design problems encountered throughout the system, thus giving each the benefit of the experience of the others.

Field Performance Tracking

While the engineering complaint routine provides a great deal of useful information about field problems, additional quantitative information is often required, particularly for high-volume products such as circuit packs and customer-premises equipment. For these products, detailed feedback is obtained through special field studies (frequently conducted with the assistance of field representatives) and by means of formal *Product Performance Surveys* (PPSs), which continuously monitor the quality of critical products.

There may be several dozen tracking studies being conducted concurrently under the aegis of the Quality Assurance Center. These special studies are frequently implemented, with the cooperation of one or more Bell operating companies, when important new products are first introduced to the field. Quantitative data are gathered for a long enough period of time, under actual operating conditions, either to confirm that the quality level is satisfactory or to identify specific performance deficiencies. Such studies are especially critical in the presence of decreasing intervals between design concept and sale to the Bell operating companies. In addition, the field studies may identify the need for changes in the audit.

Product Performance Surveys provide continuous monitoring to supplement the normal laboratory testing, field trials, and appraisal studies associated with the introduction of new or modified equipment. The basic functions of PPSs are shown in Figure 18-19. Unforeseen quality

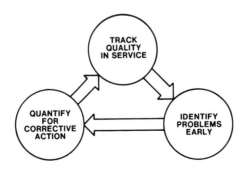

Figure 18-19. Product Performance Survey functions.

problems can appear as a result of well-intentioned cost reductions, design changes, process modifications, and material substitutions. Quality control and quality assurance at the manufacturing or repair stage can prevent quality problems only if the consequences of these changes are anticipated and if measurement techniques are feasible and economically justified.

By providing extensive information on in-service reliability and initially defective units, the PPS quantifies product quality as seen by the Bell operating companies. The continuous tracking of about one million telephone sets in service and hundreds of thousands of installations, for example, has resulted in the collection of extensive detailed defect data. Analyses of these data have provided the means for quickly and accurately identifying problems. Since the PPS is a continuing activity, the modified product can be monitored to ensure the validity of the "fixes" implemented, and the cycle repeated as shown in Figure 18-19.

In a sense, Figure 18-19 summarizes the nature of all the major quality assurance functions included under auditing and monitoring: They are on-going activities intended to identify quality problems, call them to the attention of the appropriate organizations, and then verify that the corrective actions have been effective.

AUTHORS

K. J. Cohen
B. L. Hanson
C. E. Johnson
C. H. King
D. C. Krupka
V. O. Mowery
R. Sherman

References and Additional Reading

ASQC   American Society for Quality Control
BLR   *Bell Laboratories Record*
BSTJ   *The Bell System Technical Journal*
CCITT   Comité Consultatif International
        Télégraphique et Téléphonique
EIA   Electronics Industries Association
FCC   Federal Communications Commission
ICC   International Conference on Communications
ICCC   International Conference on Computer
        Communications
IEEE   Institute of Electrical and Electronics
        Engineers
ISO   International Organization for Standardization
USITA   United States Independent Telephone
        Association

CHAPTER 1. STRUCTURE AND ACTIVITIES

References

AT&T. 1983. *1982 Statistical Report*. New York.

Bell Laboratories. 1982. *Facts About Bell Laboratories*. 12th ed. Short Hills, NJ.

———. 1977. *Engineering and Operations in the Bell System*. Murray Hill, NJ.

Fagen, M. D., ed. 1975/1978. *A History of Engineering and Science in the Bell System:* vol. 1, *The Early Years (1875-1925);* vol. 2, *National Service in War and Peace (1925-1975).* Murray Hill, NJ: Bell Laboratories.

Lustig, L. K., ed. 1981. *Impact: A Compilation of Bell System Innovations in Science and Engineering.* 2nd. ed. rev. Short Hills, NJ: Bell Laboratories.

Mueser, R., ed. 1979. *Bell Laboratories Innovation in Telecommunications.* Murray Hill, NJ: Bell Laboratories. (A second edition, updated, is scheduled for publication in 1984.)

USITA. 1983. *Independent Telephone Statistics for the Year 1982.* Vol. 1, 1982 ed. Washington, DC.

CHAPTER 2. SERVICES

Reference

AT&T Long Lines. 1982. *Basic Packet Switching Service.* Illustrative Tariff FCC No. 270. November 15.

CHAPTER 3. INTRODUCTION TO THE NETWORK

Reference

Skoog, R. A., ed. 1980. *The Design and Cost Characteristics of Telecommunications Networks.* Murray Hill, NJ: Bell Laboratories.

Additional Reading

Manhire, L. M. 1978. Physical and Transmission Characteristics of Customer Loop Plant. *BSTJ* **57**:35-59.

Noweck, H. E. 1961. The Versatility of *TOUCH-TONE* Calling. *BLR* **39**:312-316.

CHAPTER 4. NETWORK STRUCTURES AND PLANNING

References

AT&T Long Lines. 1982. *The World's Telephones—A Statistical Compilation as of January, 1981.* Morris Plains, NJ.

Skoog, R. A., ed. 1980. *The Design and Cost Characteristics of Telecommunications Networks.* Murray Hill, NJ: Bell Laboratories.

Additional Reading

Bell Laboratories. 1978. *BSTJ,* vol. **57**, no. 4 (April). A special issue on the loop plant.

Katz, S. S.; Lifchus, I. M.; and Skeer, M. H. 1979. A Sophisticated Switched Service. *BLR* **57**:38-45.

Lutz, K. J.; Pecsvaradi, T.; and Waninski, J. E. 1983. The Integrated Special Services Network. Paper read at the IEEE ICC '83. Boston.

CHAPTER 5. TRAFFIC

References

Barnes, D. H. 1976. Extreme Value Engineering of Small Switching Offices. *Proceedings of the 8th International Teletraffic Congress,* Melbourne, 10-17 November. Paper no. 242.

Cooper, R. B. 1981. *Introduction to Queueing Theory.* 2nd ed. New York: North Holland Press.

Elsner, W. B. 1977. A Descent Algorithm for the Multihour Sizing of Traffic Networks. *BSTJ* **56**:1405-1430.

Feller, W. 1966. *An Introduction to Probability Theory and Its Applications.* Vol. 2, 2nd ed. New York: John Wiley & Sons.

———. 1968. *An Introduction to Probability Theory and Its Applications.* Vol. 1, 3rd ed. New York: John Wiley & Sons.

Fuchs, E., and Jackson, P. E. 1970. Estimates of Distributions of Random Variables for Certain Computer Communications Traffic Models. *Communications of the Association for Computing Machinery* **13**:752-757.

Hill, D. W., and Neal, S. R. 1976. Traffic Capacity of a Probability-Engineered Trunk Group. *BSTJ* **55**:831-842.

Wilkinson, R. I. 1956. Theories for Toll Traffic Engineering in the USA. *BSTJ* **35**:421-514.

CHAPTER 6. TRANSMISSION

References

AT&T. 1977. *Telecommunications Transmission Engineering.* 2nd ed. 3 vols. Lisle, IL: Bell System Center for Technical Education.

Bell Laboratories. 1982. *Transmission Systems for Communications.* 5th ed. Holmdel, NJ.

Greene, E. S. 1962. *Principles of Physics.* New York: Prentice-Hall.

Lucky, R. W.; Salz, J.; and Weldon, E. J., Jr. 1968. *Principles of Data Communication.* New York: McGraw-Hill Book Company.

Martin, J. 1976. *Telecommunications and the Computer.* 2nd ed. Englewood Cliffs, NJ: Prentice-Hall.

Additional Reading

Abate, J. E.; Rosenberger, J. R.; and Yin, M. 1981. Keeping the Integrated Services Digital Network in Sync. *BLR* **59**:217-220.

AT&T. 1980. *Notes on the Network.* Section 7.

Campbell, L. W., Jr. 1970. The PAR Meter: Characteristics of a New Voiceband Rating System. *IEEE Transactions on Communications Technology* **COM-18**:147-153.

Cavanaugh, J. R.; Hatch, R. W.; and Sullivan, J. L. 1976. Models for the Subjective Effects of Loss, Noise, and Talker Echo on Telephone Connections. *BSTJ* **55**:1319-1371.

Jacobs, I. 1979. Lightwave Communications Begins Regular Service. *BLR* **57**:298-304.

Lewinski, D. A. 1964. A New Objective for Message Circuit Noise. *BSTJ* **43**:719-740.

CHAPTER 7. SWITCHING

Additional Reading

Inose, H. 1979. *Introduction to Digital Integrated Communications Systems.* Tokyo: University of Tokyo.

Joel, A. E., Jr. 1977. What Is Telecommunication Switching? *Proceedings of the IEEE* **65**:1237-1253.

————. 1979. Digital Switching—How It Has Developed. *IEEE Transactions on Communications* **COM-27**:948-959.

Marcus, M. J. 1977. The Theory of Connecting Networks and Their Complexity: A Review. *Proceedings of the IEEE* **65**:1263-1271.

McDonald, J. C., ed. 1983. *Fundamentals of Digital Switching.* New York/London: Plenum Press.

Pearce, J. G. 1981. *Telecommunication Switching.* New York/London: Plenum Press.

Talley, D. 1975. *Basic Electronic Switching for Telephone Systems.* Rochelle Park, NJ: Hayden.

————. 1979. *Basic Telephone Switching Systems.* 2nd ed. Rochelle Park, NJ: Hayden.

CHAPTER 8. SIGNALING AND INTERFACES

References

AT&T. 1978. *Interconnection Specification for Digital Cross-Connects.* AT&T Technical Reference no. 34. Issue 2. Basking Ridge, NJ.

————. 1980a. *Local Switching System General Requirements (LSSGR).* Preliminary issue. Bell System Technical Reference PUB 48501.

————. 1980b. *Operations Systems Network Communications Protocol Specification BX.25.* Bell System Technical Reference PUB 54001. Issue 2.

————. 1983. *Description of the Analog Voiceband Interface Between the Bell System Local Exchange Lines and Terminal Equipment.* Bell System Technical Reference PUB 61100.

CCITT. 1981a. *Data Transmission at 48 Kilobits Per Second Using 60-108 kHz Group Band Circuits. Recommendation V.35.* Yellow Book. Vol. 8.1. Geneva.

————. 1981b. *Interface Between Data Terminal Equipment (DTE) and Data Circuit-Terminating Equipment (DCE) for Terminal Operating in the Packet Mode on Public Data Networks. Recommendation X.25.* Yellow Book. Vol. 8.2. Geneva.

————. 1981c. *Terminal and Transmit Call Control Procedures and Data Transfer System on International Circuits Between Packet-Switched Data Networks. Recommendation X.75.* Yellow Book. Vol. 8.3. Geneva.

————. 1983. *Reference Model of Open Systems Interconnection for CCITT Applications, Draft Recommendation X.200.* CCITT Circular Letter no. 58.

EIA. 1969. *Interface Between Data Terminal Equipment and Data Communication Equipment Employing Serial Binary Data Exchange. Standard RS-232-C.*

————. 1977/1980. *General Purpose 37-Position and 9-Position Interface for Data Terminal Equipment and Data Circuit-Terminating Equipment Employing Serial Binary Data Interchange. Standard RS-449.* With Addendum 1.

FCC. 1975. *Engineering and Operations Supplement,* FCC Docket 20099.

————. 1977/1980. *Part 68 Rules and Regulations.* Vol. 10.

ISO. 1983. *Data processing — Open systems interconnection — Basic reference model. Organization Standard 7498.*

Additional Reading

AT&T. 1977. *Telecommunications Transmission Engineering.* 2nd ed. 3 vols. Lisle, IL: Bell System Center for Technical Education. Vol. 2, Section 1.

Bell Laboratories. 1978. *BSTJ*, vol. **57**, no. 2 (February). A special issue on common-channel interoffice signaling.

————. 1982. *BSTJ*, vol. **61**, no. 7, part 3 (September). A special issue on the stored-program control network.

CCITT. 1981. *Common Channel Signaling System No. 7. Recommendations Q.701-Q.707, Q.721-Q.725.* Yellow Book. Vol 6.6. Geneva.

CHAPTER 9. TRANSMISSION SYSTEMS

Reference

Bell Laboratories. 1982. *Transmission Systems for Telecommunications,* 5th ed. Holmdel, NJ.

Additional Reading

Adams, W. B., and Bailey, C. C. 1976. Identifying the Needs of Rural Networks. *BLR* **54**:184-188.

Anderson, C. D.; Gleason, R. F.; Hutchinson, P. T.; and Runge, P. K. 1980. An Undersea Communication System Using Fiberguide Cables. *Proceedings of the IEEE* **68**:1299-1303.

AT&T. 1977. *Telecommunications Transmission Engineering.* 2nd ed. 3 vols. Lisle, IL: Bell System Center for Technical Education.

Bangert, J. T. 1973. L5—A "Jumbojet" Coaxial System. *BLR* **51**:290-299.

Berger, U. S. 1973. The Old TD2 Becomes the New TD2C. *BLR* **51**:278-284.

Bergholm, J. O., and Koliss, P. P. 1972. Serving Area Concept—A Plan for Now with a Look to the Future. *BLR* **50**:212-216.

Billhardt, R. A.; Guarneri, P. J.; Madigan, T. W.; and Tentarelli, K. D. 1980. The Digital Carrier Trunk: a "Smart" Move Toward Increased ESS Capabilities. *BLR* **58**:341-345.

Bleisch, G. W., and Michaud, W. P., Jr. 1971. The A-6 Channel Bank: Putting New Technologies to Work. *BLR* **49**:251-254.

Brolin, S.; Cho, Y.-S.; Michaud, W. P., Jr.; and Williamson, D. H. 1980. Inside the New Digital Subscriber Loop System. *BLR* **58**:110-116.

Clark, M., and Porter, G. R. 1976. The Metallic Facility Terminal: Special Help for Special Services. *BLR* **54**:215-219.

Colton, J. R. 1980. Cross-connections—DACS Makes Them Digital. *BLR* **58**:248-255.

Danielson, W. E. 1975. Exchange Area and Local Loop Transmission. *BLR* **53**:40-49.

Dodds, S., and Mitchell, W. J. 1979. From Analog to Digital for the No. 4 ESS—Economically. *BLR* **57**:288-292.

Drechsler, R. C. 1980. DACS Cross-Connects—And That's Just the Beginning. *BLR* **58**:305-311.

Graczyk, J. F.; Mackey, E. T.; and Maybach, W. J. 1975. T1C Carrier: The T1 Doubler. *BLR* **53**:256-263.

Haury, P. T., and Romeiser, M. B. 1976. T1 Goes Rural. *BLR* **54**:178-183.

Hull, T. R., and LeBlanc, R. E. 1979. New Life for Short-Haul Analog Routes. *BLR* **57**:270-274.

Hymel, D. P., and Walker, J. D. 1980. The Difference with DIF? Size, Intelligence—and Cost. *BLR* **58**:346-349.

Jacobs, I. 1980. Lightwave Communications—Yesterday, Today, and Tomorrow. *BLR* **58**:2-10.

Johannes, V. I. 1976. The Evolving Digital Network. *BLR* **54**:268-273.

Markle, R. E. 1978. Single Sideband Triples Microwave Radio Route Capacity. *BLR* **56**:104-110.

O'Neill, E. F. 1975. Radio and Long-Haul Transmission. *BLR* **53**:50-59.

Sipress, J. M. 1975. T4M: New Superhighway for Metropolitan Communications. *BLR* **53**:352-359.

CHAPTER 10. NETWORK SWITCHING SYSTEMS

Additional Reading

Almquist, R. P.; Carney, D. L.; and Estvander, R. A. 1977. 1A ESS: Newest, Largest-Capacity Local Switch Cuts Over Early. *BLR* **55**:15-20

Bell Laboratories. 1964. *BSTJ*, vol. **43**, no. 5 (September). A special issue on the 1*ESS* system.

———. 1965. *BLR*, vol. **43**, no. 6 (June). A special issue on the 1*ESS* system.

———. 1969. *BSTJ*, vol. **48**, no. 6 (October). A special issue on the 2*ESS* system.

———. 1970. *BSTJ*, vol. **49**, no. 10 (December). A special issue on TSPS No. 1.

———. 1974. *BSTJ*, vol. **53**, no. 1 (January). A special issue on the Automatic Intercept System.

———. 1977a. *BSTJ*, vol. **56**, no. 2 (February). A special issue on the 1A Processor.

———. 1977b. *BSTJ*, vol. **56**, no. 7 (September). A special issue on the 4*ESS* system.

———. 1977c. *BLR*, vol. **55**, no. 11 (December). A special issue on the 4*ESS* system.

———. 1979. *BSTJ*, vol. **58**, no. 6, part 1 (July-August). A special issue on TSPS No. 1.

———. 1981. *BSTJ*, vol. **60**, no. 6, part 2 (July-August). A special issue on the 4*ESS* system.

———. 1982. *BSTJ*, vol. **61**, no. 4 (April). A special issue on the 10A Remote Switching System.

———. 1983a. *BSTJ*, vol. **62**, no. 1, part 2 (January). A special issue on the 3*B*20D Computer and DMERT Operating System.

———. 1983b. *BSTj*, vol. **62**, no. 3, part 3 (March). A special issue on TSPS No. 1B.

Bruce, R. A.; Giloth, P. K.; and Siegel, E. H., Jr. 1979. No. 4 ESS: A Continuing Evolution. *BLR* **57**:154-161.

Byrne, C. J., and Pilkinton, D. C. 1976. Towards Automated Local Billing. *BLR* **54**:104-109.

Fagan, M. D., ed. 1975. *A History of Engineering and Science in the Bell System:* vol. 1, *The Early Years (1875-1925).* Murray Hill, NJ: Bell Laboratories. Chapter 6, pp. 467-714.

Foster, R. W. 1977. No. 3 ESS Improves Telephone Service for Country Customers. *BLR* **55**:230-235.

Garraty, W. C. 1975. Introducing CAMA Features to No. 1 ESS. *BLR* **53**:395-399.

International Switching Symposium. 1981. No. 5 ESS—Overview, System Architecture, Hardware Design, Software Design. Montreal, 21-25 September. Vol. **3**, pp. 31A-1,1—31A-4,6.

Joel, A. E., Jr., et al. 1982. *A History of Engineering and Science in the Bell System: Switching Technology (1925-1975).* Murray Hill, NJ: Bell Laboratories.

Johnson, J. W.; Kennedy, J. C.; and Warner, J. C. 1981. No. 5 ESS—Serving the Present, Serving the Future. *BLR* **59**:290-293.

Kleber, J. J., and Perkinson, W. B. 1980. No. 1A AMARC: At the Center of Billing Data Collection. *BLR* **58**:236-243.

Lehder, W. E., Jr., and Whitemyer, J. G. 1979. Introduction of AMA to No. 3 ESS. *BLR* **57**:174-178.

Mandigo, P. D. 1976. No. 2B ESS: New Features from a More Efficient Processor. *BLR* **54**:304-309.

Neville, S. M., and Royer, R. D. 1976. Controlling Large Electronic Switching Systems. *BLR* **54**:30-33.

Richards, P. C. and Herndon, J. A. 1973. No. 2 ESS: An Electronic Switching System for the Suburban Community. *BLR* **51**:130-135.

Sevcik, R. W., and Smith, D. P. 1980. Custom Calling Comes to Clarksville. *BLR* **58**:63-68.

Telesis. 1978. Vol. 5, no. 10 (August). A special issue on the *DMS*-10 digital community office.

Winckelmann, W. A. 1968. Automatic Intercept Service. *BLR* **46**:138-143.

CHAPTER 11. CUSTOMER-SERVICES EQUIPMENT AND SYSTEMS

References

Balkovic, M. D.; Klancer, H. W.; Klare, S. W.; and McGruther, W. G. 1971. 1969-70 Connection Survey: High-Speed Voiceband Data Transmission Performance on the Switched Telecommunications Network. *BSTJ* **50**:1349-1384.

Kretzmer, E. R. 1973. The New Look in Data Communications. *BLR* **51**:258-265.

Larsen, A. B., and Brown, E. F. 1980. Continuous Presence Video Conferencing at 1.5/6 Mbps. *Teleconferencing and Interactive Media*, pp. 391-398. [Madison]:University of Wisconsin.

Lucky, R. W.; Salz, J.; and Weldon, E. J., Jr., 1968. *Principles of Data Communication.* New York: McGraw-Hill Book Company.

Netravali, A. N., and Robbins, J. D. 1979. Motion-Compensated Television Coding: Part I. *BSTJ* **58**:631-670.

Additional Reading

Arnold, T. F., and Toy, W. N. 1981. Inside the 3B-20 Processor. *BLR* **59**:66-71.

Bell Laboratories. 1979. *BSTJ*, vol. **58**, no. 1 (January). A special issue on Advanced Mobile Phone Service.

Borison, V. S. 1975. Transaction Telephone Gets the Facts at the Point of Sale. *BLR* **53**:376-383.

Brophy, F. J.; Honnold, G. H.; and Thayer, S. T. 1981. *DATAPHONE* II Service—New Standard for Data Communications. *BLR* **59**:248-252.

Burns, H. S., and Loosme, O. 1981. An Inside Look at *DATAPHONE* II Service—Part II: Diagnostic Tools. *BLR* **59**:268-271.

Busala, A. 1960. Fundamental Considerations in the Design of a Voice-Switched Speakerphone. *BSTJ* **39**:265-294.

Carpenter, J. R.; Dennis, T. M.; Holzman, L. N.; and Tong, K. 1981. An Inside Look at *DATAPHONE* II Service—Part I: Data Sets. *BLR* **59**:264-268.

Daugherty, T. H. 1970. Digitally Companded Delta Modulation for Voice Transmission. IEEE International Symposium on Circuit Theory, *Digest of Technical Papers*, pp. 17-18.

Douglas, V. A. 1964. The MJ Mobile Radio Telephone System. *BLR* **42**:382-389.

Ehlinger, J. C., and Stubblefield, R. W. 1983. No. 1 PSS: Number 1 Packet-Switching System—Service Capabilities and Architecture. Paper read at the IEEE ICC '83. Boston.

Ham, J. H., and West, F. 1963. A *TOUCH-TONE* Caller for Station Sets. *IEEE Transactions on Communication and Electronics*, no. **65**:17-24.

Handler, G. J., and Snowden, R. L. 1982. Planning Packet Transport Capabilities in the Bell System. Paper read at the 6th ICCC. London.

Huff, D. L., and Pennotti, R. J. 1980. Mobile Phone Service Moves Ahead. *BLR* **58**:91-96.

Inglis, A. H., and Tuffnell, W. L. 1951. An Improved Telephone Set. *BSTJ* **30**:239-270.

Levy, R. P., and Magee, F. R. 1983. An Internal Packet Network Protocol and Buffer Management Scheme for an X.25 Based Network. Paper read at IEEE INFOCOM '83. San Diego.

Mansell, J. J., and Stubbs, R. D., II. 1982. Systems Engineering Considerations in the Bell System Packet-Switching Networks. Paper read at the 6th ICCC. London.

Matlack, R. C., and Render, D. J. 1977. *DATAPHONE* Select-a-Station Service: Improved Transmission Facilities for Alarm Systems. *BLR* **55**:243-247.

Mounts, F. W. 1969. A Video Encoding System With Conditional Picture-Element Replenishment. *BSTJ* **48**:2545-2554.

O'Brien, J. A. 1978. Final Tests Begin for Mobile Telephone System. *BLR* **56**:171-174.

Rodriguez, E. J. 1982. Architectural Considerations in Bell System Packet-Switching Networks. Paper read at the 6th ICCC. London.

CHAPTER 12. COMMON SYSTEMS

Additional Reading

AT&T. 1981. *New Equipment-Building System (NEBS)—General Equipment Requirements.* Bell System Technical Reference PUB 51001. Issue 2, rev.

Pferd, W. 1979. The Evolution and Special Features of Bell System Telephone Equipment Buildings. *BSTJ* **58**:427-466.

CHAPTER 13. OVERVIEW OF TELEPHONE COMPANY OPERATIONS

Reference

Martz, L. M., and Osofsky, A. J. 1982. Operations Impact of New Technology: A Case Study of the Digital Access and Cross-Connect System. IEEE ICC '82, *Conference Record.* Philadelphia.

CHAPTER 15. OPERATIONS PLANNING

Additional Reading

Johnson, D. W.; Varma, G. K.; and Waninski, J. E. 1982. Network Operations Planning for the Bell System. IEEE ICC '82, *Conference Record.* Philadelphia.

CHAPTER 16. EVALUATION OF SERVICE AND PERFORMANCE

References

Cavanaugh, J. R.; Hatch, R. W.; and Sullivan, J. L. 1976. Models for the Subjective Effects of Loss, Noise, and Talker Echo on Telephone Connections. *BSTJ* **55**:1319-1371.

Hester, S. 1982. Taking the Pulse of the Network. *BLR* **60**:70-74.

McDade, S. 1979. Measurement for Measurement's Sake, Past Tense. *Bell Telephone Magazine* **58** (February):31-32.

Militzer, K. H. 1980. Macro versus Micro Input/Output Ratios. *Management Review* **69** (June):8-15.

Additional Reading

Alexander, A. A.; Gryb, R. M.; and Nast, D. W. 1960. Capabilities of the Telephone Network for Data Transmission. *BSTJ* **39**:431-476.

Balkovic, M. D.; Klancer, H. W.; Klare, S. W.; and McGruther, W. G. 1971. 1969-70 Connection Survey: High-Speed Voiceband Data Transmission Performance on the Switched Telecommunications Network. *BSTJ* **50**:1349-1384.

Duffy, F. P., and Thatcher, T. W., Jr. 1971. 1969-70 Connection Survey: Analog Transmission Performance on the Switched Telecommunications Network. *BSTJ* **50**:1311-1347.

Duffy, F. P.; McNees, G. K.; Nåsell, I; and Thatcher, T. W., Jr. 1975. Echo Performance of Toll Telephone Connections in the United States. *BSTJ* **54**:209-243.

Fennick, J. H., and Nåsell, I. 1966. The 1963 Survey of Impulse Noise on Bell System Carrier Facilities. *IEEE Transactions on Communications Technology* **COM-14**:520-525.

Fleming, H. C., and Hutchinson, R. M., Jr. 1971. 1969-70 Connection Survey: Low-Speed Data Transmission Performance on the Switched Telecommunications Network. *BSTJ* **50**:1385-1405.

Gresh, P. A. 1969. Physical and Transmission Characteristics of Customer Loop Plant. *BSTJ* **48**:3337-3386.

Kessler, J. E. 1971. The Transmission Performance of Bell System Toll Connecting Trunks. *BSTJ* **50**:2741-2777.

Nâsell, I. 1964. The 1962 Survey of Noise and Loss on Toll Connections. *BSTJ* **43**:697-718.

——. 1968. Some Transmission Characteristics of Bell System Toll Connections. *BSTJ* **47**:1001-1018.

Nâsell, I.; Ellison, C. R., Jr.; and Holmstrom, R. 1968. The Transmission Performance of Bell System Intertoll Trunks. *BSTJ* **47**:1561-1613.

Spang, T. C. 1976. Loss-Noise-Echo Study of the Direct Distance Dialing Network. *BSTJ* **55**:1-36.

CHAPTER 18. EVOLUTION OF PRODUCTS AND SERVICES

References

AT&T. 1983. *Purchased Products Quality Handbook.* Bell System Information Publication IP 10450.

Belfi, J. L.; Glenn, F. W.; and Keyser, C. J. 1978. Quality Assurance for Components. *BLR* **56**:296-302.

Bell Labs News. March 1, 1982.

Comella, W. K. 1978. Quality Assurance for ESS: Advanced Measures for Advanced Technology. *BLR* **56**:288-295.

Dodge, H. F. 1928. A Method of Rating Manufactured Product. *BSTJ* **7**:350-368.

Dodge, H. F., and Torrey, M. N. 1956. A Check Inspection and Demerit Rating Plan. *Industrial Quality Control* **13**:5-12.

Dreyfuss, H. 1959. *The Measure of Man; Human Factors in Design.* New York: Whitney Library of Design.

French, W. L., and Godfrey, B. 1978. Quality Assurance: The New Audit for Cables. *BLR* **56**:276-281.

Fuchs, E., and Howard, B. T. 1978. Quality Assurance: Tradition and Change. *BLR* **56**:226-231.

Godfrey, B., and Hoadley, B. 1978. Statistical Methods in Quality Assurance. *BLR* **56**:233-237.

Green, P. E., and Tull, D. S. 1978. *Research for Marketing Decisions*. 4th ed. Englewood Cliffs, NJ: Prentice-Hall.

Hiering, V. S., and Hooke, J. A. 1978. Product Performance Surveys: Field Tracking for Station Sets. *BLR* **56**:238-240.

Hoadley, B. 1981a. The Quality Measurement Plan (QMP). *BSTJ* **60**:215-273.

———. 1981b. The Universal Sampling Plan. *Transactions of the 35th Annual Quality Congress*, May 27-29.

Kennedy, K. W., and Bates, C. 1965. *Development of Design Standards for Ground Support Consoles*. Aerospace Medical Research Laboratories report no. AMRL-TR-65-163. Wright-Patterson Air Force Base, Ohio.

Mowery, V. O. 1978. A New Look at Customer Premises Products. *BLR* **56**:271-275.

Peters, R. A., and Karraker, I. O. 1975. Why You Can Trust Your Telephone. *Industrial Research* **17**:47-51.

Scientific American. 1977. A special issue on microelectronics. Vol. 237 (September).

Shewhart, W. A. 1926. Correction of Data for Errors of Averages Obtained from Small Samples. *BSTJ* **5**:308-319.

Urban, G. L., and Hauser, J. R. 1980. *Design and Marketing of New Products*. Englewood Cliffs, NJ: Prentice-Hall.

Van Cott, H. P., and Kinkade, R. G. 1972. *Human Engineering Guide to Equipment Design*. Washington: American Institutes for Research.

Woodson, W. E., and Conover, D. W. 1964. *Human Engineering Guide for Equipment Designers*. 2nd ed. Berkeley: University of California Press.

Additional Reading

AT&T. 1977. *Engineering Economy: A Manager's Guide to Economic Decision Making.* 3rd ed. New York: McGraw-Hill Book Company.

Dodge, H. F., and Romig, H. G. 1929. A Method of Sampling Inspection. *BSTJ* 8:613-631.

————. 1959. *Sampling Inspection Tables: Single and Double Sampling.* 2nd ed. New York: John Wiley & Sons.

Fagen, M. D., ed. 1975. *A History of Engineering and Science in the Bell System:* vol. 1, *The Early Years (1875-1925).* Murray Hill, NJ: Bell Laboratories. Chapter 9.

Hoadley, B. 1979. An Empirical Bayes Approach to Quality Assurance. *33rd Annual Technical Conference Transactions of ASQC,* pp. 257-263.

Hughes, G. D. 1978. *Marketing Management: A Planning Approach.* Reading, MA: Addison-Wesley.

Kolter, P. 1980. *Marketing Management: Analysis, Planning and Control.* 4th ed. Englewood Cliffs, NJ: Prentice-Hall.

Liebesman, B. S. 1979. The Use of MIL-STD 105D to Control Average Outgoing Quality. *Journal of Quality Technology* 11:36-43.

McCarthy, E. J. 1978. *Basic Marketing: A Managerial Approach.* 6th ed. Homewood, IL: Richard D. Irwin.

McCormick, E. J. 1964. *Human Factors Engineering.* 2nd ed. New York: McGraw-Hill Book Company.

Murphy, R. B. 1956. The Role of Quality Assurance in the Bell System. *BLR* 34:241-245.

Patterson, E. G. D. 1956. An Overall Quality Assurance Plan. *Industrial Quality Control* 12:32-37.

Phadke, M. S. 1979. Sequential Empirical Bayes Sampling via Microcomputers. *33rd Annual Technical Conference Transactions of ASQC,* pp. 833-841.

Shewhart, W. A. 1927. Quality Control. *BSTJ* 6:722-735.

———. 1931. *Economic Control of Quality of Manufactured Product.* New York: D. Van Nostrand Co.

———. 1958. Nature and Origin of Standards of Quality. *BSTJ* **37**:1-22.

Wagner, H. M. 1975. *Principles of Operations Research.* 2nd ed. Englewood Cliffs, NJ: Prentice-Hall.

Glossary

Words set in SMALL CAPS are defined elsewhere in this glossary. A few of the definitions in the glossary have been taken, with permission, from the sources listed below. A bracketed letter code that corresponds to the specific source follows each of these definitions.

DST   *McGraw-Hill Dictionary of Scientific and Technical Terms.* 1974. New York: McGraw-Hill.

IEC/ITU   Definitions agreed upon by Working Group C of the International Electrotechnical Commission (IEC)/International Telecommunications Union (ITU) Joint Coordination Group. 1982.

IEEE   *IEEE Standard Dictionary of Electrical and Electronics Terms.* 1977. New York: Institute of Electrical and Electronics Engineers.

TC   Martin, J. 1976. *Telecommunications and the Computer.* 2nd ed. Englewood Cliffs, NJ: Prentice-Hall, Inc.

Address. (1) A sequence of numbers that identifies the telephone or other customer-premises EQUIPMENT to which a call is directed. The *address* is usually a 7- or 10-digit number, depending on whether the destination is inside or outside the NUMBERING PLAN AREA from which the call originated. (2) Digital information (a combination of bits) that identifies a location in a storage device or equipment unit.

Addressing. Specifying to the NETWORK the destination of a call.

Address Signal. A SIGNAL used to convey call destination information such as telephone STATION NUMBER, CENTRAL OFFICE CODE, and area code. Some forms of *address signals* are called *pulses*; for example, dial pulses, multifrequency pulses.

Alerting. Generating an audio (or visual) SIGNAL to indicate that a call is being made.

Alerting Signal. A SIGNAL sent to a customer, a PRIVATE BRANCH EXCHANGE, or a SWITCHING SYSTEM to indicate an incoming call. A common form is the signal that rings a bell in the station set being called.

All-Number Calling. The system of telephone numbering that uses only numbers and replaces the 2-letter plus 5-number (2L+5N) NUMBERING PLAN. *All-number calling* offers more usable combinations of numbers than the 2L+5N numbering plan and is now the nationwide standard.

Alternate Routing. A means of selectively distributing TRAFFIC over a number of routes that ultimately lead to the same destination.

Amplitude/Frequency Distortion. Distortion in which the relative magnitudes of the different frequency components of a SIGNAL are altered during amplification or TRANSMISSION over a CHANNEL. [DST]

Amplitude Modulation (AM). One way to modify a SIGNAL to make it "carry" information. The amplitude of a CARRIER SIGNAL is modified in accordance with the amplitude of the information signal.

Analog Channel. A TRANSMISSION path that accepts a band of frequencies and is compatible with the transmission of ANALOG SIGNALS.

Analog Signal. A SIGNAL, such as voice or music, that varies in a continuous manner. An *analog signal* may be contrasted with a DIGITAL SIGNAL, which represents only discrete states.

Answer Delay. The time from the beginning of RINGING until the called station or an operator answers.

Area Code. *See* NUMBERING PLAN AREA.

Attempt. A call initiation or bid for service in which at least one digit is received by the originating SWITCHING SYSTEM. For some purposes, an *attempt* is said to occur as soon as an originating station goes OFF-HOOK and causes some response in the originating switching system.

Attempts per Circuit per Hour (ACH). A running count of the number of TRUNK requests (new calls). Used in network management as an indicator of calling pressure. *See* CONNECTIONS PER CIRCUIT PER HOUR.

Attenuation. A decrease in SIGNAL amplitude during TRANSMISSION from one point to another, usually expressed in DECIBELS (dB).

Automatic Identified Outward Dialing (AIOD). An arrangement provided with CENTREX service whereby stations associated with the service can be identified automatically when originating TOLL calls or for other purposes.

Automatic Intercept Center (AIC). A centrally located set of equipments that is a part of an AUTOMATIC INTERCEPT SYSTEM. STORED-PROGRAM CONTROL is used to inform the calling customer (by means of a recorded or electronically assembled announcement) why connection to the called number cannot be completed.

Automatic Intercept System (AIS). A type of TRAFFIC service system consisting of one or more AUTOMATIC INTERCEPT CENTERS and a centralized intercept bureau for handling INTERCEPT CALLS.

Automatic Message Accounting (AMA). The automatic collection, recording, and processing of information relating to calls for billing purposes.

Automatic Number Identification (ANI). The automatic identification of a calling station by a SWITCHING SYSTEM, usually for AUTOMATIC MESSAGE ACCOUNTING.

Automatic Voice Network (AUTOVON). A PRIVATE NETWORK serving the Department of Defense. *AUTOVON* employs automatic switching and handles both voice and data TRAFFIC. It is worldwide; the continental United States portion is known as *CONUS AUTOVON.*

Average Busy Season Busy Hour (ABSBH) Load. The expected OFFERED LOAD for which a TRUNK GROUP is engineered. Estimated by averaging the measured loads for a given peak hour of the day (BUSY HOUR) during each weekday of the peak season (BUSY SEASON). Normally, measurements of load on 20 weekdays are averaged.

Bandlimited. Applied to a SIGNAL or a CHANNEL that is limited in frequency content.

Bandpass Filter. A CIRCUIT designed to allow only frequencies within a specific range to pass. The cutoff frequencies must be finite and nonzero. The band of frequencies between the cutoff frequencies is called the *passband.*

Bandwidth. The range of frequencies that can be transmitted by a communications CHANNEL, a TRANSMISSION FACILITY, or a transmission medium, expressed in hertz (Hz).

Baseband Channel. A CHANNEL that carries a SIGNAL faithfully without requiring MODULATION, in contrast to a *passband channel*.

Baseband Signal. The original form of a SIGNAL, unchanged by MODULATION.

Bell Administrative Network Communications System (BANCS). A computer system consisting of modules that form a message SWITCHING NETWORK for business communications. *BANCS* handles computer-to-computer, interactive terminal-to-computer, and store-and-forward output DISTRIBUTION traffic.

Bit Rate. The speed at which digital signals are transmitted, usually expressed in bits per second (bps).

Blocked Calls Cleared (BCC). Designation for a queuing system in which demands (calls) that find no idle servers leave the system immediately. Commonly used to model systems for which waiting positions are not provided, such as TRUNK GROUPS.

Blocked Calls Delayed (BCD). Designation for a queuing system in which demands (calls) that find no idle servers wait until an idle server becomes available (that is, they never give up). Commonly used to model systems for which customers are not overly impatient, such as digit receivers.

Blocking. The inability of the calling party to be connected to the called party because either (1) all suitable TRUNK paths are busy or (2) a path between a given inlet and any suitable free outlet of the SWITCHING NETWORK of a SWITCHING SYSTEM is unavailable.

Bridged Tap. A cable pair connected in parallel with a customer LOOP. The connection (*tap*) may occur at the CENTRAL OFFICE or at some point along a cable route.

Bridging Connection. A parallel CONNECTION that draws some of the SIGNAL energy from a CIRCUIT, often with imperceptible effect on the circuit's normal operation.

Broadband Channel. A transmission CHANNEL with a BANDWIDTH wider than that required for transmitting voice SIGNALS, for example, 48 kilohertz (kHz), 240 kHz.

Business Information Systems Programs (BISP). One of many computer-based systems for performing voluminous business and administrative operations associated with the provision of telephone service by an OPERATING COMPANY.

Busy Hour. The hour(s) of the day during which TRAFFIC normally peaks. Hours during which there are peaks for abnormal reasons (holidays, special events) are not considered. Traffic systems are typically sized for *busy hour* demand levels.

Busy Season. A period during the year when BUSY HOURS are at their highest.

Busy Season Busy Hour (BSBH). The hour of the business day that, on the average, is the busiest hour during the BUSY SEASON.

Busy Tone. An audible SIGNAL indicating that a call cannot be completed because the called LINE is busy. The tone is applied 60 times per minute.

Cable Entrance Facility (CEF). The entrance area in a telephone EQUIPMENT building for all types of OUTSIDE PLANT cables carrying subscriber LINES and interoffice TRANSMISSION FACILITIES. The typical *cable entrance facility* is a vault-like, below-grade area, 15 feet high and 12 feet wide, that runs the length of the building directly under the main DISTRIBUTING FRAME(S).

Capacity Expansion. A NETWORK PLANNING activity that determines types, sizes, locations, and timing of SWITCHING SYSTEM and TRANSMISSION FACILITY installations.

Carried Load. The average number of busy servers in a TRAFFIC system. In BLOCKED-CALLS-DELAYED systems, *carried load* equals OFFERED LOAD since all calls are served eventually. In BLOCKED-CALLS-CLEARED systems, *carried load* is less than offered load since some calls are denied service.

Carrier-Derived Channel. One of a number of CHANNELS provided by a CARRIER SYSTEM.

Carrier Frequency. A sinusoidal waveform with constant amplitude that undergoes MODULATION by an information SIGNAL to shift the information signal frequencies to a higher frequency band.

Carrier Frequency Shift. The change in frequencies between transmitting and receiving terminals in a nonsynchronized CARRIER SYSTEM.

Carrier Signal. A SIGNAL suitable for MODULATION by an information signal. It may be a CARRIER FREQUENCY, or it may be a stream of constant-amplitude pulses as in PULSE-CODE MODULATION.

Carrier System. A TRANSMISSION system in which one or more CHANNELS of information are processed and converted to a form suitable for the transmission medium used by the system. Common types of *carrier systems* are *frequency division*, in which each information channel occupies an assigned portion of the frequency spectrum, and *time division*, in which each information channel uses the transmission medium for periodic, assigned time intervals.

Centralized Automatic Message Accounting (CAMA). A process using centrally located EQUIPMENT (including a switchboard or a TRAFFIC SERVICE POSITION), associated with a tandem or TOLL switching office, for automatically recording billing data for customer-dialed extracharge calls originating from several local CENTRAL OFFICES. A tape record is processed at an electronic data-processing center. *See* TANDEM SWITCHING SYSTEM.

Centralized Automatic Reporting on Trunks (CAROT). An OPERATIONS SYSTEM that automatically schedules tests of TRUNKS, performs the tests, and analyzes and records the results. *CAROT* also performs tests of trunks on demand to verify trouble and repair conditions.

Centralized Automatic Trouble Locating and Analysis System (CATLAS). An OPERATIONS SYSTEM designed as a maintenance aid for Stored-Program Control Systems (SPCSs). When trouble is detected and diagnosed, *CATLAS* automates trouble location procedures that identify faulty circuit packs. *See* TOTAL NETWORK DATA SYSTEMS.

Centralized System for Analysis and Reporting (CSAR). An OPERATIONS SYSTEM that measures the accuracy, timeliness, and completeness of the TOTAL NETWORK DATA SYSTEM data flow and the consistency of its record bases.

Central Office. Usually used to refer to a *LOCAL SWITCHING SYSTEM* that connects LINES to lines and lines to TRUNKS. It may be more generally applied to any *network switching system*. The term is sometimes used loosely to refer to a telephone company building in which a SWITCHING SYSTEM is located and to include other EQUIPMENT (such as transmission system terminals) that may be located in such a building.

Central Office Code. A 3-digit identification under which up to 10,000 station numbers are subgrouped. EXCHANGE AREA boundaries are associated with the *central office code*, which accordingly has billing significance. Several *central office codes* may be served by a CENTRAL OFFICE.

Central Office Equipment Engineering System (COEES). A time-sharing OPERATIONS SYSTEM that assists in the planning and engineering of local switching EQUIPMENT.

Centrex. A service for customers with many stations that permits station-to-station dialing, one listed directory number for the customer, DIRECT INWARD DIALING to a particular station, and station identification on outgoing calls. The switching functions are performed in a CENTRAL OFFICE.

Channel. A TRANSMISSION path between two points. May refer to a 1-way path or, when paths in the two directions of transmission are always associated, to a 2-way path. It is usually the smallest subdivision of a transmission system by means of which a single type of communications service (that is, a voice, teletypewriter, or data *channel*) is provided.

Channel Bank. TERMINAL EQUIPMENT used to combine (MULTIPLEX) CHANNELS on a frequency-division or time-division basis. Voice channels are combined into 12- or 24-channel groups.

Circuit. Frequently used interchangeably with CHANNEL to designate a communications path between two or more points. Other meanings include: (1) a configuration of interconnected NETWORK equipment that provides a TRANSMISSION capability and (2) a closed path through which current can flow.

Circuit Administration Center (CAC). An OPERATIONS CENTER that administers the TRUNK network. The functions of the *CAC* include: (1) determining demand and BUSY-SEASON trunk requirements and issuing MESSAGE TRUNK orders to provide the required trunks, (2) developing forecasts of trunk requirements for the NETWORK for 1 to 5 years, and (3) network routing.

Circuit Design. The OPERATING COMPANY process that specifies the types of NETWORK equipment required to be interconnected to satisfy a functional capability.

Circuit Order. The document used to transmit engineering design of a TRUNK or SPECIAL-SERVICES CIRCUIT for the PUBLIC SWITCHED TELEPHONE NETWORK to the department that implements the design.

Circuit Order Control (COC). A component system of the TRUNKS INTEGRATED RECORDS KEEPING SYSTEM (TIRKS) that controls telephone company MESSAGE TRUNK, SPECIAL-SERVICES, and CARRIER SYSTEM orders. Reports produced by *COC* provide management with the current status of CIRCUIT ORDERS. *COC* provides data to other TIRKS component systems to update the assignment status of EQUIPMENT, FACILITIES, and CIRCUITS as orders are processed.

Circuit Provision Center (CPC). An OPERATIONS CENTER that assigns EQUIPMENT and FACILITIES and prepares and distributes work documents for MESSAGE TRUNK circuits, designed SPECIAL-SERVICES CIRCUITS, and CARRIER SYSTEMS. It also generates and maintains CIRCUIT RECORDS and inventory and assignment records for all interoffice facilities and equipment.

Circuit Provisioning. The OPERATING COMPANY process that responds to needs for TRUNKS and SPECIAL-SERVICES CIRCUITS. It includes CIRCUIT DESIGN, assignment of specific components, and generation of work documents for the required installation work.

Circuit Record. The OPERATING COMPANY document that records the specific configuration of EQUIPMENT assigned to a CIRCUIT.

Circuit Routing. A NETWORK PLANNING activity that determines the most efficient configuration of TRANSMISSION FACILITIES to provide the required CIRCUITS.

Circuit Switched Digital Capability (CSDC). A NETWORK capability that provides a high-speed, digital path over portions of the PUBLIC SWITCHED TELEPHONE NETWORK. *CSDC* provides services such as audiographics TELECONFERENCING and bulk data transport.

Coaxial Cable. A type of cable made up of coaxial units, or tubes. Each tube contains an inner conductor that is centered within an outer conductor through the use of insulating disks spaced about 1 inch apart. The outer conductor forms a cylinder around the disks. Each cable contains from four to twenty-two of these coaxial tubes.

Common Carrier. A supplier in an industry that undertakes to "carry" goods, services, or people from one point to another for the public in general or for specified classes of the public. In telecommunications, such carriage relates to the provision of TRANSMISSION capabilities over the telecommunications NETWORK. A *common carrier* that offers communications services to the public is subject to regulation by federal and state regulatory commissions.

Common-Channel Interoffice Signaling (CCIS). A SIGNALING system, developed for use between SWITCHING SYSTEMS with STORED-PROGRAM CONTROL, in which all of the signaling information for one or more groups of TRUNKS is transmitted over a dedicated high-speed data LINK, rather than on a per-trunk basis. *CCIS* can reduce call-setup time and save money compared to individual trunk signaling.

Common Control. An automatic arrangement in which items of control equipment in a SWITCHING SYSTEM are shared; they are associated with a given call only during the periods required to accomplish the control functions. All Bell System crossbar and ELECTRONIC SWITCHING SYSTEMS have *common control*.

Common-Control Switching Arrangement (CCSA). An arrangement in which SWITCHING for a PRIVATE NETWORK is provided by one or more COMMON-CONTROL switching systems. The SWITCHING SYSTEMS may be shared by several private networks and may also be shared with the PUBLIC SWITCHED TELEPHONE NETWORK. This service provides uniform dialing to customers who use a NETWORK of dedicated TRANSMISSION FACILITIES between geographically dispersed locations.

Common-Language Equipment Identification (CLEI) Code. An alphanumeric code that identifies NETWORK equipment by family, subfamily, and type.

Common-Language Location Identification (CLLI) Code. A 10-character designation that identifies any physical location within the Bell System.

Common System. A system that provides common power, INTERCONNEC-TION, or environmental support for NETWORK elements associated with TRANSMISSION and SWITCHING; for example, power systems, DISTRIBUTING FRAMES, and equipment building systems.

Common Update/Equipment (CU/EQ). A master record base that stores the TRAFFIC measurement requests generated by OPERATING COMPANY personnel.

Common Update/Trunking (CU/TK). A record base system that contains trunking NETWORK information and other information required by TRUNK SERVICING and TRUNK FORECASTING SYSTEMS.

Community Dial Office (CDO). A small, electromechanical SWITCHING SYS-TEM that serves a separate EXCHANGE AREA and, ordinarily, has no operating or maintenance force located in its own building; operation is handled and maintenance is directed from a remote location.

Compandor. An abbreviation of *compressor-expandor,* a device used to compress the range of talker volumes at the input to a CARRIER SYSTEM (in particular, to increase low-level talker volumes) and to expand the received volumes at the output of the carrier system (to provide the complementary function and to make the transmission system transparent). This technique improves the SIGNAL-TO-NOISE RATIO for low-level talkers and provides a substantially reduced received NOISE level during the so-called quiet intervals.

Computer Subsystem (CSS). The computer system used by a SWITCHING CONTROL CENTER (SCC) or by several SCCs. It interfaces with the *ESS* SWITCHING SYSTEMS supported by SCCs.

Computer System for Mainframe Operations (COSMOS). A WIRE CENTER administration system operating in real time for subscriber services. The objective of COSMOS is to minimize congestion and long cross-connects on main DISTRIBUTING FRAMES while maintaining load balance across the SWITCHING equipment in the wire center.

Concentrated Range Extension with Gain (CREG). A design method that enables the increased utilization of finer gauge cables in the LOOP plant through the use of switched range extension shared by many customers.

Concentration. (1) Applies to a SWITCHING NETWORK (or portion of one) that has more inputs than outputs. (2) In a TRAFFIC NETWORK, combining calls arriving on many LINES or TRUNKS to convey them more efficiently to other TRANSMISSION or SWITCHING equipment. (3) Locating as much EQUIPMENT as possible at a given place to achieve economies in such things as building costs, power arrangements, and maintenance.

Connection. (1) Generally, in terms of a telephone *connection*, a 2-way voiceband CIRCUIT completed between two points by means of one or more SWITCHING SYSTEMS. It contains two LOOPS and may contain one or more TRUNKS. (2) A point where a junction of two or more conductors is made.

Connections per Circuit per Hour (CCH). A running count of the number of TRUNK connections between SWITCHING SYSTEMS. Used in NETWORK management as an indicator of switching congestion. *See* ATTEMPTS PER CIRCUIT PER HOUR.

Coordinate Network. A SWITCHING NETWORK connecting stages of COORDINATE SWITCHES.

Coordinate Switch. A switch with contacts or crosspoints arranged in a matrix, or gridlike, structure. The crosspoints are usually fine-motion electromechanical elements or electronic switching elements.

CORNET. A PRIVATE NETWORK serving Western Electric and Bell Laboratories. *CORNET* (a contraction of *corporate network*) is an example of an ENHANCED PRIVATE SWITCHED COMMUNICATION SERVICE.

Country Code. The 1-, 2-, or 3-digit number that, in the world NUMBERING PLAN, identifies each country or integrated numbering plan in the world. The initial digit is always the WORLD ZONE NUMBER. Any subsequent digits in the code further define the designated geographic area (normally identifying a specific country). On an international call, the *country code* is dialed before the national number.

Craft Force. OPERATING COMPANY personnel with specialized technical training who install and maintain EQUIPMENT.

Crossbar Switch. A form of COORDINATE SWITCH and the basic element of any crossbar system. A *crossbar switch* is a relay mechanism consisting of ten horizontal paths and ten or twenty vertical paths. Any horizontal path can be connected to any vertical path by means of magnets. A 2-stage operation is used to close any crosspoint. First, a selecting magnet shifts all selecting fingers in a horizontal row; then a holding magnet shifts a vertical actuating card to close the selected contacts.

Crosstalk. Interference in a communications CHANNEL caused by a SIGNAL traveling in an adjacent channel. Telephone *crosstalk* may be intelligible or unintelligible to the parties engaged in conversation.

Customer-Line Signaling. The interaction between the customer and the SWITCHING SYSTEM that serves the customer.

Customer-Premises Equipment. *See* EQUIPMENT.

Cutover. A brief interval in an overall conversion period when operation actually changes from an existing to a new system or system configuration. In some cases, the change occurs almost instantaneously and may be called a *flash cut.*

Data Circuit. A NETWORK equipment configuration that provides a capability for data services.

DATAPHONE **Digital Service.** A service in which calls are placed over the PUBLIC SWITCHED TELEPHONE NETWORK in the normal manner or automatically, and after a connection is established, DATA TERMINALS are connected at both ends for exchange of data. The term applies to PRIVATE-LINE SERVICE as well.

Data Set. Equipment that converts SIGNALS (usually DIGITAL SIGNALS) from data processors or other TERMINAL EQUIPMENT into signals suitable for TRANSMISSION over telephone lines and controls the connection. *Data sets* can be transmitters, receivers, or both. That portion of a *data set* that converts terminal signals for transmission (*modulator*) and received line signals for delivery to the terminal (*demodulator*) is called a MODEM (a contraction of *modulator/demodulator*). The terms *data set* and *modem* are often used interchangeably.

Data Terminal. A device that is used with a computer system for data input and output. If it is situated at a location remote from the computer system, it requires data TRANSMISSION. Examples of *data terminals* include teletypewriters and magnetic tape readers. The term also applies to devices for terminal-to-terminal communications.

Data Under Voice (DUV). An arrangement for transmitting 1.544-megabits per second (Mbps) pulse streams in the BANDWIDTH available underneath the portion of the baseband used for voice CHANNELS on existing microwave systems.

Decibel (dB). A logarithmic measure of the ratio between two powers:

$$dB = 10 \ \log_{10} \frac{P_2}{P_1}.$$

Named for Alexander Graham Bell.

Demarcation Point. In the INTERCONNECTION environment, the physical and electrical boundary between EQUIPMENT or FACILITIES provided by an OTHER COMMON CARRIER and Bell System facilities.

Demodulation. The process of restoring a SIGNAL to its original form at the receiving end of a TRANSMISSION system.

Dial. The part of a station set that generates a coded SIGNAL to control the CENTRAL OFFICE switching equipment in accordance with the digits dialed. It may be either a rotary or pushbutton device. (*See* STATION EQUIPMENT.) The term is sometimes used as an adjective, as in *dial administration*, the process of short-term rearranging and performance monitoring in a central office SWITCHING SYSTEM.

DIAL-IT **Network Communications Service.** Any of several services in which customers dial advertised telephone numbers to reach an announcement. Examples of *DIAL-IT* services are Public Announcement Service (Sports-Phone, Dial-a-Joke, etc.) and Media Stimulated Calling (telephone voting, telethons, and media promotions).

Dial Tone. An audible tone sent from an automatic SWITCHING SYSTEM to a customer to indicate that the EQUIPMENT is ready to receive dial SIGNALS.

Dial-Tone Delay. A measure of time required to provide DIAL TONE to customers. This measures one aspect of the performance of a SWITCHING SYSTEM.

Differential Phase-Shift Keying (DPSK). A MODULATION technique for transmitting digital information in which that information is conveyed by selecting discrete phase changes of the CARRIER SIGNAL. At the receiving end, phase changes are detected by comparing the phase of each SIGNAL element with the phase of the preceding signal element.

Digital Carrier Trunk (DCT). An internal INTERFACE that combines certain T-carrier TRANSMISSION functions and electronic SWITCHING SYSTEM control functions.

Digital Channel. A transmission CHANNEL that carries SIGNALS in digital form.

Digital Data System (DDS). A nationwide, PRIVATE-LINE, synchronous data communications NETWORK formed by interconnecting digital TRANSMISSION FACILITIES and providing special maintenance and testing capabilities. Customer CHANNELS operate at 2.4, 4.8, 9.6, or 56 kilobits per second (kbps).

Digital Signal. A SIGNAL that has a limited number of discrete states prior to TRANSMISSION. A *digital signal* may be contrasted with an ANALOG SIGNAL, which varies in a continuous manner and may be said to have an infinite number of states.

Digital Signal (DS) Level. One of several TRANSMISSION rates in the TIME-DIVISION MULTIPLEX hierarchy. For example, the DS1 level is 1.544 megabits per second (Mbps).

Digital System Cross-Connect. An internal INTERFACE that acts as a central point for cross-connecting, rearranging, patching, and testing digital EQUIPMENT and FACILITIES.

Digital Transmission. A mode of TRANSMISSION in which all information is transmitted in digital form, that is, as a serial stream of pulses. Any ANALOG SIGNAL—such as voice—can be converted into a DIGITAL SIGNAL.

Digroup. A digitally multiplexed group of twenty-four CHANNELS. *Digroup* usually refers to the T1 carrier line SIGNAL of 1.544 megabits per second (Mbps); however, the term is also used to refer to the digital CHANNEL BANK that provides the 24-channel MULTIPLEXING function.

Direct Distance Dialing (DDD). The automatic establishment of TOLL calls in response to SIGNALS from the dialing device of the originating customer.

Direct Inward Dialing (DID). A feature that permits incoming calls to stations served by a PRIVATE BRANCH EXCHANGE or by a CENTREX to be dialed directly; the call need not go through an attendant.

Direct Services Dialing Capability (DSDC). A set of service-independent NETWORK capabilities that will allow the creation of specific services to meet specific customer needs. The capabilities are provided by a set of *primitives* in a SWITCHING SYSTEM that can be summoned into use with any service. Examples of primitives are: route the call, make a billing record, play an announcement.

Direct Trunk. A TRUNK between two class 5 offices (END OFFICES) in the PUBLIC SWITCHED TELEPHONE NETWORK hierarchy.

Distributing Frame. A manually operated hardware system used to interconnect NETWORK elements (OUTSIDE PLANT cables, SWITCHING and TRANSMISSION equipment, etc.) to provide telecommunications services.

Distribution. In a SWITCHING NETWORK, *distribution* refers to the capability of connecting an input to any one of several outputs. In a TRAFFIC NETWORK, *distribution* refers to separating calls on incoming TRUNK GROUPS at a tandem TOLL OFFICE and recombining them on other outgoing trunk groups.

Double-Sideband Amplitude Modulation (DSBAM). Amplitude MODULATION in which the modulated wave is accompanied by both of the sidebands resulting from modulation; the upper sideband corresponds to the sum of the carrier and modulation frequencies, whereas the lower side band corresponds to the difference between the carrier and modulation frequencies. [*DST*]

Dual-Tone Multifrequency (DTMF). A means of SIGNALING that uses a simultaneous combination of one of a lower group of frequencies and one of a higher group of frequencies to represent each digit or character.

Echo. A wave that has been reflected or otherwise returned with sufficient magnitude and delay to be perceived in some manner as a wave distinct from that directly transmitted. *Note: Echoes* are frequently measured in DECIBELS (dBs) relative to the directly transmitted wave. *See* TALKER ECHO. [IEEE]

Echo Canceler. A device that detects transmitted speech signals, generates a SIGNAL that is a replica of the ECHO, and subtracts this signal from the actual echo, thereby canceling the echo.

Echo Suppressor. A device that detects speech signals transmitted in either direction on a 4-wire CIRCUIT and introduces LOSS in the opposite direction of speech TRANSMISSION to suppress echoes.

Economic CCS (ECCS). The load that should be carried on the last TRUNK of a HIGH-USAGE GROUP to minimize the total cost of routing the offered TRAFFIC, assuming that overflow from the high-usage route is offered to an alternate route engineered to meet an objective BLOCKING probability. *See* CARRIED LOAD.

Economic Evaluation. Analyzing the economic impact on a company's overall financial position of designing, producing, or implementing a product or service.

800 Service. Also called *INWATS*. *See* WIDE AREA TELECOMMUNICATIONS SERVICES.

Electromechanical Switching System. An automatic SWITCHING SYSTEM in which the control functions are performed principally by devices, such as relays and servos, that are electrically operated and have mechanical motion.

Electronic Switching System. A class of modern SWITCHING SYSTEMS in which the control functions are performed principally by electronic devices.

Electronic Tandem Switching (ETS). A PRIVATE NETWORK service that provides customers with a uniform NUMBERING PLAN and numerous call-routing features. The *electronic tandem switching* functions are furnished by the SWITCHING equipment that provides PRIVATE BRANCH EXCHANGE or CENTREX service.

Electronic Translator. The equipment in No. 4A and No. 5A Crossbar Systems that functions primarily to translate ADDRESS codes, by means of electronic circuitry and STORED-PROGRAM CONTROL, into information required by the system to select an available route toward the CENTRAL OFFICE of the called customer. *See* CROSSBAR SWITCH.

E&M Lead Signaling. A specific form of INTERFACE between a SWITCHING SYSTEM and a TRUNK in which the SIGNALING information is transferred across the interface via 2-state voltage conditions on two leads, each with a ground return, separate from the leads used for MESSAGE information. The message and signaling information are combined (and separated) by a signaling system appropriate for application to the TRANSMISSION FACILITY. The term *E&M lead signaling* is used also in some SPECIAL-SERVICES applications.

End Office. A local SWITCHING office where LOOPS are terminated for purposes of INTERCONNECTION to each other and to TRUNKS. *End offices* are designated class 5 offices.

End Office Toll Trunk. A HIGH-USAGE TRUNK from an END OFFICE that carries TOLL traffic only. It may be either a DIRECT TRUNK to another end office or a trunk to a TOLL CENTER in another toll center area (that is, not the toll center on which the end office homes).

Engineered Capacity. The highest load level for a TRUNK GROUP or a SWITCHING SYSTEM at which SERVICE OBJECTIVES are met.

Engineering and Administrative Data Acquisition System (EADAS). An OPERATIONS SYSTEM in which TRAFFIC data are measured at SWITCHING SYSTEMS by electronic devices, transmitted to a centrally located minicomputer, and recorded on magnetic tape in a format that is suitable for computer processing and analysis.

Engineering and Administrative Data Acquisition System/Network Management (EADAS/NM). An OPERATIONS SYSTEM that monitors SWITCHING SYSTEMS and TRUNK GROUPS that have been designated by network managers. *EADAS/NM* reports existing or anticipated congestion on a display board at operating company NETWORK MANAGEMENT CENTERS.

Engineering Period. A particular time period during which service is measured and compared to the objective GRADE OF SERVICE. Sufficient EQUIPMENT must therefore be provided (*engineered*) to meet SERVICE OBJECTIVES during this period.

Enhanced Private Switched Communication Service (EPSCS). A PRIVATE NETWORK service that, like the COMMON-CONTROL SWITCHING ARRANGEMENT, provides a uniform dialing plan for customers with geographically dispersed locations. *EPSCS* offers 4-wire TRANSMISSION (to improve transmission quality) within the private network and a Customer Network Control Center, which can be used by the customer to control some network operations and to obtain private network usage and status information.

EPLANS **Computer Program Service.** Software systems used by OPERATING COMPANY engineering and related personnel to support their planning, recordkeeping, implementation, scheduling, ordering, NETWORK performance evaluation, network characterization, and other similar activities. The programs are Western Electric products and are offered as time-share or batch-run computer services by Western Electric or, in some cases, are run in telephone company data centers.

Equipment. A term broadly applied to hardware components that includes customer-premises equipment (CPE) and SWITCHING and TRANSMISSION (and other) components located in telephone company buildings.

Equipment and Facility Recovery. The reduction of capital investment for additional EQUIPMENT accomplished by providing accurate records of what capital equipment is already available to meet a current need.

Equipment-to-Equipment Interface. Any INTERFACE between EQUIPMENT units on a user premises that is not considered a NETWORK interface.

Erlang. A dimensionless unit of TRAFFIC intensity used to express the average number of calls underway or the average number of devices in use. One *erlang* corresponds to the continuous occupancy of one traffic path. Traffic in *erlangs* is the sum of the holding times of paths divided by the period of measurement. The term *erlang* can be used to express the capacity of a system; for example, a TRUNK GROUP of thirty TRUNKS, which, in a theoretical peak sense, might carry 30 *erlangs* of traffic, would have a typical capacity of perhaps 25 *erlangs* averaged over an hour. Named for A. K. Erlang, the founder of the traffic theory.

Erlang B. One of the basic TRAFFIC models and related formulas used in the Bell System. The assumptions are POISSON input, negative exponential holding times, and BLOCKED CALLS CLEARED. Used for TRUNK engineering. Also called *Erlang's Loss Formula*.

Erlang C. One of the basic TRAFFIC models and related formulas used in the Bell System. The assumptions are POISSON input, negative exponential holding times, and BLOCKED CALLS DELAYED. Used for COMMON-CONTROL engineering. Also called *Erlang's Delay Formula*.

Error-Second. A measurement of system performance for digital TRANSMISSION FACILITIES. An *error-second* is a 1-second interval during which one or more bit errors occur.

Exchange Area. Traditionally, an area within which there is a single uniform set of charges for telephone service. An *exchange area* may be served by a number of CENTRAL OFFICES. A call between any two points within an exchange area is a local call.

Expanded 800 Service. An improvement over the basic 800 SERVICE, which uses DIRECT SERVICES DIALING CAPABILITY to provide customers with more options in defining service areas and determining the treatment a call receives.

Expansion. The term applied to a SWITCHING NETWORK (or portion of one) that has more outputs than inputs.

Facilities Assignment and Control System (FACS). An on-line data-processing system that maintains inventories and provides assign-ment of OUTSIDE PLANT and CENTRAL OFFICE facilities.

Facilities Network. The aggregate of TRANSMISSION systems, SWITCHING SYS-TEMS, and STATION EQUIPMENT; it supports a large number of traffic NET-WORKS.

Facility. Any one of the elements of physical telephone plant that are needed to provide service. Thus, SWITCHING SYSTEMS, cables, and microwave radio TRANSMISSION systems are examples of *facilities*. *Facil-ity* is sometimes used in a more restricted sense to mean TRANSMISSION FACILITY.

Facility and Equipment Planning System (FEPS). A component system of the TRUNKS INTEGRATED RECORDS KEEPING SYSTEM (TIRKS) that supports the current planning process. *FEPS* helps planners use information in the TIRKS data bases along with forecasts of future growth to allo-cate existing inventories efficiently, to determine future EQUIPMENT and FACILITY requirements, and to update planning designs.

Federal Communications Commission (FCC). A board of commissioners, appointed by the president of the United States under the Communi-cations Act of 1934, having the power to regulate interstate and foreign communications originating in the United States by wire and radio.

Feeder Route. A network of LOOP cable extending from a WIRE CENTER into a segment of the area served by the wire center.

Field Representative. A member of the Bell Laboratories Quality Assurance Center responsible for monitoring the operation of Bell System telecommunications EQUIPMENT in the field and participating in field evaluations and reliability studies.

Final Group. A TRUNK GROUP that acts as a final (last-chance) route for TRAFFIC. Traffic can overflow to a *final group* from HIGH-USAGE GROUPS that are busy. Calls blocked on a *final group* are not offered to another route.

Final Trunk. A TRUNK in a FINAL GROUP.

Flat Rate. A rate-setting principle for local service in which customers in a specific group or area are all charged the same rate for local calling regardless of the number of local calls they make or the length of the calls.

Focused Overload. Abnormal calling from many points to one point; for example, after a disaster or when a radio or television station encourages mass calling.

Forecasting. A NETWORK PLANNING activity that provides estimates of future demands for existing and new services.

Foreign Exchange (FX) Service. A service that provides a CIRCUIT between a customer's MAIN STATION or PRIVATE BRANCH EXCHANGE and a CENTRAL OFFICE other than the one that normally serves the EXCHANGE AREA in which the customer is located.

Foreign Numbering Plan Area (FNPA). Any NUMBERING PLAN AREA (NPA) outside the boundaries of the home NPA.

Frame. (1) A segment of an ANALOG SIGNAL or DIGITAL SIGNAL that has a repetitive characteristic in that corresponding elements of successive *frames* represent the same things. Examples are a television *frame*, which represents a complete scan of a picture, or a TELEMETRY *frame*, which represents values of a number of parameters in a specific order. In a TIME-DIVISION MULTIPLEX system, a *frame* is a sequence of time slots, each containing a sample from one of the CHANNELS served by the multiplex system; the *frame* is repeated at the sampling rate, and each channel occupies the same sequence position in successive *frames*. (2) An assembly of EQUIPMENT units.

Framing Bit. A non-information-carrying bit introduced into a bit stream to facilitate the separation of characters at the receiving end of a TRANSMISSION.

Frequency Content. The band of frequencies or specific frequency components contained in a SIGNAL. For example, the frequency content of a voiceband signal includes components between 200 and 3500 hertz (Hz).

Frequency-Division Multiplex (FDM). A method of providing a number of simultaneous CHANNELS over a common TRANSMISSION path by using a different frequency band for the transmission of each channel.

Frequency Modulation (FM). One way to modify a SIGNAL to make it "carry" information. The frequency of a CARRIER SIGNAL is modified in accordance with the amplitude of the information signal.

Frequency-Shift Keying (FSK). A MODULATION technique for transmitting digital information having two, or possibly more, discrete states. Each of the discrete states is represented by an associated frequency. The most common form is binary *FSK*, which uses two frequencies to represent the two states.

Fundamental Planning. A NETWORK PLANNING activity that develops long-range plans for changes and growth in NETWORK structure.

Gain. An increase in SIGNAL power during TRANSMISSION from one point to another, usually expressed in DECIBELS (dBs). Also called *amplification*. [TC]

General Trade. A Bell System term for manufacturers and suppliers of telecommunications EQUIPMENT other than Western Electric.

Generic Program. A set of instructions for an electronic SWITCHING SYSTEM that is the same for all offices using that type of system. Detailed differences for each individual office are listed in a separate parameter table.

Grade of Service (GOS). (1) An estimate of customer satisfaction with a particular aspect of service (such as NOISE or ECHO). It combines the distribution of subjective opinions of a representative group of people with the distribution of performance for the particular aspect being graded. For example, with a specified distribution of noise, 95 percent of the people may judge the noise performance to be good or better; the noise *grade of service* is then said to be 95 percent good or better. (2) In traffic NETWORKS, the proportion of calls that receive no service (BLOCKING) or poor service (long delay).

High-Usage (HU) Group. A TRUNK GROUP between two SWITCHING SYSTEMS that is designed for high average occupancy. To provide an overall acceptable probability of BLOCKING, calls blocked on a *high-usage group* are offered to other routes.

High-Usage (HU) Trunk. A TRUNK in a HIGH-USAGE GROUP.

Home Numbering Plan Area (HNPA). The NUMBERING PLAN AREA within which the calling line appears at a local (class 5) switching office (END OFFICE).

Human Factors. A scientific discipline that takes human behavioral characteristics and physical capabilities into account when designing products to be used or tasks to be performed.

Hundred Call Seconds (CCS) Per Hour. A unit of TRAFFIC used to express the average number of calls in progress or the average number of devices in use. Numerically, it is 36 times the traffic expressed in ERLANGS.

Hybrid. A NETWORK having four ports and designed so that when the ports are properly terminated, the SIGNAL input to any particular port splits equally between the two adjacent ports with essentially no signal coupled to the opposite port. *Hybrids* are used to couple 4-wire CIRCUITS to 2-wire circuits.

Impulse Noise. Short-duration, high-amplitude bursts, or spikes, of NOISE energy, much greater than the normal peaks of MESSAGE CIRCUIT NOISE on a transmission CHANNEL.

Inband Tone Signaling. SIGNALING that uses the same path as a MESSAGE and in which the signaling frequencies are in the same band used for the message.

Independent Telephone Company. A telephone company, not affiliated with the Bell System, that has its own "independent" territory. In 1981, there were more than 1450 *independent telephone companies* in the United States.

Individual Circuit Analysis (ICAN) Program. Part of the TOTAL NETWORK DATA SYSTEM that detects ELECTROMECHANICAL SWITCHING SYSTEM equipment faults by identifying abnormal load patterns on individual CIRCUITS within a circuit group. These faults, for example, include defective circuits that prevent customer calls from being completed. *ICAN* produces reports used by the NETWORK ADMINISTRATION CENTER.

Intercept Calls. Calls directed by a customer to an improper telephone number that are redirected to an operator or a recording. The caller is told why the call could not be completed and, if possible, given the correct number.

Intercity Facility Relief Planning System (IFRPS). An OPERATIONS SYSTEM that aids in FACILITY planning for a TOLL network. Its function is similar to the METROPOLITAN AREA TRANSMISSION FACILITY ANALYSIS PROGRAM.

Intercom Calling. Intralocation calling; calls between stations on the same customer premises.

Interconnection. The direct electrical connection, acoustical coupling, or inductive coupling of user-premises TERMINAL EQUIPMENT (including terminal equipment that is a part of a separate communications system) to the telephone NETWORK. It also includes the direct electrical connection of OTHER COMMON CARRIER facilities to the telephone network.

Interface. A common boundary between two systems or pieces of EQUIPMENT where they are joined.

Interface Specification. A technical requirement that must be met at an INTERFACE.

International Telecommunications Satellite Consortium (INTELSAT). An international organization established in 1964 to govern a global commercial communications satellite system to provide communications between many countries. Membership is in excess of eighty countries. The Communications Satellite Corporation (COMSAT) acts as manager for *INTELSAT* and also represents the United States.

Interoffice Call. A call between two SWITCHING SYSTEMS.

Interoffice Facilities Network. Part of the nationwide FACILITIES NETWORK, consisting of interoffice TRANSMISSION FACILITIES, TANDEM SWITCHING SYSTEMS, and LOCAL SWITCHING SYSTEMS. *See* **LOCAL FACILITIES NETWORK.**

Interoffice Trunk Signaling. The exchange of call-handling information between SWITCHING offices within the NETWORK.

Intertoll Trunk. A TRUNK between two TOLL OFFICES.

Intraoffice Call. A call involving only one SWITCHING SYSTEM.

Jumpers. Temporary wires used on a DISTRIBUTING FRAME to cross-connect the TERMINATION points of cables from particular EQUIPMENT and FACILITIES to provide service to a customer.

Key Telephone Set (KTS). A telephone set with buttons, or keys, located on or near the telephone, used as part of a KEY TELEPHONE SYSTEM.

Key Telephone System. An arrangement of KEY TELEPHONE SETS and associated circuitry, located on a customer's premises, that provides combinations of certain voice communications arrangements such as call pickup, call hold, call line status lamp signals, and INTERCONNECTION among on-premises stations without the need for connection through the CENTRAL OFFICE or PRIVATE BRANCH EXCHANGE.

Large-Scale Integrated (LSI) Circuit. An integrated CIRCUIT containing 100 gates or more on a single chip, resulting in an increase in the scope of the function performed by a single device.

License Contract. The legal agreement that governed the relationship between AT&T and a Bell OPERATING COMPANY. Each *license contract* described the reciprocal services, licenses, and privileges that existed between the parties. All *license contracts* terminated with divestiture of the Bell operating companies.

Line. (1) From a SWITCHING viewpoint, the LOOP-, STATION EQUIPMENT-, and CENTRAL OFFICE—associated EQUIPMENT assigned to a customer. (2) From a TRANSMISSION viewpoint, the transmission path between a customer's station equipment and a SWITCHING SYSTEM. In this sense, it is also called a *LOOP*. (3) In CARRIER SYSTEMS, the portion of a transmission system between two terminal locations. The *line* includes the transmission media and associated line REPEATERS. (4) The side of a piece of CENTRAL OFFICE equipment that connects to or toward the OUTSIDE PLANT; the other side of the equipment is called the *drop side*. (5) A family of equipment or apparatus designed to provide a variety of styles, a range of sizes, or a choice of service features, for example, a "product line."

Linear Distortion. Distortion resulting from a CHANNEL having a linear filter characteristic different from an ideal linear LOW-PASS or BANDPASS FILTER; in particular, amplitude-versus-frequency characteristics that are not flat over the passband, and phase-versus-frequency characteristics that are not linear over the passband. *See* BANDPASS FILTER.

Line Build-Out (LBO) Network. Amplifiers (REPEATERS) in a cable TRANSMISSION system may be designed to compensate for distortion over a specific length of cable. When the length of cable between amplifiers is less than that for which the amplifier is designed, one or more *line build-out networks* are used to bring the distortion to approximately the design level.

Line Fill. The ratio of assigned CIRCUITS to total capacity in a FACILITY or EQUIPMENT unit.

Link. A TRANSMISSION FACILITY in the telecommunications NETWORK.

Load Balance System (LBS). A CENTRAL OFFICE reporting system and part of the TOTAL NETWORK DATA SYSTEM. *LBS* helps assure network administrators that TRAFFIC loads in SWITCHING SYSTEMS are uniformly distributed. For example, reports generated by *LBS* are used to determine the "lightly loaded" line groups to which new subscriber LINES can be assigned.

Load Balancing. Procedures mainly used for equalizing the TRAFFIC load on customer TERMINATION groups in SWITCHING NETWORKS.

Load Lost. In a TRAFFIC system, the portion of OFFERED LOAD that is not served because all servers are busy and all waiting positions (if any are provided) are occupied.

Local Automatic Message Accounting (LAMA). A process using EQUIPMENT located in a local office for automatically recording billing data for MESSAGE RATE SERVICE calls (bulk billing) and for customer-dialed station-to-station TOLL calls. The tape record is sent to and processed at an electronic data-processing center.

Local Facilities Network. Part of the nationwide FACILITIES NETWORK, consisting of local SWITCHING SYSTEMS located at WIRE CENTERS and the loop TRANSMISSION FACILITIES through which customers are connected to local switching systems. Local switching systems are considered the conceptual boundary between the *local facilities network* and the INTEROFFICE FACILITIES NETWORK and, in this sense, belong to both networks. *See* INTEROFFICE FACILITIES NETWORK.

Local Switching System. A SWITCHING SYSTEM that performs END OFFICE (class 5) functions. *Local switching systems* connect customer LINES directly to other customer lines, or customer lines to TRUNKS.

Long-Haul Trunk. A FINAL TRUNK or HIGH-USAGE TRUNK that interconnects two regions in the PUBLIC SWITCHED TELEPHONE NETWORK hierarchy.

Long-Route Design. A codification of design practices used to plan customer LOOPS that exceed the resistance design limit of the serving CENTRAL OFFICE.

Loop. A CHANNEL between a customer's terminal and a CENTRAL OFFICE. A *loop* may also be called a *LINE*.

Loop Assignment Center (LAC). An OPERATIONS CENTER that assigns customer LOOP facilities, telephone numbers, CENTRAL OFFICE line EQUIPMENT, and miscellaneous central office equipment.

Loop-Reverse-Battery. A method of SIGNALING over interoffice TRUNKS in which dc changes, including directional changes associated with battery reversal, are used for SUPERVISION. This technique provides 2-way signaling on 2-wire trunks; however, a trunk can be seized at only one end; that is, it cannot be seized at the office at which battery is applied. Also called *reverse battery signaling.*

Loop Signaling. A method of SIGNALING over dc CIRCUIT paths that utilizes the metallic LOOP formed by the LINE or TRUNK conductors and terminating circuits.

Loss. In the Bell System, the term refers to *insertion loss,* a quantity that represents a specific relationship between the input and output of a NETWORK (for example, a customer connection or a CIRCUIT). The basic *insertion loss* calculation determines the difference in DECIBELS (dBs) between power applied to a load directly and power applied to a load through a network.

Loss Objective. An objective for the amount of LOSS that can be tolerated in NETWORK components and still maintain a satisfactory GRADE OF SERVICE.

Loudness Loss. A measure used to express the LOSS of communications paths in a manner that reflects loudness perception. For partial and overall telephone CIRCUITS, *loudness loss* is the ratio of suitably weighted output SIGNAL levels to input signal levels. (The signals may be electric or acoustic.)

Low-Pass Filter. A filter having a single TRANSMISSION band extending from zero to some finite cutoff frequency.

Main Station. A telephone that is connected directly to a CENTRAL OFFICE by either an individual or shared LINE. *Main stations* include the principal telephone of each party on a party line. They do not include telephones that are manually or automatically connected to a central office through a PRIVATE BRANCH EXCHANGE or extension telephones (that is, telephones that have been added to an individual or shared line to extend telephone service to other parts of the subscriber's home or business premises).

Marker (Crossbar). The heart of COMMON-CONTROL crossbar CENTRAL OFFICE equipment. *Markers* perform the following functions in a No. 5 Crossbar SWITCHING SYSTEM: (1) determine terminal locations of calling LINES, incoming TRUNKS bidding for service, called lines, and outgoing trunks in the EQUIPMENT; (2) determine the proper route for the call, establish the connection within the office, and pass routing information to the senders; (3) determine the calling line class of service and provide charge classification; (4) recognize line busy, trouble, intercept, and vacant-line conditions; and (5) call in a trouble recorder when necessary. *See* CROSSBAR SWITCH.

Market. The set of actual or potential buyers of a product or service.

Marketing. Those activities that influence the flow of goods and services from producers to consumers.

Marketing Mix. The set of variables that can be adjusted to attract MARKETS to a product or service. These include variables associated with the product itself and those related to price, product distribution, and promotion of the product.

Market Segmentation. The division of a MARKET into submarkets differing in such characteristics as customer needs and buying behavior.

Master Control Center (MCC). A FRAME of EQUIPMENT in an ELECTRONIC SWITCHING SYSTEM office with lamps that show the current state of the office equipment and with keys for operating controls.

Measurement Plan. A compilation of service and performance measurements for operational units (for example, OPERATIONS CENTERS).

Message. (1) In telephone communications, a successful call attempt that is answered by the called party and followed by some minimum period of CONNECTION. (2) In data communications, a set of information, typically digital and in a specific code such as the American Standard Code for Information Interexchange (ASCII), to be carried from a source to a destination. A header, with ADDRESS and other information regarding handling, may be considered part of or separate from the *message*.

Message Circuit Noise. The short-term average NOISE level as measured with a 3A noise measuring set or its equivalent. This set includes frequency weighting and time constants to make the set most sensitive to noise that will impair TRANSMISSION quality in telephone CIRCUITS used for speech.

Message Rate Service. Telephone service for which a charge is made in accordance with a measured amount of usage, referred to as *message units.*

Message Telecommunications Service (MTS). Non-PRIVATE-LINE intrastate and interstate long-distance telephone service.

Message Trunk. A TRUNK carrying MESSAGE TELECOMMUNICATIONS SERVICE traffic on the PUBLIC SWITCHED TELEPHONE NETWORK.

Metropolitan Area Transmission Facility Analysis Program (MATFAP). A program that aids in facility planning for metropolitan NETWORKS. Using various measures, *MATFAP* analyzes the alternatives available for future TRANSMISSION FACILITIES and EQUIPMENT and identifies what transmission plant is needed at various locations and when it will be needed. *MATFAP* also determines the economic consequences of selecting particular facilities and/or equipment and of selecting particular routes and provides the least-cost assignment of CIRCUITS to each facility as a guide to the CIRCUIT-PROVISIONING process.

Microprocessor. The control and processing portion of a microcomputer built with LARGE-SCALE INTEGRATED CIRCUITRY, usually on one chip. *Microprocessors* can handle both arithmetic and logic functions under control of a program stored in a memory chip.

Minimum-Cost Routing. A CIRCUIT-ROUTING scheme that determines a path through the NETWORK for each point-to-point demand for each year, so that, when point-to-point demands are provided on these paths and the resulting CAPACITY EXPANSION problem is solved, the total cost of TRANSMISSION FACILITIES is minimized.

Mobile Telephone Service. One of a class of services that uses radio CHANNELS to provide telephone service to customers on the move. *Mobile telephone services* include land mobile telephone service, *BELLBOY* personal signaling set paging service, air/ground service, marine radiotelephone services, and high-speed train telephone services.

Modem. A contraction of the words *modulator* and *demodulator* signifying an EQUIPMENT unit that performs both of these functions. *See* **DATA SET.**

Modular Engineering. The expression of EQUIPMENT quantities (typically, TRUNK GROUPS) as an integer multiple of some basic unit, the unit depending on physical constraints. Reflects the reality that certain equipment items can only be added in multiples, rather than one item at a time.

Modulation. The process by which the amplitude, frequency, or phase of a CARRIER SIGNAL is varied in accordance with one of the characteristics of an information SIGNAL.

Multifrequency (MF) Signaling. An inband, interoffice, ADDRESS SIGNALING method in which ten decimal digits and five auxiliary SIGNALS are each represented by selecting two frequencies out of the following group: 700, 900, 1100, 1300, 1500, and 1700 hertz (Hz).

Muldem. A contraction of the words *multiplexer* and *demultiplexer* signifying an EQUIPMENT unit that performs both of these functions.

Multiplex(ing). The EQUIPMENT or process for combining a number of individual CHANNELS into a common frequency band or into a common bit stream for TRANSMISSION. The converse equipment or process for separating into individual channels is called *demultiplex(ing)*.

Narrowband Channel. A transmission CHANNEL with a BANDWIDTH narrower than that required for transmitting voice SIGNALS.

Nationwide Rate Averaging. A rate-setting method that has been used to establish rates for interstate telephone service; like services and like calling distances carry the same charges even though the costs of interstate service to the COMMON CARRIER may differ for routes in low-density, high-cost areas and for routes in high-density, low-cost areas.

Negative Exponential Distribution. A probability distribution function used in many queuing models to describe the distribution of the length of completed telephone calls.

Network. (1) The FACILITIES NETWORK is the aggregate of TRANSMISSION systems, SWITCHING SYSTEMS, and STATION EQUIPMENT; it supports a large number of traffic *networks*. (2) A TRAFFIC NETWORK is an arrangement of CHANNELS, such as LOOPS and TRUNKS; associated SWITCHING arrangements; and station equipment, designed to handle a specific body of TRAFFIC. A traffic *network* is a subset of the facilities *network*. (3) An electrical/electronic CIRCUIT, usually packaged as a single piece of apparatus or on a printed circuit pack. Examples are a transformer *network* and an equalization *network*. (4) The switching stages and associated INTERCONNECTIONS of a switching system are collectively called the SWITCHING NETWORK.

Network Administration Center (NAC). An OPERATIONS CENTER with administrative responsibility for local and TANDEM SWITCHING SYSTEMS. The functions performed by an *NAC* include data administration, daily surveillance of service and load in the SWITCHING NETWORK, EQUIPMENT utilization, machine and TRUNK GROUP performance analysis and monitoring, and participation in Job Contact Committees associated with equipment additions, replacements, and rearrangements.

Network Channel-Terminating Equipment. EQUIPMENT located on user premises that is a part of the telephone NETWORK facility.

Network Control Point (NCP). A NODE in the STORED-PROGRAM CONTROL network that connects to a SIGNAL TRANSFER POINT in the COMMON-CHANNEL INTEROFFICE SIGNALING (CCIS) network. The *NCP's* associated data base contains customer-service data accessed by SIGNALS routed over the CCIS network and used to support extended NETWORK services.

Network Data Collection Center (NDCC). An OPERATIONS CENTER that administers NETWORK data collection. Primarily, the *NDCC* supervises the operation and maintenance of the Engineering and Administrative Data Acquisition System (EADAS), 1A EADAS central control units, data LINKS, and data collection apparatus.

Network Interface (NI). In the INTERCONNECTION environment, the physical and electrical boundary between two separately owned telecommunications capabilities. It also serves as the boundary for administrative and maintenance activities.

Network Management Center (NMC). An OPERATIONS CENTER responsible for surveillance and control of TRAFFIC flow in a specific geographic area. Control is an ongoing activity in response to OVERLOADS, especially from peak day calling, mass calling, NETWORK system failure, or network rearrangements. The *NMC* also plans strategies for potential overload situations.

Network Operations Center (NOC). The OPERATIONS CENTER, located in Bedminster, New Jersey, that oversees and coordinates management of the North American message NETWORK. It monitors the status of the intertoll network among the regions and coordinates the use of interregional SWITCHING SYSTEM and trunking network capacity that is temporarily spare. The *NOC* also directs the network management of international TRAFFIC flows for the United States and monitors the status of TOLL facilities and the effect of any problems on the interregional intertoll network. In case of problems in the toll facility network, the NOC sets restoration priorities and coordinates restoration activities.

Network Operations Center System (NOCS). The primary support system for the NETWORK OPERATIONS CENTER located in Bedminster, New Jersey. NOCS provides near real-time surveillance of major SWITCHING SYSTEMS and their associated trunking at a nationwide level.

Network Planning. A multifaceted discipline that encompasses the functions involved in planning the evolution and implementation of the nationwide NETWORK, designing and engineering the configuration of the network, and managing the total network investments.

Network Service Center (NSC). An OPERATIONS CENTER responsible for ensuring the overall quality of NETWORK service, keeping management informed about levels and trends in the quality of service provided by the network, and stimulating improvement activities when instances of substandard service are identified.

New Equipment-Building System Standards (NEBS). A set of integrated specifications for both telecommunications EQUIPMENT and equipment buildings.

No. 5 Crossbar Central Office Equipment Reports (5XB COER). A CENTRAL OFFICE equipment reporting system for No. 5 Crossbar offices. The system analyzes TRAFFIC data in support of NETWORK administration and design functions.

Node. A SWITCHING office or FACILITY junction point in the telecommunications NETWORK.

Noise. An unwanted disturbance introduced in a communications CIRCUIT. It may partially or completely obscure the information content of a desired SIGNAL. On telephone circuits, *noise* may be an annoyance during quiet intervals as well as when speech is present.

Nonlinear Distortion. A type of distortion in which the output SIGNAL amplitude does not have the desired linear relationship to the input signal amplitude.

Numbering Plan. A numbering system for a switched telephone NETWORK that identifies each MAIN STATION by a unique ADDRESS that is convenient, readily understandable, and similar in format to that of other main stations connected to the network.

Numbering Plan Area (NPA). The familiar *area code*, defining a geographic division within which telephone directory numbers are subgrouped. In North America, a 3-digit $N0/1X$ code is assigned to denote each *NPA*, where:

$$N = \text{any digit 2 through 9}$$
$$X = \text{any digit 0 through 9}$$
$$0/1 = 0 \text{ or } 1.$$

Occupancy. The fraction of time that a CIRCUIT or an EQUIPMENT unit is in use, expressed as a decimal. Numerically, it is the ERLANGS carried per circuit. *Occupancy* typically includes both MESSAGE time and call-setup time.

Offered Load. The demand placed on a TRAFFIC system, defined by the product of two parameters: the average rate at which customers place demands on the system and the average length of time they require service.

Off-Hook. Station switchhook contacts closed, resulting in LINE current, or whatever SUPERVISION condition is indicative of the in-use or request-for-service state.

One-Way Trunk. A TRUNK that can be seized for use by the SWITCHING equipment at one end only. Once a trunk is seized, 2-way TRANSMISSION may occur.

On-Hook. Station switchhook contacts open or whatever SUPERVISION condition is indicative of the EQUIPMENT-idle state.

Operating Company. A regulated telephone company whose primary business is providing telephone service to customers.

Operations Center. A group of people, reporting to a common manager, who perform a set of closely related functions for a specific geographic area, group of customers, or service.

Operations Planning. An ongoing activity to ensure that changes in the roles and responsibilities of people are linked to changes in the telephone business and that operations-related functions are assigned to people and OPERATIONS SYSTEMS in ways that realize potentials for greater efficiency and better customer service.

Operations Process. The sequence of interactions between OPERATIONS SYSTEMS and OPERATIONS CENTERS that are required to perform a particular operation, such as adding a TRUNK to the NETWORK or maintaining a SWITCHING SYSTEM.

Operations System. A computer-based system that OPERATING COMPANY employees use to support operations activities. An *operations system* does not directly provide telecommunications service to customers, but supports operating company personnel in the performance of their duties, such as testing TRUNKS or maintaining SWITCHING SYSTEMS.

Operations Systems Network. The collection of OPERATIONS SYSTEMS, communications terminals at OPERATIONS CENTERS, and the switched and direct communications LINKS interconnecting them and connecting them to a variety of telecommunications systems (for example, an ELECTRONIC SWITCHING SYSTEM).

Operator Code. A code of the form 1XX, 11XX, or 11XXX that allows outward TOLL operators to reach inward, directory assistance, or other operators in distant cities.

Operator Number Identification (ONI). Operator identification of a calling station, usually for billing purposes, when automatic identification from a local office is not available.

Operator Services. A variety of services normally performed by operators. These include completing or helping customers to complete

TOLL calls and assistance calls; preparing billing inputs on those calls; providing directory assistance; intercepting and helping customers with calls to changed or nonworking numbers; providing SPECIAL SERVICES, such as person-to-person, coin, calling card, collect, mobile, *PICTUREPHONE* MEETING SERVICE, and conference calls; and giving on-the-job consultation to business customers.

Optical Fiber. A glass fiber that provides a 1-way path for light SIGNALS. Used in lightguide cable.

Other Common Carrier (OCC). A telecommunications COMMON CARRIER authorized by the FEDERAL COMMUNICATIONS COMMISSION (FCC) to provide a variety of services. The FCC refers to these carriers as *domestic satellite carriers, miscellaneous common carriers,* and *specialized common carriers.*

Outpulsing. Sending ADDRESS or other SIGNALING information over a LINE or TRUNK.

Outside Plant. The part of the telephone system that is located physically outside of telephone company buildings. *Outside plant* includes cables, supporting structures, and certain EQUIPMENT items; it does not include microwave towers, antennas, and cable system REPEATERS.

Outstate Facility Network Planning System (OFNPS). An interactive computer system that aids in facility planning for rural NETWORKS. It is similar in function to the METROPOLITAN AREA TRANSMISSION FACILITY ANALYSIS PROGRAM.

Overflow. A count of all calls that are offered to a TRUNK GROUP but are not carried (*see* **PEG COUNT**); usually measured for one hour.

Overload. (1) In TRANSMISSION, a load greater than that which a device is designed to handle; may cause overheating of power-handling components and distortion in SIGNAL circuits. *[DST]* (2) For telecommunications TRAFFIC, an *overload* is an increase in OFFERED LOAD beyond the capacity for which NETWORK components (for example, TRUNKS and SWITCHING SYSTEMS) are engineered.

Packet. A group of bits that is switched as an integral unit. Typically, a *packet* contains data, destination and origination information, and control information, arranged in a particular format.

Packet-Switching Network. A NETWORK that is designed to transport and switch data in PACKET form.

Pair Gain. Referring to a system that uses digital or analog carrier techniques to serve many customers over a few pairs between the CENTRAL OFFICE and a remote electronic terminal.

Peakedness. A telecommunications TRAFFIC term signifying the ratio of the variance of the load to the mean of the load. For random traffic (POISSON arrivals, negative exponential holding times), the *peakedness* of the load is 1.0, that is, the variance equals the mean. For traffic overflowing a HIGH-USAGE GROUP, *peakedness* exceeds 1, reflecting the fact that OVERFLOW occurs in "bursts" or peaks.

Peak Load. Denotes a higher-than-average quantity of TRAFFIC; usually expressed for a 1-hour period and as any of several functions of the observing interval, such as peak hour during a day, average of daily peak hours over a 20-day interval, maximum of average hourly traffic over a 20-day interval. Significantly higher *peak loads* occur infrequently as the result of catastrophes and on Mother's Day and Christmas Day.

Peg Count. A count of all calls offered to a TRUNK GROUP, usually measured for 1 hour. As applied to units of SWITCHING SYSTEMS with COMMON-CONTROL, *peg count*, or *carried peg count*, means the number of calls actually handled.

Performance Measurement. A measurement intended to reflect whether an operational unit (for example, an OPERATIONS CENTER) is meeting objectives or whether NETWORK service characteristics and performance parameters are at levels required to meet SERVICE OBJECTIVES.

Performance Objective. A statement of the operational objective to be met by a component of the NETWORK or the network as a whole. For example, the MESSAGE CIRCUIT NOISE objective for customer LOOPS is stated in terms of an upper limit of 20 DECIBELS above reference noise, C-message weighted (dBrnC).

Per-Trunk Signaling. A method of SIGNALING in which the SIGNALS pertaining to a particular call are transmitted over the same TRUNK that carries the call. Interoffice signaling other than COMMON-CHANNEL INTEROFFICE SIGNALING falls into this category.

PICTUREPHONE **Meeting Service.** A service, supported by High-Speed Switched Digital Service, that allows people in two distant, specially equipped rooms to hold a fully interactive audio and video conference.

Plant. All of the FACILITIES (such as land, buildings, machinery, apparatus, instruments, and fixtures) needed to provide telecommunications services. *See* **OUTSIDE PLANT.**

Plug-In Inventory Control System/Detailed Continuing Property Records (PICS/DCPR). An OPERATIONS SYSTEM that inventories PLUG-IN UNITS of EQUIPMENT in CENTRAL OFFICES. The *DCPR* portion serves as an investment data base that supports a telephone company's accounting records for all types of central office equipment.

investment data base that supports a telephone company's accounting records for all types of central office equipment.

Plug-In Unit. A prewired modular assembly of electronics components designed to be plugged into a permanently wired receptacle. Most contemporary telecommunications EQUIPMENT consists of many *plug-in units* to perform a variety of functions.

Poisson. In TRAFFIC theory, *Poisson* refers to a distribution or a process resulting in a distribution of events such that the intervals between adjacent events are independent, random variables that are members of identical exponential distributions. Under certain conditions, the arrival of telephone calls to be routed over a TRUNK GROUP can be approximated by a *Poisson* distribution. Named after a 19th-century French mathematician.

Preferential Assignment. A method of attaining more short JUMPERS on a DISTRIBUTING FRAME than would be possible if jumpers were assigned randomly from lists of spares. The distributing frame is administered in *zones* with the objective of running jumpers between adjacent zones.

Prefix. Any SIGNAL dialed prior to the ADDRESS. *Prefixes* are used to place an address in proper context, to indicate service options, or both. An example is the prefix 0 used before an address where operator assistance or intervention is requested, such as for collect calls.

Premises Information System (PREMIS). An OPERATIONS SYSTEM that provides rapid access to information about EQUIPMENT located on customers' premises.

Present Worth. An adjustment made to amounts of money associated with different times in the future to make the amounts equivalent to present values for use in ECONOMIC EVALUATION.

Primary Center. A class 3 office in the hierarchy of TOLL switching offices. *See* **TOLL OFFICE.**

Private Branch Exchange (PBX). A private SWITCHING SYSTEM, either manual or automatic, usually serving an organization, such as a business or a government agency, and usually located on the customer's premises. Telephones served by the *PBX* are called *stations*. Calls from one station to another or to an external network such as the PUBLIC SWITCHED TELEPHONE NETWORK may be handled manually or automatically, depending on the type of *PBX*. DIRECT INWARD DIALING and AUTOMATIC IDENTIFIED OUTWARD DIALING service (formerly called CENTREX-CU) can be provided by some *PBXs*. TIE TRUNKS are commonly used between *PBX* systems of a single customer.

Private Line. A circuit used for PRIVATE-LINE SERVICE.

Private-Line Service. A service in which the customer leases a CIRCUIT, not interconnected with the PUBLIC SWITCHED TELEPHONE NETWORK, for the customer's exclusive use. The PRIVATE LINE may be used for transmission of voice, teletypewriter, data, television, etc.

Private Network. A NETWORK made up of CIRCUITS and, sometimes, SWITCHING arrangements, for the exclusive use of one customer. These networks can be nationwide in scope and typically serve large corporations or government agencies. An example of a private voiceband network is the Bell System corporate network, CORNET.

Product Life Cycle. The length of time from introduction of a product or service into the MARKET until the product or service becomes obsolete and is replaced by a newer product. It typically has four stages: introduction, growth, maturity, and decline.

Progressively Controlled Network. A SWITCHING NETWORK in which calls are set up by making a series of connections, stage by stage, based on the digits dialed.

Protector Devices. Fuse-like devices mounted on DISTRIBUTING FRAMES to guard telecommunications EQUIPMENT against spurious voltages and currents that might result from lightning strikes or power-line crosses in the OUTSIDE PLANT.

Protocol. A strict procedure for the initiation and the maintenance of data communications.

Public Communications Services. Services provided through telephones installed at locations where a public need exists, such as airports, bus and train stations, hotel lobbies, large business offices, public streets, and highways.

Public Switched Telephone Network (PSTN). The portion of the total NETWORK that supports PUBLIC SWITCHED TELEPHONE NETWORK SERVICES. It provides the capability for interconnecting virtually any home or office in the country with any other. *Public* is the key word; any equipment indeterminately shared by more than one customer is part of the *PSTN*. May also be called the *public switched network, public telephone network,* or the DIRECT DISTANCE DIALING (DDD) *network.*

Public Switched Telephone Network Services. Services provided by the PUBLIC SWITCHED TELEPHONE NETWORK including MESSAGE TELECOMMUNICATIONS SERVICE (MTS), WIDE AREA TELECOMMUNICATIONS SERVICES (WATS), and DATAPHONE DIGITAL SERVICE among others.

Public Utilities Commission (PUC). An agency charged with regulating communications services as well as other public utility services, usually within a state.

Pulse-Amplitude Modulation (PAM). A MODULATION technique in which the amplitude of each pulse is related to the amplitude of an ANALOG SIGNAL. Used, for example, in TIME-DIVISION MULTIPLEX arrangements in which successive pulses represent samples from the individual VOICEBAND CHANNELS; also used in time-division SWITCHING SYSTEMS of small and moderate size.

Pulse-Code Modulation (PCM). Conversion of an ANALOG SIGNAL, such as voice, to a digital format, ordinarily in terms of binary-coded pulses representing the quantized amplitude samples of the analog signal.

Pulse Rate. The number of pulses transmitted per unit of time. May also be called the *baud rate*. When the pulses or symbols have only two possible values (binary), the pulse rate is also the BIT RATE.

Quality Assurance. Continuous independent verification of satisfactory performance of products and services from the user's viewpoint.

Quality Assurance System. In the Bell System, a series of audits and continuous monitoring to determine the effect of QUALITY CONTROLS used by the designers and producers of a product or service.

Quality Control. A set of procedures used by designers and producers of products and services to provide sufficient control of machinery, personnel, and material necessary to meet acceptable quality criteria in an economic manner.

Regeneration. The process of receiving a DIGITAL SIGNAL and reconstructing it in a form in which the amplitudes, waveforms, and timing of the SIGNAL elements are constrained within specified limits. [IEC/ITU]

Regional Center. A class 1 office in the hierarchy of TOLL switching offices; the highest level TOLL OFFICE.

Register. A part of an automatic SWITCHING SYSTEM that receives and stores SIGNALS from a calling device or other source for interpretation and action, some of which is carried out by the *register* itself.

Remote Memory Administration System (RMAS). An OPERATIONS SYSTEM that changes translations in SWITCHING SYSTEMS.

Reorder Tone. A tone applied 120 times per minute that indicates all SWITCHING paths are busy, all TOLL trunks are busy, EQUIPMENT blockages, unassigned code dialed, or incomplete registration of digits at a tandem or TOLL OFFICE; also called *channel busy* or *fast busy tone*.

Repeater. EQUIPMENT, essentially including one or several amplifiers and/or regenerators and associated devices, inserted at a point in a TRANSMISSION. The medium may operate in one or both directions of transmission. [IEC/ITU]

Repertory Dialer. A piece of STATION EQUIPMENT that permits a user to dial telephone numbers automatically from a preprogrammed directory.

Reserve Capacity. The amount of additional capacity (beyond that needed to meet SERVICE OBJECTIVES) that should be provided in a long-term forecast to minimize the cost of underestimating the demand. Reducing *reserve capacity* in the long-term forecast generates short-term costs (for example, servicing during BUSY SEASONS). The "optional" *reserve capacity* balances the long-term capital cost against the short-term servicing cost.

Resistance Design. A design method for customer LOOPS in which an attempt is made to employ cable having the highest gauge (smallest wire) that will ensure a loop resistance less than the SIGNALING limit of the CENTRAL OFFICE serving the loop.

Revenue Accounting Office (RAO). A telephone company center utilizing large, mainframe computers for billing and other data processing. Functions performed include receipt and processing of AUTOMATIC MESSAGE ACCOUNTING (AMA) data, preparation of the customer's bill, directory preparation, MARKETING support, internal reports, and payroll and inventory management.

Ringing. The process of alerting the called party by the application of an intermittent 20-hertz (Hz) SIGNAL to the appropriate LINE; this produces a *ringing* sound at the called telephone set. When the *ringing* signal is applied to the called line, an intermittent signal called an *audible ring* is sent to the calling telephone to indicate that *ringing* is taking place.

Sectional Center. A class 2 office in the hierarchy of TOLL switching offices. *See* TOLL OFFICE.

Seize, Seizure. An action of a SWITCHING SYSTEM in selecting an outgoing TRUNK or other component for a particular call.

Separation. A rate-setting requirement that specified that the costs of providing long-distance telephone service be appropriately distributed between interstate services, which are under the jurisdiction of the FEDERAL COMMUNICATIONS COMMISSION, and intrastate services, which are under the jurisdiction of the state PUBLIC UTILITIES COMMISSIONS.

Service Code. A code, typically of the form $N11$ (N = any digit 2 through 9), that defines a connection for a service (for example, 411 for directory assistance).

Service Evaluation. The process of determining what customers and the Bell System expect of a service, setting appropriate objectives based on the expectations, and assessing compliance with objectives.

Service Measurement. A measurement reflecting aspects of operations perceivable by the customer.

Service Objective. A statement of the quality of service that is to be provided to the customer; for example, no more than 1.5 percent of customer calls should encounter a delay of more than 3 seconds for DIAL TONE during the average BUSY HOUR. *See* GRADE OF SERVICE.

Service Order. An order prepared in the commercial department of an OPERATING COMPANY, at the request of a customer, to establish a service, to change an existing service, or to terminate a service. The resultant document contains all the information required to meet the customer's needs.

Service Representative. An individual in the business office of an OPERATING COMPANY who typically deals with customers.

Serving Area Interface. A rearrangeable cross-connect point between feeder and DISTRIBUTION cables in the LOOP plant.

Settlement. An accounting procedure based on the total investment in telephone EQUIPMENT; the total investment of a company determines the base for the allowed earnings (called the *rate base*). *Settlements* define how revenue from a single call is distributed among different companies, both Bell and independent, involved in that connection.

Sidetone. The portion of the SIGNAL from a telephone transmitter that appears at the receiver of that telephone. Some *sidetone* appears to be desirable to assure the customer that the telephone is working and to help the talker adjust the level of speech.

Signal. An electrical, optical, or other representation of information for (1) MESSAGES; for example, voice, data, television; (2) NETWORK control; for example, call routing, network management; (3) internal operation of network elements; for example, timing and control of SWITCHING SYSTEMS.

Signaling. The transmission of ADDRESS, SUPERVISION, or other SWITCHING information between stations and SWITCHING SYSTEMS and between switching systems, including any information required for billing.

Signal-to-Noise Ratio. The ratio of the average SIGNAL power at any point in a TRANSMISSION path to the average NOISE power at that same point, often expressed in DECIBELS (dBs).

Signal Transfer Point (STP). A switching NODE in the COMMON-CHANNEL INTEROFFICE SIGNALING network. *STPs* operate under STORED-PROGRAM CONTROL to connect signaling LINKS to network SWITCHING SYSTEMS and other *STPs*. They may also connect directly to NETWORK CONTROL POINTS.

Single-Frequency (SF) Signaling. A method of conveying dial-pulse and SUPERVISION signals from one end of a TRUNK or LINE to the other, using the presence or absence of a single specified frequency. A 2600-hertz (Hz) tone is commonly used.

Single-Sideband Amplitude Modulation (SSBAM). AMPLITUDE MODULATION in which only one of the sidebands resulting from MODULATION is selected for TRANSMISSION by a BANDPASS FILTER. A precise and stable CARRIER FREQUENCY is inserted at the receiving terminal for DEMODULATION.

Small Office Network Data System (SONDS). An OPERATIONS SYSTEM that collects TRAFFIC data from small step-by-step offices, processes the data, and provides reports to NETWORK administrators.

Span. A collection of span LINES between two offices. The term is also used to refer to the collection of all SPAN LINES in a particular cable, all span lines on a particular route, or all span lines between two offices.

Span Line. A repeatered T1 LINE section between two CENTRAL OFFICES (not necessarily contiguous offices). A T1 CARRIER SYSTEM is made up of a tandem combination of *span lines*, plus a digital CHANNEL BANK at each terminal.

Speakerphone. An audio terminal, consisting of transmitter and loudspeaker units, used with a telephone set for TELECONFERENCING.

Special Services. Services requiring special treatment with respect to TRANSMISSION, SIGNALING, SWITCHING, billing, or customer use. Examples are PRIVATE BRANCH EXCHANGE (PBX) service; WIDE AREA TELECOMMUNICATIONS SERVICE (WATS); FOREIGN EXCHANGE (FX) SERVICE; and PRIVATE-LINE SERVICES such as CIRCUITS for voice, data, teletypewriter, and television.

Special-Services Circuit. A TRANSMISSION path used to provide SPECIAL SERVICES to a specific customer.

Standard Supply Contract. The legal agreement that has governed the relationship between Western Electric and a Bell OPERATING COMPANY. The *supply contract* required Western Electric either to manufacture or to purchase materials that the operating companies might reasonably require, which they might order from Western Electric. The *supply contract* did not, however, obligate the operating companies to purchase these materials from Western Electric. The *supply contract* terminated at the time of divestiture.

Station Equipment. EQUIPMENT that allows a customer to access the NETWORK and the available services. The most common *station equipment* is the ordinary single-line telephone set.

Station Number. The final four digits of a standard 7- or 10- digit ADDRESS that define a connection to a specific customer's line within a CENTRAL OFFICE. *See* **CENTRAL OFFICE CODE.**

Step-by-Step (SXS) System. An automatic SWITCHING SYSTEM using step-by-step switches. In most such systems, a call is extended progressively, step-by-step, to the desired terminal under direct control of pulses from a customer's DIAL or from a sender.

Stored-Program Control (SPC). A form of SWITCHING SYSTEM control in which system operations are controlled by a stored program executed by one or more processors. Operation of the system can be altered significantly by changing programs.

Stored Program Control System/Central Office Equipment Reports (SPCS/COER). A series of time-shared programs that analyzes TRAFFIC data for ELECTRONIC SWITCHING SYSTEM offices and produces reports.

Suffix. Any SIGNAL dialed after the ADDRESS. Used by operators, for example, to indicate the end of dialing.

Supervision. The constant monitoring and controlling of the status of a call.

Switching. (1) Refers to the process of connecting appropriate LINES and TRUNKS to form a desired communication path between two station sets. Included are all kinds of related functions, such as sending and receiving SIGNALS, monitoring the status of CIRCUITS, translating ADDRESSES to routing instructions, alternate routing, testing circuits for busy condition, and detecting and recording troubles. (2) Designates a field of work, such as system development, planning, or engineering, involving the application of *switching* technology in telecommunications NETWORKS. (3) Refers, in a more restricted sense, to the technology associated with any circuit that operates discretely, particularly logic and memory.

Switching Control Center (SCC). An OPERATIONS CENTER responsible for the centralized installation and maintenance of a group of SWITCHING SYSTEMS in a geographic area.

Switching Control Center System (SCCS). The Computer Subsystem (CSS) and the EQUIPMENT units that remote the MASTER CONTROL CENTER capability of an ELECTRONIC SWITCHING SYSTEM. The *SCCS* provides for the administration, control, and maintenance of electronic switching systems from central locations.

Switching Network. SWITCHING stages and their INTERCONNECTIONS within a SWITCHING SYSTEM.

Switching System. An electromechanical or electronic system for connecting LINES to lines, lines to TRUNKS, or trunks to trunks. The term includes PRIVATE BRANCH EXCHANGE *switching systems* and centrally located NETWORK *switching systems*. *See* **SWITCHING**.

System Code. A 3-digit code, usually of the form $1XX$ but including $0XX$ (X = any digit 0 through 9) assignments, available only to operators or to SWITCHING equipment for use as part of a special or modified ADDRESS to influence route selection. These codes are reserved for system-wide use; that is, they are the same across all NUMBERING PLAN AREAS.

Talker Echo. An ECHO of a talker's voice that is returned to the talker. When there is delay between the original SIGNAL and the echo, the effect is disturbing, unless the echo is attenuated to a tolerable level.

Tandem Switching System. A broad functional category representing systems that connect TRUNKS to trunks. Tandem switching divides into two applications: Those offices that connect trunks within a metropolitan area are referred to as *local tandem* offices; Those offices that connect trunks in the TOLL network portion (class 1 to class 4) of the PUBLIC SWITCHED TELEPHONE NETWORK are called *TOLL OFFICES*.

Tandem Trunk. A TRUNK that connects WIRE CENTERS through a local tandem office.

Tariff. The published rates, regulations, and descriptions governing the provision of communications service.

T-Carrier Administration System (TCAS). An OPERATIONS SYSTEM responsible for T-carrier alarms.

Teleconferencing. Voice telephone service between a group of people and one or more other groups or individuals.

Telemetry. The method or EQUIPMENT used to transmit status information such as that represented by the operation of keys or by lamp displays to a remote location.

Terminal Equipment. In the INTERCONNECTION environment, any separately housed EQUIPMENT unit or a group of equipment units located on user premises on the user side of a network INTERFACE.

Termination. (1) The points on a SWITCHING NETWORK to which a TRUNK or a LINE may be attached. (2) An item that is connected to the terminals of a CIRCUIT or piece of EQUIPMENT. (3) An impedance connected to the end of a circuit being tested.

Termination Layout Mask. A plan that reserves space on a DISTRIBUTING FRAME for different TERMINATION categories of EQUIPMENT and FACILITIES.

Thus, as the office grows, there will be room on the FRAME for the orderly addition of new terminations.

Throughput. (1) The total useful information processed or communicated during a specified time period. [TC] (2) A measure of the effective rate of TRANSMISSION of data by a communications system. [DST]

Tie Cable. Cable that interlinks DISTRIBUTING FRAMES.

Tie Trunk. A SPECIAL-SERVICES CIRCUIT connecting two PRIVATE BRANCH EXCHANGES or equivalent SWITCHING SYSTEMS.

Time-Division Multiplex (TDM). A method of serving a number of simultaneous CHANNELS over a common TRANSMISSION path by assigning the transmission path sequentially to the various channels, each assignment being for a discrete time interval.

Time-Multiplexed Switch (TMS). An element of a time-division SWITCHING NETWORK that effectively operates as a very high-speed space-division switch whose input-to-output paths can be changed in every time slot.

Time Sharing. The use of a FACILITY or piece of EQUIPMENT for more than one purpose or function or for repetition of the same function within the same overall time period. This is accomplished by interspersing or interleaving the required actions in time.

Time-Slot Interchange (TSI). An element of time-division SWITCHING that separates and switches SIGNALS from multiple calls that are presented in a TIME-DIVISION MULTIPLEXED format.

Tip and Ring Conductors. The two conductors associated with a 2-wire cable pair. The terms *tip* and *ring* derive their names from the physical characteristics of an operator's cord switchboard plug in which these two conductors terminated in the days of manual switchboards. Use of the names *tip* and *ring* has extended throughout the plant. The cord switchboard plug also had a *sleeve,* and the name is occasionally used for a third conductor associated with *tip* and *ring.*

Toll. A term describing service that is a part of public telephone service but under a TARIFF separate from the EXCHANGE AREA tariff. Also used to describe components of the FACILITIES NETWORK that are used principally for *toll* service.

Toll Center. A class 4 office in the hierarchy of toll switching offices; the lowest level TOLL OFFICE.

Toll Center Code. A 3-digit code of the form 0XX (X = any digit 0 through 9) that identifies a specific TOLL CENTER and is available only for OPERATING COMPANY use.

Toll Charge. A charge for telephone service for calls outside the designated local EXCHANGE AREA. TOLL service calls are billed individually.

Toll Connecting Trunk. A TRUNK between an END OFFICE and a TOLL OFFICE.

Toll Office. Those offices that connect TRUNKS in the TOLL network portion (class 1 to class 4) of the PUBLIC SWITCHED TELEPHONE NETWORK. *See* TANDEM SWITCHING SYSTEM.

Total Network Data System (TNDS). A coordinated family of OPERATIONS SYSTEMS. *TNDS* consists of both manual procedures and computer systems that provide OPERATING COMPANY managers with comprehensive, timely, and accurate NETWORK information. It supports OPERATIONS CENTERS responsible for administration of the trunking network, network data collection, daily surveillance of the load on the SWITCHING NETWORK, and design of local and CENTRAL OFFICE switching equipment to meet future service demands.

Total Network Operations Plan (TNOP). A Bell System operations plan that describes the OPERATIONS PROCESSES, OPERATIONS CENTERS, and OPERATIONS SYSTEMS to be used in administering and provisioning the telecommunications NETWORK in the Bell OPERATING COMPANIES.

Traffic. The flow of information or MESSAGES through the NETWORK. This information flow may be generated by telephone conversations or may be the result of providing data, audio, and video services.

Traffic Data Administration System (TDAS). Part of the TOTAL NETWORK DATA SYSTEM (TNDS) that formats and temporarily stores TRAFFIC data for other TNDS systems.

Traffic Engineering. A NETWORK PLANNING activity that determines the number and type of CHANNELS or communication paths required between SWITCHING points and the call-handling capacity of the switching points.

Traffic Network. An arrangement of CHANNELS (such as LOOPS and TRUNKS, associated SWITCHING arrangements, and STATION EQUIPMENT) designed to handle a specific body of TRAFFIC. *Traffic networks* are provided by the FACILITIES NETWORK.

Traffic Service Position (TSP). A cordless console that is associated with either a crossbar tandem office or a TRAFFIC SERVICE POSITION SYSTEM, equipped so that operators can provide assistance, if needed, on station-to-station calls, special TOLL calls, public telephone calls, and all local and toll assistance TRAFFIC. The operators provide assistance in completing these calls and ensure that correct data are recorded in the centralized AUTOMATIC MESSAGE ACCOUNTING equipment or in the Traffic Service Position System equipment. They also supervise coin deposits for calls originating at public telephones. The position is arranged for automatic display of both the calling and called numbers, as well as certain other information.

Traffic Service Position System (TSPS). A type of TRAFFIC service system, with STORED-PROGRAM CONTROL that provides for the processing and recording of special TOLL calls, public telephone toll calls, and other types of calls requiring operator assistance. It includes TRAFFIC SERVICE POSITIONS arranged in groups called *Operator Office Groups*, where operators are automatically connected in on calls to perform the functions necessary to process and record the calls correctly.

Traffic Theory. A branch of applied probability theory that produces models used to determine the capacity requirements to meet SERVICE OBJECTIVES of systems with nondeterministic demands.

Translation. The operation of converting information from one form to another. In SWITCHING SYSTEMS, the process of interpreting all or part of a destination code to determine the routing of a call.

Transmission. (1) Designates a field of work, such as EQUIPMENT development, system design, planning, or engineering, in which electrical communication technology is used to create systems to carry information over a distance. (2) Refers to the process of sending information from one point to another. (3) Used with a modifier to describe the quality of a telephone connection: good, fair, or poor *transmission*. (4) Refers to the transfer characteristic of a CHANNEL or NETWORK in general or, more specifically, to the amplitude transfer characteristic. One may hear the phrase, "*transmission* as a function of frequency."

Transmission Facility. An element of physical telephone PLANT that performs the function of TRANSMISSION; for example, a multipair cable, a COAXIAL CABLE system, or a microwave radio system.

Transmission Level Point (TLP). A specification, in DECIBELS (dBs), of the relative level at a particular point in a TRANSMISSION system as referred to a zero *transmission level point* (0 *TLP*). The *TLP* value does not specify the absolute power that will exist at that point.

Transmission Objective. Electrical performance characteristics for communication CIRCUITS, systems, and EQUIPMENTS based on both economic and technical considerations of telephone facilities and on reasonable estimates of the performance desired. Characteristics for which objectives are stated include LOSS, NOISE, ECHO, CROSSTALK, frequency shift, ATTENUATION distortion, envelope delay distortion, etc.

Trouble Ticket. A form containing either symptoms or detailed information about malfunctioning EQUIPMENT. It is given to a craftsperson whose job is to locate and repair the equipment.

Trunk. A communication CHANNEL between two SWITCHING SYSTEMS. The term *switching system* includes CENTRAL OFFICE types, TOLL switching systems, PRIVATE BRANCH EXCHANGES, KEY TELEPHONE SYSTEMS, manual and automatic switchboards, concentrators, etc.

Trunk Circuit. A CIRCUIT, part of a SWITCHING SYSTEM, associated with the connection of a TRUNK to the switching system. It serves to convert between the SIGNAL formats used internally in the switching system and those used in the TRANSMISSION circuit, and it performs logic and sometimes memory functions associated with SUPERVISION.

Trunk Forecasting System (TFS). Part of the TOTAL NETWORK DATA SYSTEM that forecasts MESSAGE TRUNK requirements for 5 years in the future.

Trunk Group. A number of TRUNKS that can be used interchangeably between two SWITCHING SYSTEMS.

Trunk Servicing System (TSS). Part of the TOTAL NETWORK DATA SYSTEM that processes TRAFFIC data from the TRAFFIC DATA ADMINISTRATION SYSTEM, computes OFFERED LOAD, and calculates TRUNK requirements. It is used by trunk administrators to maintain the MESSAGE TRUNK network.

Trunks Integrated Records Keeping System (TIRKS). An OPERATIONS SYSTEM for maintaining the inventory and assignment of the FACILITIES and EQUIPMENT used to establish TRUNKS of all kinds.

Two-Way Trunk. A TRUNK that can be seized for use by the SWITCHING equipment located at either end.

Unigauge Design. A design method for customer LOOPS that provides for the exclusive use of 26-gauge cable on all loops within 30 kilofeet of the CENTRAL OFFICE. Requires range extension equipment developed specifically for the unigauge system.

Universal Telephone Service. The goal of establishing affordable and available nationwide telephone service.

Usage-Sensitive Rate. A rate-setting principle that relates directly to customer use of EQUIPMENT and service: Those who use less, pay less.

Value-of-Service Pricing. A rate-setting principle that relates directly to the customer density in a local calling area, the frequency of use, and importance of service to the customer.

Voiceband Channel. A transmission CHANNEL with a nominal 4-kilohertz (kHz) BANDWIDTH suitable for voice TRANSMISSION.

Voice-Frequency (VF) Facility. An analog FACILITY that provides one VOICEBAND CHANNEL and carries the information in the voice-frequency band.

Volatile Memory. A computer memory in which stored information is lost if the power supply for the memory fails or is turned off.

Wide Area Telecommunications Services (WATS). A service that permits customers to make (OUTWATS) or receive (INWATS or 800 SERVICE) long-distance calls and to have them billed on a bulk basis rather than individually. *WATS* is provided within selected service areas, or bands, by means of special private-access LINES, which are connected to the PUBLIC SWITCHED TELEPHONE NETWORK through *WATS*-equipped CENTRAL OFFICES. A single access line permits inward or outward service but not both.

Wire Center. The location of one or more local SWITCHING SYSTEMS; a point at which customer LOOPS converge. May be loosely used to mean the CENTRAL OFFICE building at that location.

Wire Center Area. The area surrounding a WIRE CENTER containing all customers, other than those with FOREIGN EXCHANGE SERVICE, whose LOOPS are connected to a CENTRAL OFFICE at that wire center.

Wire Pair Cable. Cables composed of twisted pairs of wires rather than coaxial tubes, fibers, etc.

Work Package. Material sent to OPERATING COMPANY field forces that describes work to be performed.

World Numbering Plan. *See* COUNTRY CODE.

World Zone Number. A 1-digit number that, in the world numbering plan, identifies a geographic zone. The *world zone number* is the initial number in a COUNTRY CODE.

Acronyms and Abbreviations

The acronyms and abbreviations listed here reflect usage in this book. They may be used differently in other contexts.

ABSBH	average busy season busy hour	AUTOVON	automatic voice network
ACD	automatic call distributor	BANCS	Bell Administrative Network Communications System
ACH	attempts per circuit per hour	BCC	blocked calls cleared
ACP	action point	BCD	blocked calls delayed
ACTS	Automated Coin Toll Service	BDT	billing data transmitter
ACU	automatic calling unit	BISP	Business Information Systems Programs
ADS	Auxiliary Data System	BOC	Bell operating company
AIC	Automatic Intercept Center	BOSS	Billing and Order Support System
AIOD	automatic identified outward dialing	bpi	bits per inch
AIS	Automatic Intercept System	bps	bits per second
ALBO	automatic line buildout	BPSS	Basic Packet-Switching Service
ALGOL	Algorithmic Computer Language	BSBH	busy season busy hour
AM	amplitude modulation	B6ZS	bipolar with six zero substitution
AMA	automatic message accounting	BSP	Bell System Practice
AMARC	Automatic Message Accounting Recording Center	BSRFS	Bell System Reference Frequency Standard
AMAT	automatic message accounting transmitter	Btu	British thermal unit
ANI	automatic number identification	CAC	Circuit Administration Center Customer Administration Center
APS	automatic protection switching	CAMA	centralized automatic message accounting
ASCII	American Standard Code for Information Interexchange	CAMA–ONI	centralized automatic message accounting–operator number identification
ASPEN	Automatic System for Performance Evaluation of the Network	CAROT	Centralized Automatic Reporting on Trunks
ATRS	Automated Trouble Reporting System	CATLAS	Centralized Automatic Trouble Locating and Analysis System
AT&T	American Telephone and Telegraph Company		

CCH	connections per circuit per hour	CPD	central pulse distributor
CCIR	Comité Consultatif International des Radiocommunications	CPE	customer-premises equipment
		CPU	central processing unit
CCIS	common-channel interoffice signaling	CRC	Customer Record Center
		CREG	concentrated range extension with gain
CCITT	Comité Consultatif International Télégraphique et Téléphonique	CRIS	Customer Records Information System
CCS	hundred call seconds	CRS	Centralized Results System
CCSA	common-control switching arrangement	CRT	cathode-ray tube
		CSAR	Centralized System for Analysis and Reporting
CDA	call data accumulator	CSDC	circuit-switched digital capability
CDCF	cumulative discounted cash flow		
		CSO	central services organization
CDO	community dial office	CSS	Computer Subsystem
CDT	call data transmitter	CSU	channel service unit
CEF	cable entrance facility	CUCRIT	Capital Utilization Criteria
CEV	controlled environment vault	CU/EQ	Common Update/Equipment (System)
CLEI	common-language equipment identification	CU/TK	Common Update/Trunking (System)
CLLI	common-language location identification	DACS	Digital Access and Cross-Connect System
CMDF	combined main distributing frame	dB	decibel
		DBAS	Data Base Administration System
CMDS	Centralized Message Data System	dBm0	decibels with reference to a power of 0 milliwatt
CMS	Call Management System		
CNAB	Customer Name and Address Bureau	dBrnC	decibels above reference noise, using C-message weighting
CNCC	Customer Network Control Center		
		DCE	data circuit-terminating equipment
CO	central office		
COC	circuit order control	DCM	digital carrier module
codec	coder-decoder	DCP	duplex central processor
COEES	Central Office Equipment Engineering System	DCPR	Detailed Continuing Property Records (PICS/DCPR)
COER	Central Office Equipment Report		
		DCPSK	differential coherent phase-shift keying
COMSAT	Communications Satellite Corporation		
		DCT	digital carrier trunk
CONN	connector	DDC	Direct Department Calling
CONUS	continental United States	DDD	direct distance dialing
CORNET	corporate network (Bell System)	DDS	Digital Data System
		DERP	Defective Equipment Replacement Program
COSMIC	Common Systems Main Interconnection Frame System	DEW	Distant Early Warning (Line)
		DFI	digital facility interface
COSMOS	Computer System for Mainframe Operations	DIC	digital interface controller
		DID	direct inward dialing
CPC	Circuit Provision Center	DIF	digital interface frame

DIM	data in the middle	EPL	Electronic Switching System
DIU	digital interface unit		Programming Language
DLTU	digital line/trunk unit	EPROM	erasable programmable
DOC	dynamic overload control		read-only memory
DOJ	Department of Justice	EPSCS	Enhanced Private Switched
DOM	data on mastergroup		Communication Service
DOV	data over voice	ETS	electronic tandem switching
DP	demarcation point		electronic translator system
DPAC	dedicated plant assignment card	FACS	Facilities Assignment and Control System
DPP	discounted payback period	FCAP	Facility Capacity
DPSK	differential phase-shift keying	FCC	Federal Communications Commission
DR	data receive	FCG	false cross or ground
DS	digital signal	FDM	frequency-division
DSBAM	double-sideband amplitude modulation	FDX	multiplex full duplex
DSDC	direct services dialing capability	FEPS	Facility and Equipment Planning System
DSn	digital signal (level) n	FIFO	first in, first out
DSS	data station selector	FIP	facility interface processor
DSU	data service unit	5XB COER	No. 5 Crossbar Central
DSX	digital system cross-connect		Office Equipment
DT	data transmit		Reports (System)
	digroup terminal	FM	frequency modulation
DTC	digroup terminal controller	FNPA	foreign numbering plan area
DTE	data terminal equipment	FSK	frequency-shift keying
DTMF	dual-tone multifrequency	FTS	Federal Telecommunications
DTU	digroup terminal unit		Service
DUV	data under voice	FX	foreign exchange
EADAS	Engineering and Adminis-trative Data Acquisition System	GHz	gigahertz
		GND	ground
		GOS	grade of service
EADAS/NM	Engineering and Adminis-trative Data Acquisition System/Network Management	GRP MOD	group modulator
		GSAT	General Telephone and Electronics Satellite Corporation
EBCDIC	Extended Binary-Coded Decimal Interchange Code	HCSDS	High-Capacity Satellite Digital Service
ECCS	economic hundred call seconds	HCTDS	High-Capacity Terrestrial Digital Service
ECPT	electronic coin public telephone	HDX	half duplex
		HNPA	home numbering plan area
EEI	equipment-to-equipment interface	HSSDS	High-Speed Switched Digital Service
EFRAP	Exchange Feeder Route Analysis Program	HU	high usage
		Hz	hertz
EIA	Electronics Industries Association	IC	interexchange carrier
		ICAN	Individual Circuit Analysis
EISS	Economic Impact Study System	ICC	Interstate Commerce Commission
E911	Enhanced 911 (Emergency Service)	IDDD	international direct distance dialing

IDF	intermediate distributing frame	LRS	line repeater station
IEEE	Institute of Electrical and Electronics Engineers	LRSS	Long Range Switching Studies
IF	intermediate frequency	LSI	large-scale integration
IFRPS	Intercity Facility Relief Planning System	LSRP	Local Switching Replacement Planning (System)
IM	interface module	LTF	lightwave terminating frame
INTELSAT	International Telecommunications Satellite Consortium	MAS	Mass Announcement System
		MATFAP	Metropolitan Area Transmission Facility Analysis Program
INWATS	inward Wide Area Telecommunications Services	Mbps	megabits per second
IOP	input/output processor	MCC	master control center
IPLAN	Integrated Planning and Analysis	MELD	Mechanized Engineering and Layout for Distributing Frames
ipm	interruptions per minute		
IRC	international record carrier	MET	multibutton electronic telephone
IROR	internal rate of return		
ISDN	integrated services digital network	MF	multifrequency
		MFJ	Modification of Final Judgment
ISO	International Organization for Standardization	MFT	metallic facility terminal
ITU	International Telecommunication Union	MGT	mastergroup translator
		MHz	megahertz
JMX	jumbogroup multiplex	MLT	mechanical loop testing
K	kilobit	MMGT	multimastergroup translator
kbps	kilobits per second	MMX	mastergroup multiplex
kft	kilofeet	modem	modulator-demodulator
kHz	kilohertz	MSC	Media Stimulated Calling
KP	key pulse	MTSO	mobile telecommunications switching office
KSR	keyboard send-receive		
KTS	key telephone system	muldem	multiplexer-demultiplexer
LAC	Loop Assignment Center	MUX	multiplex
LAMA	local automatic message accounting	NAC	Network Administration Center
LATA	local access and transport area	NCP	network control point
		NCTE	network channel-terminating equipment
LBO	line buildout		
LBS	Load Balance System	NDCC	Network Data Collection Center
LCIE	lightguide cable interconnection equipment	NEBS	New Equipment-Building System
LED	light-emitting diode		
LF	line finder	NI	network interface
LIFO	last in, first out	NMC	Network Management Center
LLN	line link network		
LMMS	Local Message Metering System	NOC	Network Operations Center
		NOCS	Network Operations Center System
LMOS	Loop Maintenance Operations System	NORGEN	Network Operations Report Generator
LOCAP	low capacitance		
LRAP	Long Route Analysis Program	NOTIS	Network Operations Trouble Information System

NPA	numbering plan area	PREMIS/LAC	Premises Information System/Loop Assignment Center
NPV	net present value		
NSC	Network Service Center	PSAP	public safety answering point
NSCS	Network Service Center System		
NSPMP	Network Switching Performance Measurement Plan	PSK	phase-shift keying
		PSTN	public switched telephone network
NTSC	National Television Standards Committee	PUC	public utilities commission
OCC	other common carrier	QAM	quadrature-amplitude modulation
OCE	other common carrier channel equipment		
		QMP	Quality Measurement Plan
OCM	office carrier module	QSS	Quality Surveillance System
OCU	office channel unit	RAM	random-access memory
OFNPS	Outstate Facility Network Planning System	R&SE	Research and Systems Engineering
ONI	operator number identification	RAO	Revenue Accounting Office
		RASC	Residence Account Service Center
OPS	off-premises station		
OS	outstate	RBOC	regional Bell operating company
OSC	oscillator		
OSI	Open Systems Interconnection	RCC	radio common carrier
		RCM	remote carrier module
OUTWATS	outward Wide Area Telecommunications Service	RCVR	receiver
		RDES	Remote Data Entry System
PABX	private automatic branch exchange	RDS	Radio Digital System
		RDT	Radio Digital Terminal
PACE	Program for Arrangement of Cables and Equipment	REM	remote equipment module
		RF	radio frequency
PAM	pulse-amplitude modulation	RMAS	Remote Memory Administration System
P/AR	peak-to-average ratio		
PAS	Public Announcement Service	rms	root-mean-square
		RO	receive only
PBC	peripheral bus computer	ROM	read-only memory
PBX	private branch exchange	rn	reference noise
PCM	pulse-code modulation	RSM	remote switching module
PCO	peg count and overflow	RSS	remote switching system
PE	peripheral equipment	RTA	remote trunk arrangement
PIA	Plug-In Administrator	RTM	remote test module
PIC	plastic-insulated cable	R/WM	read/write memory
PICS	Plug-In Inventory Control System	SAG	Street Address Guide
		SARTS	Switched Access Remote Test System
PICS/DCPR	Plug-in Inventory Control System/Detailed Continuing Property Records		
		SCC	Switching Control Center
		SCCS	Switching Control Center System
PIN	personal identification number		
		SCU	selector control unit
PNPN	positive-negative-positive-negative (devices)	SD&D	Specific Development and Design
		SDOC	selective dynamic overload controls
PPS	Product Performance Surveys		
		SEL	selector
PREMIS	Premises Information System	SF	single frequency

SI	status indicator	TIRKS	Trunks Integrated Records
SMAS	Switched Maintenance		Keeping System
	Access System	TLN	trunk link network
SMDF	subscriber main distributing	TLP	transmission level point
	frame	TM	transverse magnetic
SMSA	standard metropolitan	TMDF	trunk main distributing
	statistical area		frame
SONDS	Small Office Network Data	TMS	time-multiplexed switch
	System	TNDS	Total Network Data System
SP	signal processor	TNOP	Total Network Operations
SPC	stored-program control		Plan
SPCS	Stored-Program Control	T1/OS	T1 (carrier) outstate
	System	TPMP	Total Network Data System
SPCS/COER	Stored-Program Control		Performance Measurement
	System/Central Office		Plan
	Equipment Reports	TRMTR	transmitter
SSAS	Station Signaling and	TSI	time-slot interchange
	Announcement Subsystem	TSORT	Transmission System
SSB	single sideband		Optimum Relief Tool
SSBAM	single-sideband amplitude	TSP	traffic service position
	modulation	TSPS	Traffic Service Position
SSTTSS	space-space-time-time-space-		System
	space (network)	TSPS/ACTS	Traffice Service Position
ST	start		System/Automated Coin
STP	signal transfer point		Toll Service
STS	space-time-space (network)	TSS	Trunk Servicing System
SXS	step-by-step	TSST	time-space-space-time
TASC	Telecommunications Alarm		(network)
	Surveillance and Control	TST	time-space-time (network)
	(System)	TSTS	time-space-time-space
TASI	Time Assignment Speech		(network)
	Interpolation (System)	TTY	teletypewriter
TCAS	T-Carrier Administration	TWT	traveling-wave tube
	System	UCD	Uniform Call Distribution
TCSP	Tandem Cross Section	USITA	United States Independent
	Program		Telephone Association
TDAS	Traffic Data Administration	USP	Universal Sampling Plan
	System	VF	voice frequency
TDM	time-division multiplex	VHF	very high frequency
TE	terminal equipment	VIU	voiceband interface unit
	transverse electric	VNL	Via Net Loss (Plan)
TELSAM	Telephone Service Attitude	VNLF	via net loss factor
	Measurement	VSB	vestigial sideband
TERM	terminal		modulation
TFLAP	T-Carrier Fault-Locating	WATS	Wide Area Telecommunica-
	Applications Program		tions Services
TFS	Trunk Forecasting System	XB	crossbar
3ACC	3A central control	XBT	crossbar tandem

Index

"Above-890" Ruling, 689, 693
Access code, 94*n*
Access line(s), 39
  engineering of, 188, 190
Access unit, 493
Accounting measures, in project evaluation, 719, 726
ACCUNET Packet Service, 71*n*
  see also Basic Packet-Switching Service
ACCUNET Reserved 1.5 Service, 73*n*
  see also High-Speed Switched Digital Service
ACCUNET T1.5 Service, 73*n*
  see also High-Capacity Terrestrial Digital Service
Acoustic signal, 97, 194
  see also Speech signal(s)
Action point (ACP), 509, 510, 512, 513
Adaptive transversal equalizer, 478–79
Address
  assignment of, by PREMIS, 620
  data base, 617
  definition of, 115
  history and evolution, 114–19
  signaling, 268, 269, 275–76, 290
Addressing, definition of, 85
  see also Signaling
Address input devices, 117
Administration, 572
  customer-service, 577–79
    billing, 577–78
    switching systems, 578–79
  network, 592–96
    data, 592–93
    equipment utilization, 594
    network management, 595–96
    office status evaluation, 594
    operator-services, 593
    problem analysis, 594
    transition management, 595

Advance Calling, human factors in design of, 739–40
Advanced Mobile Phone Service, 521–24
  cellular concept, 21, 521–23
  regulation, 524
  system operation, 523–24
  system plan, diagram, 522
  see also *AUTOPLEX* cellular radio system
Advanced Mobile Phone Service, Inc., 5, 10
Advanced Research Projects Agency, private data network of, 91
Advertising organizations, 713
Air Force, U.S., 19
Air/ground service, 74
Airline reservations
  automatic call distributor, 503, 55–56
  Basic Packet-Switching Service, 72
  DATAPHONE digital service, 70
Alarm service bureaus, use of *DATAPHONE* Select-a-station service, 538–40
Alarm systems, in telephone equipment building, 541, 567
Alerting, 85, 265, 274, 285
  electronic telephone, 469
ALGOL, 433
All-number calling, 118*n*
Allocation area(s), in local facilities network, 122–23
Allowance numbers, in quality assurance, 756
Alternate routing, 92, 110, 166, 169, 259, 630
  automatic, 106, 400
  busy hour, 173
  cancellation of, 181
  cost function for, graph, 171
  during trunk congestion, 177
  economics of, 169–72
  in ESPCS, 69
  in No. 4A Crossbar System, 400
  in private switched networks, 113

ENGINEERING AND OPERATIONS IN THE BELL SYSTEM

Text set at AT&T Bell Laboratories in Palatino Roman
on an AUTOLOGIC, Inc. APS-5 phototypesetter
using the TROFF formatting and phototypesetting program
under the **UNIX** operating system.
Display lines set on a Varityper Comp/Edit 5810 in Korinna Bold
Index set by University Graphics, Inc.
Printed by the Maple-Vail Book Manufacturing Group
on Perkins & Squier Offset Smooth text stock
Bound by the Maple-Vail Book Manufacturing Group
in James River's Kivar 9 Mulberry cover
with Lindenmeyr's Multicolor Antique Muscatel endsheets